SEVENTH EDITION

Mathematics for the Trades

A GUIDED APPROACH

Robert A. Carman

Hal M. Saunders

PEARSON

Prentice
Hall

Upper Saddle River, New Jersey
Columbus, Ohio

Library of Congress Cataloging-in-Publication Data

Carman, Robert A.
  Mathematics for the trades: a guided approach/Robert A. Carman, Hal
M. Saunders.—7th ed.
       p. cm.
Includes index.
ISBN 0-13-114525-8
1. Mathematics.   I. Saunders, Hal M. II. Title.

QA39.3.C37 2005
513′.14—dc22   2004000161

Editor in Chief: Stephen Helba
Senior Acquisitions Editor: Gary Bauer
Editorial Assistant: Natasha Holden
Production Editor: Louise N. Sette
Production Supervision: *The GTS Companies*/York, PA Campus
Design Coordinator: Diane Ernsberger
Cover Designer: Jim Hunter
Production Manager: Brian Fox
Marketing Manager: Leigh Ann Sims

This book was set in Century Schoolbook by *The GTS Companies*/York, PA Campus. It was printed and bound by Courier
Kendallville, Inc. The cover was printed by Phoenix Color Corp.

10 9 8 7 6 5 4 3 2
ISBN 0-13-114525-8

Contents

Preface

This book provides the practical mathematics skills needed in a wide variety of trade and technical areas, including electronics, auto mechanics, construction trades, air conditioning, machine technology, welding, drafting, and many other occupations. It is especially intended for students who have a poor math background and for adults who have been out of school for a time. Most of these students have had little success in mathematics, some openly fear it, and all need a direct, practical approach that emphasizes careful, complete explanations and actual on-the-job applications. This book is intended to provide practical help with real math, beginning at each student's own individual level of ability.

Features   Those who have difficulty with mathematics will find in this book several special features designed to make it most effective for them:

- Careful attention has been given to **readability.** Reading specialists have helped plan both the written text and the visual organization.
- A **diagnostic pretest** and performance **objectives** keyed to the text are included at the beginning of each unit. These clearly indicate the content of each unit and provide the student with a sense of direction.
- Each unit ends with a **problem set** covering the work of the unit.
- The **format** is clear and easy to follow. It respects the individual needs of each reader, providing immediate feedback at each step to ensure understanding and continued attention. The emphasis is on *explaining* concepts rather than simply *presenting* them. This is a practical presentation rather than a theoretical one.
- Special attention has been given to **on-the-job math skills,** using a wide variety of real problems and situations. Many problems parallel those that appear on professional and apprenticeship exams. The answers to all problems are given in the back of the book.
- A light, lively **conversational style** of writing and a pleasant, easy-to-understand visual approach are used. The use of humor is designed to appeal to students who have in the past found mathematics to be dry and uninteresting.

Seven editions and over two decades of experience with a wide variety of students indicate that this approach is successful—the book works and students learn, many of them experiencing success in mathematics for the first time.

Flexibility of use was a major criterion in the design of the book. Field testing and extensive experience with the first six editions indicate that the book can be used successfully in a variety of course formats. It can be used as a textbook in traditional lecture-oriented courses. It is very effective in situations where an instructor wishes to modify a traditional course by devoting a portion of class time to independent study. The book is especially useful in programs of individualized or self-paced instruction, whether in a learning lab situation, with tutors, with audio tapes, or in totally independent study.

Calculators   Calculators are a necessary tool for workers in trade and technical areas, and we have recognized this by using calculators extensively in the text, both in finding numerical solutions to problems, including specific keystroke sequences, and in determining the values of transcendental functions. We have taken care to first explain all concepts and problem solving without the use of the calculator and to estimate and check answers. Many realistic problems included in the exercise sets involve large numbers, repeated calculations, and large quantities of information and are ideally suited to calculator use. They are representative of actual trades situations where a calculator is needed. Detailed instruction on the use of calculators is included in special sections at the end of appropriate chapters or is integrated into the text.

Supplements

An extensive package of supplementary materials is available for teachers and students using this textbook:

- A ExamGear CD-ROM (packaged with this text) provides readers with extra practice problems.
- An Instructor's Solutions Manual containing completely worked out solutions to all exercises and providing step-by-step procedures to help students learn problem-solving skills.
- Transparency Masters for over 130 illustrations, charts, and other visual materials.
- A Test Item File of approximately 1000 problem-type, multiple-choice, and fill-in questions for a full range of testing.
- The Prentice Hall TestGen, a state-of-the-art system that provides, on CD-ROM, the test item file.
- A booklet, *How to Study Technical Math,* to illustrate methods and techniques for solving a wide range of problems in technical math.

Seventh Edition

In response to suggestions from many instructors who have used earlier versions of the text, this seventh edition includes a number of substantial revisions and improvements:

- All problem sets and examples have been revised and brought up to date. More than 100 new applied problems from the trades have been added to the exercise sets, and additional worked examples have been added throughout the text.
- Numerous changes have been made in the wording and presentation of worked examples and applied problems throughout the text. These small changes, designed to improve the text, appear on almost every page.
- Experts in trades and technical areas have been consulted to help review and bring up to date all applied problems. The text now contains nearly 1000 application problems.
- The explanation for finding the Least Common Denominator has been revised and significantly simplified.
- Following current usage, we have made more extensive use of exponents in writing units.
- All calculator examples and problems assume the use of a standard scientific calculator with a ten-digit display.
- Explanations in many step-by-step examples have been improved and their readability has been enhanced.

Acknowledgments

It is a pleasure to acknowledge the help of many people who have contributed to the development of this book. 'Lyn Carman spent countless hours interviewing trades workers, union leaders, apprentices, teachers, and training program directors so that their experience could be used in developing realistic problems, an effective format, and appropriate content. Her contributions were invaluable to us.

The staff of Prentice Hall provided outstanding assistance at every step of the development and production of this seventh edition. We are especially grateful to editors Steve Helba and Gary Bauer of Prentice Hall for their guidance and assistance and to Penny Walker and Kristin Landon for their help in moving this edition through the production process.

We are indebted to the following teachers who read preliminary versions of the text and offered many helpful suggestions:

Robert Ahntholz, Coordinator—Learning Center, Central City Occupational Center, Los Angeles, California

Dean P. Athans, East Los Angeles College, Monterey Park, California

Frances L. Brewer, Vance-Granville Community College, Henderson, North Carolina

Peter Arvanites, Rockland Community College, Suffern, New York

Gerald Barkley, Glen Oaks Community College, Centreville, Michigan

James W. Brennan, Boise State University, Boise, Idaho

Robert S. Clark, Spokane Community College, Spokane, Washington

James W. Cox, Merced College, Merced, California

Harry Craft, Spencerian College, Lexington, Kentucky

Liz Cunningham, Bellingham Technical College, Bellingham, Washington

Gordon A. DeSpain, San Juan College, Farmington, New Mexico

B. H. Dwiggins, Systems Development Engineer, Technovate, Inc., Tacoma, Washington

Kenneth R. Ebernard, Chabot College, Hayward, California

Hal Ehrenreich, North Central Technical College, Wausau, Wisconsin

Donald Fama, Chairperson, Mathematics-Engineering Science Department, Cayuga Community College, Auburn, New York

James Graham, Industrial Engineering College of Chicago, Chicago, Illinois

Mary Glenn Grimes, Bainbridge College, Bainbridge, Georgia

Ronald J. Gryglas, Tool and Die Institute, Chicago, Illinois

Vincent J. Hawkins, Chairperson, Department of Mathematics, Warwick Public Schools, Warwick, Rhode Island

Bernard Jenkins, Lansing Community College, Lansing, Michigan

Chris Johnson, Spokane Community College, Spokane, Washington

Judy Ann Jones, Madison Area Technical College, Madison, Wisconsin

Robert Kimball, Wake Technical College, Raleigh, North Carolina

J. Tad Martin, Technical College of Alamance, Haw River, North Carolina

Gregory B. McDaniel, Texas State Technical College, Texas

Sharon K. Miller, North Central Technical College, Mansfield, Ohio

David C. Mitchell, Seattle Central Community College, Seattle, Washington

Voya S. Moon, Western Iowa Technical Community College, Sioux City, Iowa

Robert E. Mullaney, Santa Barbara School District, Santa Barbara, California

Joe Mulvey, Salt Lake Community College, Salt Lake City, Utah

Jack D. Murphy, Penn College, Williamsport, Pennsylvania

R. O'Brien, SUNY–Canton, Canton, New York

Steven B. Ottmann, Southeast Community College, Lincoln, Nebraska

Emma M. Owens, Tri-County Technical College, Pendleton, South Carolina

David A. Palkovich, Spokane Community College, Spokane, Washington

William Poehler, Santa Barbara School District, Santa Barbara, California

Richard Powell, Fullerton College, Fullerton, California

Martin Prolo, San Jose City College, San Jose, California

Ilona Ridgeway, Fox Valley Technical College, Appleton, Wisconsin

Kurt Schrampfer, Fox Valley Technical College, Appleton, Wisconsin

Richard C. Spangler, Developmental Instruction Coordinator, Tacoma Community College, Tacoma, Washington

Arthur Theobald, Bergen County Vocational School, Wayne, New Jersey

Curt Vander Vere, Pennsylvania College of Technology, Williamsport, Pennsylvania

Cathy Vollstedt, North Central Technical College, Wausau, Wisconsin

Joseph Weaver, Associate Professor, State University of New York, Delhi, New York

Kay White, Walla Walla Community College, Walla Walla, Washington

Raymond E. Wilhite, Solano Community College, Vacaville, California

Edward Graper, LeBow Co., Goleta, California

This book has benefited greatly from their excellence as teachers.

Finally, through every step of the seemingly endless sequence of researching, interviewing, testing, writing, and rewriting that makes a textbook, we have benefited from the patience, understanding, and concern of our wives, 'Lyn and Chris. They have made it a better book and a more pleasant experience than we would have otherwise had.

Robert A. Carman
Hal M. Saunders

Santa Barbara, California

How to Use This Book

In this book you will find many questions, not only at the end of each chapter or section, but on every page. This textbook is designed for those who need to learn the practical math used in the trades, and who want it explained carefully and completely at each step. The questions and explanations are designed so that you can:

- Start either at the beginning or where you need to start.
- Work on only what you need to know.
- Move as fast or as slowly as you wish.
- Receive the guidance and explanation you need.
- Skip material you already understand.
- Do as many practice problems as you need.
- Test yourself often to measure your progress.

In other words, if you find mathematics difficult and you want to be guided carefully through it, this book is designed for you.

This is no ordinary book. It is not designed for browsing or casual reading. You *work* your way through it. You will be led carefully, step by step, through a worked example. After this example, you will be challenged to work through a similar problem. In the solution to these *YOUR TURN* problems, you will be guided through the explanation. Finally, you will be given a problem or a brief set of problems to check your understanding of the concepts. Answers or detailed solutions provide immediate feedback. Learning helps or hints provide additional insight.

Because we know that every person is different and has different needs, each major section of the book starts with a preview that will help you to determine the topics on which you need to work.

Most students hesitate to ask questions that nag at their understanding. They are fearful that "dumb questions" will humiliate them and reveal their lack of understanding. To relieve you of worry over dumb questions, we will ask and answer them for you. Thousands of students have taught us that "dumb questions" can produce smart students.

This textbook has been designed for students who not only will read it but will be working to achieve understanding and acquire skills.

Be alert for the following learning devices as you use this textbook.

Reference Head   Within each section of a chapter when new vocabulary is introduced or when new topics are presented, **reference headings** appear in the left margin as shown here. These signals show where the new information appears. Reference heads will be useful to you when you need to locate a word, phrase, equation, or topic, or when you need to review.

Note ▶   Every experienced teacher knows that certain mathematical concepts and procedures will present special difficulties for students. To help you with these, special notes are included in the text. A large triangle (▶) and a warning word appear in the left margin to indicate the start of the comment, and another triangle (◀)

shows when it is completed. The word **NOTE**, as used at the start of this paragraph, calls your attention to conclusions or consequences that might be overlooked. The word **CAREFUL** warns you about common mistakes many students make, and the phrase **LEARNING HELP** offers an alternative explanation or helpful hint. ◄

An important feature of this textbook is the use of many worked examples and detailed problem solutions. Each example in the text is introduced with a heading like this:

EXAMPLE

In these examples we will demonstrate in detail, step by step, how to solve a typical problem using the material being studied.

YOUR TURN

After working an example, you will be invited to work on a similar problem. To identify this second step look for the *YOUR TURN* heading.

SOLUTION

After solving a *YOUR TURN* problem, you will be guided through a solution to check your work.

PROBLEMS

Finally, when you are ready to test your understanding, a short problem set is given and followed by answers or detailed solutions.

A **check icon** appears in a problem solution to remind you to check your work.

Color, , and other $\boxed{\text{graphical aids}}$ are used to highlight the important or tricky aspects of a solution, if needed.

 The calculator is an important tool for the modern trades worker or technician, and we assume in this textbook that once you have learned the basic operations of arithmetic you will use a calculator. Problems in the exercise sets or examples in the text that involve the use of a scientific calculator are preceded by the calculator symbol shown here.

Solutions often include a display of the proper calculator key sequences. For example, the calculation

$$\frac{85.7 + (12.9)^2}{71.6}$$

would be shown as

85.7 ⊞ **12.9** $\boxed{x^2}$ ⊟ ⊡**71.6** ⊟ → 3.5210894385

**Exercises 0-1
Sample Problem Set**
At the conclusion of each section of each chapter, you will find a set of problems covering the work of that section. These will include a number of routine or drill problems as well as applications or word problems. Each applied problem begins with an indication of the trades area from which it has been taken. Many of these applications have been obtained from trades workers in these areas.

Problem Set 0   Each chapter ends with a set of problems reviewing all of the material covered in the chapter.

> Important rules, definitions, equations, or helpful hints are often placed in a box like this so that they will be easy to find.

If your approach to learning mathematics is to skim the text lightly on the way to puzzling through a homework assignment, you will have difficulty with this or any other textbook. If you are motivated to study mathematics so that you understand it and can use it correctly, this textbook is designed for you.

According to an old Spanish proverb, "the world is an ocean and he who cannot swim will sink to the bottom." A study published by the U.S. Office of Education revealed that two-thirds of the skilled and semiskilled job opportunities in today's labor market are available only to those who have an understanding of the basic principles of arithmetic, algebra, and geometry. If the modern world of work is an ocean, the skill needed to keep afloat or even swim to the top is clearly mathematics. It is the purpose of this book to help you learn these basic skills.

Now, turn to page 1 and let's begin.

R. A. C.
H. M. S.

Arithmetic of Whole Numbers

Objective	Sample Problems		For Help Go to Page

When you finish this unit you will be able to:

1. Add and subtract whole numbers.

    (a) $67 + 58$ _____ 6

    (b) $7009 + 1598$ _____

    (c) $82 - 45$ _____ 17

    (d) $4035 - 1967$ _____

    (e) $14 + 31 + 59 - 67 + 22 + 37 - 19$ _____

2. Multiply and divide whole numbers.

    (a) 64×37 _____ 24

    (b) 305×243 _____

    (c) 908×705 _____

                                34

    (d) $2006 \div 6$ _____

    (e) $7511 \div 37$ _____

3. Do word problems with whole numbers.

    Machine Technology A metal casting weighs 680 lb; 235 lb of metal is removed during shaping. What is its finished weight?   _____

_____

Name

_____

Date

_____

Course/Section

4. Determine factors and prime factors.

    (a) List all the factors of 12.   39

    (b) Write 12 as a product of its prime factors.   40

5. Use the correct order of operations with addition, subtraction, multiplication, and division.

    (a) $6 + 9 \times 3$ 44

    (b) $35 - 14 \div 7$

    (c) $56 \div 4 \times 2 + 9 - 4$

    (d) $(23 - 7) \times 24 \div (12 - 4)$

(Answers to these preview problems are given in the Appendix. Don't peek.)

If you are certain that you can work *all* these problems correctly, turn to page 49 for a set of practice problems. If you cannot work one or more of the preview problems, turn to the page indicated to the right of the problem. For those who wish to master this material with the greatest success, turn to Section 1-1 and begin to work there.

Arithmetic of Whole Numbers

Reprinted with permission of Universal Press Syndicate.

The average person a century ago used numbers to tell time, count, and keep track of money. Today, most people need to develop technical skills based on their ability to read, write, and work with numbers in order to earn a living. Although we live in an age of computers and calculators, much of the simple arithmetic used in industry, business, and the skilled trades is still done mentally or by hand. In fact, most trade and technical areas require you to *prove* that you can do the calculations by hand before you can get a job.

In the first part of this book we take a practical, how-to-do-it look at basic arithmetic: addition, subtraction, multiplication, and division, including fractions, decimal numbers, negative numbers, powers, and roots. Once we are past the basics, we will show you how to use a calculator to do such calculations. There are no quick and easy formulas here, but we do provide a lot of help for people who need to use mathematics in their daily work.

1-1 WORKING WITH WHOLE NUMBERS

The simplest numbers are the whole numbers—the numbers we use for counting the number of objects in a group. The whole numbers are 0, 1, 2, 3, . . . , and so on.

> EXAMPLE

How many letters are in the collection shown in the margin?

We counted 23. Notice that we can count the letters by grouping them into sets of ten:

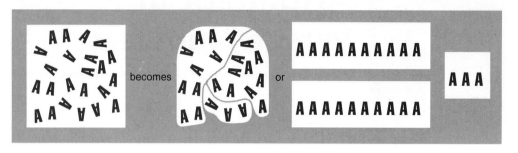

2 tens + 3 ones

20   + 3   or 23

Expanded Form   Mathematicians call this the *expanded form* of a number. For example,

$$46 = \quad 40 + 6 \qquad = 4 \text{ tens} + 6 \text{ ones}$$
$$274 = 200 + 70 + 4 = 2 \text{ hundreds} + 7 \text{ tens} + 4 \text{ ones}$$
$$305 = 300 + 5 \qquad = 3 \text{ hundreds} + 0 \text{ tens} + 5 \text{ ones}$$

Only ten numerals or number symbols—0, 1, 2, 3, 4, 5, 6, 7, 8, and 9—are needed to write any number. These ten basic numerals are called the *digits* of the number. The digits 4 and 6 are used to write 46, the number 274 is a three-digit number, and so on.

YOUR TURN

Write out the following three-digit numbers in expanded form:

(a)   362 = ____ + ____ + ____ = ____ hundreds + ____ tens + ____ ones

(b)   425 = ____ + ____ + ____ = ____ hundreds + ____ tens + ____ ones

(c)   208 = ____ + ____ + ____ = ____ hundreds + ____ tens + ____ ones

SOLUTIONS

(a)   *362 = 300 + 60 + 2 = 3 hundreds + 6 tens + 2 ones*

(b)   *425 = 400 + 20 + 5 = 4 hundreds + 2 tens + 5 ones*

(c)   *208 = 200 + 0 + 8 = 2 hundreds + 0 tens + 8 ones*

Notice that the 2 in 362 means something different from the 2 in 425 or 208. In 362 the 2 signifies two ones. In 425 the 2 signifies two tens. In 208 the 2 signifies two hundreds. Ours is a *place-value* system of naming numbers: the value of any digit depends on the place where it is located.

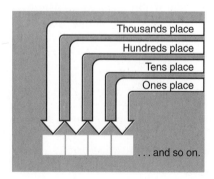

Being able to write a number in expanded form will help you to understand and remember the basic operations of arithmetic—even though you'll never find it on a blueprint or in a technical handbook.

This expanded-form idea is useful especially in naming very large numbers. Any large number given in numerical form may be translated to words by using the following diagram:

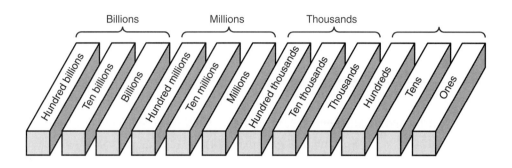

EXAMPLE

The number 14,237 can be placed in the diagram like this:

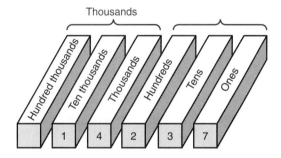

and read "fourteen thousand, two hundred thirty-seven."

EXAMPLE

The number 47,653,290,866 becomes

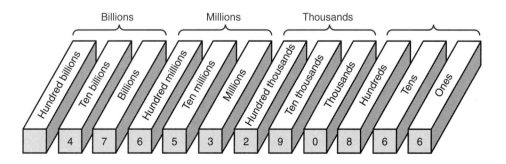

and is read "forty-seven billion, six hundred fifty-three million, two hundred ninety thousand, eight hundred sixty-six."

In each block of three digits read the digits in the normal way ("forty-seven," "six hundred fifty-three") and add the name of the block ("billion," "million"). Notice that the word "and" is not used in naming these numbers.

YOUR TURN

Use the diagram to name the following numbers.

(a) 4072

(b) 1,360,105

(c) 3,000,210

(d) 21,010,031,001

ANSWERS

(a) Four thousand, seventy-two

(b) One million, three hundred sixty thousand, one hundred five

(c) Three million, two hundred ten

(d) Twenty-one billion, ten million, thirty-one thousand, one

It is also important to be able to write numbers correctly when you hear them spoken or when they are written in words.

PROBLEMS

Read each of the following aloud and then write them in correct numerical form.

(a) Fifty-eight thousand, four hundred six

(b) Two hundred seventy-three million, five hundred forty thousand

(c) Seven thousand sixty

(d) Nine billion, six million, two hundred twenty-three thousand, fifty-eight

ANSWERS

(a) 58,406

(b) 273,540,000

(c) 7060

(d) 9,006,223,058

Addition of Whole Numbers

Adding whole numbers is fairly easy provided that you have stored in your memory a few simple addition facts. It is most important that you be able to add simple one-digit numbers mentally.

The following sets of problems in one-digit addition are designed to give you some practice. Work quickly. You should be able to answer all problems in a set in the time shown.

Practice   One-Digit Addition
A. Add.

7	5	2	5	8	2	3	8	9	7
3	6	9	7	8	5	6	7	3	6

6	8	9	3	7	2	9	9	7	4
4	5	6	5	7	7	4	9	2	7
9	2	5	8	4	9	6	4	8	8
7	6	5	9	5	5	6	3	2	3
5	6	7	7	5	6	2	3	6	9
8	7	5	9	4	5	8	7	8	8
7	5	9	4	3	8	4	8	5	7
4	9	2	6	8	6	9	4	8	8

Average time = 90 seconds
Record = 35 seconds

B. Add. Try to do all addition mentally.

2	7	3	4	2	6	3	5	9	5
5	3	6	5	7	7	4	7	6	2
4	2	5	8	9	8	4	8	3	8
6	5	4	8	6	9	7	4	8	1
2	4	2	1	8	3	1	9	4	8
7	5	9	9	8	5	6	1	6	7
1	9	3	1	7	2	9	9	8	5
9	9	1	6	9	9	8	5	3	4
2	1	4	3	6	1	2	1	3	7

Average time = 90 seconds
Record = 41 seconds

The answers are given in the Appendix.

Rounding Whole Numbers In many situations a simplified approximation of a number is more useful than its exact value. For example, the accountant for a business may calculate its total monthly revenue as $247,563, but the owner of the business may find it easier to talk about the revenue as "about $250,000." The process of approximating a number is called *rounding*. Rounding numbers comes in handy when we need to make estimates or do "mental mathematics."

A number can be rounded to any desired place. For example, $247,563 is approximately

$247,560 rounded to the nearest ten,
$247,600 rounded to the nearest hundred,
$248,000 rounded to the nearest thousand,
$250,000 rounded to the nearest ten thousand, and
$200,000 rounded to the nearest hundred thousand.

To round a whole number, follow this **step-by-step** process:

EXAMPLE

Round 247,563	to the nearest hundred thousand	to the nearest ten thousand
Step 1 Determine the place to which the number is to be rounded. Mark it on the right with a ∧.	$2 47,563 \\ \quad ∧	$24 7,563 \\ \quad ∧
Step 2 If the digit to the right of the mark is less than 5, replace all digits to the right of the mark with zeros.	$200,000	
Step 3 If the digit to the right of the mark is equal to or larger than 5, increase the digit to the left by 1 and replace all digits to the right with zeros.		$250,000

YOUR TURN

Try these for practice. Round

(a) 73,856 to the nearest thousand

(b) 64 to the nearest ten

(c) 4852 to the nearest hundred

(d) 350,000 to the nearest hundred thousand

(e) 726 to the nearest hundred

SOLUTIONS

(a) **Step 1** Place a mark to the right of the thousands place. The digit 3 is in the thousands place.

73 856
∧

Step 2 Does not apply.

Step 3 The digit to the right of the mark, 8, is larger than 5. Increase the 3 to a 4 and replace all digits to the right with zeros.

74,000

(b) **Step 1** Place a mark to the right of the tens place. The digit 6 is in the tens place.

6 4
∧

Step 2 The digit to the right of the mark, 4, is less than 5. Replace it with a zero.

60

(c) 4900   (d) 400,000   (e) 700

Doing arithmetic with one-digit numbers is very important. It is the key to any mathematical computation—even if you do the work on a calculator. Suppose that you need to find the total time spent on a job by two workers. You need to find the sum

31 hours + 48 hours = _____

Estimating   What is the first step? Start adding digits? Punch in some numbers on your trusty calculator? Rattle your abacus? None of these. The first step is to *estimate* your answer. The most important rule in any mathematical calculation is:

> Know the approximate answer to any calculation before you calculate it.

Never do an arithmetic calculation until you know roughly what the answer is going to be. Always know where you are going.

Rounding to the nearest 10 hours, the preceding sum can be estimated as 31 hours + 48 hours or approximately 30 hours + 50 hours or 80 hours, not 8 or 8000 or 800 hours. Having the estimate will keep you from making any major (and embarrassing) mistakes. Once you have a rough estimate of the answer, you are ready to do the arithmetic work.

Calculate 31 + 48 = _____

You don't really need an air-conditioned, solar-powered, talking calculator for that, do you?

You should set it up like this:

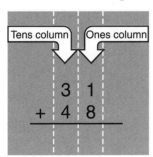

1. The numbers to be added are arranged vertically (up and down) in columns.

2. The right end or ones digits are placed in the ones column, the tens digits are placed in the tens column, and so on.

Avoid the confusion of $\begin{array}{r} 31 \\ +\ 48 \end{array}$ or $\begin{array}{r} 31 \\ +\ 48 \end{array}$

Careful ▶   Most often the cause of errors in arithmetic is carelessness, especially in simple tasks such as lining up the digits correctly. ◀

Once the digits are lined up the problem is easy.

$$\begin{array}{r} 31 \\ +\ 48 \\ \hline 79 \end{array}$$

Does the answer agree with your original estimate? Yes. The estimate, 80, is roughly equal to the actual sum, 79.

What we have just shown you is called the *guess n' check method* of doing mathematics.

1. **Estimate** the answer using rounded numbers.

2. **Work** the problem carefully.

 3. **Check** your answer against the estimate. If they disagree, repeat both steps 1 and 2. The check icon reminds you to check your work.

Most students worry about estimating, either because they won't take the time to do it or because they are afraid they might do it incorrectly. Relax. You are the only one who will know your estimate. Do it in your head, do it quickly, and make it reasonably accurate. Step 3 helps you to find incorrect answers before you finish the problem. The guess n' check method means that you never work in the dark; you always know where you are going.

Note ▶   Estimating is especially important in practical math, where a wrong answer is not just a mark on a piece of paper. An error may mean time and money lost. ◀

Here is a slightly more difficult problem:

27 lb + 58 lb = _____

Try it, then check your answer.

I called 4 + 2 "four and two" and my math teacher looked unhappy. Why?

Call 4 + 2 "four plus two." There is no math operation called "and." Your instructor gets upset because he's not sure what you are saying and he's afraid you don't know either.

SOLUTION

First, estimate the answer. 27 + 58 is roughly
30 + 60 or about 90.
The answer is about 90 lb.

Second, line up the digits in columns.

$$\begin{array}{r} 27 \\ + 58 \end{array}$$

The numbers to be added, 27 and 58 in this case, are usually called *addends.*

Third, add carefully.

$$\begin{array}{r} \overset{1}{} \\ 27 \\ + 58 \\ \hline 85 \end{array}$$

✓ **Finally,** check your answer by comparing it with the estimate. The estimate 90 lb is roughly equal to the answer 85 lb—at least you have an answer in the right ballpark.

What does that little 1 above the tens digit mean? What really happens when you "carry" a digit? Let's look at it in detail. In expanded notation,

$$\begin{aligned} 27 &\to 2 \text{ tens } + 7 \text{ ones} \\ + 58 &\to 5 \text{ tens } + 8 \text{ ones} \\ &= 7 \text{ tens } + 15 \text{ ones} \\ &= 7 \text{ tens } + 1 \text{ ten } + 5 \text{ ones} \\ &= 8 \text{ tens } \qquad + 5 \text{ ones} \\ &= 85 \end{aligned}$$

The 1 that is carried over to the tens column is really a ten.

Learning Help ▶ Trades people often must calculate exact answers mentally. To do a problem such as 27 + 58, a trick called "balancing" works nicely. Simply add 2 to the 58 to get a "round" number, 60, and subtract 2 from 27 to balance, keeping the total the same. Therefore, 27 + 58 is the same as 25 + 60, which is easy to add mentally to get 85. ◀

The following is a short set of problems. Add, and be sure to estimate your answers first. Check your answers with your estimates.

(a) 429 + 738 = _____

(b) 446 + 867 = _____

(c) 2368 + 744 = _____

(d) 409 + 72 = _____

(e) **Masonry**   Three bricklayers working together on a job each laid the following number of bricks in a day: 927, 1143, and 1065. How many bricks did all three lay that day?

SOLUTIONS

(a) **Estimate:**   Rounding each number to the nearest hundred, 400 + 700 = 1100
Line up the digits: 429
                          + 738

Calculate:

Step 1   429
            + 738   9 + 8 = 17   Write 7; carry 1 ten.
               7

Step 2   429
            + 738   1 + 2 + 3 = 6   Write 6.
              67

Step 3   429
            + 738   4 + 7 = 11   Write 11.
            1167

☑ **Check:**   The estimate 1100 and the answer 1167 are roughly equal.

(b) **Estimate:**   400 + 900 = 1300

Calculate:

Step 1   446
            + 867   6 + 7 = 13   Write 3; carry 1 ten.
               3

Step 2   446
            + 867   1 + 4 + 6 = 11   Write 1; carry 1 hundred.
              13

Step 3   446
            + 867   1 + 4 + 8 = 13   Write 13.
            1313

☑ **Check:**   The estimate 1300 and the answer 1313 are roughly equal.

(c) **Estimate:**   Rounding each number to the nearest hundred, 2400 + 700 = 3100

Calculate: 2368
                 + 744
                 3112

☑ **Check:**   The estimate 3100 and the answer 3112 are roughly equal.

(d) **Estimate:**   400 + 100 = 500

Calculate: 409
                 + 72
                 481

✓ **Check:** The estimate 500 and the answer 481 are roughly equal.

(e) **Estimate:** $900 + 1100 + 1100 = 3100$

Calculate:
$$
\begin{array}{r}
{\scriptstyle 111}\\
927\\
1143\\
+\ 1065\\
\hline
3135
\end{array}
$$

✓ **Check:** The estimate 3100 is roughly equal to the answer 3135.

> Where do the + and – signs come from?

> They were used in the fifteenth century to show that boxes of merchandise were overweight (+) or underweight (–). Within 40 years or so bookkeepers and mathematicians started using them. The + symbol came from the Latin word *et* meaning and.

Estimating answers is a very important part of any mathematics calculation, especially for the practical mathematics used in engineering, technology, and the trades. A successful builder, painter, or repairperson must make accurate estimates of job costs—business success depends on it. If you work in a technical trade, getting and keeping your job may depend on your ability to get the correct answer *every* time.

If you use a calculator to do the actual arithmetic, it is even more important to get a careful estimate of the answer first. If you plug a wrong number into the calculator, accidentally hit a wrong key, or unknowingly use a failing battery, the calculator may give you a wrong answer—lightning fast, but wrong. The estimate is your best insurance that a wrong answer will be caught immediately. Convinced?

Now, work the following problems for practice in adding whole numbers.

Exercises 1-1

Working with Whole Numbers

A. Add.

1. 47
   23
	2. 27
	   38
		3. 45
		   35
			4. 38
			   65
				5. 75
				   48

6. 26
   98
	7. 48
	   84
		8. 67
		   69
			9. 189
			   204
				10. 508
				   495

11. 684
   706
	12. 432
	   399
		13. 621
		   388
			14. 747
			    59
				15. 375
				   486

16. 4237
   1288
	17. 5076
	   4385
		18. 7907
		   1395
			19. 3785
			   7643
				20. 6709
				   9006

21. 18745	22. 40026	23. 10674	24. 9876	25. 78044
6972	7085	397	4835	97684

B. Arrange vertically and add.

1. $487 + 29 + 526 =$ _____

2. $715 + 4293 + 184 + 19 =$ _____

3. $1706 + 387 + 42 + 307 =$ _____

4. $456 + 978 + 1423 + 3584 =$ _____

5. $6284 + 28 + 674 + 97 =$ _____

6. $6842 + 9008 + 57 + 368 =$ _____

7. $322 + 46 + 5984 =$ _____

8. $7268 + 209 + 178 =$ _____

9. $5016 + 423 + 1075 =$ _____

10. $8764 + 85 + 983 + 19 =$ _____

11. $4 + 6 + 11 + 7 + 14 + 3 + 9 + 6 + 4 =$ _____

12. $12 + 7 + 15 + 16 + 21 + 8 + 10 + 5 + 30 + 17 =$ _____

13. $1 + 2 + 3 + 4 + 5 + 6 + 7 + 8 + 9 + 10 =$ _____

14. $22 + 31 + 43 + 11 + 9 + 1 + 19 + 12 =$ _____

15. $75 + 4 + 81 + 12 + 14 + 65 + 47 + 22 + 37 =$ _____

C. Writing and Rounding Whole Numbers

Write in words.

1. 357   2. 2304   3. 17,092   4. 207,630   5. 2,000,034

6. 10,007   7. 740,106   8. 5,055,550   9. 118,180,018   10. 6709

Write as numbers.

11. Three thousand six

12. Seventeen thousand twenty-four

13. Eleven thousand, one hundred

14. Three million, two thousand, seventeen

15. Four million, forty thousand, six

16. Seven hundred twenty million, ten

Round as indicated.

17. 357 to the nearest ten

18. 4386 to the nearest hundred

19. 4386 to the nearest thousand

20. 5386 to the nearest thousand

21. 225,799 to the nearest ten thousand

22. 225,799 to the nearest thousand

D. Applied Problems

1. **Electrical Technology** In setting up his latest wiring job, an electrician cut the following lengths of wire: 387, 913, 76, 2640, and 845 ft. Find the total length of wire used.

2. **Building Construction** The Acme Lumber Co. made four deliveries of 1-in. by 6-in. flooring: 3280, 2650, 2465, and 2970 fbm. What was the total number of board feet of flooring delivered? (The abbreviation for "board feet" is fbm, which is short for "feet board measure.")

3. **Machine Technology** The stockroom has eight boxes of No. 10 hexhead cap screws. How many screws of this type are in stock if the boxes contain 346, 275, 84, 128, 325, 98, 260, and 120 screws, respectively?

4. **Office Services** In calculating her weekly expenses, a contractor found that she had spent the following amounts: materials, $13,860; labor, $3854; salaried help, $942; overhead expense, $832. What was her total expense for the week?

5. **Machine Technology** The head machinist at Tiger Tool Co. is responsible for totaling time cards to determine job costs. He found that five different jobs this week took 78, 428, 143, 96, and 384 minutes each. What was the total time in minutes for the five jobs?

6. **Carpentry** On a home construction job, a carpenter laid 1480 wood shingles the first day, 1240 the second, 1560 the third, 1320 the fourth, and 1070 the fifth day. How many shingles did he lay in five days?

7. **Industrial Technology** Eight individually powered machines in a small production shop have motors using 420, 260, 875, 340, 558, 564, 280, and 310 watts each. What is the total wattage used when (a) the total shop is in operation? (b) the three largest motors are running? (c) the three smallest motors are running?

8. **Auto Mechanics** A mechanic is taking inventory of oil in stock. He has 24 quarts of 10W-30, 8 quarts of 30W, 42 quarts of 20W-50, 16 quarts of 10W-40, and 21 quarts of 20W-40. How many total quarts of oil does he have in stock?

9. **Building Construction** The Happy Helper building materials supplier has four piles of bricks containing 1250, 865, 742, and 257 bricks. What is the total number of bricks they have on hand?

10. **Machine Technology** A machinist needs the following lengths of 1-in. diameter rod: 8 in., 14 in., 6 in., 27 in., and 42 in. How long a rod is required to supply all five pieces? (Ignore cutting waste.)

11. **Landscaping** A new landscape maintenance business requires the following equipment:

    1 rear tiller for $597
    1 gas trimmer for $279
    1 mower for $299
    1 hedge trimmer for $139

    What is the total cost of this equipment?

12. **Electrician** The Radius Electronics Company orders 325 resistors, 162 capacitors, 25 integrated circuit boards, and 68 transistors. Calculate the total number of parts ordered.

13. **Electronics** When resistors are connected in series, the total resistance is the sum of the individual resistors. If resistances of 520 ohms, 1160 ohms, 49 ohms, and 1200 ohms are connected in series, calculate the total resistance.

14. **Electronics** Kirchhoff's law states that the sum of the voltage drops around a closed circuit is equal to the source voltage. $V_s = E_1 + E_2 + E_3 + E_4 + E_5$ Calculate the source voltage V_s for the circuit shown.

E. Calculator Problems

You probably own a calculator and, of course, you are eager to put it to work doing practical math calculations. In this book we include problem sets for calculator users. These problems are taken from real-life situations and, unlike most textbook problems, involve big numbers and lots of calculations. If you think that having an electronic brain-in-a-box means that you do not need to know basic arithmetic, you will be disappointed. The calculator helps you to work faster, but it will not tell you *what* to do or *how* to do it.

Detailed instruction on using a calculator with whole numbers appears on page 51.

Here are a few helpful hints for calculator users:

1. Always *estimate* your answer before doing a calculation.

2. *Check* your answer by comparing it with the estimate or by the other methods shown in this book. Be certain that your answer makes sense.

3. If you doubt the calculator (they do break down, you know), put a problem in it whose answer you know, preferably a problem like the one you are solving.

1. **Manufacturing** The following table lists the number of widget fasteners made by each of the five machines at the Ace Widget Co. during the last ten working days.

| Day | Machine | | | | | Daily Totals |
	A	B	C	D	E	
1	347	402	406	527	237	
2	451	483	312	563	316	
3	406	511	171	581	289	
4	378	413	0	512	291	
5	399	395	452	604	342	
6	421	367	322	535	308	
7	467	409	256	578	264	
8	512	514	117	588	257	
9	302	478	37	581	269	
10	391	490	112	596	310	
Machine Total						

(a) Complete the table by finding the number of fasteners produced each day. Enter these totals under the column "Daily Totals" on the right.

(b) Find the number of fasteners produced by each machine during the ten-day period and enter these totals along the bottom row marked "Machine Totals."

(c) Does the sum of the daily totals equal the sum of the machine totals?

2. Add the following as shown.

(a) $ 67429
6070
4894
137427
91006
399

(b) $216847
9757
86492
4875
386738
28104

(c) $693884
675489
47039
276921
44682
560487

(d) $4299 + $137 + $20 + $177 + $63 + $781 + $1008 + $671 = ?

3. **Office Services** Joe's Air Conditioning Installation Co. has not been successful, and he is wondering if he should sell it and move to a better location. During the first six months of the year his expenses were:

Rent $1860
Supplies $2540
Part-time helper $2100
Transportation $948

Taxes $315
Advertising $250
Miscellaneous $187

His monthly income was:

January $609
February $1151
March $1269

April $1381
May $1687
June $1638

(a) What was his total expense for the six-month period?

(b) What was his total income for the six-month period?

(c) Now turn your calculator around to learn what Joe should do about this unhappy situation.

4. **Electrical Technology** A mapper is a person employed by an electrical utility company who has the job of reading diagrams of utility installations and listing the materials to be installed or removed by engineers. Part of a typical job list might look like this:

INSTALLATION (IN FEET OF CONDUCTOR)

Location Code	No. 12 BHD (bare, hard-drawn copper wire)	#TX (triplex)	410 AAC (all-aluminum conductor)	110 ACSR (aluminum-core steel-reinforced conductor)	6B (No. 6, bare conductor)
A3	1740	40	1400		350
A4	1132		5090		2190
B1	500			3794	
B5		87	3995		1400
B6	4132	96	845		
C4		35		3258	2780
C5	3949		1385	1740	705

(a) How many total feet of each kind of conductor must the installer have to complete the job?

16

(b) How many feet of conductor are to be installed at each of the seven locations?

When you have completed these exercises, check your answers in the Appendix and then continue with Section 1-2.

1-2 SUBTRACTION OF WHOLE NUMBERS

Subtraction is the reverse of addition.

Addition: $3 + 4 = \square$

Subtraction: $3 + \square = 7$

Written this way, a subtraction problem asks the question: How much must be added to a given number to produce a required amount?

Most often, however, the numbers in a subtraction problem are written using a minus sign ($-$):

$17 - 8 = \square$ means that there is a number $\square$ such that $8 + \square = 17$

But we should remember that

$8 + 9 = 17$ or $17 - 8 = 9$ ⟵ Difference

The *difference* is the name given to the answer in a subtraction problem.

Solving simple subtraction problems depends on your knowledge of the addition of one-digit numbers.

For example, to solve the problem

$9 - 4 = $ _____

you probably go through a chain of thoughts something like this:

> Nine minus four. Four added to what number gives nine? Five? Try it: four plus five equals nine. Right.

Subtraction problems with small whole numbers will be easy for you if you know your addition tables.

EXAMPLE

Here is a more difficult subtraction problem:

$47 - 23 = $ _____

The **first** step is to estimate the answer—remember?

$47 - 23$ is roughly $50 - 20$ or 30.

The difference, your answer, will be about 30—not 3 or 10 or 300.

The **second** step is to write the numbers vertically as you did with addition. Be careful to keep the ones digits in line in one column, the tens digits in a second column, and so on.

$$\begin{array}{r} 4\ 7 \\ -2\ 3 \end{array}$$ Notice that the larger number is written above the smaller number.

Once the numbers have been arranged in this way, the difference may be written by performing the following two steps:

Step 1

$$\begin{array}{r} 4\ 7 \\ -2\ 3 \\ \hline 4 \end{array}$$ ones digits: $7 - 3 = 4$

$$\begin{array}{r} 4\ \ 7 \\ -2\ \ 3 \\ \hline 2\ \ 4 \end{array}$$ tens digits: $4 - 2 = 2$

 The difference is 24, which agrees roughly with our estimate.

EXAMPLE

With some problems it is necessary to rewrite the larger number before the problem can be solved. Let's try this one:

$64 - 37 =$ _____

First, estimate the answer. Rounding to the nearest ten, $64 - 37$ is roughly $60 - 40$ or 20.

Second, arrange the numbers vertically in columns.
$$\begin{array}{r} 64 \\ -37 \end{array}$$

Because 7 is larger than 4 we must "borrow" one ten from the 6 tens in 64. We are actually rewriting 64 (6 tens + 4 ones) as 5 tens + 14 ones. In actual practice our work would look like this:

Step 1
$$\begin{array}{r} 6\ \ 4 \\ -3\ \ 7 \\ \hline \end{array}$$

Step 2
$$\begin{array}{r} {\scriptstyle 5}\ \ {\scriptstyle 14} \\ \not{6}\ \ \not{4} \\ -3\ \ 7 \\ \hline 7 \end{array}$$ ⟵ Borrow one ten, change the 6 in the tens place to 5, change 4 to 14, subtract $14 - 7 = 7$.

Step 3
$$\begin{array}{r} {\scriptstyle 5}\ \ {\scriptstyle 14} \\ \not{6}\ \ \not{4} \\ -3\ \ 7 \\ \hline 2\ \ 7 \end{array}$$

$\boxed{5 - 3 = 2}$ $\boxed{14 - 7}$

 Double-check subtraction problems by adding the answer and the smaller number; their sum should equal the larger number.

Step 4 Check:
$$\begin{array}{r} 37 \\ +27 \\ \hline 64 \end{array}$$

Learning Help ▶ If you need to get an exact answer to a problem such as $64 - 37$ mentally, add or subtract to make the smaller number, 37, a "round" number. In this case, add 3 to make it 40. Since we're subtracting, we want the *difference,* not the *total,* to be the same or balance. Therefore, we also add 3 to the 64 to get 67. The problem becomes $64 - 37 = (64 + 3) - (37 + 3) = 67 - 40$. Subtracting a round number is easy mentally: $67 - 40 = 27$. ◀

Can you do a problem like 37 – 64?

In this section we are considering only subtraction of a smaller number from a larger number. See Chapter 6 for information on that problem.

Try these problems for practice.

(a)　71
　　−39

(b)　263
　　−127

(c)　426
　　−128

(d)　902
　　−465

SOLUTIONS

(a) **Estimate:** $70 - 40 = 30$

Step 1　　**Step 2**

Borrow one ten from 70,
change the 7 in the tens place to 6,
change the 1 in the ones place to 11.

11 − 9 = 2　Write 2

6 − 3 = 3　Write 3

☑ The answer 32 is approximately equal to the estimate 30. As a shortcut, mentally add 1 to each number,

$$71 - 39 = 72 - 40$$

then subtract.

$$72 - 40 = 32$$

(b) **Estimate:** $260 - 130 = 130.$

Step 1　　**Step 2**

Borrow one ten from 60,
change the 6 in the tens place to 5,
change the 3 in the ones place to 13.

13 − 7 = 6　Write 6

5 − 2 = 3　Write 3

2 − 1 = 1　Write 1

☑ The answer is approximately equal to the estimate.

(c) **Estimate:** $400 - 100 = 300$

Step 1　　**Step 2**　　**Step 3**

In this case we borrow twice.
Borrow one ten from the 20 in
426 and make 16. Then borrow
one hundred from the 400 in 426
to make 11 in the tens place.

16 − 8 = 8　Write 8

11 − 2 = 9　Write 9

3 − 1 = 2　Write 2

☑ The answer 298 is approximately equal to the estimate 300.

(d) **Estimate:** $900 - 500 = 400$

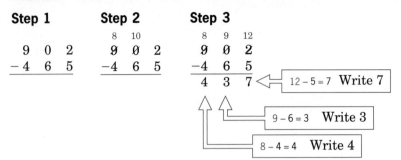

Step 1	Step 2	Step 3
	$\overset{8\quad 10}{}$	$\overset{8\quad 9\quad 12}{}$

$$
\begin{array}{r}
9\ \ 0\ \ 2 \\
-4\ \ 6\ \ 5 \\
\hline
\end{array}
\qquad
\begin{array}{r}
\overset{8\ \ 10}{\cancel{9}}\ \ 0\ \ 2 \\
-4\ \ 6\ \ 5 \\
\hline
\end{array}
\qquad
\begin{array}{r}
\overset{8\ \ 9\ \ 12}{\cancel{9}\ \ \cancel{0}}\ \ 2 \\
-4\ \ 6\ \ 5 \\
\hline
4\ \ 3\ \ 7
\end{array}
$$

$12 - 5 = 7$ Write 7

$9 - 6 = 3$ Write 3

$8 - 4 = 4$ Write 4

☑ The answer 437 is roughly equal to the estimate 400.

In problem (d) we first borrow one hundred from 900 to get a 10 in the tens place. Then we borrow one 10 from the tens place to get a 12 in the ones place.

EXAMPLE

Let's work through a few examples of subtraction problems involving zero digits.

(a) $400 - 167 = ?$

Step 1	Step 2	Step 3	Check

$$
\begin{array}{r}
4\ \ 0\ \ 0 \\
-1\ \ 6\ \ 7 \\
\hline
\end{array}
\qquad
\begin{array}{r}
\overset{3\ \ 10}{\cancel{4}\ \ \cancel{0}}\ \ 0 \\
-1\ \ 6\ \ 7 \\
\hline
\end{array}
\qquad
\begin{array}{r}
\overset{3\ \ 9\ \ 10}{\cancel{4}\ \ \cancel{0}\ \ \cancel{0}} \\
-1\ \ 6\ \ 7 \\
\hline
2\ \ 3\ \ 3
\end{array}
\qquad
\begin{array}{r}
1\ \ 6\ \ 7 \\
+2\ \ 3\ \ 3 \\
\hline
4\ \ 0\ \ 0
\end{array}
$$

Do you see in steps 2 and 3 that we have rewritten 400 as $300 + 90 + 10$?

(b) $5006 - 2487 = ?$

Step 1	Step 2	Step 3	Step 4	Check

$$
\begin{array}{r}
5\ \ 0\ \ 0\ \ 6 \\
-2\ \ 4\ \ 8\ \ 7 \\
\hline
\end{array}
\quad
\begin{array}{r}
\overset{4\ \ 10}{\cancel{5}\ \ \cancel{0}}\ \ 0\ \ 6 \\
-2\ \ 4\ \ 8\ \ 7 \\
\hline
\end{array}
\quad
\begin{array}{r}
\overset{4\ \ 9\ \ 10}{\cancel{5}\ \ \cancel{0}\ \ \cancel{0}}\ \ 6 \\
-2\ \ 4\ \ 8\ \ 7 \\
\hline
\end{array}
\quad
\begin{array}{r}
\overset{4\ \ 9\ \ 9\ \ 16}{\cancel{5}\ \ \cancel{0}\ \ \cancel{0}\ \ \cancel{6}} \\
-2\ \ 4\ \ 8\ \ 7 \\
\hline
2\ \ 5\ \ 1\ \ 9
\end{array}
\quad
\begin{array}{r}
2\ \ 4\ \ 8\ \ 7 \\
+2\ \ 5\ \ 1\ \ 9 \\
\hline
5\ \ 0\ \ 0\ \ 6
\end{array}
$$

Here is an example involving repeated borrowing.

(c) $24632 - 5718 = ?$

Step 1	Step 2	Step 3	Step 4	Check

$$
\begin{array}{r}
2\ \ 4\ \ 6\ \ 3\ \ 2 \\
-\ \ 5\ \ 7\ \ 1\ \ 8 \\
\hline
\end{array}
\quad
\begin{array}{r}
\overset{2\ \ 12}{2\ \ 4\ \ 6\ \ \cancel{3}\ \ \cancel{2}} \\
-\ \ 5\ \ 7\ \ 1\ \ 8 \\
\hline
1\ \ 4
\end{array}
\quad
\begin{array}{r}
\overset{3\ \ 16\ \ 2\ \ 12}{2\ \ 4\ \ \cancel{6}\ \ \cancel{3}\ \ \cancel{2}} \\
-\ \ 5\ \ 7\ \ 1\ \ 8 \\
\hline
9\ \ 1\ \ 4
\end{array}
\quad
\begin{array}{r}
\overset{1\ \ 13\ \ 16\ \ 2\ \ 12}{2\ \ \cancel{4}\ \ \cancel{6}\ \ \cancel{3}\ \ \cancel{2}} \\
-\ \ 5\ \ 7\ \ 1\ \ 8 \\
\hline
1\ \ 8\ \ 9\ \ 1\ \ 4
\end{array}
\quad
\begin{array}{r}
5\ \ 7\ \ 1\ \ 8 \\
+1\ \ 8\ \ 9\ \ 1\ \ 4 \\
\hline
2\ \ 4\ \ 6\ \ 3\ \ 2
\end{array}
$$

Any subtraction problem that involves borrowing should always be checked in this way. It is very easy to make a mistake in this process.

Chap. 1 Arithmetic of Whole Numbers

Why do we use = for "equals"?

The mathematician Robert Recorde invented it in 1557. He decided that two equal length parallel lines were as equal as anything available.

Now check your progress on subtraction in Exercises 1-2.

Exercises 1-2

Subtraction of Whole Numbers

A. Subtract.

1. 13 $\underline{7}$	2. 12 $\underline{5}$	3. 8 $\underline{6}$	4. 8 $\underline{0}$	5. 11 $\underline{7}$	6. 16 $\underline{7}$
7. 10 $\underline{7}$	8. 5 $\underline{5}$	9. 12 $\underline{9}$	10. 11 $\underline{8}$	11. 10 $\underline{2}$	12. 14 $\underline{6}$
13. 12 $\underline{3}$	14. 15 $\underline{6}$	15. 9 $\underline{0}$	16. 14 $\underline{5}$	17. 9 $\underline{6}$	18. 11 $\underline{5}$
19. 15 $\underline{7}$	20. 12 $\underline{7}$	21. 13 $\underline{6}$	22. 18 $\underline{0}$	23. 16 $\underline{9}$	24. 12 $\underline{4}$
25. 0 $\underline{0}$	26. 13 $\underline{8}$	27. 17 $\underline{9}$	28. 18 $\underline{9}$	29. 14 $\underline{8}$	30. 16 $\underline{8}$
31. 15 $\underline{9}$	32. 17 $\underline{8}$	33. 14 $\underline{9}$	34. 15 $\underline{8}$	35. 13 $\underline{9}$	36. 12 $\underline{8}$

B. Subtract.

1. 40 $\underline{27}$	2. 78 $\underline{49}$	3. 51 $\underline{39}$	4. 36 $\underline{17}$	5. 42 $\underline{27}$	6. 52 $\underline{16}$
7. 65 $\underline{27}$	8. 46 $\underline{17}$	9. 84 $\underline{38}$	10. 70 $\underline{48}$	11. 34 $\underline{9}$	12. 56 $\underline{18}$
13. 546 $\underline{357}$	14. 409 $\underline{324}$	15. 476 $\underline{195}$	16. 330 $\underline{76}$	17. 504 $\underline{96}$	18. 747 $\underline{593}$
19. 400 $\underline{127}$	20. 803 $\underline{88}$	21. 632 $\underline{58}$	22. 438 $\underline{409}$	23. 6218 $\underline{3409}$	24. 6084 $\underline{386}$
25. 13042 $\underline{524}$	26. 57022 $\underline{980}$	27. 5007 $\underline{266}$	28. 10000 $\underline{386}$	29. 48093 $\underline{500}$	30. 27004 $\underline{4582}$

C. Applied Problems

1. **Painting and Decorating** In planning for a particular job, a painter buys $486 worth of materials. When the job is completed, she returns five empty paint drums for a credit of $27. What was the net amount of her bill?

2. **Building Construction** How many square feet (sq ft) of plywood remain from an original supply of 8000 sq ft after 5647 sq ft is used?

3. **Welding** A storage rack at the Tiger Tool Company contains 3540 ft of 1-in. stock. On a certain job 1782 ft is used. How much is left?

4. **Welding** Five pieces measuring 26, 47, 38, 27, and 32 cm are cut from a steel bar that was 200 cm long. Allowing for a total of 1 cm for waste in cutting, what is the length of the piece remaining?

5. **Office Services** Taxes on a group of factory buildings owned by the Ace Manufacturing Company amounted to $875,977 eight years ago. Taxes on the same buildings last year amounted to $1,206,512. Find the increase in taxes.

6. **Office Services** In order to pay their bills, the owners of Edwards Plumbing Company made the following withdrawals from their bank account: $42, $175, $24, $217, and $8. If the original balance was $3610, what was the amount of the new balance?

7. **Manufacturing** Which total volume is greater, four drums containing 72, 45, 39, and 86 liters, or three drums containing 97, 115, and 74 liters? By how much is it greater?

8. **Machine Technology** Determine the missing dimension (L) in the following drawings.

(a)

(b)

(c)

(d)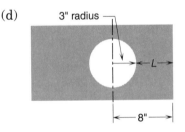

9. **Auto Mechanics** A service department began the day with 238 gallons of coolant. During the day 64 gallons were used. How many gallons remained at the end of the day?

10. **Building Construction** A truck loaded with rocks weighs 14,260 lb. If the truck weighs 8420 lb, how much do the rocks weigh?

11. **Printing** A press operator has a total of 22,000 impressions to run for a job. If the operator runs 14,250 the first day, how many are left to run?

12. **Plumbing**   In the following plumbing diagram, find pipe lengths A and B.

Problem 12

13. **Electronics**   A *potentiometer* is a device that acts as a variable resistor, allowing a resistance to change from 0 ohms to some maximum value. If a 20,000-ohm potentiometer is set to 6500 ohms (Ω) as shown, calculate the resistance R.

D. Calculator Problems

1. **Plumbing**   The Karroll Plumbing Co. has 10 trucks and, for the month of April, the following mileage was recorded on each.

Truck No.	Mileage at Start	Mileage at End
1	58352	60027
2	42135	43302
3	76270	78007
4	40006	41322
5	08642	10002
6	35401	35700
7	79002	80101
8	39987	40122
9	10210	11671
10	71040	73121

Find the mileage traveled by each truck during the month of April and the total mileage of all vehicles.

2. Which sum is greater?

987654321		123456789
87654321		123456780
7654321		123456700
654321		123456000
54321	or	123450000
4321		123400000
321		123000000
21		120000000
1		100000000

3. **Office Services**   If an electrician's helper earns $28,245 per year and she pays $3814 in withholding taxes, what is her take-home pay?

4. **Office Services**   The revenue of the Smith Construction Company for the year is $837,672 and the total expenses are $720,867. Find the difference, Smith's profit, for that year.

5. Balance the following checking account record.

Date	Deposits	Withdrawals	Balance
7/1			$6375
7/3		$ 379	
7/4	$1683		
7/7	$ 474		
7/10	$ 487		
7/11		$2373	
7/15		$1990	
7/18		$ 308	
7/22		$1090	
7/26		$ 814	
8/1			A

(a) Find the new balance A.

(b) Keep a running balance by filling each blank in the balance column.

6. **Building Construction**   A water meter installed by the BetterBilt Construction Company at a work site read 9357 cubic feet on June 1 and 17,824 cubic feet on July 1. How much water did they use during the month of June?

When you have completed these exercises, check your answers in the Appendix, then turn to Section 1-3 to study the multiplication of whole numbers.

1-3 MULTIPLICATION OF WHOLE NUMBERS

In a certain football game, the West Newton Waterbugs scored five touchdowns at six points each. How many total points did they score through touchdowns? We can answer the question several ways:

1. Count points, ...

2. Add touchdowns, $6 + 6 + 6 + 6 + 6 = ?$

or

3. Multiply $5 \times 6 = ?$

We're not sure about the mathematical ability of the West Newton scorekeeper, but most people would multiply. Multiplication is a shortcut method of performing repeated addition.

How many points did they score?

In a multiplication problem the *product* is the name given to the result of the multiplication. The numbers being multiplied are the *factors* of the product.

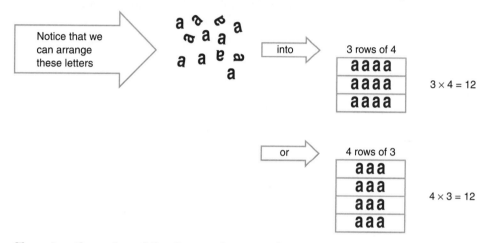

Changing the order of the factors does not change their product. This is called the *commutative* property of multiplication.

To become skillful at multiplication, you must know the one-digit multiplication table from memory. Even if you use a calculator for your work, you need to know the one-digit multiplication table to make estimates and to check your work.

YOUR TURN

Complete the table below by multiplying the number at the top by the number at the side and placing their product in the proper square. We have multiplied $3 \times 4 = 12$ and $2 \times 5 = 10$ for you.

Multiply	2	5	8	1	3	6	9	7	4
1									
7									
5	10								
4					12				
9									
2									
6									
3									
8									

Here is the completed multiplication table:

MULTIPLICATION TABLE

×	2	5	8	1	3	6	9	7	4
1	2	5	8	1	3	6	9	7	4
7	14	35	56	7	21	42	63	49	28
5	10	25	40	5	15	30	45	35	20
4	8	20	32	4	12	24	36	28	16
9	18	45	72	9	27	54	81	63	36
2	4	10	16	2	6	12	18	14	8
6	12	30	48	6	18	36	54	42	24
3	6	15	24	3	9	18	27	21	12
8	16	40	64	8	24	48	72	56	32

Multiplying by Zero and One

Notice that the product of any number and 1 is that same number. For example,

$1 \times 2 = 2$

$1 \times 6 = 6$

or even

$1 \times 753 = 753$

Zero has been omitted from the table because the product of any number and zero is zero. For example,

$0 \times 2 = 0$

$0 \times 7 = 0$

$395 \times 0 = 0$

Practice   One-Digit Multiplication

Multiply as shown. Work quickly; you should be able to answer all problems in a set correctly in the time indicated. (These times are for community college students enrolled in a developmental math course.)

A. Multiply.

6	4	9	6	3	9	7	8	2	8
2	8	7	6	4	2	0	3	7	1

6	8	5	5	2	3	9	7	3	1
8	2	9	6	5	3	8	5	6	4

7	5	4	7	4	8	6	9	8	6
4	3	9	7	2	5	7	6	8	4

5	3	5	9	9	6	1	8	4	7
4	0	5	3	9	1	1	6	4	9

Average time = 100 sec
Record = 37 sec

B. Multiply.

2	6	3	5	6	4	4	8	2	7
8	5	3	7	3	5	7	6	6	9

8	0	2	3	1	5	6	9	5	8
4	6	9	8	9	5	4	5	2	9

3	7	5	6	9	2	7	8	9	2
5	7	8	9	4	4	6	8	0	2

5	9	1	8	6	4	9	0	2	7
5	3	7	7	6	3	9	4	1	8

Average time = 100 sec
Record = 36 sec

Check your answers in the Appendix.

If you are not able to perform these one-digit multiplications quickly from memory, you should practice until you can do so. A multiplication table is given at the back of the book. Use it if you need it.

Multiplying by Larger Numbers

The multiplication of larger numbers is based on the one-digit number multiplication table.

EXAMPLE

Consider the problem:

34 × 2 = _____

First, estimate the answer: 30 × 2 = 60. The actual product of the multiplication will be about 60.

Second, arrange the factors to be multiplied vertically, with ones digits in a single column, tens digits in a second column, and so on.

$$\begin{array}{c} 3 \quad 4 \\ \times \quad 2 \end{array}$$

Finally, to make the process clear, let's write it in expanded form.

$$\begin{array}{r} 34 \\ \times 2 \end{array} \Rightarrow \begin{array}{r} 3 \text{ tens} + 4 \text{ ones} \\ \times 2 \\ \hline 6 \text{ tens} + 8 \text{ ones} = 60 + 8 = 68 \end{array}$$

 The guess 60 is roughly equal to the answer 68.

YOUR TURN

Now write the following multiplication in expanded form.

$$\begin{array}{r} 28 \\ \times 3 \end{array}$$

SOLUTION

Estimate: 30 × 3 = 90   The answer is about 90.

$$\begin{array}{r} 28 \\ \times 3 \end{array} \Rightarrow \begin{array}{r} 2 \text{ tens} + 8 \text{ ones} \\ \times 3 \\ \hline 6 \text{ tens} + 24 \text{ ones} \end{array}$$

$$= 6 \text{ tens} + 2 \text{ tens} + 4 \text{ ones}$$
$$= 8 \text{ tens} + 4 \text{ ones}$$
$$= 80 + 4$$
$$= 84$$

 90 is roughly equal to 84.

Of course, we do not normally use the expanded form; instead we simplify the work like this:

$$
\begin{array}{r}
\overset{2}{2}\ 8 \\
\times\quad 3 \\
\hline
8\ 4
\end{array}
$$

$3 \times 8 = 24$ Write 4 and carry 2 tens.
3×2 tens $= 6$ tens   6 tens $+$ 2 tens $= 8$ tens
Write 8.

PROBLEMS

Now try these problems to be certain you understand the process. Multiply as shown.

(a) $\begin{array}{r} 43 \\ \times\ 5 \\ \hline \end{array}$ (b) $\begin{array}{r} 73 \\ \times\ 4 \\ \hline \end{array}$ (c) $\begin{array}{r} 29 \\ \times\ 6 \\ \hline \end{array}$ (d) $\begin{array}{r} 258 \\ \times\ \ 7 \\ \hline \end{array}$

SOLUTIONS

(a) **Estimate:** $40 \times 5 = 200$ The answer is roughly 200.

$$
\begin{array}{r}
\overset{1}{}\ \\
4\ 3 \\
\times\quad 5 \\
\hline
2\ 1\ 5
\end{array}
$$

$5 \times 3 = 15$ Write 5; carry 1 ten.
5×4 tens $= 20$ tens
20 tens $+$ 1 ten $= 21$ tens

☑ The answer 215 is roughly equal to the estimate 200.

(b) **Estimate:** $70 \times 4 = 280$

$$
\begin{array}{r}
\overset{1}{}\ \\
7\ 3 \\
\times\quad 4 \\
\hline
2\ 9\ 2
\end{array}
$$

$4 \times 3 = 12$ Write 2; carry 1 ten.
4×7 tens $= 28$ tens
28 tens $+$ 1 ten $= 29$ tens

☑ The answer 292 is roughly equal to the estimate 280.

(c) **Estimate:** $30 \times 6 = 180$

$$
\begin{array}{r}
\overset{5}{}\ \\
2\ 9 \\
\times\quad 6 \\
\hline
1\ 7\ 4
\end{array}
$$

$6 \times 9 = 54$ Write 4; carry 5 tens.
6×2 tens $= 12$ tens
12 tens $+$ 5 tens $= 17$ tens

☑ The answer 174 is roughly equal to the estimate 180.

(d) **Estimate:** $300 \times 7 = 2100$

$$
\begin{array}{r}
\overset{4}{}\ \overset{5}{}\ \\
2\ 5\ 8 \\
\times\qquad 7 \\
\hline
1\ 8\ 0\ 6
\end{array}
$$

$7 \times 8 = 56$ Write 6; carry 5 tens.
7×5 tens $= 35$ tens
35 tens $+$ 5 tens $= 40$ tens   Write 0; carry 4 hundreds.
7×2 hundreds $= 14$ hundreds
14 hundreds $+$ 4 hundreds $= 18$ hundreds

☑ The answer and estimate are roughly equal.

Learning Help ▶ Multiplying a two- or three-digit number by a one-digit number can be done mentally. For example, think of 43×5 as 40×5 plus 3×5, or $200 + 15 = 215$. For a problem such as 29×6, think 30×6 minus 1×6 or $180 - 6 = 174$. Tricks like this are very useful on the job when neither paper and pencil nor a calculator is at hand. ◀

MULTIPLICATION SHORTCUTS

There are hundreds of quick ways to multiply various numbers. Most of them are quick only if you are already a math whiz. If you are not, the shortcuts will confuse more than help you. Here are a few that are easy to do and easy to remember.

1. To multiply by 10, attach a zero to the right end of the first factor. For example,

    $34 \times 10 = 340$

    $256 \times 10 = 2560$

    Multiplying by 100 or 1000 is similar.

    $34 \times 100 = 3400$

    $256 \times 1000 = 256000$

2. To multiply by a number ending in zeros, carry the zeros forward to the answer. For example,

    $$\begin{array}{r} 26 \\ \times\, 20 \end{array} \Rightarrow \begin{array}{r} 26 \\ \times\, 20 \\ \hline 520 \end{array}$$
    Multiply 26×2 and attach the zero on the right. The product is 520.

    $$\begin{array}{r} 34 \\ \times\, 2100 \end{array} \Rightarrow \begin{array}{r} 34 \\ \times\, 2100 \\ \hline 34\ \ \ \\ 68\ \ \ \ \\ \hline 71400 \end{array}$$

3. If both factors end in zeros, bring all zeros forward to the answer.

    $$\begin{array}{r} 230 \\ \times\, 200 \end{array} \Rightarrow \begin{array}{r} 230 \\ 200 \\ \hline 46000 \end{array}$$
    Attach three zeros to the product of 23×2.

    $$\begin{array}{r} 1000 \\ \times\ \ 100 \\ \hline 100,000 \end{array}$$
    This kind of multiplication is mostly a matter of counting zeros.

Two-Digit Multiplications

Calculations involving two-digit multipliers are done in a similar way.

EXAMPLE

To multiply

$$\begin{array}{r} 89 \\ \times\, 24 \end{array}$$

First, estimate the answer. Rounding each number to the nearest ten, $90 \times 20 = 1800$.

Second, multiply by the ones digit 4.

$$
\begin{array}{r}
\overset{3}{} \\
8\,9 \\
\times\,2\,4 \\
\hline
3\,5\,6
\end{array}
$$

$4 \times 9 = 36$ Write 6; carry 3 tens.
$4 \times 8 \text{ tens} = 32 \text{ tens}$
$32 \text{ tens} + 3 \text{ tens} = 35 \text{ tens}$

Third, multiply by the tens digit 2.

$$
\begin{array}{r}
\overset{1}{} \\
\overset{3}{} \\
8\,9 \\
\times\,2\,4 \\
\hline
3\,5\,6 \\
1\,7\,8 \leftarrow
\end{array}
$$

$2 \times 9 = 18$ Write 8; carry 1.
$2 \times 8 = 16$
$16 + 1 = 17$
└─ Leave a blank space here, since we are actually multiplying $89 \times 20 = 1780$.

Fourth, add the products obtained.

$$
\begin{array}{r}
3\,5\,6 \\
1\,7\,8 \\
\hline
2\,1\,3\,6
\end{array}
$$

✓ **Finally,** check it. The estimate and the answer are roughly the same, or at least in the same ballpark.

Notice that the product in the third step, 178, is written one digit space over from the product from the second step. When we multiplied 2×9 to get 18 in step 3, we were actually multiplying $20 \times 9 = 180$, but the zero in 180 is usually omitted to save time.

YOUR TURN

Try these:

(a) $\begin{array}{r} 64 \\ \times\,37 \\ \hline \end{array}$

(b) $\begin{array}{r} 327 \\ \times\,145 \\ \hline \end{array}$

(c) $\begin{array}{r} 342 \\ \times\,102 \\ \hline \end{array}$

(d)   **Office Services**   The Good Value hardware store sold 168 tool boxes at \$47 each. What was their total income on this item?

SOLUTIONS

(a)   **Estimate:** $60 \times 40 = 2400$

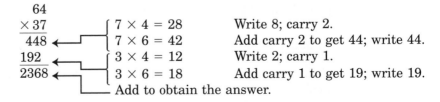

$$
\begin{array}{r}
64 \\
\times\,37 \\
\hline
448 \leftarrow \\
192 \leftarrow \\
\hline
2368 \leftarrow
\end{array}
$$

$7 \times 4 = 28$ Write 8; carry 2.
$7 \times 6 = 42$ Add carry 2 to get 44; write 44.
$3 \times 4 = 12$ Write 2; carry 1.
$3 \times 6 = 18$ Add carry 1 to get 19; write 19.
└─ Add to obtain the answer.

(b) **Estimate:** $300 \times 150 = 45,000$

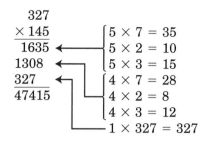

$$
\begin{array}{r}
327 \\
\times 145 \\
\hline
1635 \\
1308 \\
327 \\
\hline
47415
\end{array}
$$

$5 \times 7 = 35$	Write 5; carry 3.
$5 \times 2 = 10$	Add carry 3 to get 13; write 3, carry 1.
$5 \times 3 = 15$	Add carry 1 to get 16; write 16.
$4 \times 7 = 28$	Write 8, carry 2.
$4 \times 2 = 8$	Add carry 2 to get 10; write 0, carry 1.
$4 \times 3 = 12$	Add carry 1 to get 13; write 13.
$1 \times 327 = 327$	

The product is 47,415.

(c) **Estimate:** $300 \times 100 = 30,000$

$$
\begin{array}{r}
342 \\
\times 102 \\
\hline
684 \\
000 \\
342 \\
\hline
34884
\end{array}
$$

$2 \times 342 = 684$
$0 \times 342 = 000$
$1 \times 342 = 342$

The product is 34,884.

(d) To find the total, multiply 168 by 47.

$$
\begin{array}{r}
168 \\
\times \ 47 \\
\hline
1176 \\
672 \\
\hline
7896
\end{array}
$$

The total income is $7896.

Note ▶ Be very careful when there are zeros in the multiplier; it is very easy to misplace one of those zeros. Do not skip any steps, and be sure to estimate your answer first. ◀

THE SEXY SIX

Here are the six most often missed one-digit multiplications:

Inside digits

$9 \times 8 = 72$
$9 \times 7 = 63$
$9 \times 6 = 54$
$8 \times 7 = 56$
$8 \times 6 = 48$
$7 \times 6 = 42$

It may help you to notice that in these multiplications the "inside" digits, such as 8 and 7, are consecutive and the digits of the answer add to nine: $7 + 2 = 9$. This is true for *all* one-digit numbers multiplied by 9.

Be certain that you have these memorized.

(There is nothing very sexy about them, but we did get your attention, didn't we?)

Go to Exercises 1-3 for a set of practice problems on the multiplication of whole numbers.

Exercises 1-3

Multiplication of Whole Numbers

A. Multiply.

1. 7
  6
	2. 7
	  8
		3. 6
		  8
			4. 8
			  9
				5. 9
				  7
					6. 29
					   3

7. 9
  6
	8. 4
	  7
		9. 5
		  9
			10. 12
			   7
				11. 37
				   8
					12. 24
					   7

13. 72
   8
	14. 47
	   9
		15. 64
		   5
			16. 39
			   4
				17. 58
				   5
					18. 94
					   6

19. 32
  13
	20. 46
	  14
		21. 72
		  11
			22. 68
			  16
				23. 54
				  26
					24. 17
					   9

25. 47
   6
	26. 77
	   4
		27. 48
		  15
			28. 64
			  27
				29. 90
				  56
					30. 86
					  83

31. 34
  57
	32. 66
	  25
		33. 59
		  76
			34. 29
			  32
				35. 78
				  49
					36. 94
					  95

B. Multiply.

1. 305
  123
	2. 2006
	   125
		3. 8043
		    37
			4. 809
			   47
				5. 3706
				   102

6. 708
   58
	7. 684
	   45
		8. 2043
		   670
			9. 2008
			   198
				10. 563
				   107

11. 809
   9
	12. 609
	   7
		13. 500
		   50
			14. 542
			  600
				15. 7009
				   504

16. 407
   22
	17. 316
	   32
		18. 514
		   62
			19. 807
			  111
				20. 560
				  203

C. Practical Problems

1. **Plumbing**   A plumber receives $65 per hour. How much is he paid for 40 hours of work?

2. **Electrical Technology**   What is the total length of wire on 14 spools if each spool contains 150 ft?

3. **Building Construction**   How many total lineal feet of redwood are there in 65 2-in. by 4-in. boards each 20 ft long?

4. **Automotive Services**   An auto body shop does 17 paint jobs at a special price of $859 each. How much money does the shop receive from these jobs?

5. **Office Services**   Three different-sized boxes of envelopes contain 50, 100, and 500 envelopes, respectively. How many envelopes total are there in 18 boxes of the first size, 16 of the second size, and 11 of the third size?

6. **Machine Technology**   A machinist needs 25 lengths of steel each 9 in. long. What is the total length of steel that he needs? No allowance is required for cutting.

7. **Machine Technology**   The Ace Machine Company advertises that one of its machinists can produce 2 parts per hour. How many such parts can 27 machinists produce if they work 45 hours each?

8. **Building Construction** What is the horizontal distance in inches covered by 12 stair steps if each is 11 in. wide?

9. **Auto Mechanics** Five bolts are needed to mount a tire. How many bolts are needed to mount all four tires on 60 cars?

10. **Electrician** A voltmeter has a 4000 ohms per volt rating. This means that, for example, on the 3-volt scale, the meter reads 3×4000 or 12,000 ohms resistance. Find the resistance on the 50-volt range.

11. **Printing** A printer finds that nine 6-in. by 8-in. cards can be cut from a 19-in. by 25-in. sheet. How many such cards can be cut from 850 of the sheets?

12. **Industrial Technology** The Hold Tite fastener plant produces 738,000 cotter pins per shift. If it operates two shifts per day, five days each week, how many cotter pins will the plant produce in six weeks?

13. **Machine Technology** If a machine produces 16 screws per minute, how many screws will it produce in 24 hours?

14. **Electronics** For a transistor amplifier circuit, the collector current is calculated by multiplying the gain by the base current. For a circuit with a gain of 72, and base current 210 microamps, calculate the collector current.

15. **Roofing** In the Happy Homes development cedar shingles are laid to expose 5 in. per course. If 23 courses of shingles are needed to cover a surface, how long is the surface?

16. **Electronics** A 78-ohm resistor R draws 3 amps of current i in a given circuit. Calculate the voltage drop E across the resistor by multiplying resistance by current.

D. Calculator Problems

1. **Office Services** If an investment company started in business with $1,000,000 of capital on January 1 and, because of mismanagement, lost an average of $873 every day for a year, what would be its financial situation on the last day of the year?

2. Which of the following pay schemes gives you the most money over a one-year period?

   (a) $100 per day

   (b) $700 per week

   (c) $400 for the first month and a $400 raise each month

   (d) 1 cent for the first two-week pay period, 2 cents for the second two-week period, 4 cents for the third two-week period, and so on, the pay doubling each two weeks.

3. Multiply.

(a) $12{,}345{,}679 \times 9 =$
     $12{,}345{,}679 \times 18 =$
     $12{,}345{,}679 \times 27 =$

(b) $15{,}873 \times 7 =$
     $15{,}873 \times 14 =$
     $15{,}873 \times 21 =$

(c) $1 \times 1 =$
     $11 \times 11 =$
     $111 \times 111 =$
     $1111 \times 1111 =$
     $11111 \times 11111 =$

(d) $6 \times 7 =$
     $66 \times 67 =$
     $666 \times 667 =$
     $6666 \times 6667 =$
     $66666 \times 66667 =$

Can you see the pattern in each of these?

4. **Machine Technology** In the Aztec Machine Shop there are nine lathes each weighing 2285 lb, five milling machines each weighing 2570 lb, and three drill presses each weighing 395 lb. What is the total weight of these machines?

5. **Manufacturing** The Omega Calculator Company makes five models of electronic calculators. The following table gives the hourly production output. Find the weekly (five eight-hour days) production costs for each model.

Model	Alpha	Beta	Gamma	Delta	Tau
Cost of production of each model	$6	$17	$32	$49	$78
Number produced during typical hour	117	67	29	37	18

Check your answers in the Appendix.

1-4 DIVISION OF WHOLE NUMBERS

Division is the reverse of multiplication. It enables us to separate a given quantity into equal parts. The mathematical phrase $12 \div 3$ is read "twelve divided by three," and it asks us to separate a collection of 12 objects into 3 equal parts. The mathematical phrases

$$12 \div 3 \qquad 3\overline{)12} \qquad \frac{12}{3} \qquad \text{and} \qquad 12/3$$

all represent division, and they are all read "twelve divided by three."

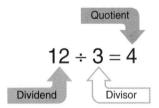

In this division problem, 12, the number being divided, is called the *dividend;* 3, the number used to divide, is called the *divisor;* and 4, the result of the division, is called the *quotient,* from a Latin word meaning "how many times."

One way to perform division is to reverse the multiplication process.

$24 \div 4 = \square$ means that $4 \times \square = 24$

If the one-digit multiplication tables are firmly in your memory, you will recognize immediately that $\square = 6$.

YOUR TURN

Try these.

$35 \div 7 = \underline{\hspace{1.5cm}}$ $30 \div 5 = \underline{\hspace{1.5cm}}$ $18 \div 3 = \underline{\hspace{1.5cm}}$

$28 \div 4 = \underline{\hspace{1.5cm}}$ $42 \div 6 = \underline{\hspace{1.5cm}}$ $63 \div 9 = \underline{\hspace{1.5cm}}$

$45 \div 5 = \underline{\hspace{1.5cm}}$ $56 \div 7 = \underline{\hspace{1.5cm}}$ $72 \div 8 = \underline{\hspace{1.5cm}}$

$70 \div 10 = \underline{\hspace{1.5cm}}$

ANSWERS

$35 \div 7 = 5$ $30 \div 5 = 6$ $18 \div 3 = 6$

$28 \div 4 = 7$ $42 \div 6 = 7$ $63 \div 9 = 7$

$45 \div 5 = 9$ $56 \div 7 = 8$ $72 \div 8 = 9$

$70 \div 10 = 7$

You should be able to do all these quickly by working backward from the one-digit multiplication tables.

Careful ▶   Division by zero has no meaning in arithmetic. (Any number) $\div$ 0 or $\dfrac{\text{(any number)}}{0}$ is not defined. But $0 \div$ (any number) $= \dfrac{0}{\text{(any number)}} = 0$ ◀

How do we divide numbers that are larger than 9×9 and therefore not in the multiplication table? Obviously, we need a better procedure.

EXAMPLE

Here is a step-by-step explanation of the division of whole numbers:

Divide $96 \div 8 = \underline{\hspace{1.5cm}}$.

First, *estimate* the answer: $100 \div 10$ is about 10. The quotient or answer will be about 10.

Second, arrange the numbers in this way:

$8\overline{)96}$

Notice that the order in which the numbers are written is reversed. With the $\div$ symbol, the divisor (8) is on the right. With the $\overline{)}$ symbol, the divisor is on the left.

Third, divide using the following step-by-step procedure.

Step 1 $8\overline{)\,\overset{1}{96}}$ 8 into 9? Once. Write 1 in the answer space above the 9.

Step 2 $8\overline{)\,\overset{1}{96}}$ Multiply $8 \times 1 = 8$ and write the product 8 under the 9.
$\phantom{8\overline{)\,}}8$

Step 3
$$8\overline{)96}$$
$$\underline{-8\downarrow}$$
$$16$$
Subtract $9 - 8 = 1$ and write 1. Bring down the next digit, 6.

Step 4
$$8\overline{)96}$$
$$\underline{-8}$$
$$16$$
8 into 16? Twice. Write 2 in the answer space above the 6.

Step 5
$$8\overline{)96}$$
$$\underline{-8}$$
$$16$$
$$-16$$
Multiply $8 \times 2 = 16$ and write the product 16 under the 16.

Step 6
$$8\overline{)96}$$
$$\underline{-8}$$
$$16$$
$$\underline{-16}$$
$$0$$
Subtract $16 - 16 = 0$. Write 0.

$\leftarrow$ The remainder is zero. 8 divides into 96 exactly 12 times.

✓ **Finally,** *check* your answer. The answer 12 is roughly equal to the original estimate 10. As a second check, multiply the divisor 8 and the quotient 12. Their product should be the original dividend number, $8 \times 12 = 96$, which is correct.

▶ YOUR TURN

Practice this step-by-step division process by finding $112 \div 7$.

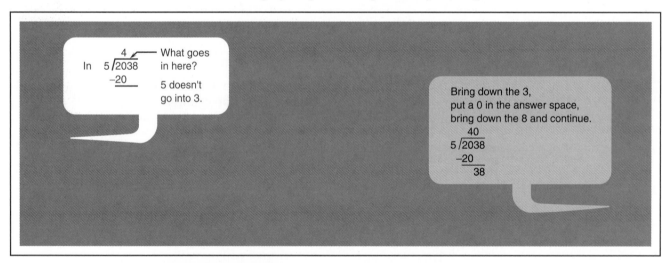

◀ SOLUTION

Estimate: $112 \div 7$ is roughly $100 \div 5$ or 20. The answer will be roughly 20.

$$7\overline{)112}$$ with quotient 016
$$\underline{-7\downarrow}$$
$$42$$
$$\underline{-42}$$
$$0$$

Step 1 7 into 1. Won't go, so put a zero in the answer place.

Step 2 7 into 11. Once. Write 1 in the answer space as shown.

Step 3 $7 \times 1 = 7$. Write 7 below 11.

Step 4 $11 - 7 = 4$. Write 4. Bring down the 2.

Step 5 7 into 42. Six. Write 6 in the answer space.

Step 6 $7 \times 6 = 42$. Subtract 42. The remainder is zero.

✓ The answer 16 is roughly equal to our original estimate of 20. Check again by multiplying. $7 \times 16 = 112$. Finding a quick estimate of the answer will help you avoid making mistakes.

Careful ▶ Notice that once the first digit of the answer has been obtained, there will be an answer digit for every digit of the dividend.

$$16$$
$$\Updownarrow$$
$$7\overline{)112} \ \blacktriangleleft$$

EXAMPLE

$203 \div 7 = ?$

Estimate: $203 \div 7$ is roughly $210 \div 7$ or 30. The answer will be roughly 30.

$$
\begin{array}{r}
029 \\
7\overline{)\ 203} \\
-14\downarrow \\
\hline
63 \\
-\ 63 \\
\hline
0
\end{array}
$$

Step 1 7 into 2. Won't go, so put a zero in the answer space.

Step 2 7 into 20. Twice. Write 2 in the answer space as shown.

Step 3 $7 \times 2 = 14$. Write 14 below 20.

Step 4 $20 - 14 = 6$. Write 6. Bring down the 3.

Step 5 7 into 63. Nine times. Subtract 63. The remainder is zero.

✓ The answer is 29, roughly equal to the original estimate of 30. Double-check by multiplying: $7 \times 29 = 203$.

The remainder is not always zero, of course.

EXAMPLE

To calculate $153 \div 4$,

$$
\begin{array}{r}
38 \quad \text{Quotient} \\
\text{Divisor} \Longrightarrow 4\overline{)153} \quad \text{Dividend} \\
-12 \\
\hline
33 \\
-32 \\
\hline
1 \quad \text{Remainder}
\end{array}
$$

Write the answer as $38r1$ to indicate that the quotient is 38 and the remainder is 1.

✓ To check an answer with a remainder, first multiply the quotient by the divisor, then add the remainder. In this example,

$38 \times 4 = 152$

$152 + 1 = 153$ which checks.

YOUR TURN

Ready for some guided practice? Try these problems.

(a) $3174 \div 6$ (b) $206 \div 6$ (c) $59 \div 8$

(d) $5084 \div 31$ (e) $341 \div 43$

(f) **Plumbing** Pete the Plumber billed $7548 for the labor on a large project. If his regular rate is $37 per hour, how many hours did he work?

SOLUTIONS

(a) $3174 \div 6 =$ _____

Estimate: $6 \times 500 = 3000$ The answer will be roughly 500.

$$\begin{array}{r} 0529 \\ 6\overline{)\ 3174} \\ -30 \\ \hline 17 \\ -\ 12 \\ \hline 54 \\ -\ 54 \\ \hline 0 \end{array}$$

Step 1 6 into 3? No. Write a zero. 6 into 31? 5 times. Write 5 above the 1.

Step 2 $6 \times 5 = 30$. Subtract $31 - 30 = 1$.

Step 3 Bring down 7. 6 into 17? Twice. Write 2 in the answer.

Step 4 $6 \times 2 = 12$. Subtract $17 - 12 = 5$.

Step 5 Bring down 4. 6 into 54? 9 times. Write 9 in the answer.

Step 6 $6 \times 9 = 54$. Subtract $54 - 54 = 0$.

Quotient = 529

The answer is 529.

☑ The estimate, 500, and the answer, 529, are roughly equal.

Double-check: $6 \times 529 = 3174$.

(b) $206 \div 6 =$ _____

Estimate: $6 \times 30 = 180$ The answer is about 30.

$$\begin{array}{r} 034 \\ 6\overline{)\ 206} \\ -18 \\ \hline 26 \\ -\ 24 \\ \hline 2 \end{array}$$

Step 1 6 into 2? No. Write a zero. 6 into 20? Three times. Write 3 above the zero in the answer.

Step 2 $6 \times 3 = 18$. Subtract $20 - 18 = 2$.

Step 3 Bring down 6. 6 into 26? 4 times. Write 4 in the answer.

Step 4 $6 \times 4 = 24$. Subtract $26 - 24 = 2$. The remainder is 2.

Quotient = 34

The answer is 34 with a remainder of 2.

☑ The estimate 30 and the answer are roughly equal.

Double-check: $6 \times 34 = 204$ $204 + 2 = 206$.

(c) $59 \div 8 =$ _____

Estimate: $7 \times 8 = 56$. The answer will be about 7.

$$\begin{array}{r} 7 \\ 8\overline{)\ 59} \\ -56 \\ \hline 3 \end{array}$$

8 into 5? No. There is no need to write the zero.

The quotient is 7 with a remainder of 3.

☑ Check it, and double-check by multiplying and adding the remainder.

(d) $5084 \div 31 =$ _____

Estimate: This is roughly about the same as $5000 \div 30$ or $500 \div 3$ or about 200. The quotient will be about 200.

$$\begin{array}{r} 164 \\ 31\overline{)\,5084} \\ -31 \\ \hline 198 \\ -186 \\ \hline 124 \\ -\,124 \\ \hline 0 \end{array}$$

Step 1 31 into 5: No. 31 into 50? Yes, once. Write 1 above the zero.

Step 2 $31 \times 1 = 31$. Subtract $50 - 31 = 19$.

Step 3 Bring down 8. 31 into 198? (That is, about the same as 3 into 19.) Yes, 6 times; write 6 in the answer.

Step 4 $31 \times 6 = 186$. Subtract $198 - 186 = 12$.

Step 5 Bring down 4. 31 into 124? (That is, about the same as 3 into 12.) Yes, 4 times. Write 4 in the answer.

Step 6 $31 \times 4 = 124$. Subtract $124 - 124 = 0$.

The quotient is 164.

☑ The estimate is reasonably close to the answer.

Double-check: $31 \times 164 = 5084$.

Notice that in step 3 it is not at all obvious how many times 31 will go into 198. Again, you must make an educated guess and check your guess as you go along.

(e) $341 \div 43 = $ _____

Estimate: This is roughly $300 \div 40$ or $30 \div 4$ or about 7. The answer will be about 7 or 8.

$$\begin{array}{r} 7 \\ 43\overline{)\,341} \\ -301 \\ \hline 40 \end{array}$$

Your first guess would probably be that 43 goes into 341 8 times (try 4 into 34), but $43 \times 8 = 344$, which is larger than 341. The quotient is 7 with a remainder of 40.

☑ Check it.

(f) To determine the total bill, Pete multiplied his hourly rate by the number of hours. Therefore, if we know the total and the hourly rate, we reverse this process and divide the total by the hourly rate to find the number of hours.

Estimate: $7000 \div 30$ or $700 \div 3$ is about 200.

$$\begin{array}{r} 204 \\ 37\overline{)\,7548} \\ 74 \\ \hline 148 \\ 148 \end{array}$$

On the second step notice that 37 does not go into 14. So we must write a zero in the tens place of the answer, above the 4, and bring down the 8.

The quotient is 204. Pete worked a total of 204 hours on the project.

☑ The answer 204 and the estimate 200 are roughly equal.

Double-check: $37 \times 204 = 7548$.

Factors It is sometimes useful in mathematics to be able to write any whole number as a product of other numbers. If we write

$6 = 2 \times 3$ 2 and 3 are called **factors** of 6.

Of course, we could also write

$6 = 1 \times 6$ and see that 1 and 6 are also factors of 6.

The factors of 6 are 1, 2, 3, and 6.

The factors of 12 are 1, 2, 3, 4, 6, and 12.

The factors of 30 are 1, 2, 3, 5, 6, 10, 15, and 30.

Any number is exactly divisible by its factors; that is, every factor divides the number with zero remainder.

List all the factors of each of the following whole numbers.

1. 4   2. 10   3. 24   4. 18   5. 14
6. 32   7. 20   8. 26   9. 44   10. 90

ANSWERS

1. 1, 2, 4   2. 1, 2, 5, 10   3. 1, 2, 3, 4, 6, 8, 12, 24

4. 1, 2, 3, 6, 9, 18   5. 1, 2, 7, 14   6. 1, 2, 4, 8, 16, 32

7. 1, 2, 4, 5, 10, 20   8. 1, 2, 13, 26   9. 1, 2, 4, 11, 22, 44

10. 1, 2, 3, 5, 6, 9, 10, 15, 18, 30, 45, 90

Primes

For some numbers the only factors are 1 and the number itself. For example, the factors of 7 are 1 and 7 because $7 = 1 \times 7$. There are no other numbers that divide 7 with remainder zero. Such numbers are known as prime numbers. A **prime number** is one for which there are no factors other than 1 and the prime itself.

Here is a list of the first few prime numbers.

2, 3, 5, 7, 11, 13, 17, 19, 23, 29, 31, 37, 41, 43, 47

(Notice that 1 is not listed. All prime numbers have two different factors: 1 and the number itself. The number 1 has only 1 factor—itself.)

Prime Factors

The **prime factors** of a number are those factors that are prime numbers. For example, the prime factors of 6 are 2 and 3. The prime factors of 30 are 2, 3, and 5. You will see later that the concept of prime factors is useful when working with fractions.

To find the prime factors of a number a **factor-tree** is often helpful.

EXAMPLE

Find the prime factors of 132.

First, write the number 132 and draw two branches below it.

Second, beginning with the smallest prime in our list, 2, test for divisibility. $132 = 2 \times 66$. Write the factors below the branches.

Next, repeat this procedure on the nonprime factor, 66, so that $66 = 2 \times 33$.

Finally, continue dividing until all branches end with a prime.

Therefore, $132 = 2 \times 2 \times 3 \times 11$.

Check by multiplying.

Write each whole number as a product of its prime factors. If the given number is a prime, label it with a letter P.

1. 9   2. 11   3. 15   4. 3   5. 17

6. 21   7. 23   8. 29   9. 52   10. 50

ANSWERS

1. 3×3 2. 11 P   3. 3×5 4. 3 P   5. 17 P

6. 3×7 7. 23 P   8. 29 P   9. $2 \times 2 \times 13$ 10. $2 \times 5 \times 5$

Exercises 1-4 provide some practice in the division of whole numbers.

SOME DIVISION "TRICKS OF THE TRADE"

1. If a number is exactly divisible by 2, that is, divisible with zero remainder, it will end in an even digit, 0, 2, 4, 6, or 8.

   Example: We know that 374 is exactly divisible by 2 since it ends in the even digit 4.

   $$\frac{187}{2\overline{)374}}$$

2. If a number is exactly divisible by 3, the sum of its digits is exactly divisible by 3.

   Example: The number 2784 has the sum of digits $2 + 7 + 8 + 4 = 21$. Since 21 is exactly divisible by 3, we know that 2784 is also exactly divisible by 3.

   $$\frac{928}{3\overline{)2784}}$$

3. A number is exactly divisible by 4 if its last two digits are exactly divisible by 4 or are both zero.

   Example: We know that the number 3716 is exactly divisible by 4 since 16 is exactly divisible by 4.

   $$\frac{929}{4\overline{)3716}}$$

4. A number is exactly divisible by 5 if its last digit is either 5 or 0.

   Example: The numbers 875, 310, and 33,195 are all exactly divisible by 5.

5. A number is exactly divisible by 8 if its last three digits are divisible by 8.

   Example: The number 35,120 is exactly divisible by 8 since the number 120 is exactly divisible by 8.

   (continued)

6. A number is exactly divisible by 9 if the sum of its digits is exactly divisible by 9.

   Example: The number 434,673 has the sum of digits 27. Since 27 is exactly divisible by 9, we know that 434,673 is also exactly divisible by 9.

$$\overset{48297}{9\overline{)434673}}$$

7. A number is exactly divisible by 10 if it ends with a zero.

   Example: The numbers 230, 7380, and 100,200 are all exactly divisible by 10.

   Can you combine the tests for divisibility by 2 and 3 to get a rule for divisibility by 6?

Exercises 1-4

Division of Whole Numbers

A. Divide.

1. $63 \div 7$	2. $92 \div 8$	3. $72 \div 0$
4. $37 \div 5$	5. $71 \div 7$	6. $6 \div 6$
7. $\dfrac{32}{4}$	8. $\dfrac{28}{7}$	9. $\dfrac{54}{9}$
10. $245 \div 7$	11. $167 \div 7$	12. $228 \div 4$
13. $310 \div 6$	14. $3310 \div 3$	15. $\dfrac{147}{7}$
16. $7\overline{)364}$	17. $6\overline{)222}$	18. $4\overline{)201}$
19. $322 \div 14$	20. $382 \div 19$	21. $936 \div 24$
22. $700 \div 28$	23. $730 \div 81$	24. $\dfrac{901}{17}$
25. $31\overline{)682}$	26. $27\overline{)1724}$	27. $42\overline{)371}$

B. Divide.

1. $61\overline{)7320}$	2. $33\overline{)303}$	3. $16\overline{)904}$
4. $2001 \div 21$	5. $2016 \div 21$	6. $1000 \div 7$
7. $2000 \div 9$	8. $2400 \div 75$	9. $14\overline{)4275}$
10. $71\overline{)6005}$	11. $53\overline{)6307}$	12. $3\overline{)9003}$
13. $7\overline{)3507}$	14. $6\overline{)48009}$	15. $6\overline{)3624}$
16. $3\overline{)62160}$	17. $15\overline{)3000}$	18. $67\overline{)3354}$
19. $24\overline{)2596}$	20. $47\overline{)94425}$	21. $38\overline{)22800}$
22. $231\overline{)14091}$	23. $411\overline{)42020}$	24. $603\overline{)48843}$
25. $111\overline{)11111}$	26. $102\overline{)2004}$	27. $405\overline{)7008}$

C. For each of the following whole numbers, (a) list all its factors, and (b) write it as a product of its prime factors.

1. 6   2. 16   3. 19

4. 27   5. 40   6. 48

D. Applied Problems

1. **Machine Technology** A machinist has a piece of bar stock 243 in. long. If she must cut nine equal pieces, how long will each piece be? (Assume no waste to get a first approximation.)

2. **Carpentry** How many braces 32 in. long can be cut from a piece of lumber 192 in. long?

3. **Building Construction** If subflooring is laid at the rate of 85 sq ft per hour, how many hours will be needed to lay 1105 sq ft?

4. **Building Construction** The illustration shows a stringer for a short flight of stairs. Determine the missing dimensions H and W.

5. **Building Construction** How many joists spaced 16 in. o.c. (on center) are required for a floor 36 ft long? (Add one joist for a starter.)

6. **Carpentry** If rafters are placed 24 in. o.c., how many are required for both sides of a gable roof that is 48 ft long? Add one rafter on each side for a starter.

7. **Carpentry** A stairway in a house is to be built with 18 risers. If the distance from the top of the first floor to the top of the second floor is 10 ft 6 in., what is the height of each step?

8. **Manufacturing** What is the average horsepower of six engines having the following horsepower ratings: 385, 426, 278, 434, 323, and 392? (*Hint:* To find the average, add the six numbers and divide their sum by 6.)

9. **Plumbing** Mr. Martinez, owner of the Maya Restaurant, hired a plumber who worked for 18 hours installing some new equipment. The total bill for wages and supplies used was $1400. If the cost for materials was $176, calculate the plumber's hourly wage.

10. A truck traveling at an average rate of 54 miles per hour covered a distance of 486 miles in one run. Allowing 2 hours for refueling and meals, how long did it take to complete the trip?

11. **Auto Mechanics** A mechanic needs to order 480 spark plugs. If there are ten plugs in a box, how many boxes does he need?

12. **Metalworking** One cubic foot of a certain alloy weighs 375 lb. How many cubic feet does a 4875-lb block occupy?

13. **Printing** A printer needs 13,500 sheets of paper for a job. If there are 500 sheets in a ream, how many reams does the printer need to get from the stockroom?

14. **Wastewater Technology** A water tank contains 18,000 gallons. If 500 gallons per day are used from this tank, how long will it last?

E. Calculator Problems

1. Calculate the following.

(a) $\dfrac{1347 \times 46819}{3}$ (b) $\dfrac{76459 + 93008 + 255}{378}$

(c) $\dfrac{4008 + 408 + 48}{48}$ (d) $\dfrac{9909 + 9090}{3303}$

W

H

4'8"

84"

Problem 4

2. **Masonry** If 6-in. clay brick is on sale at \$29 per hundred, calculate the cost of the 14,000 bricks needed on a construction project.

3. **Manufacturing** The specified shear strength of a $\frac{1}{8}$-in. 2117-T3 (AD) rivet is 344 lb. If it is necessary to assure a strength of 6587 lb to a riveted joint in an aircraft structure, how many rivets must be used?

4. A train of 74 railway cars weighed a total of 9,108,734 lb. What was the average weight of a car?

5. **Manufacturing** A high-speed stamping machine can produce small flat parts at the rate of 96 per minute. If an order calls for 297,600 parts to be stamped, how many hours will it take to complete the job? (60 minutes = 1 hour.)

6. **Painting and Decorating** Following are four problems from the national apprentice examination for painting and decorating contractors.

   (a) $11,877,372 \div 738$ (b) $87,445,005 \div 435$

   (c) $1,735,080 \div 760$ (d) $206,703 \div 579$

7. **Building Construction** Mike, the master mason, is constructing a wall. From similar jobs he estimates that he can lay an average of 115 bricks an hour. How long will it take him to lay 4830 bricks?

Check your answers in the Appendix, then turn to Section 1-5 to study another important topic in the arithmetic of whole numbers.

1-5 ORDER OF OPERATIONS

What is the value of the arithmetic expression

$3 \times 4 + 2 = ?$

If we multiply first, $(3 \times 4) + 2 = 12 + 2 = 14$

but if we add first, $3 \times (4 + 2) = 3 \times 6 = 18$

To avoid any possible confusion in situations like this, mathematicians have adopted a standard order of operations for arithmetic calculations. When two or more of the four basic arithmetic operations (addition, subtraction, multiplication, and division) are combined in the same calculation, follow these rules.

Rule 1 Perform any calculations shown inside parentheses first.

Rule 2 Perform all multiplications and divisions next, working from left to right.

Rule 3 Perform additions and subtractions last, working from left to right.

> EXAMPLE

$3 \times 4 + 2 = 12 + 2$ Multiply first, Rule 2. $3 \times 4 = 12$

$ = 14$ Add last, Rule 3.

> YOUR TURN

Perform the following calculations, using the order of operation rules.

(a) $5 + 8 \times 7$ (b) $12 \div 6 - 2$

(c) $(26 - 14) \times 2$ (d) $12 \times (6 + 2) \div 6 - 7$

(a) $5 + 8 \times 7 = 5 + 56$ Do the multiplication first, Rule 2. $(8 \times 7 = 56)$

              $= 61$ Then add, Rule 3.

(b) $12 \div 6 - 2 = 2 - 2$ First divide, Rule 2. $(12 \div 6 = 2)$

             $= 0$ Then subtract, Rule 3.

(c) $(26 - 14) \times 2 = 12 \times 2$ Perform the operation in parentheses first, Rule 1. $(26 - 14 = 12)$

              $= 24$ Then multiply, Rule 2.

(d) $12 \times (6 + 2) \div 6 - 7 = 12 \times 8 \div 6 - 7$ Work inside parentheses first, Rule 1. $(6 + 2 = 8)$

              $= 96 \div 6 - 7$ Multiply, Rule 2. $(12 \times 8 = 96)$

              $= 16 - 7$ Divide, Rule 2. $(96 \div 6 = 16)$

              $= 9$ Subtract, Rule 3.

Notice in Exercise (d) that the multiplication and division without parentheses are performed working left to right.

Using parentheses helps guard against confusion. For example, the calculation $12 - 4 \div 2$ should be written $12 - (4 \div 2)$ to avoid making a mistake.

PROBLEMS

Perform the following calculations, using the three rules of order of operations.

(a) $240 \div (18 + 6 \times 2) - 2$ (b) $6 \times 5 + 14 \div (6 + 8) - 3$

(c) $6 + 2 \times 3 \div (7 - 4) + 5$

(d) **Construction** A remodeling job requires 128 sq ft of countertops. Two options are being considered. The more expensive option is to use all Corian at $50 per sq ft. The less expensive option is to use 76 sq ft of granite at $57 per sq ft and 52 sq ft of laminate at $22 per sq ft. Write out a mathematical statement to calculate the difference in cost between these two options. Then calculate this difference.

SOLUTIONS

(a) $240 \div (18 + 6 \times 2) - 2 = 240 \div (18 + 12) - 2$ Perform the multiplication inside the parentheses first. $(6 \times 2 = 12)$

              $= 240 \div 30 - 2$ Perform the sum inside the parentheses next. $(18 + 12 = 30)$

              $= 8 - 2$ Next divide, Rule 2. $(240 \div 30 = 8)$

              $= 6$ Finally, subtract, Rule 3.

(b) $6 \times 5 + 14 \div (6 + 8) - 3 = 6 \times 5 + 14 \div 14 - 3$ Perform the operations inside parentheses first.

              $= 30 + 14 \div 14 - 3$ Next, multiply, then divide, going left to right. $(14 \div 14 = 1)$

              $= 30 + 1 - 3$ Finally, do the addition and subtraction left to right, Rule 3.

              $= 28$

(c) $6 + 2 \times 3 \div (7 - 4) + 5 = 6 + 2 \times 3 \div 3 + 5$ Parentheses first, Rule 1.
$\qquad\qquad\qquad\qquad\qquad = 6 + 6 \div 3 + 5$ Multiply, then divide, left to right,
$\qquad\qquad\qquad\qquad\qquad = 6 + 2 + 5$ Rule 2. $(6 \div 3 = 2)$
$\qquad\qquad\qquad\qquad\qquad = 13$ Add last, Rule 3.

(d)   The more expensive option is $128 \times \$50 = \$6400$

The less expensive option is $(76 \times \$57) + (52 \times \$22)$

$\qquad\qquad\qquad\qquad\qquad\qquad\qquad = \$4332 + \$1144$

$\qquad\qquad\qquad\qquad\qquad\qquad\qquad = \$5476$

The difference is $\$6400 - \$5476 = \$924$.

When division is written with a fraction bar or *vinculum,* all calculations above the bar or below the bar should be done before the division.

EXAMPLE

$\dfrac{24 + 12}{6 - 2}$ should be thought of as $\dfrac{(24 + 12)}{(6 - 2)}$

First, simplify the top and bottom to get $\dfrac{36}{4}$.

Then, divide to obtain the answer $\dfrac{36}{4} = 9$.

Note ▶   This is very different from the calculation

$\dfrac{24}{6} + \dfrac{12}{2} = 4 + 6 = 10$

where we use Rules 2 and 3, doing the divisions first and the additions last. ◀

Exercises 1-5

Order of Operations

A.   Perform all operations in the correct order.

1. $2 + 8 \times 6$	2. $20 - 3 \times 2$
3. $40 - 20 \div 5$	4. $16 + 32 \div 4$
5. $16 \times 3 + 9$	6. $2 \times 9 - 4$
7. $48 \div 8 - 2$	8. $64 \div 16 + 8$
9. $(5 + 9) \times 3$	10. $(18 - 12) \div 6$
11. $24 \div (6 - 2)$	12. $9 \times (8 + 3)$
13. $16 + 5 \times (3 + 6)$	14. $8 + 3 \times (9 - 4)$
15. $(23 + 5) \times (12 - 8)$	16. $(17 - 9) \div (6 - 2)$
17. $6 + 4 \times 7 - 3$	18. $24 - 8 \div 2 + 6$
19. $5 \times 8 + 6 \div 6 - 12 \times 2$	20. $24 \div 8 - 14 \div 7 + 8 \times 6$
21. $2 \times (6 + 4 \times 9)$	22. $54 \div (8 - 3 \times 2)$
23. $(4 \times 3 + 8) \div 5$	24. $(26 \div 2 - 5) \times 4$
25. $8 - 4 + 2$	26. $24 \div 6 \times 2$
27. $18 \times 10 \div 5$	28. $22 + 11 - 7$

29. $12 - 7 - 3$

30. $48 \div 6 \div 2$

31. $12 - (7 - 3)$

32. $18 \div (3 \times 2)$

33. $\dfrac{36}{9} + \dfrac{27}{3}$

34. $\dfrac{36 - 27}{9 - 6}$

35. $\dfrac{44 + 12}{11 - 3}$

36. $\dfrac{44}{11} + \dfrac{12}{3}$

37. $\dfrac{6 + 12 \times 4}{15 - 3 \times 2}$

38. $\dfrac{36 - (7 - 4)}{5 + 3 \times 2}$

39. $\dfrac{12 + 6}{3 + 6} + \dfrac{24}{6} - 6 \div 6$

40. $8 \times 5 - \dfrac{2 + 4 \times 12}{18 - 4 \times 2} + 72 \div 9$

B. Applied Problems

1. **Painting and Decorating** A painter ordered 3 gallons of antique white latex paint for $18 a gallon and 5 gallons of white enamel for $20 a gallon. Write out a mathematical statement giving the total cost of the paint. Calculate the cost.

2. **Landscape Architecture** On a certain landscaping job, Steve charged a customer $468 for labor and $90 each for eight flats of plants. Write a single mathematical statement giving the total cost of this job, then calculate the cost.

3. **Electrical Technology** An electrician purchased 12 dimmer switches at $11 each and received a $9 credit for each of the five power outlet panels he returned. Write out a mathematical statement that gives the amount of money he spent, then calculate this total.

4. **Auto Mechanics** At the beginning of the day on Monday, the parts department has on hand 520 spark plugs. Mechanics in the service department estimate they will need about 48 plugs per day. A new shipment of 300 will arrive on Thursday. Write out a mathematical statement that gives the number of spark plugs on hand at the end of the day on Friday. Calculate this total.

C. Calculator Problems

1. $462 + 83 \times 95$

2. $425 \div 25 + 386$

3. $7482 - 1152 \div 12$

4. $1496 - 18 \times 13$

5. $(268 + 527) \div 159$

6. $2472 \times (1169 - 763)$

7. $612 + 86 \times 9 - 1026 \div 38$

8. $12 \times 38 + 46 \times 19 - 1560 \div 24$

9. $3579 - 16 \times (72 + 46)$

10. $273 + 25 \times (362 + 147)$

11. $864 \div 16 \times 27$

12. $973 - (481 + 327)$

13. $(296 + 18 \times 48) \times 12$

14. $(27 \times 18 - 66) \div 14$

15. $\dfrac{3297 + 1858 - 493}{48 \times 16 - 694}$

16. $\dfrac{391}{17} + \dfrac{4984}{89} - \dfrac{1645}{47}$

Check your answers in the Appendix, then turn to Problem Set 1 for practice on the arithmetic of whole numbers, with many practical applications of this mathematics.

PROBLEM SET 1

Arithmetic of Whole Numbers

Answers are given in the Appendix.

A. **Perform the arithmetic as shown.**

1. $24 + 69$
2. $38 + 45$
3. $456 + 72$
4. $43 + 817$
5. $396 + 538$
6. $2074 + 906$
7. $43 - 28$
8. $93 - 67$
9. $734 - 85$
10. $315 - 119$
11. $543 - 348$
12. $3401 - 786$
13. 376×4
14. 489×7
15. 67×21
16. 45×82
17. 207×63
18. 314×926
19. 5236×44
20. 4018×392
21. $259 \div 7$
22. $1704 \div 8$
23. $42\overline{)2394}$
24. $34\overline{)2108}$
25. $1440 \div 160$
26. $11309 \div 263$
27. $\dfrac{1314}{73}$
28. $\dfrac{23 \times 51}{17}$
29. $\dfrac{36 \times 91}{13 \times 42}$
30. $(18 + 5 \times 9) \div 7$
31. $120 - 40 \div 8$
32. $32 \div 4 + 16 \div 2 \times 4$
33. $3 \times 4 - 15 \div 3$
34. $\begin{array}{r} 139 \\ 407 \\ + \ 81 \\ \hline \end{array}$
35. $\begin{array}{r} 308 \\ 793 \\ + \ 144 \\ \hline \end{array}$
36. $\begin{array}{r} 194 \\ 271 \\ + \ 368 \\ \hline \end{array}$

B. **For each of the following whole numbers, (a) list all its factors, and (b) write each as a product of its prime factors.**

1. 8
2. 28
3. 31
4. 35
5. 36
6. 42

C. **Practical Problems**

1. **Electrical Technology**   From a roll of No. 12 wire, an electrician cut the following lengths: 6, 8, 20, and 9 ft. How many feet of wire did he use?
2. **Machine Technology**   A machine shop bought 14 steel rods of $\frac{7}{8}$-in.-diameter steel, 23 rods of $\frac{1}{2}$-in. diameter, 8 rods of $\frac{1}{4}$-in. diameter, and 19 rods of 1-in. diameter. How many rods were purchased?
3. **Interior Design**   In estimating the floor area of a house, Jean listed the rooms as follows: living room, 346 sq ft; dining room, 210 sq ft; four bedrooms, 164 sq ft each; two bathrooms, 96 sq ft each; kitchen, 208 sq ft; and family room, 280 sq ft. What is the total floor area of the house?

Name

Date

Course/Section

4. **Metalworking**   A metal casting weighs 680 lb, and 235 lb of metal is removed during shaping. What is its finished weight?

5. **Electrical Technology**   Roberto has 210 ft of No. 14 wire left on a roll. If he cuts it into pieces 35 ft long, how many pieces will he get?

6. **Building Construction**   How many hours will be required to install the flooring of a 2160 sq ft house if 90 sq ft can be installed in 1 hour?

7. **Construction**   The weights of seven cement platforms are 210, 215, 245, 217, 220, 227, and 115 lb. What is the average weight per platform? (Add up and divide by 7.)

8. **Interior Design**   A room has 26 square yards (sq yd) of floor space. If carpeting costs $29 per sq yd, what will it cost to carpet the room?

9. A high-definition TV set can be bought for $350 down and 12 payments of $116 each. What is its total cost?

10. **Electrical Technology**   A 500-ft spool of No. 14-2TF wire is being used on an electrical wiring job. If 248 ft of wire is used on the job, how much wire is left on the spool?

11. Texas has more miles of highway than any other state. In a recent survey, Texas had a paved road mileage of 300,507; California was next with 166,973 miles; Illinois was third with 138,246 miles; and Kansas was fourth with 133,963. What was the total paved road mileage for these four states?

12. **Office Services**   During the first three months of the year, the Print Rite Company reported the following sales of printers:

    January $55,724
    February $47,162
    March $62,473

    What is their sales total for this quarter of the year?

13. **Fire Protection**   A fire control pumping truck can move 156,000 gallons of water in 4 hours of continuous pumping. What is the flow rate (gallons per minute) for this truck?

14. **Building Construction**   A pile of lumber contains 170 boards 10 ft long, 118 boards 12 ft long, 206 boards 8 ft long, and 19 boards 16 ft long.
    (a) How many boards are in the pile?
    (b) Calculate the total length, or linear feet, of lumber in the pile.

15. **Marine Technology**   If the liquid pressure on a surface is 17 psi (pounds per square inch), what is the total force on a surface area of 167 square inches?

16. **Auto Mechanics**   A rule of thumb useful to auto mechanics is that a car with a pressure radiator cap will boil at a temperature of $(3 \times \text{Cap Rating}) + 212$ degrees. What temperature will a Dino V6 reach with a cap rating of 17?

17. **Machine Technology**   If it takes 45 minutes to cut the teeth on a gear blank, how many hours will be needed for a job that requires cutting 32 such gear blanks?

18. **Electrician**   An 8-ft lighting track holds six fixtures, each of which requires a bulb. Each track costs $20, each fixture costs $9, and each bulb costs $4. Write out a mathematical statement that gives the total cost for three fully equipped tracks, then calculate this total.

19. **Manufacturing**   A gallon is a volume of 231 cubic inches. If a storage tank holds 380 gallons of oil, what is the volume of the tank in cubic inches?

20. **Machine Technology**   The Ace Machine Tool Co. received an order for 15,500 flanges. If two dozen flanges are packed in a box, how many boxes are needed to ship the order?

21. **Manufacturing**   At 8 A.M. the revolution counter on a diesel engine reads 460089. At noon it reads 506409. What is the average rate, in revolutions per minute, for the machine?

22. **Fire Protection**   The pressure reading coming out of a pump is 164 lb. It is estimated that every 50-ft section of hose reduces the pressure by 7 lb. What will the estimated pressure be at the nozzle end of nine 50-ft sections?

23. **Carpentry** A fireplace mantel 6 ft long is to be centered along a wall 18 ft long. How far from each end of the wall should the ends of the mantel be?

24. **Electronics** When capacitors are connected in parallel in a circuit as shown, the total capacitance can be calculated by adding the individual capacitor values C_1, C_2, C_3, C_4, and C_5. Find the total capacitance of this circuit, in picofarads (pF).

Using a Calculator, I: Whole-Number Calculations

Since its introduction in the early 1970s, the handheld electronic calculator has quickly become an indispensable tool in modern society. From clerks, carpenters, and shoppers to technicians, engineers, and scientists, people in virtually every occupation now use calculators to perform mathematical tasks. With this in mind, we have included in this book special instructions, examples, and exercises demonstrating the use of a calculator wherever it is appropriate.

To use a calculator intelligently and effectively in your work, you should remember the following:

1. Whenever possible, make an estimate of your answer before entering a calculation. Then check this estimate against your calculator result. You may get an incorrect answer by accidentally pressing the wrong keys or by entering numbers in an incorrect sequence.

2. Always try to check your answers to equations and word problems by substituting them back into the original problem statement.

3. Organize your work on paper before using a calculator, and record any intermediate results that you may need later.

4. Round your answer whenever necessary.

This is the first of several sections, spread throughout the text, designed to teach you how to use a calculator with the mathematics taught in this course. In addition to these special sections, the solutions to many worked examples include the appropriate calculator key sequences and displays. Most exercise sets contain a special section giving calculator problems.

How to Select a Calculator There are many types of calculators available. They differ in size, shape, color, cost, and most important, in how they work and what they can do. To use a calculator efficiently with the mathematical operations covered in this book, you will need a scientific calculator with the following characteristics.

* A display of ten digits.
* At least one memory (look for keys marked $\boxed{\text{x→M}}$, $\boxed{\text{STO}}$, or $\boxed{\text{M+}}$).
* Keys for calculating square root $\boxed{\sqrt{\;}}$, powers $\boxed{x^2}$ $\boxed{y^x}$, trigonometric functions $\boxed{\sin}$, $\boxed{\cos}$, $\boxed{\tan}$, inverse trigonometric functions $\boxed{\sin^{-1}}$, $\boxed{\cos^{-1}}$, $\boxed{\tan^{-1}}$, as well as the four basic arithmetic operations $\boxed{+}$ $\boxed{-}$ $\boxed{\times}$ $\boxed{\div}$. Calculators that can perform all these functions usually include additional functions that may be useful in more advanced courses.
* Algebraic logic.

Becoming Familiar with Your Calculator   First, check out the machine. You'll find an on–off switch somewhere (unless it is solar powered), a display, and a keyboard with both *numerical* (0, 1, 2, 3, . . . , 9) and basic *function* (+, −, ×, ÷, =, ·) keys usually arranged like this:

7	8	9	÷
4	5	6	×
1	2	3	−
0	·	=	+

Graphing calculators use an ENTER key instead of [=].

Next, locate the *clear* [c], *clear entry* [CE], and memory keys. There are usually three memory keys, one for storing the displayed information [x→M] or [STO], one for retrieving information from memory [RM] or [RCL], and one for adding the number in the display to the number in memory [M+] or [SUM]. Finally, look for the parentheses keys [(] and [)].

The *clear* key [c] "clears" all entries from the work space of the calculator, and causes the number zero to appear in the display. Pressing the clear key means that you want to start the entire calculation over or begin a new calculation.

Keyboard   **Display**

[c]   *0.*

The *clear entry* key [CE] "clears" the display only, leaves the work space unchanged, and causes a zero to appear in the display. Pressing the clear entry key [CE] means that you wish to remove the last entry only.

Keyboard   **Display**

[CE]   *0.*

Basic Operations   Every number is entered into the calculator digit by digit, left to right. For example, to enter the number 438, simply press the 4, 3, and 8 keys and the display will read   *438.*   . If you make an error in entering the number, press either the clear [c] or clear entry [CE] key and begin again. Note that the calculator does not know if the number you are entering is 4, 43, 438, or something even larger, until you stop entering numerical digits and press a function key such as [+], [−], [×], [÷] or [=].

> **EXAMPLE**

To add 438 + 266

first, estimate the answer: 400 + 300 = 700

then, enter this sequence of keys,

4 3 8 [+] **2 6 6** [=] →   *704.*

Notice that the calculator replaced the display of 438 with the display of 266 after the [+] key was pressed, and that the sum was displayed immediately after the [=] key was pressed.

All four basic operations work the same way.

Try the following problems for practice.

(a) 789 − 375

(b) 246 × 97

(c) 2314 ÷ 26

(d) 46129 + 8596

ANSWERS

(a) **7 8 9** $\boxed{-}$ **3 7 5** $\boxed{=}$ → *414.*

(b) **2 4 6** $\boxed{\times}$ **9 7** $\boxed{=}$ → *23862.*

(c) **2 3 1 4** $\boxed{\div}$ **2 6** $\boxed{=}$ → *89.*

(d) **4 6 1 2 9** $\boxed{+}$ **8 5 9 6** $\boxed{=}$ → *54725.*

Don't forget to estimate the answer before using the calculator.

Combined Operations A calculator can operate on only two numbers at a time, but very often you need to use the result of one calculation in a second calculation or in a long string of calculations. Because the answer to any calculation remains in the work space of the calculator, you can carry out a very complex chain of calculations without ever stopping the calculator to write down an intermediate step.

EXAMPLE

To add

387 + 942 + 876 + 88

first, estimate: 400 + 900 + 900 + 100 = 2300

then, use the calculator as follows:

3 8 7 $\boxed{+}$ **9 4 2** $\boxed{+}$ **8 7 6** $\boxed{+}$ **8 8** $\boxed{=}$ → *2293.*

Notice that it was not necessary to press the $\boxed{=}$ key until the very end, and that the display showed the running total every time you hit the $\boxed{=}$ key.

In Section 1-5 you saw that when two or more operations are combined in the same calculation a standard order of operations must be followed. Fortunately, if your calculator uses algebraic logic, you do not need to worry about entering such a calculation in the correct order. Scientific calculators with algebraic logic automatically use the correct order of operations. To test your calculator, enter this sequence:

2 $\boxed{+}$ **3** $\boxed{\times}$ **4** $\boxed{=}$ → *14.*

If your calculator displays this answer, it is using algebraic logic.

Notice that when you pressed the $\boxed{\times}$ key the calculator did not display the sum of 2 and 3. Because a multiplication must be done first, the calculator waited to perform the multiplication before doing the addition.

YOUR TURN

Enter each of the following calculations as written.

(a) 480 − 1431 ÷ 53

(b) 72 × 38 + 86526 ÷ 69

(c) 2478 − 726 + 598 × 12

(d) 271440 ÷ 48 × 65

(a) $4\,8\,0\;\boxed{-}\;1\,4\,3\,1\;\boxed{\div}\;5\,3\;\boxed{=}\;\rightarrow$ ▨▨▨▨ *453.*

(b) $7\,2\;\boxed{\times}\;3\,8\;\boxed{+}\;8\,6\,5\,2\,6\;\boxed{\div}\;6\,9\;\boxed{=}\;\rightarrow$ ▨▨▨▨ *3990.*

(c) $2\,4\,7\,8\;\boxed{-}\;7\,2\,6\;\boxed{+}\;5\,9\,8\;\boxed{\times}\;1\,2\;\boxed{=}\;\rightarrow$ ▨▨▨▨ *8928.*

(d) $2\,7\,1\,4\,4\,0\;\boxed{\div}\;4\,8\;\boxed{\times}\;6\,5\;\boxed{=}\;\rightarrow$ ▨▨▨▨ *367575.*

Parentheses and Memory As you have already learned, parentheses, brackets, and other grouping symbols are used in a written calculation to signal a departure from the standard order of operations. Fortunately, every scientific calculator has parentheses keys. To perform a calculation with grouping symbols, either enter the calculation inside the parentheses first and press $\boxed{=}$, or enter the calculation as it is written but use the parentheses keys.

EXAMPLE

To enter $12 \times (28 + 15)$ use either the sequence

$2\,8\;\boxed{+}\;1\,5\;\boxed{=}\;\boxed{\times}\;1\,2\;\boxed{=}\;\rightarrow$ ▨▨▨▨ *516.*

or

$1\,2\;\boxed{\times}\;\boxed{(}\;2\,8\;\boxed{+}\;1\,5\;\boxed{)}\;\boxed{=}\;\rightarrow$ ▨▨▨▨ *516.*

Learning Help ▶ In complex problems it is useful to do the calculation both ways as a way of checking your answer. ◀

When division problems are given in terms of fractions, parentheses are implied but not shown.

EXAMPLE

The problem

$$\frac{48 + 704}{117 - 23} \text{ is equal to } (48 + 704) \div (117 - 23)$$

The fraction bar acts as a grouping symbol. The addition and subtraction in the top and bottom of the fraction must be performed before the division. Again you may use either parentheses or the $\boxed{=}$ key to simplify the top half of the fraction, but after pressing $\boxed{=}$ you must use parentheses to enter the bottom half. Here are the two possible sequences:

$\boxed{(}\;4\,8\;\boxed{+}\;7\,0\,4\;\boxed{)}\;\boxed{\div}\;\boxed{(}\;1\,1\,7\;\boxed{-}\;2\,3\;\boxed{)}\;\boxed{=}\;\rightarrow$ ▨▨▨▨ *8.*

or

$4\,8\;\boxed{+}\;7\,0\,4\;\boxed{=}\;\boxed{\div}\;\boxed{(}\;1\,1\,7\;\boxed{-}\;2\,3\;\boxed{)}\;\boxed{=}\;\rightarrow$ ▨▨▨▨ *8.*

Press $\boxed{=}$, or the calculator will
divide into 704 rather than into $48 + 704$.

The memory keys provide a third option for this problem. Memory is used to store a result for later use. In this case, we can calculate the value of the bottom first, store it in memory, then calculate the top and divide by the contents of memory. The entire sequence looks like this:

1 1 7 ⊟ **2 3** ⊜ [STO]   **4 8** ⊞ **7 0 4** ⊜ ⊡ [RCL] ⊜ → █████ 8.

Calculate   Store it   Calculate   Divide by
the bottom. in memory.   the top.   the bottom
                                            in memory.

YOUR TURN

Here are a few practice problems involving grouping symbols. Use the method indicated.

(a) $(4961 - 437) \div 52$ Do not use the parentheses keys.

(b) $56 \times (38 + 12 \times 17)$ Use the parentheses keys.

(c) $2873 - (56 + 83) \times 16$ Use the parentheses keys.

(d) $\dfrac{263 \times 18 - 41 \times 12}{18 \times 16 - 17 \times 11}$ Use memory.

(e) $12 \times 16 - \dfrac{7 + 9 \times 17}{81 - 49} + 12 \times 19$ Use parentheses.

(f) Repeat problem (e) using the memory to store the value of the fraction, then doing the calculation left to right.

ANSWERS

(a) **4 9 6 1** ⊟ **4 3 7** ⊜ ⊡ **5 2** ⊜ → █████ 87.

(b) **5 6** ⊠ ⦅ **3 8** ⊞ **1 2** ⊠ **1 7** ⦆ ⊜ → █████ 13552.

(c) **2 8 7 3** ⊟ ⦅ **5 6** ⊞ **8 3** ⦆ ⊠ **1 6** ⊜ → █████ 649.

(d) **1 8** ⊠ **1 6** ⊟ **1 7** ⊠ **1 1** ⊜ [STO] **2 6 3** ⊠ **1 8** ⊟ **4 1** ⊠ **1 2**
    ⊜ ⊡ [RCL] ⊜ → █████ 42.

(e) **1 2** ⊠ **1 6** ⊟ ⦅ **7** ⊞ **9** ⊠ **1 7** ⦆ ⊡ ⦅ **8 1** ⊟ **4 9** ⦆ ⊞ **1 2**
    ⊠ **1 9** ⊜ → █████ 415.

(f) **7** ⊞ **9** ⊠ **1 7** ⊜ ⊡ ⦅ **8 1** ⊟ **4 9** ⦆ ⊜ [STO] **1 2** ⊠ **1 6** ⊟ [RCL]
    ⊞ **1 2** ⊠ **1 9** ⊜ → █████ 415.

Fractions

Objective	Sample Problems	For Help Go to Page

When you finish this unit you will be able to:

1. Work with fractions.

(a) Write as a mixed number $\frac{31}{4}$. _____ 62

(b) Write as an improper fraction $3\frac{7}{8}$. _____ 63

Write as an equivalent fraction.

(c) $\frac{5}{16} = \frac{?}{64}$ _____ 64

(d) $1\frac{3}{4} = \frac{?}{32}$ _____

(e) Write in lowest terms $\frac{10}{64}$. _____ 65

(f) Which is larger, $1\frac{7}{8}$ or $\frac{5}{3}$? _____ 68

2. Multiply and divide fractions.

(a) $\frac{7}{8} \times \frac{5}{32}$ _____ 70

(b) $4\frac{1}{2} \times \frac{2}{3}$ _____

(c) $\frac{3}{5}$ of $1\frac{1}{2}$ _____

(d) $\frac{3}{4} \div \frac{1}{2}$ _____ 75

(e) $2\frac{7}{8} \div 1\frac{1}{4}$ _____

(f) $4 \div \frac{1}{2}$ _____

Name _____

Date _____

Course/Section

3. Add and subtract fractions.

(a) $\dfrac{7}{16} + \dfrac{3}{16}$ _____   81

(b) $1\dfrac{3}{16} + \dfrac{3}{4}$ _____

(c) $\dfrac{3}{4} - \dfrac{1}{5}$ _____

(d) $4 - 1\dfrac{5}{16}$ _____

(Answers to these preview problems are given in the Appendix.)

If you are certain that you can work *all* these problems correctly, turn to page 95 for a set of practice problems. If you cannot work one or more of the preview problems, turn to the page indicated to the right of the problem. For those who wish to master this material with the greatest success, turn to Section 2-1 and begin work there.

2 Fractions

**"Fool! This is an eleven-sixteenths.
... I asked for a five-eighths!"**

2-1 WORKING WITH FRACTIONS

The word *fraction* comes from a Latin word meaning "to break," and fraction numbers are used when we need to break down standard measuring units into smaller parts. Anyone doing practical mathematics problems will find that fractions are used in many different trade and technical areas. Unfortunately, fraction arithmetic usually *must* be done with pencil and paper, since not all calculators handle fractions directly.

Look at the following piece of lumber. What happens when we break it into equal parts?

2 parts, each one-half of the whole

3 parts, each one-third of the whole

4 parts, each one-fourth of the whole

Suppose we divide this area [⬚] into fifths by drawing vertical lines.

Notice that the five parts or "fifths" are equal in area. [⬚⬚⬚⬚⬚]

A fraction is normally written as the division of two whole numbers: $\frac{2}{3}$, $\frac{3}{4}$, or $\frac{9}{16}$. One of the five equal areas below would be "one-fifth" or $\frac{1}{5}$ of the entire area.

EXAMPLE

How would you label this portion of the area? ⬛⬛⬛⬜⬜ = ?

⬛⬛⬛⬜⬜ $\dfrac{3}{5} = \dfrac{3 \text{ shaded parts}}{5 \text{ total parts}}$

The fraction $\frac{3}{5}$ implies an area equal to three of the original parts.

$$\frac{3}{5} = 3 \times \left(\frac{1}{5}\right)$$

There are three equal parts, and the name of each part is $\frac{1}{5}$ or one-fifth.

YOUR TURN

In this collection of letters, HHHHPPT, what fraction are Hs?

SOLUTION

Fraction of Hs $= \dfrac{\text{number of Hs}}{\text{total number of letters}} = \dfrac{4}{7}$ (read it "four sevenths")

The fraction of Ps is $\frac{2}{7}$, and the fraction of Ts is $\frac{1}{7}$.

Numerator   The two numbers that form a fraction are given special names to simplify talking about them. In the fraction $\frac{3}{5}$ the upper number 3 is called the *numerator* from the Latin *numero* meaning "number." It is a count of the number of parts.

Denominator   The lower number 5 is called the *denominator* from the Latin *nomen* or "name." It tells us the name of the part being counted. The numerator and denominator are called the *terms* of the fraction.

Learning Help ▶   A handy memory aid is to remember that the denominator is the "down part"—D for down. ◀

YOUR TURN

A paperback book costs $6 and I have $5. What fraction of its cost do I have? Write the answer as a fraction.

numerator = _____ , denominator = _____

ANSWER

$ \$ \$ \$ \$ \$ \$ $ $5 is $\dfrac{5}{6}$ of the total cost.

$\underbrace{}_{5}$ ∟numerator = 5, denominator = 6

Chap. 2 Fractions

Complete the following sentences by writing in the correct fraction.

(a) If we divide a length into eight equal parts, each part will be _____ of the total length.

(b) Then three of these parts will represent _____ of the total length.

(c) Eight of these parts will be _____ of the total length.

(d) Ten of these parts will be _____ of the total length.

ANSWERS

(a) $\dfrac{1}{8}$ (b) $\dfrac{3}{8}$ (c) $\dfrac{8}{8}$ (d) $\dfrac{10}{8}$

Proper Fraction

The original length is used as a standard for comparison, and any other length—smaller or larger—can be expressed as a fraction of it. A *proper fraction* is a number less than 1, as you would suppose a fraction should be. It represents a quantity less than the standard. For example, $\frac{1}{2}$, $\frac{2}{3}$, and $\frac{17}{20}$ are all proper fractions. Notice that for a proper fraction, the numerator is less than the denominator—the top number is less than the bottom number in the fraction.

Improper Fraction

An *improper fraction* is a number greater than 1 and represents a quantity greater than the standard. If a standard length is 8 in., a length of 11 in. will be $\frac{11}{8}$ of the standard. Notice that for an improper fraction the numerator is greater than the denominator—top number greater than the bottom number in the fraction.

YOUR TURN

Circle the proper fractions in the following list.

$$\frac{3}{2} \quad \frac{3}{4} \quad \frac{7}{8} \quad \frac{5}{4} \quad \frac{15}{12} \quad \frac{1}{16} \quad \frac{35}{32} \quad \frac{7}{50} \quad \frac{65}{64} \quad \frac{105}{100}$$

ANSWERS

You should have circled the following proper fractions: $\frac{3}{4}$, $\frac{7}{8}$, $\frac{1}{16}$, $\frac{7}{50}$. In each fraction the numerator (top number) is less than the denominator (bottom number). Each of these fractions represents a number less than 1.

The improper fraction $\frac{7}{3}$ can be shown graphically as follows:

Unit standard = ☐☐☐ is equal to 1

$\frac{1}{3}$ = ☐

then, $\frac{7}{3}$ = ▭▭▭▭▭▭▭ (seven; count 'em)

We can rename this number by regrouping.

$\dfrac{1}{3}$ $\dfrac{1}{3}$ $\dfrac{1}{3}$ equals 1

$\dfrac{1}{3}$ $\dfrac{1}{3}$ $\dfrac{1}{3}$ equals 1

$\dfrac{1}{3}$ $\dfrac{7}{3} = 2 + \dfrac{1}{3}$ or $2\dfrac{1}{3}$

Mixed Numbers

A *mixed number* is an improper fraction written as the sum of a whole number and a proper fraction.

$$\frac{7}{3} = 2 + \frac{1}{3} \text{ or } 2\frac{1}{3}$$

We usually omit the $+$ sign and write $2 + \frac{1}{3}$ as $2\frac{1}{3}$, and read it as "two and one-third." The numbers $1\frac{1}{2}$, $2\frac{2}{5}$, and $16\frac{2}{3}$ are all written as mixed numbers.

To write an improper fraction as a mixed number, divide numerator by denominator and form a new fraction as shown:

Improper fraction $\left\{ \dfrac{13}{5} = 13 \div 5 = 2 \right.$ with remainder 3 $2\dfrac{3}{5}$ $\left. \right\}$ Mixed number

Remainder   Quotient   Same denominator

$$\frac{13}{5} = 5\overline{)13} \quad\begin{array}{c}2 \leftarrow \text{Quotient}\\ \\ \underline{10}\\ 3 \leftarrow \text{Remainder}\end{array} \quad = \quad 2\frac{3}{5}$$

YOUR TURN

Now you try it. Rename $\frac{23}{4}$ as a mixed number. $\frac{23}{4} =$ _____

SOLUTION

$\dfrac{23}{4} = 23 \div 4 = 5$ with remainder $3 \longrightarrow 5\dfrac{3}{4}$

If in doubt, check your work with a diagram like this:

■■■■■■■■■■■■■■■■■■■■■■■
$\underbrace{}_{23}$ = ▦▦ $\left. \right\}$ 5 rows of 4

■ ■ ■   3 remaining

PROBLEMS

Now try these for practice. Write each improper fraction as a mixed number.

(a) $\dfrac{9}{5}$ (b) $\dfrac{13}{4}$ (c) $\dfrac{27}{8}$ (d) $\dfrac{31}{4}$ (e) $\dfrac{41}{12}$ (f) $\dfrac{17}{2}$

ANSWERS

(a) $\dfrac{9}{5} = 1\dfrac{4}{5}$ (b) $\dfrac{13}{4} = 3\dfrac{1}{4}$ (c) $\dfrac{27}{8} = 3\dfrac{3}{8}$

(d) $\dfrac{31}{4} = 7\dfrac{3}{4}$ (e) $\dfrac{41}{12} = 3\dfrac{5}{12}$ (f) $\dfrac{17}{2} = 8\dfrac{1}{2}$

The reverse process, rewriting a mixed number as an improper fraction, is equally simple.

Work in a clockwise direction, first multiply, $5 \times 2 = 10$
then add the numerator $10 + 3 = 13$

$10 + 3 = 13$

$5 \times 2 = 10$

Same denominator

Graphically, $2\dfrac{3}{5} = 1 + 1 + \dfrac{3}{5}$

$1 = \qquad = \dfrac{5}{5}$

$1 = \qquad = \dfrac{5}{5}$

$\dfrac{3}{5} = \qquad = \dfrac{3}{5}$

or $\dfrac{13}{5}$; (count them)

YOUR TURN

Now you try it. Rewrite these mixed numbers as improper fractions.

(a) $3\dfrac{1}{5}$ (b) $4\dfrac{3}{8}$ (c) $1\dfrac{1}{16}$ (d) $5\dfrac{1}{2}$ (e) $15\dfrac{3}{8}$ (f) $9\dfrac{3}{4}$

ANSWERS

(a) **Step 1** $3 \times 5 = 15$

     Step 2 $15 + 1 = 16 \leftarrow$ The new numerator

     Step 3 $3\dfrac{1}{5} = \dfrac{16}{5}$ ✓ The original denominator

(b) $4\dfrac{3}{8} = \dfrac{35}{8}$ (c) $1\dfrac{1}{16} = \dfrac{17}{16}$ (d) $5\dfrac{1}{2} = \dfrac{11}{2}$

(e) $15\dfrac{3}{8} = \dfrac{123}{8}$ (f) $9\dfrac{3}{4} = \dfrac{39}{4}$

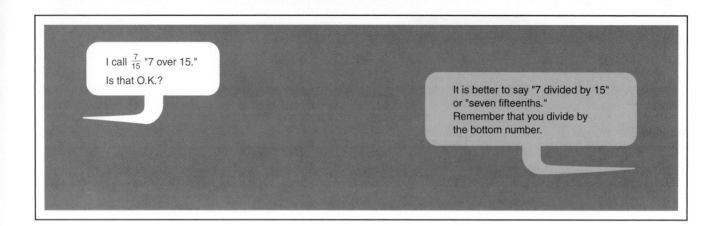

Equivalent Fractions

$\frac{1}{2}$

$\frac{2}{4}$

Two fractions are said to be *equivalent* if they represent the same number. For example, $\frac{1}{2} = \frac{2}{4}$ since both fractions represent the same portion of some standard amount.

There is a very large set of fractions equivalent to $\frac{1}{2}$.

$$\frac{1}{2} = \frac{2}{4} = \frac{3}{6} = \frac{4}{8} = \frac{5}{10} = \cdots = \frac{46}{92} = \frac{61}{122} = \frac{1437}{2874} \quad \text{and so on}$$

Each fraction is the same number, and we can use these fractions interchangeably in any mathematics problem.

To obtain a fraction equivalent to any given fraction, multiply the original numerator and denominator by the same nonzero number.

EXAMPLE

Rewrite the fraction $\frac{2}{3}$ as an equivalent fraction with denominator 15.

$$\frac{2}{3} = \frac{2 \times \boxed{5}}{3 \times \boxed{5}} = \frac{10}{15}$$

⬆

Multiply top and bottom by 5

YOUR TURN

Rewrite the fraction $\frac{3}{4}$ as an equivalent fraction with denominator equal to 20.

$$\frac{3}{4} = \frac{?}{20}$$

$$\frac{3}{4} = \frac{3 \times \square}{4 \times \square} = \frac{?}{20} \qquad 4 \times \square = 20, \text{ so } \square \text{ must be 5.}$$

$$= \frac{3 \times \boxed{5}}{4 \times \boxed{5}} = \frac{15}{20}$$

The number value of the fraction has not changed; we have simply renamed it.

PROBLEMS

Practice with these.

(a) $\dfrac{5}{8} = \dfrac{?}{32}$ 　　　　(b) $\dfrac{7}{16} = \dfrac{?}{48}$ 　　　　(c) $1\dfrac{2}{3} = \dfrac{?}{12}$

SOLUTIONS

(a) $\dfrac{5}{8} = \dfrac{5 \times 4}{8 \times 4} = \dfrac{20}{32}$ 　　　　(b) $\dfrac{7}{16} = \dfrac{7 \times 3}{16 \times 3} = \dfrac{21}{48}$

(c) $1\dfrac{2}{3} = \dfrac{5}{3} = \dfrac{5 \times 4}{3 \times 4} = \dfrac{20}{12}$

What is an equivalent fraction?

Fractions that have the same value but are written differently are equivalent.
$1\frac{1}{2} = \frac{3}{2} = \frac{6}{4} = \frac{18}{12}$
. . . are equivalent.

Writing in Lowest Terms Very often in working with fractions you will be asked to *write a fraction in lowest terms.* This means to replace it with the most simple fraction in its set of equivalent fractions. To write $\frac{15}{30}$ in its lowest terms means to replace it by $\frac{1}{2}$. They are equivalent.

$$\frac{15}{30} = \frac{15 \div 15}{30 \div 15} = \frac{1}{2}$$

We have divided both the top and bottom of the fraction by 15.

In general, you would write a fraction in lowest terms as follows:

EXAMPLE

Write $\dfrac{30}{48}$ in lowest terms.

First, find the largest number that divides both top and bottom of the fraction exactly.

$$30 = 5 \times 6$$

$$48 = 8 \times 6$$

In this case, the factor 6 is the largest number that divides both parts of the fraction exactly.

Second, eliminate this common factor by dividing.

$$\frac{30}{48} = \frac{30 \div 6}{48 \div 6} = \frac{5}{8}$$

The fraction $\frac{5}{8}$ is the simplest fraction equivalent to $\frac{30}{48}$. No whole number greater than 1 divides both 5 and 8 exactly.

EXAMPLE

Write $\dfrac{90}{105}$ in lowest terms.

$$\frac{90}{105} = \frac{90 \div 15}{105 \div 15} = \frac{6}{7}$$

This process of eliminating a common factor is usually called *canceling*. When you cancel a factor, you divide both top and bottom of the fraction by that number. We would write $\dfrac{90}{105}$ as $\dfrac{6 \times \cancel{15}}{7 \times \cancel{15}} = \dfrac{6}{7}$. When you "cancel," you must divide by a *pair* of common factors, one factor in the numerator and the same factor in the denominator.

EXAMPLE

In roof construction, the *pitch* or steepness of a roof is defined as the fraction

$$\text{Pitch} = \frac{\text{rise}}{\text{run}}$$

Rise 8 ft

Run 24 ft

The **rise** is the increase in height and the **run** is the corresponding horizontal distance. Typically, roofers express the pitch as the amount of rise per foot, or per 12 inches, of run. Therefore, rather than write the pitch as a fraction in lowest terms, they express it as an equivalent fraction with a denominator of 12. For the roof shown,

$$\text{Pitch} = \frac{\text{Rise}}{\text{Run}} = \frac{8\,\text{ft}}{24\,\text{ft}} = \frac{8 \div \boxed{2}}{24 \div \boxed{2}} = \frac{4}{12}$$

Divide by 2 ... to get a denominator of 12

Write the following fractions in lowest terms.

(a) $\dfrac{6}{8}$ (b) $\dfrac{12}{16}$ (c) $\dfrac{2}{4}$ (d) $\dfrac{4}{12}$

(e) $\dfrac{15}{84}$ (f) $\dfrac{21}{35}$ (g) $\dfrac{12}{32}$

(h) **Roofing** Calculate the pitch of a storage shed roof having a rise of 6 ft over a run of 36 ft.

SOLUTIONS

(a) $\dfrac{6}{8} = \dfrac{6 \div 2}{8 \div 2} = \dfrac{3}{4}$ (b) $\dfrac{12}{16} = \dfrac{12 \div 4}{16 \div 4} = \dfrac{3}{4}$

(c) $\dfrac{2}{4} = \dfrac{2 \div 2}{4 \div 2} = \dfrac{1}{2}$ (d) $\dfrac{4}{12} = \dfrac{4 \div 4}{12 \div 4} = \dfrac{1}{3}$

(e) $\dfrac{15}{84} = \dfrac{15 \div 3}{84 \div 3} = \dfrac{5}{28}$ (f) $\dfrac{21}{35} = \dfrac{21 \div 7}{35 \div 7} = \dfrac{3}{5}$

(g) $\dfrac{12}{32} = \dfrac{12 \div 4}{32 \div 4} = \dfrac{3}{8}$ (h) Pitch $= \dfrac{6 \text{ ft}}{36 \text{ ft}} = \dfrac{6 \div 3}{36 \div 3} = \dfrac{2}{12}$

EXAMPLE

This one is a little tricky. Write $\dfrac{6}{3}$ in lowest terms.

$\dfrac{6}{3} = \dfrac{6 \div 3}{3 \div 3} = \dfrac{2}{1}$ or simply 2

Any whole number may be written as a fraction by using a denominator equal to 1.

$3 = \dfrac{3}{1}$ $4 = \dfrac{4}{1}$ and so on.

Writing numbers in this way will be helpful when you learn to do arithmetic with fractions.

Comparing Fractions   If you were offered your choice between $\frac{2}{3}$ of a certain amount of money and $\frac{5}{8}$ of it, which would you choose?

EXAMPLE

Which is the larger fraction, $\frac{2}{3}$ or $\frac{5}{8}$?

Can you decide? Rewriting the fractions as equivalent fractions will help.

To compare two fractions, rename each by changing them to equivalent fractions with the same denominator.

$$\frac{2}{3} = \frac{2 \times 8}{3 \times 8} = \frac{16}{24} \qquad \text{and} \qquad \frac{5}{8} = \frac{5 \times 3}{8 \times 3} = \frac{15}{24}$$

Now compare the new fractions: $\frac{16}{24}$ is greater than $\frac{15}{24}$ and therefore $\frac{2}{3}$ is larger than $\frac{5}{8}$.

Learning Help ▶   1.   The new denominator is the product of the original ones ($24 = 8 \times 3$).

2.   Once both fractions are written with the same denominator, the one with the larger numerator is the larger fraction. (16 of the fractional parts is more than 15 of them.) ◀

YOUR TURN

Which of the following quantities is the larger?

(a) $\frac{3}{4}$ in. or $\frac{5}{7}$ in.

(b) $\frac{7}{8}$ or $\frac{19}{21}$

(c) 3 or $\frac{40}{13}$

(d) $1\frac{7}{8}$ lb or $\frac{5}{3}$ lb

(e) $2\frac{1}{4}$ ft or $\frac{11}{6}$ ft

(f) $\frac{5}{16}$ or $\frac{11}{36}$

SOLUTIONS

(a) $\frac{3}{4} = \frac{21}{28}, \frac{5}{7} = \frac{20}{28}; \frac{21}{28}$ is larger than $\frac{20}{28}$, so $\frac{3}{4}$ in. is larger than $\frac{5}{7}$ in.

(b) $\frac{7}{8} = \frac{147}{168}, \frac{19}{21} = \frac{152}{168}; \frac{152}{168}$ is larger than $\frac{147}{168}$, so $\frac{19}{21}$ is larger than $\frac{7}{8}$.

(c) $3 = \frac{39}{13}; \frac{40}{13}$ is larger than $\frac{39}{13}$, so $\frac{40}{13}$ is larger than 3.

(d) $1\frac{7}{8} = \frac{15}{8} = \frac{45}{24}, \frac{5}{3} = \frac{40}{24}; \frac{45}{24}$ is larger than $\frac{40}{24}$, so $1\frac{7}{8}$ lb is larger than $\frac{5}{3}$ lb.

(e) $2\frac{1}{4} = \frac{9}{4} = \frac{54}{24}, \frac{11}{6} = \frac{44}{24}; \frac{54}{24}$ is larger than $\frac{44}{24}$, so $2\frac{1}{4}$ ft is larger than $\frac{11}{6}$ ft.

(f) $\frac{5}{16} = \frac{180}{576}, \frac{11}{36} = \frac{176}{576}; \frac{180}{576}$ is larger than $\frac{176}{576}$, so $\frac{5}{16}$ is larger than $\frac{11}{36}$.

Now turn to Exercises 2-1 for some practice in working with fractions.

Exercises 2-1

Working with Fractions

A. Write as an improper fraction.

1. $2\frac{1}{3}$ 2. $7\frac{1}{2}$ 3. $8\frac{3}{8}$ 4. $1\frac{1}{16}$ 5. $2\frac{7}{8}$

6. 2 7. $2\frac{2}{3}$ 8. $4\frac{3}{64}$ 9. $4\frac{5}{6}$ 10. $1\frac{13}{16}$

B. Write as a mixed number.

1. $\frac{17}{2}$ 2. $\frac{8}{5}$ 3. $\frac{11}{8}$ 4. $\frac{40}{16}$ 5. $\frac{3}{2}$

6. $\frac{11}{3}$ 7. $\frac{100}{6}$ 8. $\frac{4}{3}$ 9. $\frac{80}{32}$ 10. $\frac{5}{2}$

C. Write in lowest terms.

1. $\frac{12}{16}$ 2. $\frac{4}{6}$ 3. $\frac{6}{16}$ 4. $\frac{18}{4}$ 5. $\frac{4}{10}$

6. $\frac{35}{30}$ 7. $\frac{24}{30}$ 8. $\frac{10}{4}$ 9. $4\frac{3}{12}$ 10. $\frac{34}{32}$

11. $\frac{42}{64}$ 12. $\frac{10}{35}$ 13. $\frac{15}{36}$ 14. $\frac{45}{18}$ 15. $\frac{38}{24}$

D. Complete.

1. $\frac{7}{8} = \frac{?}{16}$ 2. $\frac{3}{4} = \frac{?}{16}$ 3. $\frac{1}{8} = \frac{?}{64}$

4. $\frac{3}{8} = \frac{?}{64}$ 5. $1\frac{1}{4} = \frac{?}{16}$ 6. $2\frac{7}{8} = \frac{?}{32}$

7. $3\frac{3}{5} = \frac{?}{10}$ 8. $1\frac{1}{16} = \frac{?}{32}$ 9. $1\frac{40}{60} = \frac{?}{3}$

10. $4 = \frac{?}{6}$ 11. $2\frac{5}{8} = \frac{?}{16}$ 12. $2\frac{5}{6} = \frac{?}{12}$

E. Which is larger?

1. $\frac{3}{5}$ or $\frac{4}{7}$ 2. $\frac{3}{2}$ or $\frac{13}{8}$ 3. $1\frac{1}{2}$ or $1\frac{3}{7}$

4. $\frac{3}{4}$ or $\frac{13}{16}$ 5. $\frac{7}{8}$ or $\frac{5}{6}$ 6. $2\frac{1}{2}$ or $1\frac{11}{8}$

7. $1\frac{2}{5}$ or $\frac{6}{4}$ 8. $\frac{3}{16}$ or $\frac{25}{60}$ 9. $\frac{13}{5}$ or $\frac{5}{2}$

10. $3\frac{1}{2}$ or $2\frac{7}{4}$ 11. $\frac{3}{8}$ or $\frac{5}{12}$ 12. $1\frac{1}{5}$ or $\frac{8}{7}$

F. Practical Problems

1. **Carpentry** Maria, an apprentice carpenter, measured the length of a 2 by 4 as $15\frac{6}{8}$ in. Express this measurement in lowest terms.

2. **Electrical Technology** An electrical light circuit in John's welding shop had a load of 2800 watts. He changed the circuit to ten 150-watt bulbs and six 100-watt bulbs. What fraction represents a comparison of the new load with the old load?

Problem 5

3. The numbers $\frac{22}{7}$, $\frac{19}{6}$, $\frac{47}{15}$, $\frac{25}{8}$, and $\frac{41}{13}$ are all reasonable approximations to the number π. Which is the largest approximation? Which is the smallest approximation?

4. **Sheet Metal Technology** Which is thicker, a $\frac{3}{16}$-in. sheet of metal or a $\frac{13}{64}$-in. fastener?

5. **Plumbing** Is it possible to have a $\frac{7}{8}$-in. pipe with an inside diameter of $\frac{29}{32}$-in.?

6. **Sheet Metal Technology** Fasteners are equally spaced on a metal vent cover, with nine spaces between fasteners covering 24 in. Write the distance between spaces as a mixed number.

$$\frac{24}{9} = ?$$

7. **Printing** A printer has 15 rolls of newsprint in the warehouse. What fraction of this total will remain if six rolls are used?

8. **Machine Technology** A machinist who had been producing 40 parts per day increased the output to 60 parts per day by going to a faster machine. How much faster is the new machine? Express your answer as a mixed number.

9. **Landscaping** Before it can be used, a 12-ounce container of liquid fertilizer must be mixed with 48 ounces of water. What fraction of fertilizer is in the final mixture?

10. **Roofing** A ridge beam rises 18 inches over a horizontal run of 72 inches. Calculate the pitch of the roof and express it as a fraction with a denominator of 12.

When you have had the practice you need, check your answers in the Appendix.

2-2 MULTIPLICATION OF FRACTIONS

The simplest arithmetic operation with fractions is multiplication and, happily, it is easy to show graphically. The multiplication of a whole number and a fraction may be illustrated this way.

$$3 \times \frac{1}{4} = \frac{1}{4} + \frac{1}{4} + \frac{1}{4} = \frac{3}{4} \qquad \text{three segments each } \frac{1}{4} \text{ unit long.}$$

Any fraction such as $\frac{3}{4}$ can be thought of as a product: $\qquad 3 \times \frac{1}{4}$

The product of two fractions can also be shown graphically.

$$\frac{1}{2} \times \frac{1}{3} \qquad \text{means} \qquad \frac{1}{2} \text{ of } \frac{1}{3}$$

| The whole | $\frac{1}{3}$ of the whole | $\frac{1}{2}$ of $\frac{1}{3}$ of the whole | $\frac{1}{6}$ of the whole |

The product $\frac{1}{2} \times \frac{1}{3}$ is

$$\frac{1}{2} \times \frac{1}{3} = \frac{1}{6} = \frac{1 \text{ shaded area}}{6 \text{ equal areas in the square}}$$

In general, we calculate this product as

$$\frac{1}{2} \times \frac{1}{3} = \frac{1 \times 1}{2 \times 3}$$

Multiply the numerators (top)

Multiply the denominators (bottom)

$$\frac{1}{2} \times \frac{1}{3} = \frac{1}{6}$$

The product of two fractions is a fraction whose numerator is the product of their numerators and whose denominator is the product of their denominators.

EXAMPLE

Multiply $\frac{5}{6} \times \frac{2}{3}$.

$$\frac{5}{6} \times \frac{2}{3} = \frac{5 \times 2}{6 \times 3} = \frac{10}{18} = \frac{5 \times \overset{1}{2}}{9 \times 2} = \frac{5}{9}$$

Note ▶ Always write your answer in lowest terms. In this problem you probably recognized that both 10 and 18 were evenly divisible by 2, so you canceled out that common factor. It will save you time and effort if you cancel common factors, such as the 2 above, *before* you multiply this way:

$$\frac{5 \times \overset{1}{2}}{\underset{3}{\cancel{6}} \times 3} = \frac{5}{9} \blacktriangleleft$$

EXAMPLE

Whole Number Times a Fraction
When multiplying a whole number by a fraction, first write the whole number as a fraction.

Step 1 Rewrite 6 as a fraction with denominator 1.

$$6 \times \frac{1}{8} = \frac{6}{1} \times \frac{1}{8}$$

Step 2 Divide by common factors and multiply.

$$= \frac{\overset{3}{\cancel{6}} \times 1}{1 \times \underset{4}{\cancel{8}}} = \frac{3}{4}$$

EXAMPLE

Mixed Number Times a Fraction

Step 1 Write the mixed number $2\frac{1}{2}$ as the improper fraction $\frac{5}{2}$.

$$2\frac{1}{2} \times \frac{3}{4} = \frac{5}{2} \times \frac{3}{4}$$

Step 2 Multiply.

$$= \frac{5 \times 3}{2 \times 4} = \frac{15}{8}$$

Step 3 Write the product as a mixed number.

$$= 1\frac{7}{8}$$

Mixed Number Times a Mixed Number

Step 1 Write as improper fractions.

$$3\frac{1}{3} \times 2\frac{1}{4} = \frac{10}{3} \times \frac{9}{4}$$

Step 2 Multiply.

$$= \frac{\overset{5}{\cancel{10}} \times \overset{3}{\cancel{9}}}{\underset{1}{\cancel{3}} \times \underset{2}{\cancel{4}}}$$

Step 3 Write the answer as a mixed number.

$$= \frac{15}{2} = 7\frac{1}{2}$$

Note ▶ In many word problems the words "of" or "product of" appear as signals that you are to multiply. For example, the phrase "one-half of 16" means

$$\frac{1}{2} \times 16 = \frac{1}{2} \times \frac{16}{1} = \frac{16}{2} = 8,$$ and the phrase "the product of $\frac{2}{3}$ and $\frac{1}{4}$"

should be translated as $\frac{2}{3} \times \frac{1}{4} = \frac{2}{12} = \frac{1}{6}$. ◀

Now test your understanding with these problems. Multiply as shown. Change any mixed numbers to improper fractions *before* you multiply.

(a) $\frac{7}{8}$ of $\frac{2}{3}$ = _____

(b) $\frac{8}{12} \times \frac{3}{16}$ = _____

(c) $\frac{3}{32}$ of $\frac{4}{15}$ = _____

(d) $\frac{15}{4} \times \frac{9}{10}$ = _____

(e) $\frac{3}{2}$ of $\frac{2}{3}$ = _____

(f) $1\frac{1}{2} \times \frac{2}{5}$ = _____

(g) $4 \times \frac{7}{8}$ = _____

(h) $3\frac{5}{6} \times \frac{3}{10}$ = _____

(i) $1\frac{4}{5} \times 1\frac{3}{4}$ = _____

(j) **Plumbing** Polly the Plumber needs six lengths of PVC pipe each $26\frac{3}{4}$ in. long. What total length of pipe will she need?

Remember to write your answer in lowest terms.

(a) (*Hint:* "Of" means multiply.)

$$\frac{7}{8} \times \frac{2}{3} = \frac{7 \times \cancel{2}}{(4 \times \cancel{2}) \times 3} = \frac{7}{4 \times 3} = \frac{7}{12}$$

Eliminate common factors before you multiply.

Your work will look like this when you learn to do these operations mentally:

$$\frac{7}{\underset{4}{\cancel{8}}} \times \frac{\overset{1}{\cancel{2}}}{3} = \frac{7}{12}$$

(b) $\frac{8}{12} \times \frac{3}{16} = \frac{\overset{1}{\cancel{8}} \times \overset{1}{\cancel{3}}}{(4 \times \underset{1}{\cancel{3}}) \times (8 \times \underset{1}{\cancel{2}})} = \frac{1}{4 \times 2} = \frac{1}{8}$

$$\text{or } \frac{\overset{1}{\cancel{8}}}{\underset{4}{\cancel{12}}} \times \frac{\overset{1}{\cancel{3}}}{\underset{2}{\cancel{16}}} = \frac{1}{8}$$

(c) $\dfrac{\overset{1}{\cancel{3}}}{\underset{8}{\cancel{32}}} \times \dfrac{\overset{1}{\cancel{4}}}{\underset{5}{\cancel{15}}} = \dfrac{1}{40}$

(d) $\dfrac{\overset{3}{\cancel{15}}}{4} \times \dfrac{9}{\underset{2}{\cancel{10}}} = \dfrac{27}{8} = 3\dfrac{3}{8}$

(e) $\dfrac{\overset{1}{\cancel{3}}}{\underset{1}{\cancel{2}}} \times \dfrac{\overset{1}{\cancel{2}}}{\underset{1}{\cancel{3}}} = 1$

(f) $1\dfrac{1}{2} \times \dfrac{2}{5} = \dfrac{3}{2} \times \dfrac{\overset{1}{\cancel{2}}}{5} = \dfrac{3}{5}$

If you don't remember how to change a mixed number to an improper fraction see page 63.

(g) $4 \times \dfrac{7}{8} = \dfrac{4}{1} \times \dfrac{7}{\underset{2}{\cancel{8}}} = \dfrac{7}{2} = 3\dfrac{1}{2}$

(h) $3\dfrac{5}{6} \times \dfrac{3}{10} = \dfrac{23}{\underset{2}{\cancel{6}}} \times \dfrac{\overset{1}{\cancel{3}}}{10} = \dfrac{23}{20} = 1\dfrac{3}{20}$

(i) $1\dfrac{4}{5} \times 1\dfrac{3}{4} = \dfrac{9}{5} \times \dfrac{7}{4} = \dfrac{63}{20} = 3\dfrac{3}{20}$

(j) Total length of pipe = number of pieces × length of each piece

$$= 6 \times 26\dfrac{3}{4}$$

$$= 6 \times \dfrac{107}{4}$$

$$= \dfrac{\overset{3}{\cancel{6}}}{1} \times \dfrac{107}{\underset{2}{\cancel{4}}}$$

$$= \dfrac{321}{2} = 160\dfrac{1}{2} \text{ in.}$$

Exercises 2-2

Multiplication of Fractions

A. Multiply and write the answer in lowest terms.

1. $\dfrac{1}{2} \times \dfrac{1}{4}$ 2. $\dfrac{2}{5} \times \dfrac{2}{3}$ 3. $\dfrac{4}{5} \times \dfrac{1}{6}$ 4. $6 \times \dfrac{1}{2}$

5. $\dfrac{8}{9} \times 3$ 6. $\dfrac{11}{12} \times \dfrac{4}{15}$ 7. $\dfrac{8}{3} \times \dfrac{5}{12}$ 8. $\dfrac{7}{8} \times \dfrac{13}{14}$

9. $\dfrac{12}{8} \times \dfrac{15}{9}$ 10. $\dfrac{4}{7} \times \dfrac{49}{2}$ 11. $4\dfrac{1}{2} \times \dfrac{2}{3}$ 12. $6 \times 1\dfrac{1}{3}$

13. $2\dfrac{1}{6} \times 1\dfrac{1}{2}$ 14. $\dfrac{5}{7} \times 1\dfrac{7}{15}$ 15. $4\dfrac{3}{5} \times 15$ 16. $10\dfrac{5}{6} \times 3\dfrac{3}{10}$

17. $34 \times 2\dfrac{3}{17}$ 18. $7\dfrac{9}{10} \times 1\dfrac{1}{4}$ 19. $11\dfrac{6}{7} \times \dfrac{7}{8}$

20. $18 \times 1\dfrac{5}{27}$ 21. $\dfrac{1}{2} \times \dfrac{1}{2} \times \dfrac{1}{2}$ 22. $1\dfrac{4}{5} \times \dfrac{2}{3} \times \dfrac{1}{4}$

23. $\dfrac{1}{4} \times \dfrac{2}{3} \times \dfrac{2}{5}$ 24. $2\dfrac{1}{2} \times \dfrac{3}{5} \times \dfrac{8}{9}$ 25. $\dfrac{2}{3} \times \dfrac{3}{2} \times 2$

B. Find.

1. $\frac{1}{2}$ of $\frac{1}{3}$ 2. $\frac{1}{4}$ of $\frac{3}{8}$ 3. $\frac{2}{3}$ of $\frac{3}{4}$ 4. $\frac{7}{8}$ of $\frac{1}{2}$

5. $\frac{1}{2}$ of $1\frac{1}{2}$ 6. $\frac{3}{4}$ of $1\frac{1}{4}$ 7. $\frac{5}{8}$ of $2\frac{1}{10}$ 8. $\frac{5}{3}$ of $1\frac{2}{3}$

9. $\frac{4}{3}$ of $\frac{3}{4}$ 10. $\frac{3}{5}$ of $1\frac{1}{6}$ 11. $\frac{7}{8}$ of $1\frac{1}{5}$ 12. $\frac{3}{5}$ of 4

13. $\frac{7}{16}$ of 6 14. $\frac{5}{16}$ of $1\frac{1}{7}$ 15. $\frac{3}{8}$ of $2\frac{2}{3}$ 16. $\frac{15}{16}$ of $1\frac{3}{5}$

C. Practical Problems

1. **Building Construction** Find the width of floor space covered by 38 boards with $3\frac{5}{8}$-in. exposed surface each.

2. **Building Construction** There are 14 risers in the stairs from the basement to the first floor of a house. Find the total height of the stairs if the risers are $7\frac{1}{8}$ in. high.

3. **Roofing** Shingles are laid so that 5 in. or $\frac{5}{12}$ ft is exposed in each layer. How many feet of roof will be covered by 28 courses?

4. **Carpentry** A board $5\frac{3}{4}$ in. wide is cut to three-fourths of its original width. Find the new width.

5. **Carpentry** What length of 2-in. by 4-in. material will be required to make six bench legs each $28\frac{1}{4}$ in. long?

6. **Electrical Technology** Find the total length of 12 pieces of wire each $9\frac{3}{16}$ in. long.

7. **Auto Services** If a car averages $22\frac{3}{4}$ miles to a gallon of gas, how many miles can it travel on 14 gallons of gas?

8. **Machine Technology** What is the shortest bar that can be used for making six chisels each $6\frac{1}{8}$ in. in length?

9. **Manufacturing** How many pounds of grease are contained in a barrel if a barrel holds $46\frac{1}{2}$ gallons, and a gallon of grease weighs $7\frac{2}{3}$ lb?

10. **Plumbing** A plumber cut eight pieces of copper tubing from a coil. Each piece is $14\frac{3}{4}$ in. long. What is the total length used?

→| |← Pitch $\frac{1"}{24}$

Problem 11

11. **Machine Technology** How far will a nut advance if it is given 18 turns on a $\frac{1}{4}$-in. 20-NF (National Fine thread) bolt? (*Hint:* The designation 20-NF means that the nut advances $\frac{1}{20}$ in. for each complete turn.)

12. **Building Construction** What width of floor space can be covered by 48 boards each with $4\frac{3}{8}$ in. of exposed surface?

13. **Drafting** If $\frac{3}{8}$ in. on a drawing represents 1 ft, how many inches on the drawing will represent 26 ft?

14. **Manufacturing** What is the volume of a rectangular box with interior dimensions $12\frac{1}{2}$ in. long, $8\frac{1}{8}$ in. wide, and $4\frac{1}{4}$ in. deep? (*Hint:* Volume = length × width × height.)

15. **Machine Technology** How long will it take to machine 45 pins if each pin requires $6\frac{3}{4}$ minutes? Allow 1 minute per pin for placing stock in the lathe.

16. **Manufacturing** There are 231 cu in. in a gallon. How many cubic inches are needed to fill a container with a rated capacity of $4\frac{1}{3}$ gallons?

17. **Printing**   In a print shop, 1 unit of labor is equal to $\frac{1}{6}$ hour. How many hours are involved in 64 units of work?

18. **Carpentry**   Find the total width of 36 2-by-4s if the finished width of each board is actually $3\frac{1}{2}$ in.

19. **Photography**   A photograph must be reduced to four-fifths of its original size to fit the space available in a newspaper. Find the length of the reduced photograph if the original was $6\frac{3}{4}$ in. long.

20. **Printing**   A bound book weighs $1\frac{5}{8}$ lb. How many pounds will 20 cartons of 12 books each weigh?

21. **Printing**   There are 6 picas in 1 in. If a line of type is $3\frac{3}{4}$ in. long, what is this length in picas?

22. **Wastewater Technology**   The normal daily flow of raw sewage into a treatment plant is 32 MGD (million gallons per day). Because of technical problems one day, the plant had to cut back to three-fourths of its normal intake. What was this reduced flow?

23. **Masonry**   What is the height of 12 courses of $2\frac{1}{4}$-in. bricks with $\frac{3}{8}$-in. mortar joints? (12 rows of brick and 11 mortar joints)

24. **Plumbing**   To find the degree measure of the bend of a pipe fitting, multiply the fraction of bend by 360 degrees. What is the degree measure of a $\frac{1}{8}$ bend? A $\frac{1}{5}$ bend? A $\frac{1}{6}$ bend?

25. **Plumbing**   A drain must be installed with a grade of $\frac{1}{8}$ in. of vertical drop per foot of horizontal run. How much drop will there be for 26 ft of run?

Problem 26

26. **Machine Technology**   The center-to-center distance between consecutive holes in a strip of metal is $\frac{5}{16}$ in. What is the total distance x between the first and last centers as shown in the figure?

27. **Sheet Metal Technology**   The allowance for a wired edge on fabricated metal is $2\frac{1}{2}$ times the diameter of the wire. Calculate the allowance for a wired edge if the diameter of the wire is $\frac{3}{16}$ in.

Check your answers in the Appendix, then continue in Section 2-3.

2-3   DIVISION OF FRACTIONS

Addition and multiplication are both reversible arithmetic operations. For example,

2×3 and 3×2 both equal 6

$4 + 5$ and $5 + 4$ both equal 9

The order in which you add or multiply is not important. This reversibility is called the *commutative* property of addition and multiplication.

In division this type of exchange is not allowed, and because it is not allowed many people find division very troublesome. In the division of fractions it is very important that you set up the problem correctly.

The phrase "8 divided by 4" can be written as

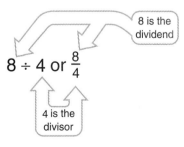

In the previous problem you were being asked to divide a set of eight objects into sets of four objects.

EXAMPLE

In the division $5 \div \frac{1}{2}$, which number is the divisor?

The divisor is $\frac{1}{2}$.

The division $5 \div \frac{1}{2}$, read "5 divided by $\frac{1}{2}$," asks how many $\frac{1}{2}$-unit lengths are included in a length of 5 units.

Division answers the question: How many of the divisor are in the dividend?

$8 \div 4 = \square$ asks you to find how many 4s are in 8.
It is easy to see that $\square = 2$.

$5 \div \frac{1}{2} = \square$ asks you to find how many $\frac{1}{2}$s are in 5. Do you see that $\square = 10$?

There are ten $\frac{1}{2}$-unit lengths contained in the 5-unit length.

Using a drawing like this to solve a division problem is difficult and clumsy. We need a simple rule.

Reciprocal   We can simplify the division of fractions with the help of a new concept called the **reciprocal**. The reciprocal of a fraction is obtained by switching its numerator and denominator. (This is often called *inverting* the fraction.)

EXAMPLE

The reciprocal of $\frac{5}{6}$ is $\frac{6}{5}$.

The reciprocal of $\frac{1}{4}$ is $\frac{4}{1}$ or 4.

The reciprocal of 8 or $\frac{8}{1}$ is $\frac{1}{8}$.

To find the reciprocal of a mixed number, first convert it to an improper fraction.

The reciprocal of $2\frac{3}{5}$ or $\frac{13}{5}$ is $\frac{5}{13}$.

YOUR TURN

Find the reciprocal of each of the following numbers.

(a) $\frac{2}{5}$ (b) $\frac{7}{8}$ (c) $\frac{1}{6}$ (d) 7   (e) $2\frac{1}{5}$

(a) $\dfrac{5}{2}$ (b) $\dfrac{8}{7}$ (c) $\dfrac{6}{1}$ or 6   (d) $7 = \dfrac{7}{1}$ so its reciprocal is $\dfrac{1}{7}$.

(e) $2\dfrac{1}{5}$ is $\dfrac{11}{5}$ and its reciprocal is $\dfrac{5}{11}$.

We can now use the following simple rule to divide fractions.

> To divide by a fraction, multiply by its reciprocal.

EXAMPLE

$5 \div \dfrac{1}{2} = ?$

$$5 \div \underset{\text{Multiply}}{\overset{\text{reciprocal}}{\dfrac{1}{2}}} = 5 \times \dfrac{2}{1} = \dfrac{10}{1} = 10 \text{ as shown graphically on the previous page}$$

EXAMPLE

$\dfrac{2}{5} \div \dfrac{4}{3} = ?$

$$\underset{\text{The reciprocal}}{\overset{\text{Multiply}}{\dfrac{2}{5} \div \dfrac{4}{3}}} = \dfrac{2}{5} \times \dfrac{3}{4} = \dfrac{\overset{1}{\cancel{2}} \times 3}{5 \times \underset{2}{\cancel{4}}} = \dfrac{3}{10}$$

Careful ▶ When dividing fractions, never attempt to cancel common factors until *after* converting the division to multiplication by the reciprocal. ◀

EXAMPLE

$\dfrac{5}{8} \div 3\dfrac{3}{4} = ?$

$$\dfrac{5}{8} \div 3\dfrac{3}{4} = \dfrac{5}{8} \div \dfrac{15}{4} = \dfrac{\overset{1}{\cancel{5}}}{\underset{2}{\cancel{8}}} \times \dfrac{\overset{1}{\cancel{4}}}{\underset{3}{\cancel{15}}} = \dfrac{1}{6}$$

We have converted division problems that are difficult to picture into simple multiplication.

The final, and very important, step in every division is checking the answer. To check, multiply the divisor by the quotient and compare this answer with the original fraction or dividend.

If $\dfrac{2}{5} \div \dfrac{4}{3} = \dfrac{3}{10}$ then $\dfrac{4}{3} \times \dfrac{3}{10}$ should equal $\dfrac{2}{5}$.

$$\dfrac{\overset{2}{\cancel{4}}}{\underset{1}{\cancel{3}}} \times \dfrac{\overset{1}{\cancel{3}}}{\underset{5}{\cancel{10}}} = \dfrac{2}{5}$$

YOUR TURN

Divide:

$$\dfrac{7}{8} \div \dfrac{3}{2} = \text{_____}$$

SOLUTION

Reciprocal

$$\dfrac{7}{8} \div \dfrac{3}{2} = \dfrac{7}{8} \times \dfrac{2}{3} = \dfrac{7 \times \overset{1}{\cancel{2}}}{\underset{4}{\cancel{8}} \times 3} = \dfrac{7}{12} \qquad \boxed{\checkmark} \quad \dfrac{\overset{1}{\cancel{3}}}{2} \times \dfrac{7}{\underset{4}{\cancel{12}}} = \dfrac{7}{8}$$

Multiply

Learning Help ▶ The chief source of confusion for many people in dividing fractions is deciding which fraction to invert. It will help if you:

1. Put every division problem in the form

    (dividend) ÷ (divisor)

    then find the reciprocal of the divisor, and finally, multiply to obtain the quotient.

2. Check your answer by multiplying. The product

    (divisor) × (quotient or answer)

    should equal the dividend. ◀

PROBLEMS

Here are a few problems to test your understanding.

(a) $\dfrac{2}{5} \div \dfrac{3}{8}$

(b) $\dfrac{7}{40} \div \dfrac{21}{25}$

(c) $3\dfrac{3}{4} \div \dfrac{5}{2}$

(d) $4\dfrac{1}{5} \div 1\dfrac{4}{10}$

(e) $3\dfrac{2}{3} \div 3$

(f) Divide $\dfrac{3}{4}$ by $2\dfrac{5}{8}$.

(g) Divide $1\dfrac{1}{4}$ by $1\dfrac{7}{8}$.

(h) **Carpentry** How many sheets of plywood, each $\dfrac{3}{4}$ in. thick, are in a stack 18 in. high?

Work carefully and check each answer.

(a) $\dfrac{2}{5} \div \dfrac{3}{8} = \dfrac{2}{5} \times \dfrac{8}{3} = \dfrac{16}{15} = 1\dfrac{1}{15}$

The answer is $1\dfrac{1}{15}$. ☑ $\dfrac{3}{\overset{}{8}_{1}} \times \dfrac{\overset{2}{16}}{\overset{}{15}_{5}} = \dfrac{2}{5}$

(b) $\dfrac{7}{40} \div \dfrac{21}{25} = \dfrac{\overset{1}{7}}{\overset{}{40}_{8}} \times \dfrac{\overset{5}{25}}{\overset{}{21}_{3}} = \dfrac{5}{24}$ ☑ $\dfrac{\overset{7}{21}}{\overset{}{25}_{5}} \times \dfrac{\overset{1}{5}}{\overset{}{24}_{8}} = \dfrac{7}{40}$

(c) $3\dfrac{3}{4} \div \dfrac{5}{2} = \dfrac{15}{4} \div \dfrac{5}{2} = \dfrac{\overset{3}{15}}{\overset{}{4}_{2}} \times \dfrac{\overset{1}{2}}{\overset{}{5}_{1}} = \dfrac{3}{2} = 1\dfrac{1}{2}$ ☑ $\dfrac{5}{2} \times \dfrac{3}{2} = \dfrac{15}{4} = 3\dfrac{3}{4}$

(d) $4\dfrac{1}{5} \div 1\dfrac{4}{10} = \dfrac{21}{5} \div \dfrac{14}{10} = \dfrac{\overset{3}{21}}{\overset{}{5}_{1}} \times \dfrac{\overset{2}{10}}{\overset{}{14}_{2}} = \dfrac{6}{2} = 3$

☑ $1\dfrac{4}{10} \times 3 = \dfrac{\overset{7}{14}}{\overset{}{10}_{5}} \times \dfrac{3}{1} = \dfrac{21}{5} = 4\dfrac{1}{5}$

(e) $3\dfrac{2}{3} \div 3 = \dfrac{11}{3} \div \dfrac{3}{1} = \dfrac{11}{3} \times \dfrac{1}{3} = \dfrac{11}{9} = 1\dfrac{2}{9}$

☑ $3 \times \dfrac{11}{9} = \dfrac{3}{1} \times \dfrac{11}{\overset{}{9}_{3}} = \dfrac{11}{3} = 3\dfrac{2}{3}$

(f) $\dfrac{3}{4} \div 2\dfrac{5}{8} = \dfrac{3}{4} \div \dfrac{21}{8} = \dfrac{\overset{1}{3}}{4} \times \dfrac{\overset{2}{8}}{\overset{}{21}_{7}} = \dfrac{2}{7}$ ☑ $2\dfrac{5}{8} \times \dfrac{2}{7} = \dfrac{\overset{3}{21}}{\overset{}{8}_{4}} \times \dfrac{\overset{1}{2}}{7} = \dfrac{3}{4}$

(g) $1\dfrac{1}{4} \div 1\dfrac{7}{8} = \dfrac{5}{4} \div \dfrac{15}{8} = \dfrac{\overset{1}{5}}{\overset{}{4}_{1}} \times \dfrac{\overset{2}{8}}{\overset{}{15}_{3}} = \dfrac{2}{3}$

☑ $1\dfrac{7}{8} \times \dfrac{2}{3} = \dfrac{\overset{5}{15}}{\overset{}{8}_{4}} \times \dfrac{\overset{1}{2}}{3} = \dfrac{5}{4} = 1\dfrac{1}{4}$

(h) A question of the form "How many X are in Y?" tells us we must divide Y by X. In this problem, divide 18 in., the total height, by $\frac{3}{4}$ in., the thickness of each sheet.

$18 \div \dfrac{3}{4} = \dfrac{\overset{6}{18}}{1} \times \dfrac{4}{\overset{}{3}_{1}} = 24$ There are 24 sheets of plywood in the stack.

WHY DO WE USE THE RECIPROCAL WHEN WE DIVIDE FRACTIONS?

Notice that $8 \div 4 = 2$ and $8 \times \dfrac{1}{4} = \dfrac{\overset{2}{8}}{1} \times \dfrac{1}{\underset{1}{4}} = 2$.

Dividing by a number gives the same result as multiplying by its reciprocal. The following shows why this is so.

The division $8 \div 4$ can be written $\dfrac{8}{4}$. Similarly, $\dfrac{1}{2} \div \dfrac{2}{3}$ can be written $\dfrac{\frac{1}{2}}{\frac{2}{3}}$.

To simplify this fraction, multiply by $\dfrac{\frac{3}{2}}{\frac{3}{2}}$ (which is equal to 1).

$$\dfrac{\frac{1}{2}}{\frac{2}{3}} = \dfrac{\frac{1}{2} \times \boxed{\frac{3}{2}}}{\frac{2}{3} \times \boxed{\frac{3}{2}}} = \dfrac{\frac{1}{2} \times \frac{3}{2}}{1} = \dfrac{1}{2} \times \dfrac{3}{2}$$

$$\dfrac{2}{3} \times \dfrac{3}{2} = \dfrac{2 \times 3}{3 \times 2} = \dfrac{6}{6} = 1$$

Therefore, $\dfrac{1}{2} \div \dfrac{2}{3} = \dfrac{1}{2} \times \dfrac{3}{2}$, the reciprocal of the divisor $\dfrac{2}{3}$.

Turn to Exercises 2-3 for a set of practice problems on dividing fractions.

Exercises 2-3

Division of Fractions

A. Divide and write the answer in lowest terms.

1. $\dfrac{5}{6} \div \dfrac{1}{2}$ $=1.6667$ 2. $6 \div \dfrac{2}{3}$ $=9$ 3. $\dfrac{5}{12} \div \dfrac{4}{3}$ $=0.3125$ 4. $8 \div \dfrac{1}{4}$ $=32$

5. $\dfrac{6}{16} \div \dfrac{3}{4}$ 0.5 6. $\dfrac{1}{2} \div \dfrac{1}{2}$ 1 7. $\dfrac{3}{16} \div \dfrac{6}{8}$ $.25$ 8. $\dfrac{3}{4} \div \dfrac{5}{16}$ 2.5

9. $1\dfrac{1}{2} \div \dfrac{1}{6}$ $=9$ 10. $6 \div 1\dfrac{1}{2}$ 4 11. $3\dfrac{1}{7} \div 2\dfrac{5}{14}$ $=1.454$ 12. $3\dfrac{1}{2} \div 2$ 1.75

13. $6\dfrac{2}{5} \div 5\dfrac{1}{3}$ 1.2 14. $10 \div 1\dfrac{1}{5}$ 8.333 15. $8 \div \dfrac{1}{2}$ 16 16. $\dfrac{2}{3} \div 6$ 0.111

17. $\dfrac{12}{\frac{2}{3}}$ $=18$ 18. $\dfrac{\frac{3}{4}}{\frac{7}{8}}$ $=0.8571$ 19. $\dfrac{5}{\frac{2}{3}}$ 7.5 20. $\dfrac{1\frac{1}{2}}{2\frac{1}{2}}$ $.6$

21. $\dfrac{5}{16} \div \dfrac{3}{8}$ 0.8333 22. $\dfrac{7}{12} \div \dfrac{2}{3}$ $.875$ 23. $\dfrac{7}{32} \div 1\dfrac{3}{4}$ 0.125 24. $1\dfrac{2}{3} \div 1\dfrac{1}{4}$ 1.333

B. Practical Problems

1. **Drafting** How many feet are represented by a 4-in. line if it is drawn to a scale of $\dfrac{1}{2}$ in. $= 1$ ft?

2. **Drafting** If $\dfrac{1}{4}$ in. on a drawing represents 1 ft 0 in., then $3\dfrac{1}{2}$ in. on the drawing will represent how many feet?

3. **Carpentry**   How many boards $4\frac{5}{8}$ in. wide will it take to cover a floor 222 in. wide?

4. **Building Construction**   How many supporting columns $88\frac{1}{2}$ in. long can be cut from six pieces each 22 ft long? (*Hint:* Be careful of units.)

5. **Building Construction**   How many pieces of $\frac{1}{2}$-in. plywood are there in a stack 42 in. high?

6. **Drafting**   If $\frac{1}{4}$ in. represents 1 ft 0 in. on a drawing, how many feet will be represented by a line $10\frac{1}{8}$ in. long?

7. **Masonry**   If we allow $2\frac{5}{8}$ in. for the thickness of a course of brick, including mortar joints, how many courses of brick will there be in a wall $47\frac{1}{4}$ in. high?

8. **Plumbing**   How many lengths of pipe $2\frac{5}{8}$ ft long can be cut from a pipe 21 ft long?

9. **Machine Technology**   How many pieces $6\frac{1}{4}$ in. long can be cut from 35 metal rods each 40 in. long? Disregard waste.

10. **Machine Technology**   The architectural drawing for a room measures $3\frac{5}{8}$ in. by $4\frac{1}{4}$ in. If $\frac{1}{8}$ in. is equal to 1 ft on the drawing, what are the actual dimensions of the room?

11. **Printing**   How many full $3\frac{1}{2}$-in. sheets can be cut from $24\frac{3}{4}$-in. stock?

12. **Machine Technology**   The feed on a boring mill is set for $\frac{1}{32}$ in. How many revolutions are needed to advance the tool $3\frac{3}{8}$ in.?

13. **Machine Technology**   If the pitch of a thread is $\frac{1}{18}$ in., how many threads are needed for the threaded section of a pipe to be $2\frac{1}{2}$ in. long?

14. **Building Construction**   The floor area of a room on a house plan measures $3\frac{1}{2}$ in. by $4\frac{5}{8}$ in. If the drawing scale is $\frac{1}{4}$ in. represents 1 ft, what is the actual size of the room?

15. **Building Construction**   The family room ceiling of a new Happy Home is to be taped and mudded. A $106\frac{1}{2}$ sq ft area has already been done, and this represents $\frac{3}{8}$ of the job. How large is the area of the finished ceiling?

16. **Plumbing**   A pipe fitter needs to divide a pipe $32\frac{5}{8}$ inches long into three pieces of equal length. Calculate the length of each piece.

Check your answers in the Appendix, then continue in Section 2-4.

2-4   ADDITION AND SUBTRACTION OF FRACTIONS

Addition   At heart, adding fractions is a matter of counting:

$$\frac{1}{5}+\frac{3}{5}=\frac{1+3}{5}=\frac{4}{5}$$

$\frac{1}{5}$ ▨☐☐☐☐   1 fifth

$+$

$\frac{3}{5}$ ▨▨▨☐☐   + 3 fifths

$=$

$\frac{4}{5}$ ▨▨▨▨☐   = 4 fifths, count them.

Add $\dfrac{1}{8} + \dfrac{3}{8} = $ _____

This is easy to see with measurements:

$$\frac{1}{8}" + \frac{3}{8}" = \frac{4}{8}" \text{ or } \frac{1}{2}"$$

EXAMPLE

Add $\frac{2}{7} + \frac{3}{7} = $ _____

$$\frac{2}{7} + \frac{3}{7} = \frac{2+3}{7} = \frac{5}{7}$$

$\frac{2}{7}$ ⬜ 2 sevenths
+
$\frac{3}{7}$ ⬜ + 3 sevenths
=
$\frac{5}{7}$ ⬜ 5 sevenths or $\frac{5}{7}$

Like Fractions

Fractions having the same denominator are called *like* fractions. In the preceding problem, $\frac{2}{7}$ and $\frac{3}{7}$ both have denominator 7 and are like fractions. Adding like fractions is easy: *first,* add the numerators to find the numerator of the sum and *second,* use the denominator the fractions have in common as the denominator of the sum.

$$\frac{2}{9} + \frac{5}{9} = \frac{2+5}{9}$$ ⬅ Add numerators
⬅ Same denominator
$$= \frac{7}{9}$$

EXAMPLE

Adding three or more like fractions is easy: $\frac{3}{12} + \frac{1}{12} + \frac{5}{12} = ?$

$$\frac{3}{12} + \frac{1}{12} + \frac{5}{12} = \frac{3+1+5}{12} = \frac{9}{12} = \frac{3}{4}$$

Notice that we write the sum in lowest terms.

YOUR TURN

Try these problems for exercise.

(a) $\frac{1}{8} + \frac{3}{8}$

(b) $\frac{7}{9} + \frac{5}{9}$

(c) $2\frac{1}{5} + 3\frac{3}{5}$

(d) $\frac{1}{7} + \frac{4}{7} + \frac{5}{7} + 1\frac{2}{7} + \frac{8}{7}$

(e) $2 + 3\frac{1}{2}$

(f) $3\frac{1}{8} + 2\frac{3}{8}$

(a) $\dfrac{1}{8} + \dfrac{3}{8} = \dfrac{1+3}{8} = \dfrac{4}{8} = \dfrac{1}{2}$

(b) $\dfrac{7}{9} + \dfrac{5}{9} = \dfrac{7+5}{9} = \dfrac{12}{9} = \dfrac{4}{3} = 1\dfrac{1}{3}$

(c) $2\dfrac{1}{5} + 3\dfrac{3}{5} = 2 + 3 + \dfrac{1}{5} + \dfrac{3}{5} = 5 + \dfrac{4}{5} = 5\dfrac{4}{5}$

(d) $\dfrac{1}{7} + \dfrac{4}{7} + \dfrac{5}{7} + 1\dfrac{2}{7} + \dfrac{8}{7} = \dfrac{1}{7} + \dfrac{4}{7} + \dfrac{5}{7} + \dfrac{9}{7} + \dfrac{8}{7} = \dfrac{1+4+5+9+8}{7} = \dfrac{27}{7} = 3\dfrac{6}{7}$

(e) $2 + 3\dfrac{1}{2} = 2 + 3 + \dfrac{1}{2} = 5\dfrac{1}{2}$. Remember: $3\dfrac{1}{2}$ means $3 + \dfrac{1}{2}$.

(f) $3\dfrac{1}{8} + 2\dfrac{3}{8} = (3 + 2) + \left(\dfrac{1}{8} + \dfrac{3}{8}\right) = 5 + \dfrac{4}{8} = 5 + \dfrac{1}{2} = 5\dfrac{1}{2}$

This last problem can also be done by rewriting each of the mixed numbers as an improper fraction before adding.

$3\dfrac{1}{8} + 2\dfrac{3}{8} = \dfrac{25}{8} + \dfrac{19}{8} = \dfrac{44}{8} = \dfrac{11}{2}$ Write in lowest terms.

$\qquad\qquad\qquad = 5\dfrac{1}{2}$ Rewrite as a mixed number.

Note ▶ If the addition is done using improper fractions, large and unwieldy numerators may result. Be careful. ◀

Unlike Fractions How do we add fractions whose denominators are not the same? For example, how do we add $\dfrac{2}{3} + \dfrac{3}{4}$?

The problem is to find a name for this new number. One way to find it is to change these fractions to equivalent fractions with the same denominator.

$\dfrac{2}{3} = \dfrac{2 \times 4}{3 \times 4} = \dfrac{8}{12}$ We chose $\dfrac{4}{4}$ as a multiplier because it changes $\dfrac{2}{3}$ to an equivalent fraction with a denominator of 12.

$\dfrac{3}{4} = \dfrac{3 \times 3}{4 \times 3} = \dfrac{9}{12}$ We chose $\dfrac{3}{3}$ as a multiplier because it changes $\dfrac{3}{4}$ to an equivalent fraction with a denominator of 12.

The two fractions now have the same denominator.

Note ▶ You should remember that we discussed equivalent fractions on page 64. Return for a quick review if you need it. ◀

EXAMPLE

Add $\dfrac{2}{3} + \dfrac{3}{4}$ using the equivalent fractions.

$\dfrac{2}{3} + \dfrac{3}{4} = \dfrac{8}{12} + \dfrac{9}{12} = \dfrac{17}{12} = 1\dfrac{5}{12}$

We change the original fractions to equivalent fractions with the same denominator and then add as before.

Least Common Denominator How do you know what number to use as the new denominator? In general, you cannot simply guess at the best new denominator. We need a method for finding it from the denominators of the fractions to be added. The new denominator is called the *least common denominator,* abbreviated LCD.

Suppose that we want to add the fractions $\frac{1}{8} + \frac{5}{12}$.

The first step is to find the LCD of the denominators 8 and 12. To find the LCD follow this procedure.

Step 1 Write each denominator as a product of its prime factors. If you need to review this concept, see page 40.

$$8 = 2 \cdot 2 \cdot 2 \qquad 12 = 2 \cdot 2 \cdot 3$$

Step 2 To form the LCD, write each factor that appears in either denominator, then repeat it for the most number of times it appears in any one denominator.

$$\text{LCD} = \underbrace{2 \cdot 2 \cdot 2} \cdot 3 \qquad = \qquad 24$$

↳ There are ↱ There is
3 factors of 1 factor of
2 in 8. 3 in 12.

The LCD of 8 and 12 is 24. This means that 24 is the smallest number that is exactly divisible by both 8 and 12.

EXAMPLE

Find the LCD of 12 and 45.

Step 1 Find the prime factors of each number.

$$12 = 2 \cdot 2 \cdot 3 \qquad 45 = 3 \cdot 3 \cdot 5$$

Step 2 The LCD must contain the factors 2, 3, and 5. The factor 2 occurs twice in 12, the factor 3 occurs twice in 45, and the factor 5 occurs just once in 45.

$$\text{LCD} = 2 \cdot 2 \cdot 3 \cdot 3 \cdot 5 = 180$$

The number 180 is the smallest number that is exactly divisible by both 12 and 45.

YOUR TURN

Use the method described to find the LCD of the numbers 28 and 42.

SOLUTION

Step 1 Find the prime factors of each number.

$$28 = 2 \cdot 2 \cdot 7 \qquad 42 = 2 \cdot 3 \cdot 7$$

Step 2 The factors 2, 3, and 7 all appear. The factor 2 appears at most twice (in 28) and both 3 and 7 appear at most just once in any single number. Therefore the LCD is

$$2 \cdot 2 \cdot 3 \cdot 7 = 84$$

A CALCULATOR METHOD FOR FINDING THE LCD

Here is an alternative method for finding the LCD that you might find easier than the method just described. Follow these two steps.

Step 1 Choose the larger denominator and write down a few multiples of it.

> **Example:** To find the LCD of 12 and 15, first write down a few of the multiples of the larger number, 15. The multiples are 15, 30, 45, 60, 75, and so on.

Step 2 Test each multiple until you find one that is exactly divisible by the smaller denominator.

> **Example:** 15 is not exactly divisible by 12. 30 is not exactly divisible by 12. 45 is not exactly divisible by 12. 60 *is* exactly divisible by 12. The LCD is 60.

This method of finding the LCD has the advantage that you can use it with an electronic calculator. For this example the calculator steps would look like this:

15 ÷ **12** = → *1.25* *Not* a whole number; therefore, *not* exactly divisible by 12.

15 × **2** ÷ **12** = → *2.5* Second multiple: answer is *not* a whole number.

15 × **3** ÷ **12** = → *3.75* Third multiple: answer is *not* a whole number.

15 × **4** ÷ **12** = → *5.* Fourth multiple: answer *is* a whole number; therefore, the LCD is 4 × 15 or 60.

With smaller numbers, this process may be done mentally.

Ready for more practice in finding LCDs?

YOUR TURN

Find the LCD of 4, 10, and 15.

SOLUTION

Step 1 Write the prime factors of all three numbers.

$$4 = 2 \cdot 2 \qquad 10 = 2 \cdot 5 \qquad 15 = 3 \cdot 5$$

Step 2 The factors 2, 3, and 5 all appear and therefore must be included in the LCD. The factor 2 appears at most twice (in 4), and the factors 3 and 5 both appear at most just once in any single number. Therefore the LCD is

$$2 \cdot 2 \cdot 3 \cdot 5 = 60$$

This means that 60 is the smallest number that is exactly divisible by 4, 10, and 15.

Practice by finding the LCD for each of the following sets of numbers.

(a) 2 and 4   (b) 8 and 4   (c) 6 and 3

(d) 5 and 4   (e) 9 and 15   (f) 15 and 24

(g) 4, 5, and 6   (h) 4, 8, and 12   (i) 12, 15, and 21

ANSWERS

(a) 4   (b) 8   (c) 6   (d) 20   (e) 45

(f) 120   (g) 60   (h) 24   (i) 420

In order to use the LCD to add fractions, rewrite the fractions with the LCD as the new denominator, then add the new equivalent fractions.

EXAMPLE

Add $\dfrac{1}{6} + \dfrac{5}{8}$.

First, find the LCD. The LCD of 6 and 8 is 24.

Next, rewrite the two fractions with denominator 24.

$$\frac{1}{6} = \frac{1 \times 4}{6 \times 4} = \frac{4}{24}$$

$$\frac{5}{8} = \frac{5 \times 3}{8 \times 3} = \frac{15}{24}$$

Finally, add the new equivalent fractions.

$$\frac{1}{6} + \frac{5}{8} = \frac{4}{24} + \frac{15}{24} = \frac{19}{24}$$

YOUR TURN

Add $\dfrac{3}{8} + \dfrac{1}{10}$.

SOLUTION

The LCD of 8 and 10 is 40.

$$\frac{3}{8} = \frac{3 \times 5}{8 \times 5} = \frac{15}{40}$$

$$\frac{1}{10} = \frac{1 \times 4}{10 \times 4} = \frac{4}{40}$$

$$\frac{3}{8} + \frac{1}{10} = \frac{15}{40} + \frac{4}{40} = \frac{19}{40}$$

Ready for some practice at this?

Find the LCD, rewrite the fractions, and add.

(a) $\dfrac{1}{2} + \dfrac{1}{4}$ (b) $\dfrac{3}{8} + \dfrac{1}{4}$ (c) $\dfrac{1}{6} + \dfrac{2}{3} + \dfrac{5}{9}$

(d) $\dfrac{3}{4} + \dfrac{1}{6}$ (e) $\dfrac{2}{5} + \dfrac{1}{4}$

(f) **Manufacturing**   The Monterey Canning Co. packs $8\frac{1}{2}$ ounces of tuna into each can. If the can itself weighs $1\frac{7}{8}$ ounces, what is the total weight of a can of Monterey tuna?

SOLUTIONS

(a) The LCD of 2 and 4 is 4.

$$\dfrac{1}{2} = \dfrac{?}{4} = \dfrac{1 \times \boxed{2}}{2 \times \boxed{2}} = \dfrac{2}{4}$$

$$\dfrac{1}{2} + \dfrac{1}{4} = \dfrac{2}{4} + \dfrac{1}{4} = \dfrac{3}{4}$$

(b) The LCD of 8 and 4 is 8.

$$\dfrac{1}{4} + \dfrac{?}{8} = \dfrac{1 \times \boxed{2}}{4 \times \boxed{2}} = \dfrac{2}{8}$$

$$\dfrac{3}{8} + \dfrac{1}{4} = \dfrac{3}{8} + \dfrac{2}{8} = \dfrac{5}{8}$$

(c) The LCD of 6, 3, and 9 is 18.

$$\dfrac{1}{6} = \dfrac{?}{18} = \dfrac{1 \times \boxed{3}}{6 \times \boxed{3}} = \dfrac{3}{18} \qquad\qquad \dfrac{2}{3} = \dfrac{?}{18} = \dfrac{2 \times \boxed{6}}{3 \times \boxed{6}} = \dfrac{12}{18}$$

$$\dfrac{5}{9} = \dfrac{?}{18} = \dfrac{5 \times \boxed{2}}{9 \times \boxed{2}} = \dfrac{10}{18}$$

$$\dfrac{1}{6} + \dfrac{2}{3} + \dfrac{5}{9} = \dfrac{3}{18} + \dfrac{12}{18} + \dfrac{10}{18} = \dfrac{25}{18} = 1\dfrac{7}{18}$$

(d) The LCD of 4 and 6 is 12.

$$\dfrac{3}{4} = \dfrac{?}{12} = \dfrac{3 \times \boxed{3}}{4 \times \boxed{3}} = \dfrac{9}{12} \qquad\qquad \dfrac{1}{6} = \dfrac{?}{12} = \dfrac{1 \times \boxed{2}}{6 \times \boxed{2}} = \dfrac{2}{12}$$

$$\dfrac{3}{4} + \dfrac{1}{6} = \dfrac{9}{12} + \dfrac{2}{12} = \dfrac{11}{12}$$

(e) The LCD of 5 and 4 is 20.

$$\dfrac{2}{5} = \dfrac{?}{20} = \dfrac{2 \times \boxed{4}}{5 \times \boxed{4}} = \dfrac{8}{20} \qquad\qquad \dfrac{1}{4} = \dfrac{?}{20} = \dfrac{1 \times \boxed{5}}{4 \times \boxed{5}} = \dfrac{5}{20}$$

$$\dfrac{2}{5} + \dfrac{1}{4} = \dfrac{8}{20} + \dfrac{5}{20} = \dfrac{13}{20}$$

(f) $8\frac{1}{2} + 1\frac{7}{8} = (8 + 1) + \left(\frac{1}{2} + \frac{7}{8}\right)$

$\qquad = 9 + \left(\frac{4}{8} + \frac{7}{8}\right)$ since $\frac{1}{2} = \frac{4}{8}$

$\qquad = 9 + \frac{11}{8} = 9 + 1 + \frac{3}{8}$

$\qquad = 10\frac{3}{8}$ oz

Note ▶ It is not necessary to find the LCD when you *multiply* fractions. ◀

Subtraction Once you have mastered the process of adding fractions, subtraction is very simple indeed. To find $\frac{3}{8} - \frac{1}{8}$, notice that the denominators are the same. We can subtract the numerators and write this difference over the common denominator.

$$\frac{3}{8} - \frac{1}{8} = \frac{3 - 1}{8} \quad \longleftarrow \boxed{\text{Subtract numerators}}$$

$$\qquad\qquad \longleftarrow \boxed{\text{Same denominator}}$$

$$= \frac{2}{8} \quad \text{or} \quad \frac{1}{4}$$

To find $\frac{3}{4} - \frac{1}{5}$, first find the LCD of 4 and 5.

The LCD of 4 and 5 is 20.

Find equivalent fractions with a denominator of 20:

$$\frac{3}{4} = \frac{?}{20} = \frac{3 \times \boxed{5}}{4 \times \boxed{5}} = \frac{15}{20}$$

$$\frac{1}{5} = \frac{?}{20} = \frac{1 \times \boxed{4}}{5 \times \boxed{4}} = \frac{4}{20}$$

Subtract the equivalent fractions:

$$\frac{3}{4} - \frac{1}{5} = \frac{15}{20} - \frac{4}{20} = \frac{15 - 4}{20} = \frac{11}{20}$$

The procedure is exactly the same as for addition.

If the fractions to be subtracted are given as mixed numbers, it is usually simplest to work with them as improper fractions.

EXAMPLE

$3\frac{1}{2} - 1\frac{1}{3} = \frac{7}{2} - \frac{4}{3}$. The LCD of 2 and 3 is 6.

$$\frac{7}{2} = \frac{7 \times \boxed{3}}{2 \times \boxed{3}} = \frac{21}{6} \qquad \frac{4}{3} = \frac{4 \times \boxed{2}}{3 \times \boxed{2}} = \frac{8}{6}$$

Then, $3\frac{1}{2} - 1\frac{1}{3} = \frac{21}{6} - \frac{8}{6} = \frac{13}{6} = 2\frac{1}{6}$

Because $\frac{1}{2}$ is greater than $\frac{1}{3}$, we could have worked with the whole-number parts separately.

$$3\frac{1}{2} - 1\frac{1}{3} = (3 - 1) + \left(\frac{1}{2} - \frac{1}{3}\right)$$

$$= 2 + \left(\frac{3}{6} - \frac{2}{6}\right)$$

$$= 2 + \frac{1}{6} = 2\frac{1}{6}$$

Arranged vertically:

$$
\begin{array}{ll}
3\frac{1}{2} \rightarrow & 3 + \frac{1}{2} \rightarrow \quad 3 + \frac{3}{6} \\[2mm]
-1\frac{1}{3} \rightarrow -\left(1 + \frac{1}{3}\right) \rightarrow \dfrac{-1 - \frac{2}{6}}{2 \quad \frac{1}{6}} & \text{or} \quad 2\frac{1}{6}
\end{array}
$$

EXAMPLE

$$5\frac{1}{4} - 2\frac{3}{8} = \frac{21}{4} - \frac{19}{8}$$

$$= \frac{42}{8} - \frac{19}{8} = \frac{23}{8} = 2\frac{7}{8}$$

If one of the fractions is a whole number, write it as a fraction first, then add or subtract as with any other fraction.

EXAMPLE

$$3 - \frac{2}{5} = \frac{3}{1} - \frac{2}{5}$$

$$= \frac{15}{5} - \frac{2}{5}$$

$$= \frac{13}{5} \quad \text{or} \quad 2\frac{3}{5}$$

YOUR TURN

Try these problems for practice in subtracting fractions.

(a) $\dfrac{7}{8} - \dfrac{5}{8}$ (b) $9\dfrac{13}{16} - 3\dfrac{1}{4}$ (c) $\dfrac{4}{5} - \dfrac{1}{6}$

(d) $6 - 2\dfrac{3}{4}$ (e) $7\dfrac{1}{6} - 2\dfrac{5}{8}$

(f) **Machine Technology** A bar $6\frac{3}{16}$ in. long is cut from a piece $24\frac{3}{8}$ in. long. If $\frac{5}{32}$ in. is wasted in cutting, what length remains?

SOLUTIONS

(a) $\dfrac{7}{8} - \dfrac{5}{8} = \dfrac{7 - 5}{8} = \dfrac{2}{8} = \dfrac{1}{4}$ (b) $9\dfrac{13}{16} - 3\dfrac{1}{4} = (9 - 3) + \left(\dfrac{13}{16} - \dfrac{1}{4}\right)$

$$= 6 + \left(\frac{13}{16} - \frac{4}{16}\right) = 6\frac{9}{16}$$

(c) The LCD of 5 and 6 is 30.

$$\frac{4}{5} = \frac{?}{30} = \frac{4 \times 6}{5 \times 6} = \frac{24}{30}$$

$$\frac{1}{6} = \frac{?}{30} = \frac{1 \times 5}{6 \times 5} = \frac{5}{30}$$

$$\text{so}\ \frac{4}{5} - \frac{1}{6} = \frac{24}{30} - \frac{5}{30}$$

$$= \frac{19}{30}$$

(d) $$6 - 2\frac{3}{4} = \frac{24}{4} - \frac{11}{4} = \frac{24 - 11}{4} = \frac{13}{4} = 3\frac{1}{4}$$

$$6 = \frac{6}{1} = \frac{6 \times 4}{1 \times 4} \qquad\qquad (2 \times 4) + 3$$

(e) The LCD of 6 and 8 is 24. $\qquad \frac{1}{6} = \frac{4}{24} \qquad \frac{5}{8} = \frac{15}{24}$

$$\text{so}\ 7\frac{1}{6} = 7\frac{4}{24} = \frac{172}{24} \quad \text{and} \quad 2\frac{5}{8} = 2\frac{15}{24} = \frac{63}{24}$$

$$7\frac{1}{6} - 2\frac{5}{8} = \frac{172}{24} - \frac{63}{24} = \frac{109}{24} = 4\frac{13}{24}$$

(f) $$24\frac{3}{8} - \frac{5}{32} - 6\frac{3}{16} = 24\frac{12}{32} - \frac{5}{32} - 6\frac{6}{32}$$

$$= (24 - 6) + \left(\frac{12}{32} - \frac{5}{32} - \frac{6}{32} \right)$$

$$= 18 + \frac{1}{32} = 18\frac{1}{32}\ \text{in.}$$

Now turn to Exercises 2-4 for a set of problems on adding and subtracting fractions.

Exercises 2-4

Addition and Subtraction of Fractions

A. Add or subtract as shown.

1. $\frac{1}{16} + \frac{3}{16}$ 2. $\frac{5}{12} + \frac{11}{12}$ 3. $\frac{5}{16} + \frac{7}{16}$

4. $\frac{2}{6} + \frac{3}{6}$ 5. $\frac{3}{4} - \frac{1}{4}$ 6. $\frac{13}{16} - \frac{3}{16}$

7. $\frac{3}{5} - \frac{1}{5}$ 8. $\frac{5}{12} - \frac{2}{12}$ 9. $\frac{5}{16} + \frac{3}{16} + \frac{7}{16}$

10. $\frac{1}{8} + \frac{3}{8} + \frac{7}{8}$ 11. $1\frac{7}{8} - \frac{3}{8}$ 12. $3\frac{9}{16} - 1\frac{5}{16}$

13. $\frac{1}{4} + \frac{1}{2}$ 14. $\frac{7}{16} + \frac{3}{8}$ 15. $\frac{5}{8} + \frac{1}{12}$

16. $\frac{5}{12} + \frac{3}{16}$ 17. $\frac{1}{2} - \frac{3}{8}$ 18. $\frac{5}{16} - \frac{3}{32}$

19. $\frac{15}{16} - \frac{1}{2}$ 20. $\frac{7}{16} - \frac{1}{32}$ 21. $\frac{3}{5} + \frac{1}{8}$

22. $\dfrac{2}{3} + \dfrac{4}{5}$ 23. $\dfrac{7}{8} - \dfrac{2}{5}$ 24. $\dfrac{4}{9} - \dfrac{1}{4}$

25. $\dfrac{1}{2} + \dfrac{1}{4} - \dfrac{1}{8}$ 26. $\dfrac{11}{16} - \dfrac{1}{8} - \dfrac{1}{3}$ 27. $1\dfrac{1}{2} + \dfrac{1}{4}$

28. $2\dfrac{7}{16} + \dfrac{3}{4}$ 29. $2\dfrac{1}{2} + 1\dfrac{5}{8}$ 30. $2\dfrac{8}{32} + 1\dfrac{1}{10}$

31. $2\dfrac{1}{3} + 1\dfrac{1}{5}$ 32. $1\dfrac{7}{8} + \dfrac{1}{4}$ 33. $4\dfrac{1}{8} - 1\dfrac{3}{4}$

34. $5\dfrac{3}{4} - 2\dfrac{1}{12}$ 35. $3\dfrac{1}{5} - 2\dfrac{1}{12}$ 36. $5\dfrac{1}{3} - 2\dfrac{2}{5}$

B. Add or subtract as shown.

1. $8 - 2\dfrac{7}{8}$ 2. $3 - 1\dfrac{3}{16}$ 3. $3\dfrac{5}{8} - \dfrac{13}{16}$

4. $\dfrac{1}{2} + \dfrac{1}{3} + \dfrac{1}{4} + \dfrac{1}{5}$ 5. $\dfrac{1}{2} + \dfrac{1}{4} + \dfrac{1}{8}$ 6. $6\dfrac{1}{2} + 5\dfrac{3}{4} + 8\dfrac{1}{8}$

7. $\dfrac{7}{8} - 1\dfrac{1}{4} + 2\dfrac{1}{2}$ 8. $1\dfrac{3}{8}$ subtracted from $4\dfrac{3}{4}$

9. $2\dfrac{3}{16}$ less than $4\dfrac{7}{8}$ 10. $6\dfrac{2}{3}$ reduced by $1\dfrac{1}{4}$

11. $2\dfrac{3}{5}$ less than $6\dfrac{1}{2}$ 12. By how much is $1\dfrac{8}{7}$ larger than $1\dfrac{7}{8}$?

C. Practical Problems

1. **Building Construction** The exterior wall of a small office building under construction is constructed of $\frac{1}{4}$-in. paneling, $\frac{5}{8}$-in. firecode sheetrock, $5\frac{3}{4}$-in. studs, $\frac{1}{2}$-in. CDX plywood sheathing, $1\frac{1}{4}$-in. insulation board, and $\frac{5}{8}$-in. exterior surfacing. Calculate the total thickness of the wall.

2. **Carpentry** A countertop is made of $\frac{5}{8}$-in. particleboard and is covered with $\frac{3}{16}$-in. laminated plastic. What width of metal edging is needed to finish off the edge?

3. **Welding** A welder needs a piece of half-inch pipe $34\frac{3}{4}$ in. long. She has a piece that is $46\frac{3}{8}$ in. long. How much must she cut off from the longer piece?

Problem 5

4. **Plumbing** If a piece of $\frac{3}{8}$-in.-I.D. (inside diameter) copper tubing measures $\frac{9}{16}$ in. O.D. (outside diameter), what is the wall thickness?

5. **Manufacturing** What is the outside diameter of tubing whose inside diameter is $1\frac{5}{16}$ in. and whose wall thickness is $\frac{1}{8}$ in.?

Problem 6

6. **Machine Technology** How long a bolt is needed to go through a piece of tubing $\frac{5}{8}$ in. long, a washer $\frac{1}{16}$ in. thick, and a nut $\frac{1}{4}$ in. thick?

7. **Office Services** Newspaper ads are sold by the column inch (c.i.). What is the total number of column inches for a month in which a plumbing contractor has had ads of $6\frac{1}{2}$, $5\frac{3}{4}$, $3\frac{1}{4}$, $4\frac{3}{4}$, and 5 c.i.?

8. **Plumbing** While installing water pipes, a plumber used pieces of pipe measuring $2\frac{3}{4}$, $4\frac{1}{8}$, $3\frac{1}{2}$, and $1\frac{1}{4}$ ft. How much pipe would remain if these pieces were cut from a 14-ft length of pipe? (Ignore waste in cutting.)

Problem 9

9. **Electrical Technology** A piece of electrical pipe conduit has a diameter of $1\frac{1}{2}$ in. and a wall thickness of $\frac{3}{16}$ in. What is its inside diameter?

Problem 12

10. **Metalworking** What is the total length of a certain machine part that is made by joining four pieces that measure $3\frac{1}{8}$, $1\frac{5}{32}$, $2\frac{7}{16}$, and $1\frac{1}{4}$ in.?

11. **Carpentry** A blueprint requires four separate pieces of wood measuring $5\frac{3}{8}$, $8\frac{1}{4}$, $6\frac{9}{16}$, and $2\frac{5}{8}$ in. How long a piece of wood is needed to cut these pieces if we allow $\frac{1}{2}$ in. for waste?

12. **Drafting** Find the missing dimension A in the drawing shown.

13. **Machine Technology** Two splice plates are cut from a piece of sheet steel that has an overall length of $18\frac{5}{8}$ in. The plates are $9\frac{1}{4}$ in. and $6\frac{7}{16}$ in. long. How much material remains from the original piece if each saw cut removes $\frac{1}{16}$ in.?

14. **Printing** A printer has $2\frac{3}{4}$ rolls of a certain kind of paper in stock. He must do three jobs that require $\frac{5}{8}$, $1\frac{1}{2}$, and $\frac{3}{4}$ roll, respectively. Does he have enough?

15. **Auto Mechanics** The wheel stagger of an automobile is the difference between the axle-to-axle lengths on the right and left sides. If this length is $101\frac{1}{4}$ in. on the right side of a particular car and is $100\frac{7}{8}$ in. on the left side, find the wheel stagger of this automobile.

Problem 16

16. **Woodworking** A cabinet 30 in. high must have a $4\frac{1}{2}$-in. base and a $1\frac{3}{4}$-in. top. How much space is left for drawers?

17. **Printing** Before it was trimmed, a booklet measured $8\frac{1}{4}$ in. high by $6\frac{3}{4}$ in. wide. If each edge of the height and one edge of the width were trimmed $\frac{1}{4}$ in. what is the finished size?

18. **Carpentry** A wall has $\frac{1}{2}$-in. paneling covering $\frac{3}{4}$-in. drywall attached to a $3\frac{3}{4}$-in. stud. What is the total thickness of the three components?

Problem 20

19. **Machine Technology** The large end of a tapered pin is $2\frac{15}{16}$ in. in diameter, while the small end is $2\frac{3}{8}$ in. in diameter. Calculate the difference to get the amount of taper.

20. **Machine Technology** Find the total length of the metal casting shown.

21. **Carpentry** A joiner is set to remove $\frac{7}{64}$ in. from the width of an oak board. If the board was $4\frac{5}{8}$ in. wide, find its width after joining once.

22. **Carpentry** A rule of thumb used in constructing stairways is that the rise and the run should always add up to 17 inches. Applying this rule, what should be the run of a stairway if the rise is $7\frac{3}{4}$ in.?

Check your answers in the Appendix.

Using a Calculator, II: Fractions

Fractions can be entered directly on most calculators, and the results of arithmetic calculations with fractions can be displayed as fractions or decimals. If your calculator has a $\boxed{a\frac{b}{c}}$ key, you may enter fractions or mixed numbers directly into your machine without using the division key.

The calculator display will indicate fractions and mixed numbers with a ⌐ symbol. For example, the fraction $\frac{3}{4}$ is shown as 3⌐4,

3 $\boxed{a\frac{b}{c}}$ **4** $\boxed{=}$ → 3⌐4

and the mixed number $1\frac{7}{8}$ is shown as 1⌐7⌐8,

1 $\boxed{a\frac{b}{c}}$ **7** $\boxed{a\frac{b}{c}}$ **8** $\boxed{=}$ → 1⌐7⌐8

Improper fractions or fractions not in lowest terms can be simplified or written in decimal form. The improper fraction $\frac{12}{7}$ can be simplified as follows:

12 $\boxed{a^b_c}$ **7** $\boxed{=}$ $\rightarrow$ ▨▨▨ *1⌐5⌐7* or $1\frac{5}{7}$

To write it as a decimal number press the fraction key again.

12 $\boxed{a^b_c}$ **7** $\boxed{=}$ $\rightarrow$ ▨▨▨ *1⌐5⌐7* $\boxed{a^b_c}$ $\rightarrow$ ▨▨ *1.714285714*

To simplify the fraction $\frac{6}{16}$

6 $\boxed{a^b_c}$ **16** $\boxed{=}$ $\rightarrow$ ▨▨▨ *3⌐8* or $\frac{3}{8}$ in lowest terms.

Perform arithmetic operations in the usual way. For example, $\frac{26}{8} - 1\frac{2}{3}$ is

26 $\boxed{a^b_c}$ **8** $\boxed{-}$ **1** $\boxed{a^b_c}$ **2** $\boxed{a^b_c}$ **3** $\boxed{=}$ $\rightarrow$ ▨▨▨ *1⌐7⌐12* $\boxed{a^b_c}$ $\rightarrow$ ▨▨ *1.583333333*

$\left(1\frac{7}{12}\right.$

Notice that pressing the $\boxed{a^b_c}$ key again will display the fraction as a decimal number.

YOUR TURN

For practice in calculating with fractions and decimals, work the following problems. Write your answer both as a fraction or mixed number and as a decimal rounded to two decimal places.

(a) $\frac{2}{3} + \frac{7}{8}$ (b) $1\frac{3}{4} - \frac{2}{5}$ (c) $\frac{25}{32} + 1.87$ (d) $8\frac{1}{5} \div 2\frac{1}{6}$

(e) $\frac{17}{20} \times \frac{1}{3}$

° ANSWERS

(a) **2** $\boxed{a^b_c}$ **3** $\boxed{+}$ **7** $\boxed{a^b_c}$ **8** $\boxed{=}$ $\rightarrow$ ▨▨ *1⌐13⌐24* $\boxed{a^b_c}$ $\rightarrow$ ▨▨ *1.541666667* or 1.54 rounded

$\left(1\frac{13}{24}\right.$

(b) **1** $\boxed{a^b_c}$ **3** $\boxed{a^b_c}$ **4** $\boxed{-}$ **2** $\boxed{a^b_c}$ **5** $\boxed{=}$ $\rightarrow$ ▨▨ *1⌐7⌐20* $\boxed{a^b_c}$ $\rightarrow$ ▨▨ *1.35*

$\left(1\frac{7}{20}\right.$

(c) **25** $\boxed{a^b_c}$ **32** $\boxed{+}$ **1.87** $\boxed{=}$ $\rightarrow$ ▨▨ *2.65125* or 2.65 rounded

(d) **8** $\boxed{a^b_c}$ **1** $\boxed{a^b_c}$ **5** $\boxed{\div}$ **2** $\boxed{a^b_c}$ **1** $\boxed{a^b_c}$ **6** $\boxed{=}$ $\rightarrow$ ▨▨ *3⌐51⌐65* $\boxed{a^b_c}$ $\rightarrow$ ▨▨ *3.784615385* or 3.78 rounded

$\left(3\frac{51}{65}\right.$

(e) **17** $\boxed{a^b_c}$ **20** $\boxed{\times}$ **1** $\boxed{a^b_c}$ **3** $\boxed{=}$ $\rightarrow$ ▨▨ *17⌐60* $\boxed{a^b_c}$ $\rightarrow$ ▨▨ *0.283333333* or 0.28 rounded

$\llcorner \frac{17}{60}$

Fractions

Answers are given in the Appendix.

A. Write as an improper fraction.

1. $1\frac{1}{8}$ 2. $4\frac{1}{5}$ 3. $1\frac{2}{3}$ 4. $2\frac{3}{16}$

5. $3\frac{3}{32}$ 6. $2\frac{1}{16}$ 7. $1\frac{5}{8}$ 8. $3\frac{7}{16}$

Write as a mixed number.

9. $\frac{10}{4}$ 10. $\frac{19}{2}$ 11. $\frac{25}{3}$ 12. $\frac{9}{8}$

13. $\frac{25}{16}$ 14. $\frac{21}{16}$ 15. $\frac{35}{4}$ 16. $\frac{7}{3}$

Write in lowest terms.

17. $\frac{6}{32}$ 18. $\frac{8}{32}$ 19. $\frac{12}{32}$ 20. $\frac{18}{24}$

21. $\frac{5}{30}$ 22. $1\frac{12}{21}$ 23. $1\frac{16}{20}$ 24. $3\frac{10}{25}$

Complete these.

25. $\frac{3}{4} = \frac{?}{12}$ 26. $\frac{7}{16} = \frac{?}{64}$ 27. $2\frac{3}{4} = \frac{?}{16}$ 28. $1\frac{3}{8} = \frac{?}{32}$

29. $5\frac{2}{3} = \frac{?}{12}$ 30. $1\frac{4}{5} = \frac{?}{10}$ 31. $1\frac{1}{4} = \frac{?}{12}$ 32. $2\frac{3}{5} = \frac{?}{10}$

Circle the larger number.

33. $\frac{7}{16}$ or $\frac{2}{15}$ 34. $\frac{2}{3}$ or $\frac{4}{7}$ 35. $\frac{13}{16}$ or $\frac{7}{8}$ 36. $1\frac{1}{4}$ or $\frac{7}{6}$

37. $\frac{13}{32}$ or $\frac{3}{5}$ 38. $\frac{2}{10}$ or $\frac{3}{16}$ 39. $1\frac{7}{16}$ or $\frac{7}{4}$ 40. $\frac{3}{32}$ or $\frac{1}{9}$

Name _____

Date _____

Course/Section _____

B. Multiply or divide as shown.

1. $\frac{1}{2} \times \frac{3}{16}$ 2. $\frac{3}{4} \times \frac{2}{3}$ 3. $\frac{7}{16} \times \frac{4}{3}$ 4. $\frac{15}{64} \times \frac{1}{12}$

5. $1\frac{1}{2} \times \frac{5}{6}$ **6.** $3\frac{1}{16} \times \frac{1}{5}$ **7.** $\frac{3}{16} \times \frac{5}{12}$ **8.** $14 \times \frac{3}{8}$

9. $\frac{3}{4} \times 10$ **10.** $\frac{1}{2} \times 1\frac{1}{3}$ **11.** $18 \times 1\frac{1}{2}$ **12.** $16 \times 2\frac{1}{8}$

13. $2\frac{2}{3} \times 4\frac{3}{8}$ **14.** $3\frac{1}{8} \times 2\frac{2}{5}$ **15.** $\frac{1}{2} \div \frac{1}{4}$ **16.** $\frac{2}{5} \div \frac{1}{2}$

17. $4 \div \frac{1}{8}$ **18.** $8 \div \frac{3}{4}$ **19.** $\frac{2}{3} \div 4$ **20.** $1\frac{1}{2} \div 2$

21. $3\frac{1}{2} \div 5$ **22.** $1\frac{1}{4} \div 1\frac{1}{2}$ **23.** $2\frac{3}{4} \div 1\frac{1}{8}$ **24.** $3\frac{1}{5} \div 1\frac{5}{7}$

C.   Add or subtract as shown.

1. $\frac{3}{8} + \frac{7}{8}$ **2.** $\frac{1}{2} + \frac{3}{4}$ **3.** $\frac{3}{32} + \frac{1}{8}$ **4.** $\frac{3}{8} + 1\frac{1}{4}$

5. $\frac{3}{5} + \frac{5}{6}$ **6.** $\frac{5}{8} + \frac{1}{10}$ **7.** $\frac{9}{16} - \frac{3}{16}$ **8.** $\frac{7}{8} - \frac{1}{2}$

9. $\frac{11}{16} - \frac{1}{4}$ **10.** $\frac{5}{6} - \frac{1}{5}$ **11.** $\frac{7}{8} - \frac{3}{10}$ **12.** $1\frac{1}{2} - \frac{3}{32}$

13. $2\frac{1}{8} + 1\frac{1}{4}$ **14.** $1\frac{5}{8} + \frac{13}{16}$ **15.** $6 - 1\frac{1}{2}$ **16.** $3 - 1\frac{7}{8}$

17. $3\frac{2}{3} - 1\frac{7}{8}$ **18.** $2\frac{1}{4} - \frac{5}{6}$ **19.** $\frac{1}{2} + \frac{1}{3} + \frac{1}{5}$ **20.** $1\frac{1}{2} + 1\frac{1}{4} + 1\frac{1}{5}$

21. $3\frac{1}{2} - 2\frac{1}{3}$ **22.** $2\frac{3}{5} - 1\frac{4}{15}$ **23.** $2 - 1\frac{3}{5}$ **24.** $4\frac{5}{6} - 1\frac{1}{2}$

D.   Practical Problems

1. **Welding**   In a welding job three pieces of 2-in. I-beam with lengths $5\frac{7}{8}$, $8\frac{1}{2}$, and $22\frac{3}{4}$ in. are needed. What is the total length of I-beam needed? (Do not worry about the waste in cutting.)

2. **Machine Technology**   How many pieces of $10\frac{5}{16}$-in. bar can be cut from a stock 20-ft bar? The metal is torch cut and allowance of $\frac{3}{16}$ in. kerf (waste) should be made for each piece. (*Hint:* 20 ft = 240 in.)

3. **Welding**   A piece of metal must be cut to a length of $22\frac{3}{8}$ in. $\pm \frac{1}{16}$ in. What are the longest and shortest acceptable lengths? (*Hint:* The symbol $\pm$ means to add $\frac{1}{16}$ in. to get the longest length and subtract $\frac{1}{16}$ in. to get the shortest length. Longest = $22\frac{3}{8}$ in. $+ \frac{1}{16}$ in. = ? Shortest = $22\frac{3}{8}$ in. $- \frac{1}{16}$ in. = ?)

4. **Auto Mechanics**   A damaged car is said to have "sway" when two corresponding diagonal measurements under the hood are different. If these diagonals are found to be $64\frac{1}{4}$ in. and $62\frac{7}{8}$ in., calculate the magnitude of the sway, the difference between these measurements.

5. **Machine Technology**   A shaft $1\frac{7}{8}$ in. in diameter is turned down on a lathe to a diameter of $1\frac{3}{32}$ in. What is the difference in diameters?

6. **Machine Technology**   A bar $14\frac{5}{16}$ in. long is cut from a piece $25\frac{1}{4}$ in. long. If $\frac{3}{32}$ in. is wasted in cutting, will there be enough left to make another bar $10\frac{3}{8}$ in. long?

7. **Manufacturing**   A cubic foot contains roughly $7\frac{1}{2}$ gallons. How many cubic feet are there in a tank containing $34\frac{1}{2}$ gallons?

8. **Manufacturing**   Find the total width of the three pieces of steel plate shown.

9. **Machine Technology**   What would be the total length of the bar formed by welding together the five pieces of bar stock shown on the next page?

$5\frac{1}{4}''$

$3\frac{5}{16}''$

$7\frac{3}{8}''$

Problem 8

Problem 9

10. **Machine Technology** The Ace Machine Shop has the job of producing 32 zinger bars. Each zinger bar must be turned on a lathe from a piece of stock $4\frac{7}{8}$ in. long. How many feet of stock will they need?

11. **Carpentry** What is the thickness of a tabletop made of $\frac{3}{4}$-in. plywood and covered with a $\frac{3}{16}$-in. sheet of glass?

12. **Building Construction** For the wooden form shown, find the lengths A, B, C, and D.

Problem 12

Problem 13

13. **Carpentry** Find the spacing x between the holes.

14. **Masonry** Find the height of the five-course (five-bricks-high) brick wall shown if each brick is $2\frac{1}{2}$ in. by $3\frac{7}{8}$ in. by $8\frac{1}{4}$ in. and all mortar joints are $\frac{1}{2}$ in.

15. **Masonry** If the wall in problem 14 has 28 stretchers (bricks laid lengthwise), what is its length?

16. **Electrical Technology** An electrical wiring job requires the following lengths of 14/2 BX cable: seven pieces each $6\frac{1}{2}$ ft long, four pieces each $34\frac{3}{4}$ in. long, and nine pieces each $19\frac{3}{8}$ in. long. What is the total length of cable needed?

17. **Printing** An invitation must be printed on card stock measuring $4\frac{1}{4}$ in. wide by $5\frac{1}{2}$ in. long. The printed material covers a space measuring $2\frac{1}{8}$ in. wide by $4\frac{1}{8}$ in. long. If the printed material is centered in both directions, what are the margins?

18. **Printing** As a rule of thumb, the top margin of a page of a book should be $\frac{2}{5}$ of the total margin, and the bottom margin should be $\frac{3}{5}$ of the total margin. If the print takes up $9\frac{1}{2}$ in. of an 11-in.-long page, what should the top and bottom margins be? (*Hint:* The total margin = 11 in. $-$ $9\frac{1}{2}$ in. = $1\frac{1}{2}$ in.)

19. **Welding** A 46-in. bar must have 9 equally spaced holes drilled through the centerline. If the centers of the two end holes are each $2\frac{1}{4}$ in. in from their respective ends, what should the center-to-center distance of the holes be? (*Hint:* There are 8 spaces between holes.)

20. **Building Construction** If an I-beam is to be $24\frac{3}{8}$ in. long with a tolerance of $\pm \frac{1}{4}$ in., find the longest and shortest acceptable lengths.

21. **Machine Technology** If a positioner shaft turns at 18 revolutions per minute, and the tool feed is $\frac{1}{16}$ in. per revolution, how long will it take to advance $7\frac{1}{2}$ in.?

22. **Sheet Metal Technology** The total allowance for both edges of a grooved seam is three times the width of the seam. Half of this total is added to each edge of the seam. Find the allowance for each edge of a grooved seam if the width of the seam is $\frac{5}{16}$ in.

Problem 14

			For Help Go to Page
Objective	**Sample Problems**		

When you finish this unit you will be able to:

1. Add, subtract, multiply, and divide decimal numbers.

(a) 5.82 + 0.096 _____ 105

(b) 3.78 − 0.989 _____ 108

(c) 27 − 4.03 _____

(d) 7.25 × 0.301 _____ 113

(e) 104.2 ÷ 0.032 _____ 117

(f) 0.09 ÷ 0.0004 _____

(g) 20.4 ÷ 6.7 (round to three decimal places) _____

2. Find averages.

Find the average of 4.2, 4.8, 5.7, 2.5, 3.6, 5.0 _____ 123

3. Work with decimal fractions.

(a) Write as a decimal number $\frac{3}{16}$ _____ 128

(b) Write as a decimal number: One hundred six and twenty-seven thousandths _____ 103

(c) Write in words 26.035 _____ 101

(d) $1\frac{2}{3}$ + 1.785 _____

(e) $4.1 \times 2\frac{1}{4}$ _____

(f) $1\frac{5}{16}$ ÷ 4.3 (round to three decimal places) _____

_____
Name

_____
Date

_____
Course/Section

4. Solve practical problems involving decimal numbers.

(a) **Electrician** Six recessed lights must be installed in a strip of ceiling. The housings cost $16.45 each, the trims cost $13.75 each, the lamps cost $15.20 each, miscellaneous hardware and wiring cost $23.45, and labor is estimated at 3.25 hours at $55 per hour. What will be the total cost of the job?

_____

(b) **Machinist** A container of 175 bolts weighs 61.3 lb. If the container itself weighs 1.8 lb, how much does each bolt weigh?

117

_____

(Answers to these preview problems are given in the Appendix.)

If you are certain that you can work *all* these problems correctly, turn to page 137 for a set of practice problems. If you cannot work one or more of the preview problems, turn to the page indicated to the right of the problem. Those who wish to master this material with the greatest success should turn to Section 3-1 and begin work there.

3 Decimal Numbers

Reprinted with permission of Universal Press Syndicate.

3-1 ADDITION AND SUBTRACTION OF DECIMALS

Place Value of Decimal Numbers

By now you know that whole numbers are written in a form based on powers of ten. A number such as

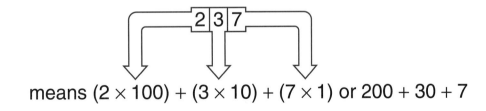

means $(2 \times 100) + (3 \times 10) + (7 \times 1)$ or $200 + 30 + 7$

This way of writing numbers can be extended to fractions. A *decimal* number is a fraction whose denominator is 10 or some multiple of 10.

A decimal number may have both a whole-number part and a fraction part. For example, the number 324.576 means

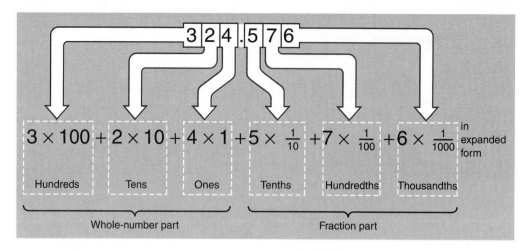

You are already familiar with this way of interpreting decimal numbers from working with money.

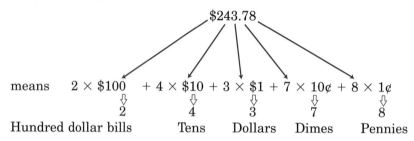

To write a decimal number in words, remember this diagram:

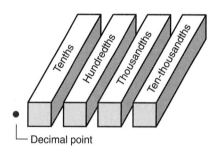

For example,

$$.6 = 6 \text{ tenths} = \frac{6}{10}$$

$$.05 = 5 \text{ hundredths} = \frac{5}{100}$$

$$.32 = 32 \text{ hundredths} = \frac{32}{100}$$

$$.004 = 4 \text{ thousandths} = \frac{4}{1000}$$

$$.267 = 267 \text{ thousandths} = \frac{267}{1000}$$

Decimal form

Fraction form

YOUR TURN

Write the following decimal numbers in words.

(a) 0.56   (b) 19.278   (c) 6.4

(d) 0.07   (e) 5.064   (f) 0.0018

ANSWERS

(a) Fifty-six hundredths

(b) Nineteen and two hundred seventy-eight thousandths

(c) Six and four tenths

(d) Seven hundredths

(e) Five and sixty-four thousandths

(f) Eighteen ten-thousandths

Learning Help ▶ The word **and** represents the decimal point. Everything preceding **and** is the whole-number part, and everything after **and** is the decimal part of the number. ◀

It is also important to be able to perform the reverse process—that is, write a decimal in numerical form if it is given in words.

EXAMPLE

To write "six and twenty-four thousandths" as a numeral,

First, write the whole-number part (six). This is the part before the "and." Follow this by a decimal point—the "and."   6.

Next, draw as many blanks as the decimal part indicates. In this case, the decimal part is "thousandths," so we allow for three decimal places. Draw three blanks.   6.__ __ __

Finally, write a number giving the decimal (twenty-four). Write it so it *ends* on the far right blank. Fill in any blank decimal places with zeros.   6.__ 2 4

6.024

YOUR TURN

Now you try it. Write each of the following as decimal numbers.

(a) Five thousandths

(b) One hundred and six tenths

(c) Two and twenty-eight hundredths

(d) Seventy-one and sixty-two thousandths

(e) Three and five hundred eighty-nine ten-thousandths

ANSWERS

(a) 0.005   (b) 100.6   (c) 2.28   (d) 71.062   (e) 3.0589

Expanded Form The decimal number .267 can be written in expanded form as

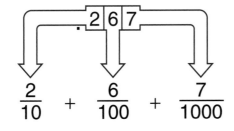

$$\frac{2}{10} \; + \; \frac{6}{100} \; + \; \frac{7}{1000}$$

Write the decimal number .526 in expanded form.

$$.526 = \frac{5}{10} + \frac{2}{100} + \frac{6}{1000}$$

We usually write a decimal number less than 1 with a zero to the left of the decimal point.

.526 would be written 0.526

.4   would be written 0.4

.001 would be written 0.001

It is easy to mistake .4 for 4, but the decimal point in 0.4 cannot be overlooked. That zero out front will help you remember where the decimal point is located.

YOUR TURN

To help get these ideas clear in your mind, write the following in expanded form.

(a) 86.42   (b) 43.607   (c) 14.5060   (d) 235.22267

ANSWERS

(a) 86.42 $= 8 \times 10 + 6 \times 1 + 4 \times \frac{1}{10} + 2 \times \frac{1}{100}$

$\quad\quad\quad = \quad 80 \quad + \quad 6 \quad + \quad \frac{4}{10} \quad + \quad \frac{2}{100}$

(b) 43.607 $= 4 \times 10 + 3 \times 1 + 6 \times \frac{1}{10} + 0 \times \frac{1}{100} + 7 \times \frac{1}{1000}$

$\quad\quad\quad\quad = \quad 40 \quad + \quad 3 \quad + \quad \frac{6}{10} \quad + \quad \frac{0}{100} \quad + \quad \frac{7}{1000}$

(c) 14.5060 $= 10 + 4 + \frac{5}{10} + \frac{0}{100} + \frac{6}{1000} + \frac{0}{10000}$

(d) 235.22267 $= 200 + 30 + 5 + \frac{2}{10} + \frac{2}{100} + \frac{2}{1000} + \frac{6}{10000} + \frac{7}{100000}$

Learning Help ▶ Notice that the denominators in the decimal fractions change by a factor of 10. For example,

3247 . 8956

3 × 1000	thousands	3000 . 0006	ten-thousandths	6 × 0.0001
2 × 100	hundreds	200 . 005	thousandths	5 × 0.001
4 × 10	tens	40 . 09	hundredths	9 × 0.01
7 × 1	ones	7 . 8	tenths	8 × 0.1

Each row changes by a factor of ten

Decimal Digits In the decimal number 86.423 the digits 4, 2, and 3 are called *decimal digits*.

The number 43.6708 has four decimal digits: 6, 7, 0, and 8.

The number 5376.2 has one decimal digit: 2.

All digits to the right of the decimal point, those that name the fractional part of the number, are decimal digits.

YOUR TURN

How many decimal digits are included in each of these numbers?

(a) 1.4   (b) 315.7   (c) 0.425   (d) 324.0075

ANSWERS

(a) one   (b) one   (c) three   (d) four

We will use the idea of decimal digits often in doing arithmetic with decimal numbers.

The decimal point is simply a way of separating the whole-number part from the fraction part. It is a place marker. In whole numbers the decimal point usually is not written, but it is understood to be there.

The whole number 2 is written 2. as a decimal.

$$2 = 2.\qquad\qquad\text{or}\qquad 324 = 324.$$

The decimal point   The decimal point

This is very important. Many people make big mistakes in arithmetic because they do not know where that decimal point should go.

Very often, additional zeros are attached to the decimal number without changing its value. For example,

$8.5 = 8.50 = 8.5000$ and so on

$6 = 6. = 6.0 = 6.000$ and so on

The value of the number is not changed, but the additional zeros may be useful, as we shall see.

Addition Because decimal numbers represent fractions with denominators equal to multiples of ten, addition is very simple.

$$2.34 = 2 + \frac{3}{10} + \frac{4}{100}$$
$$+5.23 = 5 + \frac{2}{10} + \frac{3}{100}$$
$$\overline{\; 7 + \frac{5}{10} + \frac{7}{100}} = 7.57$$

Adding like fractions

Of course, we do not need this clumsy business in order to add decimal numbers. As with whole numbers, we may arrange the digits in vertical columns and add directly.

EXAMPLE

Let's add 1.45 + 3.42.

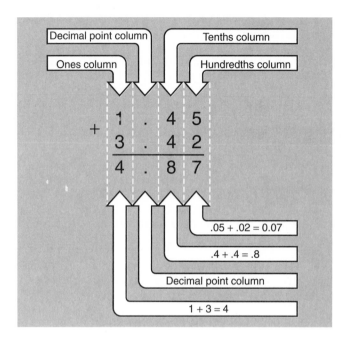

Digits of the same power of ten are placed in the same vertical column. Decimal points are always lined up vertically.

If one of the numbers is written with fewer decimal digits than the other, attach as many zeros as needed so that both have the same number of decimal digits.

$$\begin{array}{r} 2.345 \\ +1.5 \\ \hline \end{array} \quad \text{becomes} \quad \begin{array}{r} 2.345 \\ +1.500 \\ \hline \end{array}$$

Except for the preliminary step of lining up decimal points, addition of decimal numbers is exactly the same process as addition of whole numbers.

YOUR TURN

Add the following decimal numbers.

(a) $4.02 + $3.67   = _____

(b) 13.2 + 1.57   = _____

(c) 23.007 + 1.12   = _____

(d) 14.6 + 1.2 + 3.15   = _____

(e) 5.7 + 3.4   = _____

(f) $9 + $0.72 + $6.09 = _____

(g) 0.07 + 6.79 + 0.3 + 3 = _____

(h) **Sheet Metal Technology** A sheet metal worker uses 2.36, 7, 3.9, and 0.6 ounces of cleaning concentrate on successive jobs. What total amount of concentrate has he used during this time?

Arrange each sum vertically, placing the decimal points in the same column, then add as with whole numbers.

SOLUTIONS

Decimal points are lined up vertically.

(a) $4.02
$3.67
$7.69 — 0.02 + 0.07 = 0.09   Add cents.
0.0 + 0.6 = 0.6   Add 10-cent units.
4 + 3 = 7   Add dollars.

As a check, notice that the sum is roughly $4 + $4 or $8, which agrees with the actual answer. Always check your answer by first estimating it, then comparing your estimate or rough guess with the final answer.

Decimal points are in line.

(b) 13.20 ◄———— Attach a zero to provide the same number of
1.57   decimal digits as in the other addend.
14.77
↑
└———— Place answer decimal point in the same vertical line.

☑ 13 + 2 = 15, which agrees roughly with the answer.

(c)   23.007
    +   1.120 ← Attach extra zero.
       24.127

(d)   14.60
         1.20 ◄— Attach extra
     +   3.15   zeros.
        18.95

(e) ¹
    5.7
   +3.4
    9.1   0.7 + 0.4 = 1.1 Write 0.1.
                        Carry 1.
        Add: 1 + 5 + 3 = 9

(f) ¹
    $ 9.00
      0.72
      6.09
    $15.81

(g) ¹ ¹
    0.07
    6.79
    0.30 ◄┐
    3.00 ◄┴ Attach extra zeros.
   10.16

(h) ¹
    2.36
    7.00 ◄
    3.90 ◄— Attach extra
    0.60 ◄   zeros.
   13.86

Careful ▶ You must line up the decimal points vertically to be certain of getting a correct answer. ◀

Subtraction Subtraction is equally simple if you line up the decimal points carefully and attach any needed zeros before you begin work. As in the subtraction of whole numbers, you may need to "borrow" to complete the calculation.

EXAMPLE

$437.56 − $41 = _____ is

$$
\begin{array}{r}
\overset{3}{}\ \overset{13}{} \\
\$\ \ \cancel{4}\ \ \cancel{3}\ \ 7\ .5\ \ 6 \\
-\ \$\ \ \ \ \ 4\ \ 1\ .0\ \ 0 \\
\hline
\$\ \ 3\ \ 9\ \ 6\ .5\ \ 6
\end{array}
$$

Decimal points are in a vertical line.
Attach zeros (remember that $41 is $41. or $41.00).

or again   19.452 − 7.3617 = _____

$$
\begin{array}{r}
\overset{3}{}\ \overset{15}{}\ \overset{1}{}\ \overset{10}{} \\
1\ \ 9\ \ \cancel{4}\ \ \cancel{5}\ \ \cancel{2}\ \ \cancel{0} \\
-\ \ \ \ 7\ .3\ \ 6\ \ 1\ \ 7 \\
\hline
1\ \ 2\ .0\ \ 9\ \ 0\ \ 3
\end{array}
$$

Decimal points are in a vertical line.
Attach zero.

Answer decimal point is in same vertical line.

YOUR TURN

Try these problems to test yourself on the subtraction of decimal numbers.

(a) $37.66 − $14.57 = _____   (b) 248.3 − 135.921 = _____

(c) 6.4701 − 3.2   = _____   (d) 7.304 − 2.59   = _____

(e) $20 − $7.74   = _____   (f) 36 − 11.132   = _____

(g) **Machine Technology** If 0.037 in. of metal is machined from a rod exactly 10 in. long, what is the new length of the rod?

Work carefully.

SOLUTIONS

(a)
$$
\begin{array}{r}
\overset{5}{}\ \overset{16}{} \\
\$\ \ 3\ \ 7\ .\cancel{6}\ \ \cancel{6} \\
-\ \$\ \ 1\ \ 4\ .5\ \ 7 \\
\hline
\$\ \ 2\ \ 3\ .0\ \ 9
\end{array}
$$
Line up decimal points.

☑ $14.57 + $23.09 = $37.66

(b)
$$
\begin{array}{r}
\overset{7}{}\ \overset{12}{}\ \overset{9}{}\ \overset{10}{} \\
2\ \ 4\ \ 8\ .\cancel{3}\ \ \cancel{0}\ \ \cancel{0} \\
-\ 1\ \ 3\ \ 5\ .9\ \ 2\ \ 1 \\
\hline
1\ \ 1\ \ 2\ .3\ \ 7\ \ 9
\end{array}
$$
Line up decimal points.
Attach zeros.

☑ 135.921 + 112.379 = 248.300
Answer decimal point is in the same vertical line.

(c)
$$
\begin{array}{r}
6\ .4\ \ 7\ \ 0\ \ 1 \\
-\ 3\ .2\ \ 0\ \ 0\ \ 0 \\
\hline
3\ .2\ \ 7\ \ 0\ \ 1
\end{array}
$$
← Attach zeros.

☑ 3.2000 + 3.2701 = 6.4701

(d)
$$
\begin{array}{r}
\overset{6}{}\ \overset{12}{}\ \overset{10}{} \\
\cancel{7}\ .\cancel{3}\ \ \cancel{0}\ \ 4 \\
-\ 2\ .5\ \ 9\ \ 0 \\
\hline
4\ .7\ \ 1\ \ 4
\end{array}
$$
← Attach a zero.

☑ 2.590 + 4.714 = 7.304

(e)

```
     1  9  9  10
  $  2  0 .0  0  ◄────── Attach zeros.
- $     7 .7  4
  $  1  2 .2  6    ☑  $7.74 + $12.26 = $20.00
```

(f)

```
     5  9  9  10
  3  6 .0  0  0  ◄────── Attach zeros.
- 1  1 .1  3  2
  2  4 .8  6  8    ☑  11.132 + 24.868 = 36.000
```

(g)

```
     9  9  9  10
  1  0 .0  0  0  ◄────── Attach zeros.
-    0 .0  3  7
     9 .9  6  3    ☑  0.037 + 9.963 = 10.000
```

☑ Notice that each problem is checked by comparing the sum of the answer and the number subtracted with the first number. You should also start by estimating the answer. Whether you use a calculator or work it out with pencil and paper, checking your answer is important if you are to avoid careless mistakes.

Learning Help ▶ The balancing method used to add and subtract whole numbers mentally can also be used with decimal numbers.

For example, to add 12.8 and 6.3 mentally, first increase 12.8 by 0.2 to get a whole number 13. To keep the total the same, balance by subtracting 0.2 from 6.3 to get 6.1. Now 12.8 + 6.3 = 13.0 + 6.1, which is easy to do mentally. The answer is 19.1.

To subtract 34.6 − 18.7, add 0.3 to 18.7 to get the whole number 19. Balance this by adding 0.3 to 34.6 to get 34.9. Now 34.6 − 18.7 = 34.9 − 19.0. If this is still too tough to do mentally, balance again to make it 35.9 − 20.0 or 15.9. ◄

Now, for a set of practice problems on addition and subtraction of decimal numbers, turn to Exercises 3-1.

Exercises 3-1   Addition and Subtraction of Decimals

A. Write in words.

1. 0.72　　　2. 8.7　　　3. 12.36　　　4. 0.05

5. 3.072　　　6. 14.091　　　7. 3.0024　　　8. 6.0083

Write as a number.

9. Four thousandths

10. Three and four tenths

11. Six and seven tenths

12. Five thousandths

13. Twelve and eight tenths

14. Three and twenty-one thousandths

15. Ten and thirty-two thousandths

16. Forty and seven tenths

17. One hundred sixteen ten-thousandths

18. Forty-seven ten-thousandths

19. Two and three hundred seventy-four ten-thousandths

20. Ten and two hundred twenty-two ten-thousandths

B. Add or subtract as shown.

1. 14.21 + 6.8
2. 75.6 + 2.57
3. $2.83 + $12.19
4. $52.37 + $98.74
5. 0.687 + 0.93
6. 0.096 + 5.82
7. 507.18 + 321.42
8. 212.7 + 25.46
9. 45.6725 + 18.058
10. 390 + 72.04
11. 19 − 12.03
12. 7.83 − 6.79
13. $33.40 − $18.04
14. $20.00 − $13.48
15. 75.08 − 32.75
16. 40 − 3.82
17. $30 − $7.98
18. $25 − $0.61
19. 130 − 16.04
20. 19 − 5.78
21. 37 + 0.09 + 3.5 + 4.605
22. 183 + 3.91 + 45 + 13.2
23. $14.75 + $9 + $3.76
24. 148.002 + 3.4
25. 68.708 + 27.18
26. 35.36 + 4.347
27. 47.04 − 31.88
28. 180.76 − 94.69
29. 26.45 − 17.832
30. 92.302 − 73.6647
31. 6.4 + 17.05 + 7.78
32. 212.4 + 76 + 3.79
33. 26.008 − 8.4
34. 36.4 − 7.005
35. 0.0046 + 0.073
36. 0.038 + 0.00462
37. 28.7 − 7.38 + 2.9
38. 0.932 + 0.08 − 0.4
39. 6.01 − 3.55 − 0.712
40. 2.92 − 1.007 − 0.08

C. Practical Problems

1. **Machine Technology** What is the combined thickness of these five shims: 0.008, 0.125, 0.150, 0.185, and 0.005 in.?

2. **Electrical Technology** The combined weight of a spool and the wire it carries is 13.6 lb. If the weight of the spool is 1.75 lb, what is the weight of the wire?

3. **Electrical Technology** The following are diameters of some common household wires: No. 10 is 0.102 in., No. 11 is 0.090 in., No. 12 is 0.081 in., No. 14 is 0.064 in., and No. 16 is 0.051 in.

   (a) The diameter of No. 16 wire is how much smaller than the diameter of No. 14 wire?

   (b) Is No. 12 wire larger or smaller than No. 10 wire? What is the difference in their diameters?

   (c) John measured the thickness of a wire with a micrometer as 0.059 in. Assuming that the manufacturer was slightly off, what wire size did John have?

4. **Painting and Decorating** In estimating a finishing job, a painter included the following items:

Material	$ 677.85
Trucking	$ 32.80
Profit	$ 250
Labor	$1645.50
Overhead	$ 67.75

What was his total estimate for the job?

Problem 5

5. **Metalworking** Find *A*, *B*, and *C*.

6. **Plumbing** A piece of pipe 8.4 in. long is cut from a piece 40.8 in. long. What is the length of the longer piece remaining if the width of the saw cut is 0.2 in.?

7. **Machine Technology** A certain machine part is 2.345 in. thick. What is its thickness after 0.078 in. is ground off?

8. **Masonry** Find the total cost of the materials for a certain masonry job if sand cost $43.65, cement cost $114.47, a steel lintel cost $32.65, and brick cost $1476.28.

9. **Masonry** The specifications for a reinforced masonry wall called for 1.5 sq in. of reinforcing steel per square foot of cross-sectional area. If the three pieces of steel being used had cross sections of 0.125, 0.200, and 1.017 sq in., did they meet the specifications for a 1-sq-ft area?

10. **Building Construction** A plot plan of a building site showed that the east side of the house was 46.35 ft from the east lot line, and the west side of the house was 41.65 ft from the west lot line. If the lot was 156.00 ft wide along the front, how wide is the house?

11. **Auto Mechanics** A mechanic must estimate the total time for a particular servicing. He figures 0.3 hour for an oil change, 1.5 hours for a tune-up, 0.4 hour for a brake check, and 1.2 hours for air-conditioning service. What is the total number of hours of his estimate?

12. **Auto Mechanics** A heated piston measures 8.586 cm in diameter. When cold it measures 8.573 cm in diameter. How much does it expand when heated?

Problem 13

13. **Auto Mechanics** A piston must fit in a bore with a diameter of 3.3750 in. What must be the diameter of the piston given the clearance shown in the diagram?

14. **Auto Mechanics** In squaring a damaged car frame, an autobody worker measured the diagonals between the two front cross-members. One diagonal was 196.1 cm and the other was 176.8 cm. What is the difference that must be adjusted?

15. **Machine Technology** The diameter of a steel shaft is reduced 0.007 in. The original diameter of the shaft was 0.850 in. Calculate the reduced diameter of the shaft.

16. **Building Construction** A structural steel Lally column is mounted on a concrete footing as shown. Find the length of the Lally column.

D. Calculator Problems (If you need help using a calculator with decimals, turn to page 141.)

1. Balance this checkbook by finding the closing balance as of November 4.

Date	Balance	Withdrawals	Deposits
Oct. 1	$367.21		
Oct. 3		$167.05	
Oct. 4		104.97	
Oct. 8			$357.41
Oct. 16		87.50	
Oct. 18		9.43	
Oct. 20		30.09	
Oct. 22			364.85
Oct. 27		259.47	
Oct. 30		100.84	
Nov. 2		21.88	
Nov. 4	?		

2. What is the actual cost of the following car?

"Sticker price"	$21,325.00
Destination and delivery	635.00
Leather interior	875.40
5-speed automatic transmission	900.00
CD player	575.95
Moonroof	735.50
Tax and license	1602.34
Less trade-in	$ 1780.00

3. Add as shown.

(a)	(b)	(c)
0.0067	1379.4	14.07
0.032	204.5	67.81
0.0012	16.75	132.99
0.0179	300.04	225.04
0.045	2070.08	38.02
0.5	167.99	4
0.05	43.255	16.899
0.0831	38.81	7.007
0.004	19.95	4.6

(d) $0.002 + 17.1 + 4.806 + 9.9981 - 3.1 + 0.701 - 1.001 - 14 - 8.09 + 1.0101$

Check your answers in the Appendix, then continue in Section 3-2.

3-2 MULTIPLICATION AND DIVISION OF DECIMAL NUMBERS

A decimal number is really a fraction with a multiple of 10 as denominator. For example,

$$0.5 = \frac{5}{10} \qquad 0.3 = \frac{3}{10} \qquad \text{and} \qquad 0.85 = \frac{85}{100}$$

Multiplication of decimals is easy to understand if we think of it in this way:

$$0.5 \times 0.3 = \frac{5}{10} \times \frac{3}{10} = \frac{15}{100} = 0.15$$

Learning Help ▶ To estimate the product of 0.5 and 0.3, remember that if two numbers are both less than 1, their product must be less than 1. ◀

Multiplication   Of course it would be very, very clumsy and time consuming to calculate every decimal multiplication this way. We need a simpler method. Here is the procedure most often used:

Step 1   Multiply the two decimal numbers as if they were whole numbers. Pay no attention to the decimal points.

Step 2   The sum of the decimal digits in the two numbers being multiplied will give you the number of decimal digits in the answer.

EXAMPLE

Multiply 3.2 by 0.41.

Step 1   Multiply, ignoring the decimal points.

$$\begin{array}{r} 32 \\ \times\ 41 \\ \hline 1312 \end{array}$$

Step 2   Count decimal digits in each number: 3.2 has *one* decimal digit (the 2), and 0.41 has *two* decimal digits (the 4 and the 1). The total number of decimal digits in the two factors is three. The answer will have *three* decimal digits. Count over *three* digits from right to left in the answer.

1.312   three decimal digits

☑ 3.2×0.41 is roughly $3 \times \frac{1}{2}$ or about $1\frac{1}{2}$. The answer 1.312 agrees with our rough guess. Remember, even if you use a calculator to do the actual work of arithmetic, you must *always* estimate your answer first and check it afterward.

Try these simple decimal multiplications.

(a) $2.5 \times 0.5 =$ _____   (b) $0.1 \times 0.1 =$ _____

(c) $10 \times 0.6 =$ _____   (d) $2 \times 0.4 =$ _____

(e) $2 \times 0.003 =$ _____   (f) $0.01 \times 0.02 =$ _____

(g) $0.04 \times 0.005 =$ _____

(h) **Roofing** If asphalt tile weighs 1.1 lb per square foot, what is the weight of tile covering 8 sq ft?

SOLUTIONS

(a) $2.5 \times 0.5 =$ _____

First, multiply $25 \times 5 = 125$.

Second, count decimal digits.
2.5×0.5

= a total of *two* decimal digits.

Count over *two* decimal digits from the right: 1.25 two decimal digits.

The product is 1.25.

☑ $2 \times \frac{1}{2}$ is about 1, so the answer seems reasonable.

(b) 0.1×0.1 $1 \times 1 = 1$

Count over *two* decimal digits from the right. Since there are not two decimal digits in the product, attach a zero on the left: 0.01

So $0.1 \times 0.1 = 0.01$.

☑ $\frac{1}{10} \times \frac{1}{10} = \frac{1}{100}$.

Two
decimal
digits

(c) 10×0.6 $10 \times 6 = 60$

Count over *one* decimal digit from the right: 6.0

So $10 \times 0.6 = 6.0$.

Learning Help ▶ Notice that multiplication by 10 simply shifts the decimal place one digit to the right.

$10 \times 6.2 \ \ = 62$
$10 \times 0.075 = \ \ 0.75$
$10 \times 8.123 = 81.23$ and so on ◀

(d) 2×0.4 $2 \times 4 = 8$

Count over *one* decimal digit:

$2 \times 0.4 = 0.8$

(e) 2×0.003 $\qquad$ $2 \times 3 = 6$

Count over three decimal digits. Attach two zeros as place holders.

0.006 $\qquad$ $2 \times 0.003 = 0.006$

(f) 0.01×0.02 $\qquad$ $1 \times 2 = 2$

Count over four decimal digits. Attach three zeros.

0.0002 $\qquad$ $0.01 \times 0.02 = 0.0002$

Total of four decimal digits

(g) 0.04×0.005 $\qquad$ $4 \times 5 = 20$

Count over five decimal digits. Attach three zeros.

0.00020 $\qquad$ $0.04 \times 0.005 = 0.00020 = 0.0002$

Total of five decimal digits

(h) We are given the weight of 1 sq ft and we need to calculate the weight of 8 sq ft. Multiply

1.1×8 $\qquad$ $11 \times 8 = 88$

One decimal digit

The answer will have *one* decimal digit: $1.1 \times 8 = 8.8$

The weight of 8 sq ft of tile is 8.8 lb.

Learning Help ▶

1. Do not try to do this entire process mentally until you are certain you will not misplace zeros.

2. Always estimate before you begin the arithmetic, and check your answer against your estimate. ◀

Multiplication of larger decimal numbers is performed in exactly the same manner.

PROBLEMS

Try these:

(a) $4.302 \times 12.05 =$ _____

(b) $6.715 \times 2.002 =$ _____

(c) $3.144 \times 0.00125 =$ _____

(d) **Construction Technology** How much does 25.8 sq ft of heavy plastic sheeting weigh if 1 sq ft weighs 2.37 lb?

(a) **Estimate:** $4 \times 12 = 48$ The answer will be about 48.

Multiply: 4302
             $\times$ 1205
             5183910

(If you cannot do this multiplication correctly, turn to Section 1-3 for help with the multiplication of whole numbers.)

The two factors being multiplied have a total of five decimal digits (three in 4.302 and two in 12.05). Count over five decimal digits from the right in the answer.

51.83910

So that $4.302 \times 12.05 = 51.83910 = 51.8391$

☑ The answer 51.8391 is approximately equal to the estimate of 48.

(b) **Estimate:** 6.7×2 is about 7×2 or 14.

Multiply: 6.715   6.715 has *three* decimal digits
                  $\times$ 2.002   2.002 has *three* decimal digits
                 13.443430   a total of *six* decimal digits

six decimal digits

$6.715 \times 2.002 = 13.44343$

☑ The answer and estimate are roughly equal.

(c) **Estimate:** 3×0.001 is about 0.003.

Multiply: 3.144   3.144 has *three* decimal digits
                $\times$ 0.00125   0.00125 has *five* decimal digits
                .00393000   a total of *eight* decimal digits

eight decimal digits   In order to count over 8 digits we had to attach 2 zeros on the left.

$3.144 \times 0.00125 = 0.00393$

☑ The answer and estimate are roughly equal.

(d) **Estimate:** 25.8×2.37 is about 30×2 or 60 lb.

Multiply: 25.8   *one* decimal digit
                $\times$ 2.37   *two* decimal digits
                61.146   *three* decimal digits

☑ The answer and estimate are roughly equal.

Note ▶ When multiplying, you need not line up the decimal points. ◀

One nice thing about a calculator is that it counts decimal digits and automatically adjusts the answer in any calculation. But to get an estimate of the answer, you must still understand the process.

Division Division of decimal numbers is very similar to the division of whole numbers. For example,

$$6.8 \div 1.7 \qquad \text{can be written} \qquad \frac{6.8}{1.7}$$

and if we multiply both top and bottom of the fraction by 10 we have

$$\frac{6.8}{1.7} = \frac{6.8 \times \boxed{10}}{1.7 \times \boxed{10}} = \frac{68}{17}$$

But you should know how to divide these whole numbers.

$$68 \div 17 = 4$$

Therefore, $6.8 \div 1.7 = 4$.

To divide decimal numbers, use the following procedure.

Step 1 Write the divisor and dividend in standard long-division form.

Step 2 Shift the decimal point in the divisor to the right so as to make the divisor a whole number.

Step 3 Shift the decimal point in the dividend the same amount. (Attach zeros if necessary.)

Step 4 Place the decimal point in the answer space directly above the new decimal position in the dividend.

Step 5 Now divide exactly as you would with whole numbers. The decimal points in divisor and dividend may now be ignored.

$$6.8 \div 1.7 = 4$$

Example:
$6.8 \div 1.7$

$1.7\overline{)6.8}$

$1.7.\overline{)}$

$1.7.\overline{)6.8.}$

$17.\overline{)68.}$

$$17.\overline{)\underset{\displaystyle -68}{68.}}^{\displaystyle 4}$$
$$0$$

Notice in steps 2 and 3 that we have simply multiplied both divisor and dividend by 10.

If there is a remainder in step 5, we must add additional zeros to complete the division. This will be necessary in the next example.

EXAMPLE

Divide: $1.38 \div 2.4$

Let's do it step by step.

$2.4\overline{)1.38}$

$2.4.\overline{)13.8}$ Shift both decimal points one digit to the right to make the divisor (2.4) a whole number (24).

$24.\overline{)13\overset{\displaystyle .}{.}8}$ Place the decimal point in the answer space.

$\begin{array}{r} .5 \\ 24.\overline{)13.8} \\ 120 \\ \hline 18 \end{array}$ Divide as usual. 24 goes into 138 five times.

There is a remainder (18), so we must keep going.

$\begin{array}{r} .57 \\ 24.\overline{)13.80} \\ 120 \\ \hline 180 \\ 168 \\ \hline 12 \end{array}$ Attach a zero to the dividend and bring it down.

Divide 180 by 24. 24 goes into 180 seven times.

Now the remainder is 12.

$$\begin{array}{r} .575 \\ 24.\overline{)13.800} \\ \underline{12\ 0} \\ 1\ 80 \\ \underline{1\ 68} \\ 120 \\ \underline{120} \\ 0 \end{array}$$

Attach another zero and bring it down.

24 goes into 120 exactly five times.

The remainder is zero. $1.38 \div 2.4 = 0.575$

☑ $1.38 \div 2.4$ is roughly $1 \div 2$ or 0.5.

Double-Check: $2.4 \times 0.575 = 1.38$. Always multiply the answer by the divisor to double-check the work. Do this even if you are using a calculator.

Note ▶ In some problems, we may continue to attach zeros and divide yet never get a remainder of zero. We will examine problems like this later. ◀

YOUR TURN

How would you do this one?

$2.6 \div 0.052 = ?$

SOLUTION

$0.052.\overline{)2.6}$

To shift the decimal place three digits in the dividend, we must attach two zeros to its right.

$0.052.\overline{)2.600.}$ Now place the decimal point in the answer space above that in the dividend.

$$\begin{array}{r} 50. \\ 52.\overline{)2600.} \\ \underline{260} \\ 0 \\ 0 \end{array}$$ ⟵———— $5 \times 52 = 260$

$2.6 \div 0.052 = 50$ ☑ $0.052 \times 50 = 2.6$

Shifting the decimal point three digits and attaching zeros to the right of the decimal point in this way is equivalent to multiplying both divisor and dividend by 1000.

Note ▶ In this last example inexperienced students think the problem is finished after the first division because they get a remainder of zero. You must keep dividing at least until all the places up to the decimal point in the answer are filled. The answer here is 50, not 5. ◀

PROBLEMS

Try these problems.

(a) $3.5 \div 0.001$ = _____

(b) $9 \div 0.02$ = _____

(c) $0.365 \div 18.25$ = _____

(d) $8.8 \div 3.2$ = _____

(e) $7.230 \div 6$ = _____   (f) $3 \div 4$ = _____

(g) $30.24 \div 0.42$ = _____   (h) $273.6 \div 0.057$ = _____

(i) **Construction Technology** How many pieces of plywood 0.375 in. thick are in a stack 30 in. high?

SOLUTIONS

(a) $0.001.\overline{)3.500.}$

$$\begin{array}{r} 3500. \\ 1.\overline{)3500.} \end{array}$$ $3.5 \div 0.001 = 3500$

           ☑ $0.001 \times 3500 = 3.5$

(b) $0.02.\overline{)9.00.}$ Shift the decimal point two places to the right. Divide 900 by 2.

$$\begin{array}{r} 450. \\ 2.\overline{)900.} \end{array}$$ $9 \div 0.02 = 450$

         ☑ $0.02 \times 450 = 9$

Dividing a whole number by a decimal is very troublesome for most people.

(c) $18.25.\overline{)0.36.5}$

$$\begin{array}{r} .02 \\ 1825.\overline{)36.50} \\ \underline{36.50} \\ 0 \end{array}$$

1825 does not go into 365, so place a zero above the 5. Attach a zero after the 5.

1825 goes into 3650 twice. Place a 2 in the answer space above the zero.

     $0.365 \div 18.25 = 0.02$

     ☑ $18.25 \times 0.02 = 0.365$

(d) $3.2.\overline{)8.8.}$

$$\begin{array}{r} 2.75 \\ 32.\overline{)88.00} \\ \underline{64}\downarrow \\ 240 \\ \underline{224}\downarrow \\ 160 \\ \underline{160} \\ 0 \end{array}$$

$2 \times 32 = 64$ Subtract, attach a zero, and bring it down.

$7 \times 32 = 224$ Subtract, attach another zero, and bring it down.

$5 \times 32 = 160$ $8.8 \div 3.2 = 2.75$

☑   The estimated answer is $9 \div 3$ or 3.

Double-check: $3.2 \times 2.75 = 8.8$

(e) $$\begin{array}{r} 1.205 \\ 6\overline{)7.230} \\ \underline{-6}\downarrow \\ 1\,2 \\ \underline{-1\,2}\downarrow \\ 03 \\ \underline{0}\downarrow \\ 30 \\ \underline{-30} \\ 0 \end{array}$$

The divisor 6 is a whole number, so we can bring the decimal point in 7.23 up to the answer space.

☑ $1.205 \times 6 = 7.230$

$7.230 \div 6 = 1.205$

$$
\begin{array}{r}
.75 \\
\text{(f)} \quad 4\overline{)3.00} \\
2\,8 \\
\hline
20 \\
20 \\
\hline
\end{array}
\qquad \checkmark \quad 0.75 \times 4 = 3.00
$$

$$3 \div 4 = 0.75$$

(g) $\quad 0.42.\overline{)30.24}$
$$
\begin{array}{r}
72 \\
29\,4 \\
\hline
84 \\
84 \\
\hline
\end{array}
$$

$\checkmark \quad 72 \times 0.42 = 30.24$

(h) $\quad 0.057.\overline{)273.600}$
$$
\begin{array}{r}
4\,800 \\
228 \\
\hline
45\,6 \\
45\,6 \\
\hline
\end{array}
$$

$\checkmark \quad 4800 \times 0.057 = 273.6$

(i) $\quad 0.375.\overline{)30.000}$
$$
\begin{array}{r}
80 \\
30\,00 \\
\hline
0 \\
\end{array}
$$

$\checkmark \quad 80 \times 0.375 = 30$

There are 80 pieces of plywood in the stack.

If the dividend is not exactly divisible by the divisor, either stop the process after some preset number of decimal places in the answer, or round the answer. We do not generally indicate a remainder in decimal division.

ARITHMETIC "TRICKS OF THE TRADE"

1. To divide a number by 5, use the fact that 5 is one-half of 10. First, multiply the number by 2, then divide by 10 by shifting the decimal point one place to the left.

   Example: $64 \div 5 \qquad 64 \times 2 \div 10 = 128 \div 10 = 12.8$

2. To divide a number by 25, first multiply by 4, then divide by 100, shifting the decimal point two places left.

   Example: $304 \div 25 = 304 \times 4 \div 100 = 1216 \div 100 = 12.16$

3. To divide a number by 20, first divide by 2, then divide by 10.

   Example: $86 \div 20 = 86 \div 2 \div 10 = 43 \div 10 = 4.3$

4. To multiply a number by 20, first multiply by 2, then multiply by 10, shifting the decimal point one place to the right.

   Example: $73 \times 20 = 73 \times 2 \times 10 = 146 \times 10 = 1460$

When doing "mental arithmetic" like this, it is important that you start with a rough estimate of the answer.

Rounding Decimal Numbers

Calculations involving decimal numbers often result in answers having more decimal digits than are justified. When this occurs, we must round our answer. The process of rounding a decimal number is very similar to the procedure for rounding a whole number. The only difference is that after rounding to a given decimal place, all digits to the right of that place are dropped. The following examples illustrate this difference.

Round 35782.462 to the	**nearest thousand**	**nearest hundredth**
Step 1 Place a $\wedge$ mark to the right of the place to which the number is to be rounded.	35 $\wedge$ 782.462	35782.46 $\wedge$ 2
Step 2 If the digit to the right of the mark is less than 5, replace all digits to the right of the mark with zeros.		35782.460 → Drop this decimal digit zero.
If these zeros are decimal digits, discard them.		The rounded number is 35782.46
Step 3 If the digit to the right of the mark is equal to or larger than 5, increase the digit to the left by 1 and replace all digits to the right with zeros.	36 000.000 → Drop these right-end decimal zeros	
Drop all right-end decimal zeros, but keep zero placeholders.	36000 ↗ Keep these zeros as placeholders.	
	The rounded number is 36000.	

Careful ▶ Drop only the decimal zeros that are to the right of the $_\wedge$ mark. For example, to round 6.4086 to three decimal digits, we write 6.408 $_\wedge$ 6, which becomes 6.4090 or 6.409. We dropped the end zero because it was a decimal zero to the *right* of the mark. But we retained the other zero because it is needed as a placeholder. ◀

YOUR TURN

Try rounding these numbers.

(a) Round   74.238   to two decimal places.

(b) Round   8.043   to two decimal places.

(c) Round   0.07354   to three decimal places.

(d) Round   7.98   to the nearest tenth.

ANSWERS

(a) 74.238 is 74.24 to two decimal places.
(Write 74.23$_\wedge$8 and note that 8 is larger than 5, so increase the 3 to 4 and drop the last digit because it is a decimal digit.)

(b) 8.043 is 8.04 to two decimal places.
(Write 8.04$_\wedge$3 and note that 3 is less than 5 so drop it.)

(c) 0.07354 is 0.074 to three decimal places.
(Write 0.073$_\wedge$54 and note that the digit to the right of the mark is 5; therefore, change the 3 to a 4 to get 0.074$_\wedge$00. Finally, drop the digits on the right to get 0.074.)

(d) 7.98 is 8.0 to the nearest tenth.
(Write 7.9$_\wedge$8 and note that 8 is greater than 5, so increase the 9 to 0 and the 7 to 8. Drop the digit in the hundredths place.)

There are a few very specialized situations where this rounding rule is not used:

1. Some engineers use a more complex rule when rounding a number that ends in 5.

2. In business, fractions of a cent are usually rounded up to determine selling price. Three items for 25 cents or $8\frac{1}{3}$ cents each is rounded to 9 cents each.

Our rule will be quite satisfactory for most of your work in arithmetic.

HOW TO NAME DECIMAL NUMBERS

The decimal number 3,254,935.4728 should be interpreted as

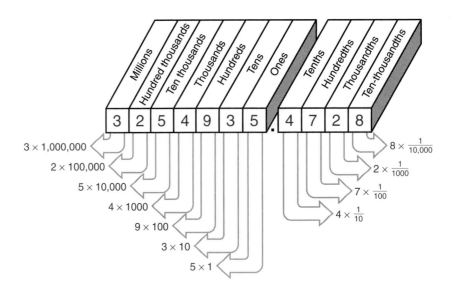

It may be read "three million, two hundred fifty-four thousand, nine hundred thirty-five, and four thousand seven hundred twenty-eight ten-thousandths."

Notice that the decimal point is read "and."

It is useful to recognize that, for example, the digit 8 represents 8 ten-thousandths or $\frac{8}{10,000}$, and the digit 7 represents 7 hundredths or $\frac{7}{100}$. Most often, however, this number is read more simply as "three million, two hundred fifty-four thousand, nine hundred thirty-five, *point* four, seven, two, eight." This way of reading the number is easiest to write, to say, and to understand.

Rounding During Division

In some division problems this process of dividing will never result in a zero remainder. At some point the answer must be rounded.

To round answers in a division problem, first continue the division so that your answer has one place more than the rounded answer will have, then round it. For example, in the division problem

$4.7 \div 1.8 = ?$

to get an answer rounded to one decimal place, first, divide to two decimal places:

$4.7 \div 1.8 = 2.61\ldots$

then round back to one decimal place:

$4.7 \div 1.8 = 2.6$ rounded

YOUR TURN

For the following problem, divide and round your answer to two decimal places.

$6.84 \div 32.7 = $ _____

Careful now.

SOLUTION

$$32.7\overline{)6.8.4}$$

$$
\begin{array}{r}
.209 \\
327.\overline{)68.400} \\
65\ 4\downarrow\downarrow \\
\hline
3\ 000 \\
2\ 943 \\
\hline
\end{array}
$$

Carry the answer to three decimal places.
Notice that two zeros must be attached to the dividend.
$2 \times 327 = 654$

$9 \times 327 = 2943$

0.209 rounded to two decimal places is 0.21.

$6.84 \div 32.7 = 0.21$ rounded to two decimal places

 $32.7 \times 0.21 = 6.867$, which is approximately equal to 6.84. (The check will not be exact because we have rounded.)

The ability to round numbers is especially important for people who work in the practical, trade, or technical areas. You will need to round answers to practical problems if they are obtained "by hand" or with a calculator. Rounding is discussed in more detail in Chapter 5.

Averages

Suppose that you needed to know the diameter of a steel connecting pin. As a careful and conscientious worker, you would probably measure its diameter several times with a micrometer, and you might come up with a sequence of numbers like this (in inches):

1.3731, 1.3728, 1.3736, 1.3749, 1.3724, 1.3750

What is the actual diameter of the pin? The best answer is to find the **average** value or **arithmetic mean** of these measurements.

$$\text{Average} = \frac{\text{sum of the measurements}}{\text{number of measurements}}$$

For the preceding problem,

$$\text{Average} = \frac{1.3731 + 1.3728 + 1.3736 + 1.3749 + 1.3724 + 1.3750}{6}$$

$$= \frac{8.2418}{6} = 1.3736333 \ldots$$

$$= 1.3736 \quad \text{rounded to four decimal places}$$

When you calculate the average of a set of numbers, the usual rule is to round the answer to the same number of decimal places as the least precise number in the set—that is, the one with the fewest decimal digits. If the numbers to be averaged are all whole numbers, the average will be a whole number.

> **EXAMPLE**

The average of 4, 6, 4, and 5 is $\frac{19}{4} = 4.75$ or 5 when rounded.

If one number of the set has fewer decimal places than the rest, round off to agree with the least precise number.

> **EXAMPLE**

The average of 2.41, 3.32, 5.23, 3.51, 4.1, and 4.12 is

$\frac{22.69}{6} = 3.78166 \ldots$ and this answer should be rounded to 3.8 to

agree in precision with the least precise number, 4.1. (We will learn more about precision in Chapter 5.)

To avoid confusion, we will usually give directions for rounding.

> **YOUR TURN**

Try it. Find the average for each of these sets of numbers.

(a) 8, 9, 11, 7, 5

(b) 0.4, 0.5, 0.63, 0.2

(c) 2.35, 2.26, 2.74, 2.55, 2.6, 2.31

> **SOLUTIONS**

(a) $\text{Average} = \frac{8 + 9 + 11 + 7 + 5}{5} = \frac{40}{5} = 8$

(b) $\text{Average} = \frac{0.4 + 0.5 + 0.63 + 0.2}{4} = \frac{1.73}{4} = 0.4324 \approx 0.4 \quad \text{rounded}$

(c) $\text{Average} = \frac{2.35 + 2.26 + 2.74 + 2.55 + 2.6 + 2.31}{6}$

$$= \frac{14.81}{6} = 2.468333 \ldots$$

$\approx 2.5 \quad$ rounded to agree in precision with 2.6, the least precise number in the set. (The symbol $\approx$ means "approximately equal to.")

Now turn to Exercises 3-2 for a set of problems on the multiplication and division of decimal numbers.

Exercises 3-2

Multiplication and Division of Decimals

A. Multiply or divide as shown.

1.	0.01×0.001	2.	10×2.15	3.	0.04×100
4.	0.3×0.3	5.	0.7×1.2	6.	0.005×0.012
7.	0.003×0.01	8.	7.25×0.301	9.	2×0.035
10.	$0.2 \times 0.3 \times 0.5$	11.	$0.6 \times 0.6 \times 6.0$	12.	$2.3 \times 1.5 \times 1.05$
13.	$3.618 \div 0.6$	14.	$3.60 \div 0.03$	15.	$4.40 \div 0.22$
16.	$6.5 \div 0.05$	17.	$0.0405 \div 0.9$	18.	$0.378 \div 0.003$
19.	$3 \div 0.05$	20.	$10 \div 0.001$	21.	$4 \div 0.01$
22.	$2.59 \div 70$	23.	$44.22 \div 6.7$	24.	$104.2 \div 0.0320$
25.	$484 \div 0.8$	26.	6.05×2.3	27.	0.0027×1.4
28.	$0.0783 \div 0.27$	29.	$0.00456 \div 0.095$	30.	$800 \div 0.25$
31.	$324 \div 0.0072$	32.	$0.08322 \div 228$	33.	$0.0092 \div 115$
34.	0.047×0.024	35.	0.0056×0.065	36.	$0.02 \times 0.06 \times 0.04$
37.	$0.008 \times 0.4 \times 0.03$	38.	123.4×0.45	39.	0.062×27.5

B. Round the following numbers as indicated.

1.	42.875 (Two decimal places)	2.	0.5728 (Nearest tenth)
3.	6.54 (One decimal place)	4.	117.6252 (Three decimal places)
5.	79.135 (Nearest hundredth)	6.	1462.87 (Nearest whole number)
7.	3.64937 (Four decimal places)	8.	19.4839 (Nearest hundredth)
9.	0.2164 (Nearest thousandth)	10.	56.826 (Two decimal places)

C. Divide and round as indicated.
Round to two decimal digits.

1.	$10 \div 3$	2.	$5 \div 6$	3.	$2.0 \div 0.19$
4.	$3 \div 0.081$	5.	$0.023 \div 0.19$	6.	$12.3 \div 4.7$
7.	$2.37 \div 0.07$	8.	$6.5 \div 1.31$		

Round to the nearest tenth.

9.	$100 \div 3$	10.	$21.23 \div 98.7$	11.	$1 \div 4$
12.	$100 \div 9$	13.	$0.006 \div 0.04$	14.	$1.008 \div 3$

Round to three decimal places.

15.	$10 \div 70$	16.	$0.09 \div 0.402$	17.	$0.091 \div 0.0014$
18.	$3.41 \div 0.257$	19.	$6.001 \div 2.001$	20.	$123.21 \div 0.1111$

D. Word Problems

1. **Painting and Decorating** A spray painting outfit is advertised for $579. It can also be bought "on time" for 24 payments of $28.50 each. How much extra do you pay by purchasing it on the installment plan?

2. If the telephone rates between Zanzibar and Timbuctou are $3.10 for the first minute and $1.25 for each additional minute, what will be the cost of an 11-minute telephone call?

3. **Metalworking** Find the average weight of five castings that weigh 17.0, 21.0, 12.0, 20.6, and 23.4 lb.

4. **Office Services** If you work $8\frac{1}{4}$ hours on Monday, 10.1 hours on Tuesday, 8.5 hours on Wednesday, 9.4 hours on Thursday, $6\frac{1}{2}$ hours on Friday, and 4.2 hours on Saturday, what is the average number of hours worked per day?

5. **Machine Technology** For the following four machine parts, find W, the number of pounds per part; C, the cost of the metal per part; and T, the total cost.

Metal Parts	Number of Inches Needed	Number of Pounds per Inch	Cost per Pound	Pounds (W)	Cost per Part (C)
A	44.5	0.38	$0.98		
B	122.0	0.19	$0.89		
C	108.0	0.08	$1.05		
D	9.5	0.32	$2.15		

6. **Machine Technology** How much does 15.7 sq ft of No. 16 gauge steel weigh if 1 sq ft weighs 2.55 lb?

7. **Building Construction** A 4-ft by 8-ft sheet of $\frac{1}{4}$-in. plywood has an area of 32 sq ft. If the weight of $\frac{1}{4}$-in. plywood is 1.5 lb/sq ft (pounds per square foot), what is the weight of the sheet?

8. **Manufacturing** Each inch of 1-in.-diameter cold-rolled steel weighs 0.22 lb. How much would a piece weigh that was 38 in. long?

9. **Welding** A truck can carry a load of 5000 lb. Assuming that it could be cut to fit, how many feet of steel beam weighing 32.6 lb/ft (pounds per foot) can the truck carry?

10. **Fire Protection** One gallon of water weighs 8.34 lb. How much weight is added to a fire truck when its tank is filled with 750 gallons of water?

11. **Electrician** Voltage values are often stated in millivolts [1 millivolt (mV) equals 0.001 volt] and must be converted to volts to be used in Ohm's law. If a voltage is given to be 75 mV, how many volts is this?

12. **Industrial Technology** An industrial engineer must estimate the cost of building a storage tank. The tank requires 208 sq ft of material at $7.29 per square foot. It also requires 4.5 hours of labor at $26.40 per hour plus 1.6 hours of labor at $19.60 per hour. Find the total cost of the tank.

13. **Sheet Metal Technology** How many sheets of metal are in a stack 5.00 in. high if each sheet is 0.0149 in. thick?

14. **Metalworking** A casting weighs 3.68 lb. How many castings are contained in a load weighing 5888.00 lb?

15. A barrel partially filled with liquid fertilizer weighs 267.75 lb. The empty barrel weighs 18.00 lb, and the fertilizer weighs 9.25 lb per gallon. How many gallons of fertilizer are in the barrel?

16. **Construction** A four-foot width of $\frac{1}{4}$-inch gypsum costs $5.23. Using this material, how much would it cost to drywall a room with 52 ft of total wall width?

17. **Construction** A room addition requires 320 sq ft of wood floor. Hardwood costs $4.59 per sq ft, while laminate costs $2.79 per sq ft. What is the total difference in cost between the two types of flooring?

18. **Electronics** The Happy Hacker electronics warehouse purchases 660 stereo adaptors, which are then packed in sets of three selling for $3.95 per set. Calculate the total income from the sale of these adaptors.

19. **Electronics** An oscillator is a device that generates an ac signal at some specified frequency. If the oscillator's output frequency is 7500 cycles per second (hertz), how many cycles does it generate in a 0.35-sec interval?

E. Calculator Problems

1. Divide. 9.87654321 ÷ 1.23456789.

   Notice anything interesting? (Divide it out to eight decimal digits.) You should be able to get the correct answer even if your calculator will not accept a nine-digit number.

2. Divide. (a) $\dfrac{1}{81}$ (b) $\dfrac{1}{891}$ (c) $\dfrac{1}{8991}$

   Notice a pattern? (Divide them out to about eight decimal places.)

3. **Metalworking** The outside diameter of a steel casting is measured six times at different positions with a digital caliper. The measurements are 4.2435, 4.2426, 4.2441, 4.2436, 4.2438, and 4.2432 in. Find the average diameter of the casting.

4. **Printing** To determine the thickness of a sheet of paper, five batches of 12 sheets each are measured with a digital micrometer. The thickness of each of the five batches of 12 is 0.7907, 0.7914, 0.7919, 0.7912, and 0.7917 mm. Find the average thickness of a single sheet of paper.

5. **Building Construction** The John Hancock Towers office building in Boston had a serious problem when it was built: When the wind blew hard, the pressure caused its windows to pop out! The only reasonable solution was to replace all 10,344 windows in the building at a cost of $6,000,000. What was the replacement cost per window? Round to the nearest dollar.

6. **Food Services** In purchasing food for his chain of burger shops, José pays $3.85 per pound for cheese to be used in cheeseburgers. He calculates that an order of 5180 lb will be enough for 165,000 cheeseburgers.

   (a) How much cheese will he use on each cheeseburger? (Round to three decimal places.)

   (b) What is the cost, to the nearest cent, of the cheese used on a cheeseburger?

7. **Plumbing** The pressure in psi (pounds per square inch) in a water system at a point 140.8 ft below the water level in a storage tank is given by the formula

   $$P = \frac{62.4 \text{ lb per cu ft} \times 140.8 \text{ ft} \times 0.43}{144 \text{ sq in. per sq ft}}$$

   Find P to the nearest tenth.

8. **Office Services** A carpet cleaner charges $48 for two rooms and $98 for five rooms. Suppose a two-room job contained 350 sq ft and a five-room job

contained 850 sq ft. Calculate the difference in cost per square foot between these two jobs. (Round to the nearest tenth of a cent.)

9. **Electronics**   Radar operates by broadcasting a high-frequency radio wave pulse that is reflected from a target object. The reflected signal is detected at the original source. The distance of the object from the source can be determined by multiplying the speed of the pulse by the elapsed time for the round trip of the pulse. A radar signal traveling at 186,000 miles per second is reflected from an object. The signal has an elapsed round trip time of 0.0001255 sec. Calculate the distance of the object from the radar antenna.

10. **Plumbing**   Normal air pressure at sea level is about 14.7 psi. Additional pressure exerted by a depth of water is given by the formula

Pressure in psi ≈ 0.434 × depth of water in ft

Calculate the total pressure on the bottom of a diving pool filled with water to a depth of 12 ft.

When you have finished these exercises, check your answers in the Appendix, then continue in Section 3-3.

3-3   DECIMAL FRACTIONS

Since decimal numbers are fractions, they may be used, as fractions are used, to represent a part of some quantity. For example, recall that

"$\frac{1}{2}$ of 8 equals 4"   means $\frac{1}{2} \times 8 = 4$

and therefore,

"0.5 of 8 equals 4"   means $0.5 \times 8 = 4$

The word *of* used in this way indicates multiplication, and decimal numbers are often called "decimal fractions."

It is very useful to be able to convert any number from fraction form to decimal form. Simply divide the top by the bottom of the fraction. If the division has no remainder, the decimal number is called a *terminating* decimal.

EXAMPLE

$\dfrac{5}{8} = \underline{\quad ? \quad}$

$$
\begin{array}{r}
.625 \\
8)\overline{5.000} \\
\underline{4\ 8} \\
20 \\
\underline{16} \\
40 \\
\underline{40} \\
0
\end{array}
$$

← Attach as many zeros as needed.

← Zero remainder; therefore, the decimal terminates or ends.

$\dfrac{5}{8} = 0.625$

If the decimal does not terminate, you may round it to any desired number of decimal digits.

$$\frac{2}{13} = \underline{\quad ? \quad}$$

$$
\begin{array}{r}
.1538 \\
13\overline{)2.0000} \\
\underline{1\ 3} \\
70 \\
\underline{65} \\
50 \\
\underline{39} \\
110 \\
\underline{104} \\
6
\end{array}
$$
← Attach zeros.

← Remainder is not equal to zero.

$$\frac{2}{13} = 0.154 \quad \text{rounded to three decimal places}$$

Convert the following fractions to decimal form and round if necessary to two decimal places.

(a) $\dfrac{4}{5}$ (b) $\dfrac{2}{3}$ (c) $\dfrac{17}{7}$ (d) $\dfrac{5}{6}$ (e) $\dfrac{7}{16}$ (f) $\dfrac{5}{9}$

(a) $\dfrac{4}{5} = \underline{\quad ? \quad}$

$$
\begin{array}{r}
.8 \\
5\overline{)4.0}
\end{array}
$$

$$\frac{4}{5} = 0.8$$

(b) $\dfrac{2}{3} = \underline{\quad ? \quad}$

$$
\begin{array}{r}
.666\ldots \\
3\overline{)2.000} \\
\underline{1\ 8} \\
20 \\
\underline{18} \\
20 \\
\underline{18} \\
2
\end{array}
$$

$$\frac{2}{3} = 0.67 \quad \text{rounded to two decimal places}$$

Notice that in order to round to *two* decimal places, we must carry the division out to at least *three* decimal digits.

(c) $\dfrac{17}{7} = \underline{\quad ? \quad}$

$$
\begin{array}{r}
2.428 \\
7\overline{)17.000} \\
\underline{14} \\
3\ 0 \\
\underline{2\ 8} \\
20 \\
\underline{14} \\
60 \\
\underline{56} \\
4
\end{array}
$$

$$\frac{17}{7} = 2.43 \quad \text{rounded to two decimal digits}$$

(d) $\dfrac{5}{6} = $ _____?_____

$$\begin{array}{r} .833\ldots \\ 6\overline{)5.000} \\ \underline{4\,8} \\ 20 \\ \underline{18} \\ 20 \\ \underline{18} \\ 2 \end{array}$$

$\dfrac{5}{6} = 0.83$ rounded to two decimal places

(e) $\dfrac{7}{16} = $ _____?_____

$$\begin{array}{r} .4375 \\ 16\overline{)7.0000} \\ \underline{6\,4} \\ 60 \\ \underline{48} \\ 120 \\ \underline{112} \\ 80 \\ \underline{80} \\ 0 \end{array}$$

$\dfrac{7}{16} = 0.4375$ or 0.44 rounded to two decimal digits

(f) $\dfrac{5}{9} = $ _____?_____

$$\begin{array}{r} .55 \\ 9\overline{)5.00} \\ \underline{4\,5} \\ 50 \\ \underline{45} \\ 5 \end{array}$$

$\dfrac{5}{9} = 0.555\ldots$ or 0.56 rounded

Repeating Decimals   Decimal numbers that do not terminate will repeat a sequence of digits. This type of decimal number is called a *repeating* decimal. For example,

$$\frac{1}{3} = 0.3333\ldots$$

where the three dots are read "and so on," and they tell us that the digit 3 continues without end.

Similarly, $\dfrac{2}{3} = 0.6666\ldots$ and $\dfrac{3}{11}$ is

$$\begin{array}{r} .2727 \\ 11\overline{)3.0000} \\ \underline{2\,2} \\ 80 \\ \underline{77} \\ 30 \\ \underline{22} \\ 80 \\ \underline{77} \\ 3 \end{array}$$

The remainder 3 is equal to the original dividend. This tells us that the decimal quotient repeats itself.

$$\frac{3}{11} = 0.272727\ldots$$

If you did the division in problem (c) with a calculator, you may have noticed that

$$\frac{17}{7} = 2.\overset{\frown}{4285714}\overset{\frown}{28571}\ldots \qquad \text{The digits 428571 repeat endlessly.}$$

Mathematicians often use a shorthand notation to show that a decimal repeats.

Write $\dfrac{1}{3} = 0.\overline{3}$ or $\dfrac{2}{3} = 0.\overline{6}$

where the bar means that the digits under the bar repeat endlessly.

$\dfrac{3}{11} = 0.\overline{27}$ means $0.272727\ldots$ and $\dfrac{17}{7} = 2.\overline{428571}$

YOUR TURN

Write $\dfrac{41}{33}$ as a repeating decimal using the "bar" notation.

SOLUTION

$$
\begin{array}{r}
1.24 \\
33\overline{)41.00} \\
33 \\
\hline
8\,0 \\
6\,6 \\
\hline
1\,40 \\
1\,32 \\
\hline
8
\end{array}
$$

These remainders are the same, so we know that further division will produce a repeat of the digits 24 in the answer.

$\dfrac{41}{33} = 1.242424\ldots = 1.\overline{24}$

Terminating Decimals   Some fractions are used so often in practical work that it is important for you to know their decimal equivalents as shown on p. 132.

PROBLEMS

Quick now, without looking at the table, fill in the blanks in the problems shown.

$\dfrac{1}{2} = \underline{\hspace{2cm}}$ $\dfrac{3}{4} = \underline{\hspace{2cm}}$ $\dfrac{3}{8} = \underline{\hspace{2cm}}$ $\dfrac{7}{8} = \underline{\hspace{2cm}}$

$\dfrac{1}{4} = \underline{\hspace{2cm}}$ $\dfrac{5}{8} = \underline{\hspace{2cm}}$ $\dfrac{3}{16} = \underline{\hspace{2cm}}$ $\dfrac{7}{16} = \underline{\hspace{2cm}}$

$\dfrac{1}{8} = \underline{\hspace{2cm}}$ $\dfrac{5}{16} = \underline{\hspace{2cm}}$ $\dfrac{9}{16} = \underline{\hspace{2cm}}$ $\dfrac{13}{16} = \underline{\hspace{2cm}}$

$\dfrac{1}{16} = \underline{\hspace{2cm}}$ $\dfrac{11}{16} = \underline{\hspace{2cm}}$ $\dfrac{15}{16} = \underline{\hspace{2cm}}$

Check your work against the table when you are finished, then turn to p. 133 and continue.

DECIMAL–FRACTION EQUIVALENTS

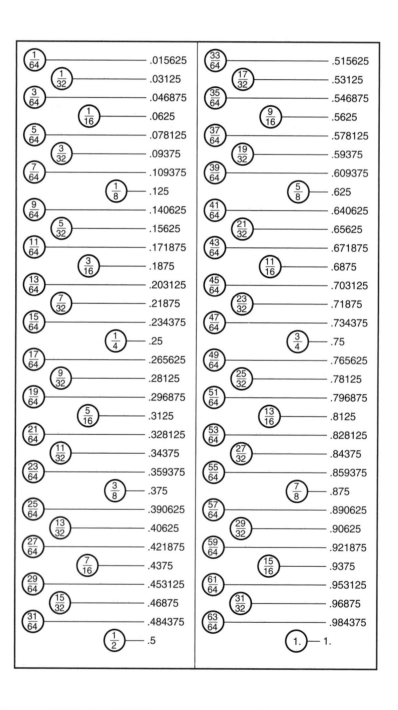

$\frac{1}{64}$ — .015625	$\frac{33}{64}$ — .515625
$\frac{1}{32}$ — .03125	$\frac{17}{32}$ — .53125
$\frac{3}{64}$ — .046875	$\frac{35}{64}$ — .546875
$\frac{1}{16}$ — .0625	$\frac{9}{16}$ — .5625
$\frac{5}{64}$ — .078125	$\frac{37}{64}$ — .578125
$\frac{3}{32}$ — .09375	$\frac{19}{32}$ — .59375
$\frac{7}{64}$ — .109375	$\frac{39}{64}$ — .609375
$\frac{1}{8}$ — .125	$\frac{5}{8}$ — .625
$\frac{9}{64}$ — .140625	$\frac{41}{64}$ — .640625
$\frac{5}{32}$ — .15625	$\frac{21}{32}$ — .65625
$\frac{11}{64}$ — .171875	$\frac{43}{64}$ — .671875
$\frac{3}{16}$ — .1875	$\frac{11}{16}$ — .6875
$\frac{13}{64}$ — .203125	$\frac{45}{64}$ — .703125
$\frac{7}{32}$ — .21875	$\frac{23}{32}$ — .71875
$\frac{15}{64}$ — .234375	$\frac{47}{64}$ — .734375
$\frac{1}{4}$ — .25	$\frac{3}{4}$ — .75
$\frac{17}{64}$ — .265625	$\frac{49}{64}$ — .765625
$\frac{9}{32}$ — .28125	$\frac{25}{32}$ — .78125
$\frac{19}{64}$ — .296875	$\frac{51}{64}$ — .796875
$\frac{5}{16}$ — .3125	$\frac{13}{16}$ — .8125
$\frac{21}{64}$ — .328125	$\frac{53}{64}$ — .828125
$\frac{11}{32}$ — .34375	$\frac{27}{32}$ — .84375
$\frac{23}{64}$ — .359375	$\frac{55}{64}$ — .859375
$\frac{3}{8}$ — .375	$\frac{7}{8}$ — .875
$\frac{25}{64}$ — .390625	$\frac{57}{64}$ — .890625
$\frac{13}{32}$ — .40625	$\frac{29}{32}$ — .90625
$\frac{27}{64}$ — .421875	$\frac{59}{64}$ — .921875
$\frac{7}{16}$ — .4375	$\frac{15}{16}$ — .9375
$\frac{29}{64}$ — .453125	$\frac{61}{64}$ — .953125
$\frac{15}{32}$ — .46875	$\frac{31}{32}$ — .96875
$\frac{31}{64}$ — .484375	$\frac{63}{64}$ — .984375
$\frac{1}{2}$ — .5	$1.$ — 1.

Convert each of the following fractions into decimal form.

(a) $\dfrac{95}{100}$ (b) $\dfrac{1}{20}$ (c) $\dfrac{7}{10}$ (d) $\dfrac{4}{1000}$ (e) $\dfrac{11}{1000}$

(f) $\dfrac{327}{10000}$ (g) $\dfrac{473}{1000}$ (h) $\dfrac{3}{50}$ (i) $\dfrac{1}{25}$

(j) $\dfrac{27}{64}$ in. (to the nearest thousandth of an inch)

ANSWERS

(a) 0.95   (b) 0.05   (c) 0.7   (d) 0.004   (e) 0.011

(f) 0.0327   (g) 0.473   (h) 0.06   (i) 0.04   (j) 0.422 in.

You will do many of these calculations more quickly if you remember that dividing by a multiple of ten is equivalent to shifting the decimal point to the left. To divide by a multiple of ten, move the decimal point to the left as many digits as there are zeros in the multiple of ten. For example,

and

HOW TO WRITE A REPEATING DECIMAL AS A FRACTION

A repeating decimal is one in which some sequence of digits is endlessly repeated. For example, $0.333\ldots = 0.\overline{3}$ and $0.272727\ldots = 0.\overline{27}$ are repeating decimals. The bar over the number is a shorthand way of showing that those digits are repeated.

What fraction is equal to $0.\overline{3}$? To answer this, form a fraction with numerator equal to the repeating digits and denominator equal to a number formed with the same number of 9s.

$0.\overline{3} = \frac{3}{9} = \frac{1}{3}$

$0.\overline{27} = \frac{27}{99} = \frac{3}{11}$ Two digits in $0.\overline{27}$; therefore, use 99 as the denominator.

$0.\overline{123} = \frac{123}{999} = \frac{41}{333}$ Three digits in $0.\overline{123}$; therefore, use 999 as the denominator.

This procedure works only when *all* of the decimal part repeats.

Now turn to Exercises 3-3 for a set of problems on decimal fractions.

Exercises 3-3

Decimal Fractions

A. Write as decimal numbers. (Round to two decimal digits if necessary.)

1. $\frac{1}{4}$	2. $\frac{2}{3}$	3. $\frac{3}{4}$	4. $\frac{2}{5}$
5. $\frac{4}{5}$	6. $\frac{5}{6}$	7. $\frac{2}{7}$	8. $\frac{4}{7}$
9. $\frac{6}{7}$	10. $\frac{3}{8}$	11. $\frac{6}{8}$	12. $\frac{1}{10}$
13. $\frac{3}{10}$	14. $\frac{2}{12}$	15. $\frac{5}{12}$	16. $\frac{3}{16}$
17. $\frac{6}{16}$	18. $\frac{9}{16}$	19. $\frac{13}{16}$	20. $\frac{3}{20}$
21. $\frac{7}{32}$	22. $\frac{13}{20}$	23. $\frac{11}{24}$	24. $\frac{39}{64}$
25. $\frac{3}{100}$	26. $\frac{213}{1000}$	27. $\frac{19}{1000}$	28. $\frac{17}{50}$

B. Calculate in decimal form. (Round to agree with the number of decimal digits in the decimal number.)

1. $2\frac{3}{5} + 1.785$	2. $\frac{1}{5} + 1.57$	3. $3\frac{7}{8} - 2.4$
4. $1\frac{3}{16} - 0.4194$	5. $2\frac{1}{2} \times 3.15$	6. $1\frac{3}{25} \times 2.08$
7. $3\frac{4}{5} \div 2.65$	8. $3.72 \div 1\frac{1}{4}$	9. $2.76 + \frac{7}{8}$
10. $16\frac{3}{4} - 5.842$	11. $3.14 \times 2\frac{7}{16}$	12. $1.17 \div 1\frac{3}{32}$

C. Practical Problems

1. **Pharmacy** If one tablet of calcium pantothenate contains 0.5 gram, how much is contained in $2\frac{3}{4}$ tablets? How many tablets are needed to make up 2.6 grams?

2. **Building Construction** Estimates of matched or tongue-and-groove (T & G) flooring—stock that has a tongue on one edge—must allow for the waste from milling. To allow for this waste, $\frac{1}{4}$ of the area to be covered must be added to the estimate when 1-in. by 4-in. flooring is used. If 1-in. by 6-in. flooring is used, $\frac{1}{6}$ of the area must be added to the estimate. (*Hint:* The floor is a rectangle. Area of a rectangle = length × width.)

   (a) A house has a floor size 28 ft long by 12 ft wide. How many square feet of 1-in. by 4-in. T & G flooring will be required to lay the floor?

   (b) A motel contains 12 units or rooms, and each room is 22 ft by 27 ft or 594 sq ft. In five rooms the builder is going to use 4-in. stock and in the other seven rooms 6-in. stock—both being T & G. How many square feet of each size will be used?

   (c) A contractor is building a motel with 24 rooms. Eight rooms will be 16 ft by 23 ft, nine rooms 18 ft by 26 ft, and the rest of the rooms 14 ft by 20 ft. How much did she pay for flooring, using 1-in. by 4-in. T & G at $3450 per 1000 sq ft?

3. **Carpentry** If a carpenter lays $10\frac{1}{2}$ squares of shingles in $4\frac{1}{2}$ days, how many squares does he do in one day?

4. **Machine Technology** A bearing journal measures 1.996 in. If the standard readings are in $\frac{1}{32}$-in. units, what is its probable standard size as a fraction?

5. **Upholstering** Complete the following invoice for upholstery fabric.

   (a) $5\frac{1}{4}$ yd @ $12.37 per yd = _____

   (b) $23\frac{3}{4}$ yd @ $16.25 per yd = _____

   (c) $31\frac{5}{6}$ yd @ $9.95 per yd = _____

   (d) $16\frac{2}{3}$ yd @ $17.75 per yd = _____

   Total _____

6. **Building Construction** A land developer purchased a piece of land containing 437.49 acres. He plans to divide it into a 45-acre recreational area and lots of $\frac{3}{4}$ acre each. How many lots will he be able to form from this piece of land?

Problem 7

7. **Machine Technology** Find the corner measurement A needed to make an octagonal end on a square bar as shown in the diagram.

8. **Auto Mechanics** A cylinder is normally $3\frac{7}{16}$ in. in diameter. It is rebored 0.040 in. larger. What is the size of the rebored cylinder? (Give your answer in decimal form.)

9. Plumbers deal with measurements in inches and common fractions of an inch, while surveyors use feet and decimal fractions of a foot. Often one trade needs to interpret the measurements of the other.

   (a) **Surveying** A drain has a run of 52 ft at a grade of $\frac{1}{8}$ in./ft. The high end of the drain has an elevation of 126.70 ft. What is the elevation at the low end?

   (b) **Plumbing** The elevation at one end of a lot is 84.25 ft, and the elevation at the other end is 71.70 ft. Express the difference in elevation in feet and inches and round to the nearest $\frac{1}{8}$ in.

10. **Plumbing** To fit a 45° connection, a plumber can approximate the diagonal length by multiplying the offset length by 1.414. Find the diagonal length of a 45° fitting if the offset length is $13\frac{11}{16}$ in. Express the answer as a fraction to the nearest $\frac{1}{16}$ in.

D. Calculator Problems

   1. **Sheet Metal Technology** In the table on page 136 the first column lists the U.S. Standard sheet metal gauge number. These tables were at one time used to give the equivalent sheet metal thickness in fractions of an inch, as shown in the second column. Today, these thicknesses are listed as decimal numbers. Complete column three to reflect this change. Round to the nearest thousandth of an inch.

   2. **Building Construction** Calculate the total cost of the following building materials.

8024 sq ft of flooring	@ $3248.75 per 1000 sq ft	_____
1825 ft of 16-ft siding	@ $1125.50 per 100 ft	_____
24,390 bricks	@ $328 per 1000	_____
19,200 sq ft of shingles	@ $42.95 per 100 sq ft	_____
	Total	_____

3. **Office Services** Machine helpers at the CALMAC Tool Co. earn $10.72 per hour. The company pays time and one-half for hours over 40 per week. Sunday hours are paid at double time. How much would you earn if you worked the following hours: Monday, 9; Tuesday, 8; Wednesday, $9\frac{3}{4}$; Thursday, 8; Friday, $10\frac{1}{2}$; Saturday, $5\frac{1}{4}$; Sunday, $4\frac{3}{4}$?

When you have finished these exercises, check your answers in the Appendix, and then turn to Problem Set 3 for practice working with decimal numbers.

Gauge No.	Fraction Thickness (in.)	Decimal Thickness (in.)	Gauge No.	Fraction Thickness (in.)	Decimal Thickness (in.)
7-0	$\frac{1}{2}$		14	$\frac{5}{64}$	
6-0	$\frac{15}{32}$		15	$\frac{9}{128}$	
5-0	$\frac{7}{16}$		16	$\frac{1}{16}$	
4-0	$\frac{13}{32}$		17	$\frac{9}{160}$	
3-0	$\frac{3}{8}$		18	$\frac{1}{20}$	
2-0	$\frac{11}{32}$		19	$\frac{7}{160}$	
0	$\frac{5}{16}$		20	$\frac{3}{80}$	
1	$\frac{9}{32}$		21	$\frac{11}{320}$	
2	$\frac{17}{64}$		22	$\frac{1}{32}$	
3	$\frac{1}{4}$		23	$\frac{9}{320}$	
4	$\frac{15}{64}$		24	$\frac{1}{40}$	
5	$\frac{7}{32}$		25	$\frac{7}{320}$	
6	$\frac{13}{64}$		26	$\frac{3}{160}$	
7	$\frac{3}{16}$		27	$\frac{11}{640}$	
8	$\frac{11}{64}$		28	$\frac{1}{64}$	
9	$\frac{5}{32}$		29	$\frac{9}{640}$	
10	$\frac{9}{64}$		30	$\frac{1}{80}$	
11	$\frac{1}{8}$		31	$\frac{7}{640}$	
12	$\frac{7}{64}$		32	$\frac{13}{1280}$	
13	$\frac{3}{32}$		—	—	—

Decimal Numbers

Answers are in the Appendix.

A. **Write in words.**

1. 0.91	**2.** 0.84	**3.** 23.164	**4.** 63.219
5. 9.3	**6.** 3.45	**7.** 10.06	**8.** 15.037

Write as decimal numbers.

9. Seven hundredths

10. Eighteen thousandths

11. Two hundred and eight tenths

12. Sixteen and seventeen hundredths

13. Sixty-three and sixty-three thousandths

14. One hundred ten and twenty-one thousandths

15. Five and sixty-three ten-thousandths

16. Eleven and two hundred eighteen ten-thousandths

B. **Add or subtract as shown.**

1. $4.39 + 18.8$	**2.** $18.8 + 156.16$
3. $\$7.52 + \$11.77$	**4.** $26 + 0.06$
5. $3.68 - 1.74$	**6.** $\$12.46 - \$8.51$
7. $104.06 - 15.80$	**8.** $16 - 3.45$
9. $264.3 + 12.804$	**10.** $0.232 + 5.079$
11. $165.4 + 73.61$	**12.** $245.94 + 7.07$
13. $116.7 - 32.82$	**14.** $4.07 - 0.085$
15. $0.42 + 1.452 + 31.8$	**16.** $\frac{1}{2} + 4.21$
17. $3\frac{1}{5} + 1.08$	**18.** $1\frac{1}{4} - 0.91$
19. $3.045 - 1\frac{1}{8}$	**20.** $8.1 + 0.47 - 1\frac{4}{5}$

Name _____

Date _____

Course/Section _____

C. Calculate as shown.

1. 0.004×0.02
2. 0.06×0.05
3. 1.4×0.6
4. 3.14×12
5. $0.2 \times 0.6 \times 0.9$
6. 5.3×0.4
7. 6.02×3.3
8. $3.224 \div 2.6$
9. $187.568 \div 3.04$
10. $0.078 \div 0.3$
11. $0.6 \times 3.15 \times 2.04$
12. $3.78 + 4.1 \times 6.05$
13. $0.008 - 0.001 \div 0.5$
14. $3.1 \times 4.6 - 2.7 \div 0.3$
15. $\dfrac{4.2 + 4.6 \times 1.2}{2.73 \div 21 + 0.41}$
16. $\dfrac{7.2 - 3.25 \div 1.3}{3.5 + 5.7 \times 4.0 - 2.8}$
17. $1.2 + 0.7 \times 2.2 + 1.6$
18. $1.2 \times 0.7 + 2.2 \times 1.6$

Round to two decimal digits.

19. $0.007 \div 0.03$
20. $3.005 \div 2.01$
21. $17.8 \div 6.4$
22. $0.0041 \div 0.019$
23. 0.0371
24. 16.8449

Round to three decimal places.

25. $0.04 \div 0.076$
26. $234.1 \div 465.8$
27. $17.6 \div 0.082$
28. $0.051 \div 1.83$
29. 27.0072
30. 1.1818

Round to the nearest tenth.

31. $0.08 \div 0.053$
32. $3.05 \div 0.13$
33. $18.76 \div 4.05$
34. $0.91 \div 0.97$
35. 47.233
36. 123.7666

D. Write as a decimal number.

1. $\frac{1}{16}$
2. $\frac{7}{8}$
3. $\frac{5}{32}$
4. $\frac{7}{16}$
5. $\frac{11}{8}$
6. $1\frac{3}{4}$
7. $2\frac{11}{20}$
8. $1\frac{1}{6}$
9. $2\frac{2}{3}$
10. $1\frac{1}{32}$
11. $2\frac{5}{16}$

Write in decimal form, calculate, round to two decimal digits.

12. $4.82 \div \frac{1}{4}$
13. $11.5 \div \frac{3}{8}$
14. $1\frac{3}{16} \div 0.62$
15. $2\frac{3}{4} \div 0.035$
16. $0.45 \times 2\frac{1}{8}$
17. $0.068 \times 1\frac{7}{8}$

E. Practical Problems

1. **Machine Technology**   A $\frac{5}{16}$-in. bolt weighs 0.43 lb. How many bolts are there in a 125-lb keg?

2. **Metalworking**   Twelve equally spaced holes are to be drilled in a metal strip $34\frac{1}{4}$-in. long, with 2 in. remaining on each end. What is the distance, to the nearest hundredth of an inch, from center to center of two consecutive holes?

3. **Plumbing**   A plumber finds the length of pipe needed to complete a bend by performing the following multiplication:

   Length needed = 0.01745 × (radius of the bend) × (angle of bend)

   What length of pipe is needed to complete a 35° bend with a radius of 16 in.?

4. **Printing**   During the first six months of the year the Busy Bee Printing Co. spent the following amounts on paper:

January $870.16	February $590.12	March $376.77
April $400.09	May $906.45	June $389.21

   What is the average monthly expenditure for paper?

5. **Machine Technology**   The *feed* of a drill is the distance it advances with each revolution. If a drill makes 310 rpm and drills a hole 2.125 in. deep in 3.25 minutes, what is the feed? Round to four decimal places.

   $$\left(Hint: \text{Feed} = \frac{\text{total depth}}{(\text{rpm}) \times (\text{time of drilling in minutes})}. \right)$$

6. **Electrical Technology**   The resistance of an armature while it is cold is 0.208 ohm. After running for several minutes, the resistance increases to 1.340 ohms. Find the increase in resistance of the armature.

7. **Painting and Decorating**   The Ace Place Paint Co. sells paint in steel drums each weighing 36.4 lb empty. If 1 gallon of paint weighs 9.06 lb, what is the total weight of a 50-gallon drum of paint?

8. **Electrical Technology**   The diameter of No. 12 bare copper wire is 0.08081 in., and the diameter of No. 15 bare copper wire is 0.05707 in. How much larger is No. 12 wire than No. 15 wire?

9. **Machine Technology**   What is the decimal equivalent of each of the following drill bit diameters?

   (a) $\frac{3}{16}$ in.   (b) $\frac{5}{32}$ in.   (c) $\frac{3}{8}$ in.   (d) $\frac{13}{64}$ in.

10. What is the closest fractional equivalent in 16ths of an inch of the following dimensions? (*Hint:* Use the table on page 132.)

    (a) 0.185 in.   (b) 0.127 in.   (c) 0.313 in.   (d) 0.805 in.

11. **Sheet Metal Technology**   The following table lists the thickness in inches of several sizes of sheet steel:

U.S. Gauge	Thickness (in.)
35	0.0075
30	0.0120
25	0.0209
20	0.0359
15	0.0673
10	0.1345
5	0.2092

(a) What is the difference in thickness between 30 gauge and 25 gauge sheet?

(b) What is the difference in thickness between seven sheets of 25 gauge and four sheets of 20 gauge sheet?

(c) What length of $\frac{3}{16}$-in.-diameter rivet is needed to join one thickness of 25 gauge sheet to a strip of $\frac{1}{4}$-in. stock? Add $1\frac{1}{2}$ times the diameter of the rivet to the length of the rivet to ensure that the rivet is long enough that a proper rivet head can be formed.

(d) What length of $\frac{5}{32}$-in. rivet is needed to join two sheets of 20 gauge sheet steel? (Don't forget to add the $1\frac{1}{2}$ times rivet diameter.)

12. **Building Construction** When the owners of the Better Builder Co. completed a small construction job, they found that the following expenses had been incurred: labor, $972.25; gravel, $86.77; sand, $39.41; cement, $280.96; and bricks, $2204.35. What total bill should they give the customer if they want to make $225 profit on the job?

13. **Building Construction** One lineal foot of 12-in. I-beam weighs 25.4 lb. What is the length of a beam that weighs 444.5 lb?

14. **Sheet Metal Technology** To determine the average thickness of a metal sheet, a sheet metal worker measures it at five different locations. His measurements are 0.0401, 0.0417, 0.0462, 0.0407, and 0.0428 in. What is the average thickness of the sheet?

15. **Drafting** A dimension in a technical drawing is given as $2\frac{1}{2}$ in. with a tolerance of ±0.025 in. What are the maximum and minimum permissible dimensions written in decimal form?

16. **Auto Mechanics** The four employees of the Busted Body Shop earned the following amounts last week: $811.76, $796.21, $808.18, and $876.35. What is the average weekly pay for the employees of the shop?

17. **Electrical Technology** If 14-gauge Romex cable sells for 19 cents per foot, what is the cost of 1210 ft of this cable?

18. **Electrical Technology** In Lucy's first week as an electrician, she earned $1593.75 for 37.5 hours of work. What is her hourly rate of pay?

19. **Welding** A welder finds that 2.1 cu ft of acetylene gas is needed to make one bracket. How much gas will be needed to make 27 brackets?

Problem 20

20. **Auto Mechanics** What is the outside diameter d of a 6.50 × 14 tire? (See the figure.)

21. **Metalworking** How many 2.34-in. spacer blocks can be cut from a 2-in. by 2-in. square bar 48 in. long? Allow $\frac{1}{4}$ in. waste for each saw cut.

22. **Auto Mechanics** At the Fixum Auto Shop, a mechanic is paid $22.50 per hour for a motor overhaul and is allotted 23 hours of labor time for the job. The mechanic completes that job in 18.5 hours, but he is still paid as if he had worked for 23 hours. Calculate his actual hourly compensation under these circumstances.

23. **Machine Technology** A machinist estimates the following times for fabricating a certain part: 0.6 hour for setup, 2.4 hours of turning, 5.2 hours of milling, 1.4 hours of grinding, and 1.8 hours of drilling. What is the total time needed to make the part?

24. **Printing** A 614-page book was printed on paper specified as 0.00175 in. thick and finished with a cover 0.165 in. thick. What was the total thickness of the bound book? (Be sure to count the cover twice and remember that there are two pages of the book for every one sheet of paper.)

140

25. **Printing** A certain type sets at an average of 14 characters to the inch. How many lines will it take to set 4325 characters to a width of 6 in.?

26. **Fire Protection** One gallon of water weighs approximately 8.34 lb. When sent to a particularly rough terrain, a Forestry Service truck is allowed to carry only 1.5 tons of weight. How many gallons of water can it carry? (1 ton = 2000 pounds.)

27. **Plumbing** To construct a certain pipe system, the following material is needed:

6 90° elbows at $1.19 each
5 tees at $0.74 each
3 couplings at $0.68 each
4 pieces of pipe each 6 ft long
6 pieces of pipe each 2 ft long
4 pieces of pipe each 1.5 ft long

The pipe is schedule 40 1-in. PVC pipe, costing $0.22 per linear foot. Find the total cost of the material for the system.

28. **Masonry** A tile setter purchases the following supplies for the day:

One bag of thin-set mortar @	$5.97 per bag.
44 sq ft of tile @	$2.25 per sq ft
One gallon of grout @	$16.97 per gallon
2 tubes of caulk @	$2.89 per tube
One container of grout sealer @	$24.77 per container
3 containers of grout and tile cleaner @	$5.99 per container
4 scrub pads @	$2.57 each
One trowel @	$3.65 each
2 packages of tile spacers @	$1.97 each
One grout bag @	$2.49 each
One grout float @	$9.45 each

What is the total cost of these items before tax is added?

29. **Auto Mechanics** One way in which the owner of an auto shop evaluates the success of his business is to keep track of "production per square foot." To calculate this indicator, he divides total revenue by the area (in square feet) of his shop. Last year, a 2260-sq-ft shop generated total revenue of $586,435. This year, the shop expanded to 2840 sq ft and generated total revenue of $734,592. During which year was the production per square foot the greater and by how much?

30. **Plumbing** The inside diameter (I.D.) and outside diameter (O.D.) of a pipe are shown in the figure. The wall thickness of the pipe is the dimension labeled t. Calculate the wall thickness of schedule 120 pipe if its I.D. is 0.599 in. and its O.D. is 1.315 in.

Problem 30

Using a Calculator, III: Decimals To perform addition, subtraction, multiplication, and division of decimal numbers on a calculator, use the same procedures outlined in Chapter 1, and press the $\boxed{\cdot}$ key to enter the decimal point.

EXAMPLE

To calculate 243.78 + 196.1 × 2.75 use the keystroke sequence

243 $\boxed{\cdot}$ **78** $\boxed{+}$ **196** $\boxed{\cdot}$ **1** $\boxed{\times}$ **2** $\boxed{\cdot}$ **75** $\boxed{=}$ → $\boxed{783.055}$

The decimal point key was shown as a special symbol here for emphasis, but in future calculator sequences in this book we will save space by not showing the decimal point as a separate key.

Notice that in the preceding calculation, the numbers and operations were entered, and the calculator followed the standard order of operations.

When you work with decimal numbers, remember that for a calculator with a ten-digit display if an answer is a nonterminating decimal or a terminating decimal with more than ten digits, the calculator will display a number rounded to ten digits.* Knowledge of rounding is especially important in such cases because the result must often be rounded further.

YOUR TURN

Work the following problems for practice and round as indicated.

(a) $28.75 + $161.49 - 37.60

(b) 2.8×0.85 (Round to the nearest tenth.)

(c) $2347.68 \div 12.9$ (Round to the nearest whole number.)

(d) $46.8 - 27.3 \times 0.49$ (Round to the nearest hundredth.)

(e) $\dfrac{16500 + 3700}{12 \times 68}$ (Round to the nearest tenth.)

(f) $(3247.9 + 868.7) \div 0.816$ (Round to the nearest ten.)

(g) $6 \div 0.07 \times 0.8 + 900$ (Round to the nearest hundred.)

ANSWERS

(a) **28.75** $\boxed{+}$ **161.49** $\boxed{-}$ **37.6** $\boxed{=}$ → $\boxed{152.64}$

Final zeros to the right of the decimal point may be omitted.

(b) **2.8** $\boxed{\times}$ **.85** $\boxed{=}$ $\boxed{2.38}$ or 2.4   rounded

We need not enter the leading zero in 0.85

(c) **2347.68** $\boxed{\div}$ **12.9** $\boxed{=}$ → $\boxed{181.9906977}$ or 182   rounded

(d) **46.8** $\boxed{-}$ **27.3** $\boxed{\times}$ **.49** $\boxed{=}$ → $\boxed{33.423}$ or 33.42   rounded

(e) **16500** $\boxed{+}$ **3700** $\boxed{=}$ $\boxed{\div}$ $\boxed{(}$ **12** $\boxed{\times}$ **68** $\boxed{)}$ → $\boxed{24.75490196}$ or 24.8 rounded

(f) **3247.9** $\boxed{+}$ **868.7** $\boxed{=}$ $\boxed{\div}$ **.816** $\boxed{=}$ → $\boxed{5044.852941}$ or 5040   rounded

(g) **6** $\boxed{\div}$ **.07** $\boxed{\times}$ **.8** $\boxed{+}$ **900** $\boxed{=}$ → $\boxed{968.5714286}$ or 1000   rounded

---

*Some calculators will truncate an answer instead of rounding it. This means that they will show the first ten digits of the result. Be sure you know which type of calculator you have.

Problem Set Chapter 3 Decimal Numbers

PREVIEW

4

Ratio, Proportion, and Percent

Objective	Sample Problems		For Help Go to Page

When you finish this unit you will be able to:

1. Calculate ratios.

    (a) Find the ratio of the pulley diameters.

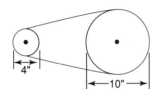

                                                       147

    (b) **Auto Mechanics** A small gasoline engine has a maximum cylinder volume of 520 cu cm and a compressed volume of 60 cu cm. Find the compression ratio. _____ 148

2. Solve proportions.

    (a) Solve for x.

$$\frac{8}{x} = \frac{12}{15}$$
                                                       151

    (b) Solve for y.

$$\frac{4.4}{2.8} = \frac{y}{9.1}$$

3. Solve problems involving proportions.

    (a) **Landscaping** A mixture of plant food must be prepared by combining three parts of a concentrate with every 16 parts water. How much water should be added to 12 ounces of concentrate? _____ 155

    (b) **Drafting** An architectural drawing of the living room of a house is $5\frac{1}{2}$ in. long and $4\frac{1}{2}$ in. wide. If the actual length of the living room is 22 ft, what is the actual width? _____ 198

(c) **Manufacturing** A large production job is completed by four machines in 6 hours. The same size job must be finished in 2 hours the next time it is ordered. How many machines should be used to accomplish this task? _____ 203

4. Write fractions and decimal numbers as percents.

    (a) Write $\frac{1}{4}$ as a percent. _____ 162

    (b) Write 0.46 as a percent. _____ 161

    (c) Write 5 as a percent. _____

    (d) Write 0.075 as a percent. _____

5. Convert percents to decimal numbers and fractions.

    (a) Write 35% as a decimal number. _____ 164

    (b) Write 0.25% as a decimal number. _____

    (c) Write 112% as a fraction. _____ 165

6. Solve problems involving percent.

    (a) Find $37\frac{1}{2}\%$ of 600. _____ 168

    (b) Find 120% of 45. _____

    (c) What percent of 80 is 5? _____ 171

    (d) 12 is 16% of what number? _____ 172

    (e) If a measurement is given as 2.778 ± 0.025 in., state the tolerance as a percent. _____ 188

    (f) **Machine Technology** If a certain kind of solder is 52% tin, how many pounds of tin are needed to make 20 lb of solder? _____

    (g) **Painting and Decorating** The paint needed for a redecorating job originally cost $144.75, but is on sale at 35% off. What is its sale price? _____ 177

    (h) **Machine Technology** A mower motor rated at 2.0 hp is found to deliver only 1.6 hp when connected to a transmission system. What is the efficiency of the transmission? _____ 187

(i) **Electronics** If the voltage in a circuit is increased from 70 volts to 78 volts, what is the percent increase in voltage? _____ 190

(Answers to these preview problems are given in the Appendix.)

If you are certain that you can work *all* these problems correctly, turn to page 213 for a set of practice problems. If you cannot work one or more of the preview problems, turn to the page indicated to the right of the problem. Those who wish to master this material with the greatest success should turn to Section 4-1 and begin work there.

4 Ratio, Proportion, and Percent

Reprinted by permission: Tribune Media Services

4-1 RATIO AND PROPORTION

Machinists, mechanics, carpenters, and other trades workers use the ideas of ratio and proportion to solve very many technical problems. The compression ratio of an automobile engine, the gear ratio of a machine, scale drawings, the pitch of a roof, the mechanical advantage of a pulley system, and the voltage ratio in a transformer are all practical examples of the ratio concept.

Ratio A *ratio* is a comparison, using division, of two quantities of the same kind, both expressed in same units. For example, the steepness of a hill can be written as the ratio of its height to its horizontal extent.

Steepness $= \frac{10 \text{ ft}}{80 \text{ ft}} = \frac{1}{8}$

The ratio is usually written as a fraction in lowest terms, and you would read this ratio as either "one-eighth" or "one to eight."

The *gear ratio* of a gear system is defined as

$$\text{Gear ratio} = \frac{\text{number of teeth on the driving gear}}{\text{number of teeth on the driven gear}}$$

EXAMPLE

Find the gear ratio of the system shown if A is the driving gear and B is the driven gear, and where A has 64 teeth and B has 16 teeth.

Gear ratio = ?

The ratio of the number of teeth on the driving gear to the number of teeth on the driven gear is

$$\text{Gear ratio} = \frac{64 \text{ teeth}}{16 \text{ teeth}} = \frac{4}{1}$$

Always reduce the fraction to lowest terms.

The gear ratio is 4 to 1. In technical work this is sometimes written as 4:1 and is read "four to one."

Typical gear ratios on a passenger car are

First gear:	3.54:1
Second gear:	1.90:1
Third gear:	1.31:1
Reverse:	3.25:1

A gear ratio of 3.54 means that the engine turns 3.54 revolutions for each revolution of the drive shaft. If the drive or rear-axle ratio is 3.72, the engine needs to make 3.54×3.72 or approximately 13.2 revolutions in first gear to turn the wheels one full turn.

Here are a few important examples of the use of ratios in practical work.

Pulley Ratios A pulley is a device that can be used to transfer power from one system to another. A pulley system can be used to lift heavy objects in a shop or to connect a power source to a piece of machinery. The ratio of the pulley diameters will determine the relative pulley speeds.

EXAMPLE

Find the ratio of the diameter of pulley A to the diameter of pulley B in the following drawing.

$$\frac{\text{Diameter of pulley } A}{\text{Diameter of pulley } B} = \frac{21 \text{ in.}}{9 \text{ in.}} \quad \Longleftarrow \boxed{\text{Same units}}$$

$$= \frac{7}{3}$$

The ratio is 7 to 3 or 7:3.

Notice that the units, inches, cancel from the ratio. A ratio is a fraction or decimal number, and it has no units.

Compression Ratio In an automobile engine there is a large difference between the volume of the cylinder space when a piston is at the bottom of its stroke and when it is at the top of its stroke. This difference in volumes is called the *engine displacement*. Automotive mechanics find it very useful to talk about the compression ratio of an automobile engine. The *compression ratio* of an engine compares the volume of the cylinder at maximum expansion to the volume of the cylinder at maximum compression.

$$\text{Compression ratio} = \frac{\text{expanded volume}}{\text{compressed volume}}$$

Expanded   Compressed

EXAMPLE

Find the compression ratio of a gasoline engine if each cylinder has a maximum volume of 475 cu cm and a minimum or compression volume of 50 cu cm.

$$\text{Compression ratio} = \frac{475 \text{ cu cm}}{50 \text{ cu cm}} = \frac{19}{2}$$

$$= 9.5$$

Compression ratios are always written so that the second number in the ratio is 1. This compression ratio would be written as $9\frac{1}{2}$ to 1.

YOUR TURN

Now, for some practice in calculating ratios, work the following problems.

(a)

Ratio of pulley diameters $= \dfrac{A}{B} =$

(b)

Gear ratio $=$

(c) **Roofing**

Pitch $= \dfrac{\text{rise}}{\text{run}} =$

(d) **Auto Mechanics** Find the rear-axle ratio of a car if the ring gear has 54 teeth and the pinion gear has 18 teeth.

$$\text{Rear-axle ratio} = \frac{\text{number of teeth on ring gear}}{\text{number of teeth on pinion gear}}$$

(e) **Auto Mechanics** A gasoline engine has a maximum cylinder volume of 47 cu in. and a compressed volume of 5.0 cu in. Find its compression ratio.

(f) **Stockbroker** Stock in Amazing.com sells for $18 per share. The company has earnings of $0.50 per share. Find its P-E or price to earnings ratio.

SOLUTIONS

(a) Ratio of pulley diameters $= \dfrac{24 \text{ cm}}{6 \text{ cm}}$ or   4 to 1.

(b) Gear ratio $= \dfrac{60 \text{ teeth}}{8 \text{ teeth}} = \dfrac{15}{2}$ or   15 to 2.

(c) Pitch $= \dfrac{6\text{-ft rise}}{24\text{-ft run}} = \dfrac{6}{24}$ or $\dfrac{3}{12}$.

The pitch is 3:12. Recall that roofers usually express pitch with a denominator of 12.

(d) Rear-axle ratio $= \dfrac{54 \text{ teeth}}{18 \text{ teeth}} = \dfrac{3}{1}$.

The rear-axle ratio is 3 to 1.

(e) Compression ratio $= \dfrac{47 \text{ cu in.}}{5.0 \text{ cu in.}} = \dfrac{47}{5.0}$ or   9.4 to 1.

(f) P-E $= \dfrac{\$18}{\$0.50} = \dfrac{\$18 \times 2}{\$0.50 \times 2} = \dfrac{\$36}{\$1}$ or   36 to 1.

Simple Equations In order to use ratios to solve a variety of problems, you must be able to solve a simple kind of algebraic equation. Consider this puzzle: "I'm thinking of a number. When I multiply my number by 3, I get 15. What is my number?" Solve the puzzle and check your solution with ours.

If you answered "five," you're correct. Think about how you worked it out. Most people take the answer 15 and do the "reverse" of multiplying by 3. That is, they divide by 3. In symbols the problem would look like this:

If $3 \times \square = 15$

then $\square = 15 \div 3 = 5$

EXAMPLE

Try another one: A number multiplied by 40 gives 2200. Find the number.

Using symbols again,

$40 \times \square = 2200$

so $\square = 2200 \div 40 = \dfrac{2200}{40} = 55$

You have just solved a simple algebraic equation. But in algebra, instead of drawing boxes to represent unknown quantities, we use letters of the alphabet. Instead of using the $\times$ symbol for multiplication, in algebra we use a raised dot $\cdot$, or we simply write the multiplier next to the letter.

> ### EXAMPLE

$$40 \times \square = 2200$$

can be written $\quad 40 \cdot N = 2200$

$$\text{or } 40N = 2200$$

To solve this type of equation, you divide 2200 by the multiplier of N, which is 40.

> ### YOUR TURN

For practice, solve these equations.

(a) $25n = 275$ (b) $6M = 96$

(c) $100x = 550$ (d) $12y = 99$

(e) $88 = 8P$ (f) $15 = 30a$

> ### SOLUTIONS

(a) $n = \dfrac{275}{25} = 11$ (b) $M = \dfrac{96}{6} = 16$

(c) $x = \dfrac{550}{100} = 5.5$ (d) $y = \dfrac{99}{12} = 8.25$

(e) Notice that P, the unknown quantity, is on the *right* side of the equation. You must divide the number by the multiplier of the unknown, so

$$P = \frac{88}{8} = 11$$

(f) Again, a, the unknown quantity, is on the right side, so

$$a = \frac{15}{30} = 0.5$$

Careful ▶   In problems like (f) some students mistakenly think that they must always divide the larger number by the smaller. That is not always correct. Remember, for an equation of this simple form, always divide by the number that multiplies the unknown. ◀

Other types of equations will be solved in Chapter 7, but only this type is needed for our work with proportions and percent.

Proportions   A *proportion* is an equation stating that two ratios are equal. For example,

$$\frac{1}{3} = \frac{4}{12} \quad \text{is a proportion}$$

Notice that the equation is true because the fraction $\frac{4}{12}$ expressed in lowest terms is equal to $\frac{1}{3}$.

Note ▶ When you are working with proportions, it is helpful to have a way of stating them in words. For example, the proportion

$$\frac{2}{5} = \frac{6}{15} \quad \text{can be stated in words as}$$

"Two is to five as six is to fifteen." ◀

When one of the four numbers in a proportion is unknown, it is possible to find the value of that number. In the proportion

$$\frac{1}{3} = \frac{4}{12}$$

notice what happens when we multiply diagonally:

$$\frac{1}{3} = \frac{4}{12} \rightarrow 1 \cdot 12 = 12 \qquad\qquad \frac{1}{3} = \frac{4}{12} \rightarrow 3 \cdot 4 = 12$$

These diagonal products are called the *cross-products* of the proportion. If the proportion is a true statement, the cross-products will always be equal. Here are more examples:

$$\frac{5}{8} = \frac{10}{16} \rightarrow 5 \cdot 16 = 80 \text{ and } 8 \cdot 10 = 80$$

$$\frac{9}{12} = \frac{3}{4} \rightarrow 9 \cdot 4 = 36 \text{ and } 12 \cdot 3 = 36$$

$$\frac{10}{6} = \frac{5}{3} \rightarrow 10 \cdot 3 = 30 \text{ and } 6 \cdot 5 = 30$$

This very important fact is called the cross-product rule.

THE CROSS-PRODUCT RULE

If $\frac{a}{b} = \frac{c}{d}$ then $a \cdot d = b \cdot c$.

EXAMPLE

Find the cross-products for the proportion $\frac{3}{5} = \frac{12}{20}$.

The cross-products are $3 \cdot 20 = 60$ and $5 \cdot 12 = 60$.

The cross-product rule can be used to solve proportions, that is, to find the value of an unknown number in the proportion. For example, in the proportion

$$\frac{x}{4} = \frac{12}{16}$$

to solve the proportion means to find the value of the unknown quantity x that makes the equation true. To do this:

First, use the cross-product rule.

If $\quad \dfrac{x}{4} = \dfrac{12}{16} \quad$ then $\quad 16 \cdot x = 4 \cdot 12 \quad$ or $\quad 16x = 48$

Then, solve this equation using the division technique.

$$x = \dfrac{48}{16} = 3$$

☑ **Finally,** check your answer by replacing x with 3 in the original proportion equation.

$$\dfrac{3}{4} = \dfrac{12}{16}$$

Check 1: Find the cross-products.

$$3 \cdot 16 = 48 \qquad \text{and} \qquad 4 \cdot 12 = 48 \qquad \text{The answer is correct.}$$

Check 2: Write $\frac{12}{16}$ in lowest terms.

$$\dfrac{12}{16} = \dfrac{4 \cdot 3}{4 \cdot 4} = \dfrac{3}{4}$$

YOUR TURN

Now try this one:

$$\dfrac{6}{N} = \dfrac{15}{10} \qquad \text{Solve this proportion for } N.$$

SOLUTION

Step 1 Apply the cross-product rule.

$$\text{If } \dfrac{6}{N} = \dfrac{15}{10} \quad \text{then} \quad 6 \cdot 10 = 15N$$
$$\text{or} \quad 15N = 60$$

Step 2 Solve the equation.

$$N = \dfrac{60}{15} = 4$$

 Step 3 Substitute 4 for N in the original proportion.

$$\dfrac{6}{4} = \dfrac{15}{10}$$

The cross-products are equal: $6 \cdot 10 = 60$ and $4 \cdot 15 = 60$.

Notice that, in lowest terms,

$$\dfrac{6}{4} = \dfrac{3}{2} \quad \text{and} \quad \dfrac{15}{10} = \dfrac{3}{2}$$

Note ▶ The unknown quantity can appear in any one of the four positions in a proportion. No matter where the unknown appears, solve the proportion the same way: write the cross-products and solve the resulting equation. ◀

PROBLEMS

Here are more practice problems. Solve each proportion.

(a) $\dfrac{28}{40} = \dfrac{x}{100}$ (b) $\dfrac{12}{y} = \dfrac{8}{50}$ (c) $\dfrac{n}{6} = \dfrac{7}{21}$

(d) $\dfrac{12}{9} = \dfrac{32}{M}$ (e) $\dfrac{Y}{7} = \dfrac{3}{4}$ (f) $\dfrac{6}{5} = \dfrac{2}{T}$

(g) $\dfrac{2\frac{1}{2}}{3\frac{1}{2}} = \dfrac{w}{2}$ (h) $\dfrac{12}{E} = \dfrac{0.4}{1.5}$

ANSWERS

(a) $\dfrac{28}{40} = \dfrac{x}{100}$ $28 \cdot 100 = 40 \cdot x$ or $40x = 2800$

$x = 2800 \div 40$

$x = 70$

(b) $\dfrac{12}{y} = \dfrac{8}{50}$ $600 = 8y$ or $y = \dfrac{600}{8} = 75$

(c) $\dfrac{n}{6} = \dfrac{7}{21}$ $21n = 42$ $n = \dfrac{42}{21} = 2$

(d) $\dfrac{12}{9} = \dfrac{32}{M}$ $12M = 288$ $M = \dfrac{288}{12} = 24$

(e) $\dfrac{Y}{7} = \dfrac{3}{4}$ $4Y = 21$ $Y = \dfrac{21}{4} = 5\frac{1}{4}$ or 5.25

(f) $\dfrac{6}{5} = \dfrac{2}{T}$ $6T = 10$ $T = \dfrac{10}{6} = 1\frac{2}{3}$

(g) $\dfrac{2\frac{1}{2}}{3\frac{1}{2}} = \dfrac{w}{2}$ $3.5w = 5$ $w = \dfrac{5}{3.5} = 1\frac{3}{7}$

(h) $\dfrac{12}{E} = \dfrac{0.4}{1.5}$ $18 = 0.4E$ $E = \dfrac{18}{0.4} = 45$

Note ▶ You may have noticed that the two steps for solving a proportion can be simplified into one step. For example, in problem (a), to solve

$\dfrac{28}{40} = \dfrac{x}{100}$ we can write the answer directly as

$x = \dfrac{28 \cdot 100}{40}$

Always divide by the number that is diagonally opposite the unknown, and multiply the two remaining numbers that are diagonally opposite each other.

This shortcut comes in handy when using a calculator. For example, with a calculator, problem (h) becomes

12 ⊠ **1.5** ⊡ **.4** ⊟ →   **45.**

The numbers 12 and 1.5 are diagonally opposite each other in the proportion.

0.4 is diagonally opposite the unknown E. ◀

Practice this one-step process by solving the proportion

$$\frac{120}{25} = \frac{6}{Q}$$

SOLUTION

$$Q = \frac{6 \cdot 25}{120} = 1.25 \qquad \text{Check it.}$$

Problem Solving Using Proportions

We can use proportions to solve a variety of problems involving ratios.

If you are given the value of a ratio and one of its terms, it is possible to find the other term. For example, if the pitch of a roof is supposed to be 2:12, and the run is 20 ft, what must be the rise?

Rise

Run

$$\text{Pitch} = \frac{\text{rise}}{\text{run}}$$

A ratio of 2:12 is equivalent to the fraction $\frac{2}{12}$; therefore, the equation becomes

$$\frac{2}{12} = \frac{\text{rise}}{20}$$

or

$$\frac{2}{12} = \frac{R}{20}$$

Find the cross-products.

$$12R = 40$$

Then solve for R.

$$R = \frac{40}{12}$$

$$R = 3\frac{1}{3} \text{ ft} \qquad \text{The rise is 3 ft 4 inches.} \ \left(\frac{1}{3} \text{ ft} = 4 \text{ in.}\right)$$

The algebra you learned earlier in this chapter will enable you to solve any ratio problem of this kind.

YOUR TURN

Try these problems.

(a) **Manufacturing**   If the gear ratio on a mixing machine is 6:1 and the driven gear has 12 teeth, how many teeth are on the driving gear?

(b) **Manufacturing**   The pulley system of an assembly belt has a pulley diameter ratio of 4. If the larger pulley has a diameter of 15 in., what is the diameter of the smaller pulley?

(c) **Auto Mechanics**   The compression ratio of a classic Datsun 280Z is 8.3 to 1. If the compressed volume of the cylinder is 36 cu cm, what is the expanded volume of the cylinder?

(d) **Construction Technology** On a certain construction job, concrete is made using a volume ratio of 1 part cement to $2\frac{1}{2}$ parts sand and 4 parts gravel. How much sand should be mixed with 3 cu ft of cement?

(e) **Sports Technology** Manager Sparky Spittoon of the Huntville Hackers wants his ace pitcher Lefty Groove to improve his strikeouts-to-walks ratio to at least 5:2. If he has 65 strikeouts so far this season, what is the maximum number of walks he should have to keep Sparky happy?

(f) **Roofing** If the pitch of a roof must be 5:12, how much rise should there be over a run of 54 ft?

SOLUTIONS

(a) Gear ratio $= \dfrac{\text{number of teeth on driving gear}}{\text{number of teeth on driven gear}}$

A ratio of 6:1 is equivalent to the fraction $\dfrac{6}{1}$.

Therefore,

$$\frac{6}{1} = \frac{x}{12}$$

or $72 = x$ The driving gear has 72 teeth.

(b) Pulley ratio $= \dfrac{\text{diameter of larger pulley}}{\text{diameter of smaller pulley}}$

$$4 = \frac{15 \text{ in.}}{D} \qquad \text{or} \qquad \frac{4}{1} = \frac{15}{D}$$

$$4D = 15$$

$$D = \frac{15}{4} \qquad \text{or} \qquad D = 3\frac{3}{4} \text{ in.}$$

(c) Compression ratio $= \dfrac{\text{expanded volume}}{\text{compressed volume}}$

$$8.3 = \frac{V}{36 \text{ cu cm}} \qquad \text{or} \qquad \frac{83}{10} = \frac{V}{36}$$

$$V = \frac{83 \cdot 36}{10} = 298.8 \text{ cu cm}$$

 83 ⊗ 36 ⊙ 10 ⊜ → ▨▨▨▨ *298.8*

In any practical situation this answer would be rounded to 300 cu cm.

(d) Ratio of cement to sand $= \dfrac{\text{volume of cement}}{\text{volume of sand}}$

$$\frac{1}{2\frac{1}{2}} = \frac{3 \text{ cu ft}}{S}$$

Therefore,

$$S = 3 \cdot 2\frac{1}{2}$$

$$S = 7\frac{1}{2} \text{ cu ft}$$

(e) $\dfrac{\text{Strikeouts}}{\text{Walks}} = \dfrac{5}{2} = \dfrac{65}{x}$

$$5x = 130$$

$$x = 26 \qquad \text{Lefty should have no more than 26 walks.}$$

(f) $\text{Pitch} = \dfrac{\text{Rise}}{\text{Run}}$

$$\dfrac{5}{12} = \dfrac{R}{54}$$

$$12R = 270$$

$$R = 22.5\ \text{ft}$$

Exercises 4-1

Ratio and Proportion

A. Complete the following tables.

1.

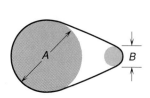

	Teeth on Driving Gear A	Teeth on Driven Gear B	Gear Ratio, $\dfrac{A}{B}$
(a)	35	5	*7-1*
(b)	12	7	*1.71+3=51 12-51*
(c)	*6*	3	2 to 1
(d)	21	*6*	$3\frac{1}{2}$ to 1
(e)	15	*45*	1 to 3
(f)	*9*	18	1 to 2
(g)	*16*	24	2:3
(h)	30	*50*	3:5
(i)	27	18	*1.5 -1*
(j)	12	30	*0.4 -1*

2.

	Diameter of Pulley A	Diameter of Pulley B	Pulley Ratio, $\dfrac{A}{B}$
(a)	16 in.	6 in.	*2.666) - 1*
(b)	15 in.	12 in.	*1.25 - 1*
(c)	*16*	8 in.	2 to 1
(d)	27 cm	*6*	4.5 to 1
(e)	*40*	10 cm	4 to 1
(f)	$8\frac{1}{8}$ in.	$3\frac{1}{4}$ in.	*2.5 - 1*
(g)	8.46 cm	11.28 cm	
(h)	20.41 cm	*6.5*	3.14 to 1
(i)	*5.4*	12.15 cm	1 to 2.25
(j)	4.45 cm	*17.8*	0.25 to 1

3.

	Rise	Run	Pitch
(a)	8 ft	6 ft	1.2333 ~1
(b)	8	24 ft	4:12
(c)	7 ft	28	3:12
(d)	14 ft 4 in.	25 ft	0.5733 ~1
(e)	9 ft	15 ft	0.6 ~1
(f)	4	20 ft	2.4:12
(g)	3 ft	20	1.8:12
(h)	~~30.5~~ 5.0833	30 ft 6 in.	2:12

$\dfrac{x}{30.5} = \dfrac{2}{12}$

$\dfrac{3}{x} = \dfrac{1.8}{12}$

B. Solve these proportion equations.

1. $\dfrac{3}{2} = \dfrac{x}{8} = 12$

2. $\dfrac{6}{R} = \dfrac{5}{72} = 86.4$

3. $\dfrac{y}{60} = \dfrac{5}{3} = 100$

4. $\dfrac{2}{15} = \dfrac{8}{H} = 60$

5. $\dfrac{5}{P} = \dfrac{30}{7} = 1.1667$

6. $\dfrac{1}{6} = \dfrac{17}{x} = 102$

7. $\dfrac{A}{2.5} = \dfrac{13}{10} = 3.25$

8. $\dfrac{27}{M} = \dfrac{3}{0.8} = 7.2$

9. $\dfrac{2}{5} = \dfrac{T}{4.5} = 1.8$

10. $\dfrac{0.12}{N} = \dfrac{2}{7}$ .42

11. $\dfrac{138}{23} = \dfrac{18}{x} = 3$

12. $\dfrac{3.25}{1.5} = \dfrac{A}{0.6}$ 1.3

13. $\dfrac{x}{34.86} = \dfrac{1.2}{8.3}$ 5.04

14. $\dfrac{2\frac{1}{2}}{R} = \dfrac{1\frac{1}{4}}{3\frac{1}{4}}$ 6.5

15. $\dfrac{2\text{ ft } 6\text{ in.}}{4\text{ ft } 3\text{ in.}} = \dfrac{L}{8\text{ ft } 6\text{ in.}}$ 5

16. $\dfrac{6.2\text{ cm}}{x} = \dfrac{1.2\text{ in.}}{11.4\text{ in.}} = 23.18?0$ in.

17. $\dfrac{3\text{ ft } 4\text{ in.}}{4\text{ ft } 2\text{ in.}} = \dfrac{3.2\text{ cm}}{x} = 1.5748$

18. $\dfrac{3\frac{1}{2}\text{ in.}}{W} = \dfrac{1.4}{0.05}$ $\dfrac{3.5}{w} = \dfrac{1.4}{0.05}$ = .05 in

C. Solve.

1. **Auto Mechanics** The compression ratio in a certain engine is 9.6 to 1. If the expanded volume of a cylinder is 48 cu in., what is the compressed volume?

2. **Painting and Decorating** If 1 gallon of paint covers 360 sq ft, how many gallons will be needed to cover 2650 sq ft with two coats? (Assume that you cannot buy a fraction of a gallon.)

3. **Machine Technology** If 28 tapered pins can be machined from a steel rod 12 ft long, how many tapered pins can be made from a steel rod 9 ft long?

4. **Carpentry** If 6 lb of nails are needed for each thousand lath, how many pounds are required for 4250 lath?

5. **Masonry** For a certain kind of plaster work, 1.5 cu yd of sand are needed for every 100 sq yd of surface. How much sand will be needed for 350 sq yd of surface?

6. **Printing** The paper needed for a printing job weighs 12 lb per 500 sheets. How many pounds of paper are needed to run a job requiring 12,500 sheets?

7. **Machine Technology** A cylindrical oil tank 8 ft deep holds 420 gallons when filled to capacity. How many gallons remain in the tank when the depth of oil is $5\frac{1}{2}$ ft?

8. **Agricultural Technology** A liquid fertilizer must be prepared by using one part of concentrate for every 32 parts of water. How much water should be added to 7 oz of concentrate?

9. **Welding** A 10-ft bar of I-beam weighs 208 lb. What is the weight of a 6-ft length?

10. **Photography** A photographer must mix a chemical in the ratio of 1 part chemical for every 7 parts of water. How many ounces of chemical should be used to make a 3-qt *total* mixture? (1 qt = 32 oz)

11. If you earn $588.80 for a 32-hour work week, how much would you earn for a 40-hour work week at the same hourly rate?

12. **Machine Technology** A machinist can produce 12 parts in 40 min. How many parts can the machinist produce in 4 hours?

13. **Auto Mechanics** The headlights on a car are set so the light beam drops 2 in. for each 25 ft measured horizontally. If the headlights are mounted 30 in. above the ground, how far ahead of the car will they hit the ground?

14. **Machine Technology** A machinist creates $2\frac{3}{4}$ lb of steel chips in fabricating 16 rods. How many pounds of steel chips will be created in producing 120 rods?

15. **Stockbroker** The annual earnings per share of the Web-site maker DotCom.com are $3.40. This stock should be fairly priced when its P–E ratio is about 30 to 1. What would be a fair price for the stock?

16. **Painting** A painter must thin some paint for use in a sprayer. If the recommended rate is $\frac{1}{2}$ pint of water per gallon of paint, how many pints of water should be added to $5\frac{1}{2}$ gallons of paint?

17. **Sports Technology** Power hitter Sammy McGwire strikes out too much to suit manager Sparky Spittoon. Sammy is promised an incentive bonus of a million dollars if he can reduce his strikeout-to-home run ratio to 3:1. If Sammy hits 56 home runs, what is the maximum number of strikeouts he can have and still earn the bonus?

18. **Painting and Decorating** The ratio 1.61803 . . . is called the *Golden Ratio* because it is the most visually pleasing ratio of the sides of a rectangle. If a rectangular painting is created with a length of 33 inches, what must be its width (the smaller dimension) if its dimensions are in a Golden Ratio? (Round to the nearest tenth of an inch.)

Check your answers in the Appendix, then turn to Section 4-2.

4-2 INTRODUCTION TO PERCENT

In many practical calculations, it is helpful to be able to compare two numbers, and it is especially useful to express the comparison in terms of a percent. Percent calculations are a very important part of any work in business, technical skills, or the trades.

The word *percent* comes from the Latin phrase *per centum* meaning "by the hundred" or "for every hundred." A number written as a percent is being compared with a second number called the standard or *base*. For example, what part of the base length is length *A?*

We could answer the question with a fraction or ratio, a decimal, or a percent. First, divide the base into 100 equal parts. Use a 100-mm metric rule to help visualize this.

Then compare length A with it.

The length of A is 40 parts out of the 100 parts that make up the base.

A is $\dfrac{40 \text{ mm}}{100 \text{ mm}}$ or 0.40 or 40% of A.

$\dfrac{40}{100} = 40\%$ 40% means 40 parts in 100 or $\dfrac{40}{100}$

YOUR TURN

For the following diagram, what part of the base length is length B?

Answer with a percent.

SOLUTION

B is $\dfrac{60}{100}$ or 60% of the base.

The compared number may be larger than the base. For example,

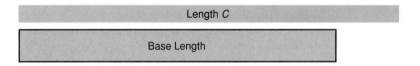

In this case, divide the base into 100 parts and extend it in length.

The length of C is 120 of the 100 parts that make up the base.

C is $\dfrac{120}{100}$ or 120 % of the base.

Ratios, decimals, and percents are all alternative ways to compare two numbers. For example, in the following drawing what fraction of the rectangle is shaded? What part of 12 is 3?

First, we can write the answer as a ratio:

$$\frac{3 \text{ shaded squares}}{12 \text{ squares}} = \frac{3}{12} = \frac{1}{4}$$

Second, by dividing 4 into 1 we can write this ratio or fraction as a decimal:

$$\frac{1}{4} = 0.25 \qquad 4\overline{)1.00}^{.25}$$

Finally, we can rewrite the fraction with a denominator of 100:

$$\frac{1}{4} = \frac{1 \times \boxed{25}}{4 \times \boxed{25}}$$

$$\frac{1}{4} = \frac{25}{100} \qquad \text{or} \qquad 25\%$$

Changing Decimal Numbers to Percents

Converting a number from fraction or decimal form to a percent is a very useful skill.

> To write a decimal number as a percent, multiply the decimal number by 100%.

EXAMPLE

$$0.60 = 0.60 \times 100\% = 60\%$$

Multiply by 100%

$$0.375 = 0.375 \times 100\% = 37.5\%$$
$$3.4 = 3.4 \times 100\% = 340\%$$
$$0.02 = 0.02 \times 100\% = 2\%$$

Notice that multiplying by 100 is equivalent to shifting the decimal point two places to the right. For example,

$$0.60 = 0.60 = 60.\% \qquad \text{or} \qquad 60\%$$

Shift decimal point two places right

$$0.056 = 0.056 = 5.6\%$$

Rewrite the following decimal numbers as percents.

(a)   0.75   (b)   1.25   (c)   0.064

(d)   3   (e)   0.05   (f)   0.004

SOLUTIONS

(a) $0.75 = 0.75 \times 100\% = 75\%$:   0.75 becomes 75%

(b) $1.25 = 1.25 \times 100\% = 125\%$ 1.25 becomes 125%

(c) $0.064 = 0.064 \times 100\% = 6.4\%$ 0.064 becomes 6.4%

(d) $3 = 3 \times 100\% = 300\%$ $3 = 3.00$ or   300%

(e) $0.05 = 0.05 \times 100\% = 5\%$ 0.05 becomes 5%

(f) $0.004 = 0.004 \times 100\% = 0.4\%$ 0.004 becomes 0.4%

Changing Fractions to Percents

> To rewrite a fraction as a percent, first change to decimal form by dividing, then multiply by 100%.

For example,

$\frac{1}{2}$ is 1 divided by 2, or $2\overline{)1.0}$ with quotient 0.5   so that

$\frac{1}{2} = 0.5 = 0.5 \times 100\% = 50\%$

with arrows indicating: Change to a decimal, then Multiply by 100%

EXAMPLE

$\frac{3}{4} = 0.75 = 0.75 \times 100\% = 75\%$

$\frac{3}{20} = 0.15 = 0.15 \times 100\% = 15\%$ since $\frac{3}{20} = 20\overline{)3.00}$ with quotient 0.15

$1\frac{7}{20} = \frac{27}{20} = 1.35 = 1.35 \times 100\% = 135\%$ since $\frac{27}{20} = 20\overline{)27.00}$ with quotient 1.35

YOUR TURN

Rewrite $\frac{5}{16}$ as a percent.

$\frac{5}{16}$ means $16\overline{)5.0000}^{0.3125}$ so that $\frac{5}{16} = 0.3\overset{\frown}{1}25 = 31.25\%$

This is often written as $31\frac{1}{4}\%$.

Some fractions cannot be converted to an exact decimal. For example, $\frac{1}{3} = 0.3333\ldots$, where the 3s continue endlessly. We can round to get an approximate percent.

$\frac{1}{3} \approx 0.3333 \approx 33.33\%$ or $33\frac{1}{3}\%$ (The sign $\approx$ means "approximately equal to.")

The fraction $\frac{1}{3}$ is roughly equal to 33.33% and exactly equal to $33\frac{1}{3}\%$.

Learning Help ▶ Here is an easy way to change certain fractions into percents. If the denominator of the fraction divides exactly into 100, change the given fraction into an equivalent fraction with a denominator of 100. The new numerator is the percent. For example,

$\frac{3}{20} = \frac{3 \times 5}{20 \times 5} = \frac{15}{100}$ so $\frac{3}{20} = 15\%$ ◀

PROBLEMS

Rewrite the following fractions as percents.

(a) $\frac{4}{5}$ (b) $\frac{2}{3}$ (c) $3\frac{1}{8}$

(d) $\frac{5}{12}$ (e) $\frac{5}{6}$ (f) $1\frac{1}{3}$

SOLUTIONS

(a) $\frac{4}{5} = 0.\overset{\frown}{80} = 80\%$

Shift the decimal point two places to the right

Change to a decimal by dividing

(b) $\frac{2}{3} \approx 0.6667 \approx 66.67\%$ or exactly $66\frac{2}{3}\%$

(c) $\frac{1}{8}$ is $8\overline{)1.000}^{0.125}$ so $3\frac{1}{8} = 3.125$

$3\frac{1}{8} = 3.1\overset{\frown}{25} = 312.5\%$

(d) $\frac{5}{12}$ is $12\overline{)5.000}^{0.4166\ldots}$ so

$\frac{5}{12} = 0.41\overset{\frown}{66}\ldots \approx 41.67\%$ or exactly $41\frac{2}{3}\%$

(e) $\dfrac{5}{6}$ is $6\overline{)5.000}$ $\overset{0.833\ldots}{\phantom{6\overline{)5.000}}}$ so

$$\frac{5}{6} \approx 0.833 \approx 83.3\% \text{ or exactly } 83\frac{1}{3}\%$$

(f) $1\dfrac{1}{3} = 1.333\ldots \approx 133.3\%$ or exactly $133\dfrac{1}{3}\%$

Changing Percents to Decimal Numbers

In order to use percent numbers when you solve practical problems, it is often necessary to change a percent to a decimal number.

> To change a percent to a decimal number, divide by 100%.

EXAMPLE

$$50\% = \frac{50\%}{100\%} = \frac{50}{100} = 0.5$$

Divide by 100%

$$6\% = \frac{6\%}{100\%} = \frac{6}{100} = 0.06$$

$$0.2\% = \frac{0.2\%}{100\%} = \frac{0.2}{100} = 0.002$$

Notice that in each of these examples division by 100% is the same as moving the decimal point two digits to the left. For example,

$$50\% = 50.\% = 0.50 \text{ or } 0.5$$

Shift the decimal point two places left

$$6\% = 06.\% = 0.06$$
$$0.2\% = 00.2\% = 0.002$$

If a fraction is part of the percent number, write it as a decimal number before dividing by 100%.

EXAMPLE

$$8\frac{1}{2}\% = 8.5\% = \frac{8.5\%}{100\%} = 0.085 \qquad \text{or} \qquad 8.5\% = 08.5\% = 0.085$$

Now try these. Write each percent as a decimal number.

(a) 4%   (b) 112%   (c) 0.5%

(d) $9\frac{1}{4}\%$ (e) 45%   (f) $12\frac{1}{3}\%$

(a) $4\% = \dfrac{4\%}{100\%} = \dfrac{4}{100} = 0.04$ since $100)\overline{4.00}^{\,0.04}$

This may also be done by shifting the decimal point: $4\% = 04.\% = 0.04$.

(b) $112\% = 112.\% = 1.12$

(c) $0.5\% = 00.5\% = 0.005$

(d) $9\frac{1}{4}\% = 9.25\% = 09.25\% = 0.0925$

(e) $45\% = 45.\% = 0.45$

(f) $12\frac{1}{3}\% \approx 12.33\% = 0.1233$

Did you notice that in problem (b) a percent greater than 100% gives a decimal number greater than 1?

$100\% = 1$

$200\% = 2$

$300\% = 3$ and so on

Changing Percents to Fractions It is often necessary to rewrite a percent number as a fraction in lowest terms.

> To write a percent number as a fraction, form a fraction with the percent number as numerator and 100 as the denominator. Then rewrite this fraction in lowest terms.

$40\% = \dfrac{40}{100} = \dfrac{40 \div 20}{100 \div 20} = \dfrac{2}{5}$

$68\% = \dfrac{68}{100} = \dfrac{68 \div 4}{100 \div 4} = \dfrac{17}{25}$

$125\% = \dfrac{125}{100} = \dfrac{125 \div 25}{100 \div 25} = \dfrac{5}{4}$ or $1\frac{1}{4}$

Write each percent as a fraction in lowest terms.

(a) 8%   (b) 65%   (c) 140%

(d) 73%   (e) 0.2%   (f) $16\frac{2}{3}\%$

SOLUTIONS

(a) $8\% = \dfrac{8}{100} = \dfrac{8 \div 4}{100 \div 4} = \dfrac{2}{25}$

(b) $65\% = \dfrac{65}{100} = \dfrac{65 \div 5}{100 \div 5} = \dfrac{13}{20}$

(c) $140\% = \dfrac{140}{100} = \dfrac{140 \div 20}{100 \div 20} = \dfrac{7}{5}$ or $1\frac{2}{5}$

(d) $73\% = \dfrac{73}{100}$ in lowest terms

(e) $0.2\% = \dfrac{0.2}{100} = \dfrac{0.2 \times 5}{100 \times 5} = \dfrac{1}{500}$

(f) $16\frac{2}{3}\% = \dfrac{16\frac{2}{3}}{100} = \dfrac{\frac{50}{3}}{100} = \dfrac{50}{3} \times \dfrac{1}{100} = \dfrac{1}{6}$

Notice in (e) that we had to multiply numerator and denominator by 5 to avoid having a decimal number within a fraction.

Here is a table of the most often used fractions with their percent equivalents.

Study the table, then go to Exercises 4-2 for a set of problems involving percent conversion.

Percent Equivalents

Percent	Decimal	Fraction	Percent	Decimal	Fraction
5%	0.05	$\frac{1}{20}$	50%	0.50	$\frac{1}{2}$
$6\frac{1}{4}\%$	0.0625	$\frac{1}{16}$	60%	0.60	$\frac{3}{5}$
$8\frac{1}{3}\%$	$0.08\overline{3}$	$\frac{1}{12}$	$62\frac{1}{2}\%$	0.625	$\frac{5}{8}$
10%	0.10	$\frac{1}{10}$	$66\frac{2}{3}\%$	$0.\overline{6}$	$\frac{2}{3}$
$12\frac{1}{2}\%$	0.125	$\frac{1}{8}$	70%	0.70	$\frac{7}{10}$
$16\frac{2}{3}\%$	$0.1\overline{6}$	$\frac{1}{6}$	75%	0.75	$\frac{3}{4}$
20%	0.20	$\frac{1}{5}$	80%	0.80	$\frac{4}{5}$
25%	0.25	$\frac{1}{4}$	$83\frac{1}{3}\%$	$0.8\overline{3}$	$\frac{5}{6}$
30%	0.30	$\frac{3}{10}$	$87\frac{1}{2}\%$	0.875	$\frac{7}{8}$
$33\frac{1}{3}\%$	$0.\overline{3}$	$\frac{1}{3}$	90%	0.90	$\frac{9}{10}$
$37\frac{1}{2}\%$	0.375	$\frac{3}{8}$	100%	1.00	$\frac{10}{10}$
40%	0.40	$\frac{2}{5}$			

Exercises 4-2

Introduction to Percent

A. Convert to a percent.

1. 0.32	2. 1	3. 0.5	4. 2.1
5. $\frac{1}{4}$	6. 3.75	7. 40	8. 0.675
9. 2	10. 0.075	11. $\frac{1}{2}$	12. $\frac{1}{6}$
13. 0.335	14. 0.001	15. 0.005	16. $\frac{3}{10}$
17. $\frac{3}{2}$	18. $\frac{3}{40}$	19. $3\frac{3}{10}$	20. $\frac{1}{5}$
21. 0.40	22. 0.10	23. 0.95	24. 0.03
25. 0.3	26. 0.015	27. 0.60	28. 7.75
29. 1.2	30. 4	31. 6.04	32. 9
33. $\frac{1}{4}$	34. $\frac{1}{5}$	35. $\frac{7}{20}$	36. $\frac{3}{8}$
37. $\frac{5}{6}$	38. $2\frac{3}{8}$	39. $3\frac{7}{10}$	40. $1\frac{4}{5}$

B. Convert to a decimal number.

1. 6%	2. 45%	3. 1%	4. 33%
5. 71%	6. 456%	7. $\frac{1}{2}$%	8. 0.05%
9. $6\frac{1}{4}$%	10. $8\frac{3}{4}$%	11. 30%	12. 2.1%
13. 800%	14. 8%	15. 0.25%	16. $16\frac{1}{3}$%
17. 7%	18. 3%	19. 56%	20. 15%
21. 1%	22. $7\frac{1}{2}$%	23. 90%	24. 0.3%
25. 150%	26. $1\frac{1}{2}$%	27. $6\frac{3}{4}$%	28. $\frac{1}{2}$%
29. $12\frac{1}{4}$%	30. $125\frac{1}{5}$%	31. 1.2%	32. 240%

C. Rewrite each percent as a fraction in lowest terms.

1. 5%	2. 20%	3. 250%	4. 16%
5. 53%	6. 0.1%	7. 92%	8. 2%
9. 45%	10. 175%	11. $8\frac{1}{3}$%	12. 37.5%
13. 24%	14. 119%	15. 0.05%	16. 15%
17. 480%	18. $83\frac{1}{3}$%	19. 62.5%	20. 1.2%

When you have had the practice you need, check your answers in the Appendix, then turn to Section 4-3.

4-3 PERCENT PROBLEMS

In all your work with percent you will find that there are three basic types of problems. These three are related to all percent problems that arise in business, technology, or the trades. In this section we show you how to solve any percent problem, and we examine each of the three types of problems.

All percent problems involve three quantities:

B, the *base* or whole or total amount, a standard used for comparison

P, the *percentage* or part being compared with the base

R, the *rate* or *percent,* a percent number

Note ▶ To avoid possible confusion between the words "percentage" and "percent," we will refer to the percentage P as the "part." ◀

For any percent problem, these three quantities are related by the proportion

$$\frac{P}{B} = \frac{R}{100}$$

For example, the proportion

$$\frac{3}{4} = \frac{75}{100}$$

can be translated to the percent statement

"Three is 75% of four."

Here, the part P is 3, the base B is 4, and the percent R is 75.

To solve any percent problem, we need to identify which of the quantities given in the problem is P, which is B, and which is R. Then we can write the percent proportion and solve for the missing or unknown quantity.

When P Is Unknown Consider these three problems:

What is 30% of 50?

Find 30% of 50.

30% of 50 is what number?

These three questions are all forms of the same problem. They are all asking you to find P, the part.

We know that 30 is the percent R because it has the % symbol attached to it. The number 50 is the base B. To solve, write the percent proportion, substituting 50 for B and 30 for R.

$$\frac{P}{50} = \frac{30}{100}$$

Now solve using cross-products and division.

$$100P = 30 \cdot 50$$

$$100P = 1500$$

$$P = \frac{1500}{100} = 15$$

15 is 30% of 50.

✓ Substitute the answer back into the proportion.

$$\frac{\boxed{15}}{50} = \frac{30}{100} \qquad \text{or} \qquad 15 \cdot 100 = 30 \cdot 50 \qquad \text{which is correct}$$

The answer is reasonable: 30% is roughly one-third, and 15 is roughly one-third of 50.

Learning Help ▶ The percent number R is easy to identify because it always has the % symbol attached to it. If you have trouble distinguishing the part P from the base B, notice that B is usually associated with the word "of" and P is usually associated with the word "is."

"is" indicates the part **P**. It is unknown in this case.

"%" indicates the percent or rate **R**.

"of" indicates the base **B**.

Therefore the proportion $\frac{P}{B} = \frac{R}{100}$ can be thought of as $\frac{\text{is}}{\text{of}} = \frac{\%}{100}$. ◀

YOUR TURN

Now try this problem to test yourself.

Find $8\frac{1}{2}\%$ of 160.

SOLUTION

Step 1 If your mental math skills are good, estimate the answer first. In this case, $8\frac{1}{2}\%$ of 160 should be a little less than 10% of 160, or 16.

Step 2 Now identify the three quantities, P, B, and R. R is obviously $8\frac{1}{2}$. The word "is" does not appear, but "of" appears with 160, so 160 is B. Therefore, P must be the unknown quantity.

Step 3 Set up the percent proportion.

$$\frac{P}{B} = \frac{R}{100} \rightarrow \frac{P}{160} = \frac{8\frac{1}{2}}{100}$$

Step 4 Solve.

$$100P = 8\frac{1}{2} \cdot 160$$
$$100P = 1360$$
$$P = 13.6$$

 Step 5 Check your answer by substituting into the original proportion equation and comparing cross-products.

Notice that the calculated answer, 13.6, is a little less than 16, our rough estimate.

Solve the following percent problems.

(a) Find 2% of 140 lb.

(b) 35% of $20 is equal to what amount?

(c) What is $7\frac{1}{4}\%$ of $1000?

(d) Calculate $16\frac{2}{3}\%$ of 66.

(e) **Carpentry** To account for waste, a carpenter needs to order 120% of the total wood used in making a cabinet. If 15 ft of white oak is actually used, how much should he order?

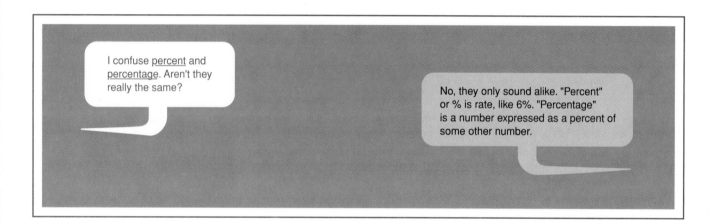

SOLUTIONS

(a) $\dfrac{P}{140} = \dfrac{2}{100}$

$100P = 2 \cdot 140 = 280$

$P = 2.8 \text{ lb}$

☑ $\dfrac{2.8}{140} = \dfrac{2}{100}$

$280 = 280$

(b) $\dfrac{P}{20} = \dfrac{35}{100}$

$100P = 700$

$P = \$7$

☑ $\dfrac{7}{20} = \dfrac{35}{100}$

$700 = 700$

(c) $\dfrac{P}{1000} = \dfrac{7\frac{1}{4}}{100}$

$100P = 7\frac{1}{4} \cdot 1000 = 7250$

$P = \$72.50$

☑ $\dfrac{72.50}{1000} = \dfrac{7\frac{1}{4}}{100}$

$7250 = 7250$

(d) $\dfrac{P}{66} = \dfrac{16\frac{2}{3}}{100}$ $16\frac{2}{3} = \dfrac{50}{3}$

$$100P = 66 \cdot \dfrac{50}{3} = \dfrac{\overset{22}{\cancel{66}}}{1} \cdot \dfrac{50}{\underset{1}{\cancel{3}}} = 22 \cdot 50$$

$$100P = 1100$$

$$P = 11 \qquad \text{Check it.}$$

 16 $\boxed{a\frac{b}{c}}$ **2** $\boxed{a\frac{b}{c}}$ **3** $\boxed{\times}$ **66** $\boxed{\div}$ **100** $\boxed{=}$ → ▨▨▨▨▨▨▨▨▨▨ *11.*

(e) The key phrases in this problem are

"120% of the total . . ."   and   "15 ft . . . is used."

We are finding 120% of 15 ft. $B = 15$ ft

$$\dfrac{P}{15} = \dfrac{120}{100}$$

$$100P = 1800$$

$$P = 18 \text{ ft}$$

Learning Help ▶ Notice that in all problems where P is to be found, we end by dividing by 100 in the last step. Recall that the quick way to divide by 100 is to move the decimal point two places to the left. In problem (c), for example, $100P = 7250$. The decimal point is after the zero in 7250. Move it two places to the left.

$$7250 \div 100 = 7250. = 72.50 \blacktriangleleft$$

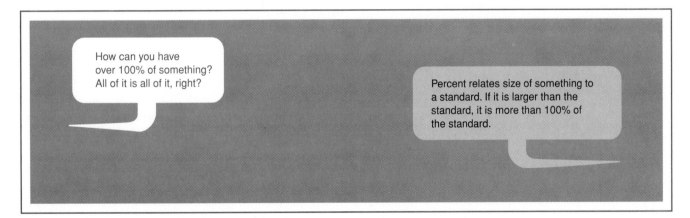

When *R* Is Unknown   Consider the following problems:

> *5 is what percent of 8?*

or   *Find what percent 5 is of 8.*

or   *What percent of 8 is 5?*

Once again, these statements represent three different ways of asking the same question. In each statement the percent R is unknown. We know this because neither of the other two numbers has a % symbol attached.

To solve this kind of problem, first identify P and B. The word "of" is associated with the base B, and the word "is" is associated with the part P.
In this case, $B = 8$ and $P = 5$.

Next, set up the percent proportion.

$$\frac{P}{B} = \frac{R}{100} \rightarrow \frac{5}{8} = \frac{R}{100}$$

Finally, solve for R.

$$8R = 5 \cdot 100 = 500$$
$$R = 62.5\%$$

Note ▶ When you solve for R, remember to include the % symbol with your answer. ◀

YOUR TURN

Now try these problems for practice.

(a) What percent of 40 lb is 16 lb?

(b) 65 is what percent of 25?

(c) Find what percent $9.90 is of $18.00.

(d) **Machine Technology** During reshaping, 6 lb of metal is removed from a casting originally weighing 80 lb. What percent of the metal is removed?

SOLUTIONS

(a) $\dfrac{16}{40} = \dfrac{R}{100}$

$$40R = 1600$$
$$R = 40\%$$

(b) $\dfrac{65}{25} = \dfrac{R}{100}$

$$25R = 6500$$
$$R = 260\%$$

(c) $\dfrac{\$9.90}{\$18.00} = \dfrac{R}{100}$

$$18R = 990$$
$$R = 55\%$$

(d) $\dfrac{6}{80} = \dfrac{R}{100}$

$$80R = 600$$
$$R = 7.5\%$$

Notice that in problem (b) the percentage is larger than the base, resulting in a percent larger than 100%. Some students mistakenly believe that the percentage is always smaller than the base. This is not always true. When in doubt use the "is–of" method to determine P and B.

Learning Help ▶ In all problems where R is unknown, we end up multiplying by 100. Recall that the quick way to do this is to move the decimal point two places to the right. If there is no decimal point, attach two zeros. ◀

When B Is Unknown This third type of percent problem requires that you find the total or base when the percent and the percentage are given. Problems of this kind can be stated as

8.7 is 30% of what number?

or *30% of what number is 8.7?*

or *Find a number such that 30% of it is 8.7.*

Solve this problem in the same way as the previous ones.

First, identify P, B, and R. Because it carries the % symbol, we know that $R = 30$. Because the word "is" is associated with the number 8.7, we know that $P = 8.7$. The base B is unknown.

Next, set up the percent proportion.

$$\frac{P}{B} = \frac{R}{100} \rightarrow \frac{8.7}{B} = \frac{30}{100}$$

Finally, solve the proportion.

$$30B = 870$$

$$B = 29 \qquad ☑ \qquad \frac{8.7}{\boxed{29}} = \frac{30}{100}$$

$$8.7(100) = 30 \cdot 29 \quad \text{which is correct}$$

YOUR TURN

Ready for a few practice problems? Try these.

(a) 16% of what amount is equal to $5.76?

(b) $2 is 8% of the cost. Find the cost.

(c) Find a distance such that $12\frac{1}{2}\%$ of it is $26\frac{1}{4}$ ft.

(d) **Office Services** The $2800 actually spent on a construction job was 125% of the original estimate. What was the original estimate?

(e) **Construction Technology** A builder told his crew that the site preparation and foundation work on a house would represent about 5% of the total building time. If it took them 21 days to do this preliminary work, how long will the entire construction take?

SOLUTIONS

(a) $\dfrac{\$5.76}{B} = \dfrac{16}{100}$

$$16B = 576$$

$$B = \$36$$

(b) $\dfrac{2}{B} = \dfrac{8}{100}$

$$8B = 200$$

$$B = \$25$$

(c) $\dfrac{26\frac{1}{4}}{B} = \dfrac{12\frac{1}{2}}{100}$

$$B = \frac{26.25 \times 100}{12.5} = 210 \text{ ft}$$

(d) $\dfrac{2800}{B} = \dfrac{125}{100}$

$$B = \frac{2800 \times 100}{125} = \$2240$$

For problem (c): **26.25** ⊗ **100** ⊘ **12.5** ⊜ → `210.`

(e) We are asked to find the total construction time, and we are told that preliminary work is 5% of the total. Since preliminary work takes 21 days, we can restate the problem as

21 is 5% of the total

or $\dfrac{21}{B} = \dfrac{5}{100}$

$$B = \frac{21 \times 100}{5} = 420 \text{ days}$$

So far we have solved sets of problems that were all of the same type. The key to solving percent problems is to be able to identify correctly the quantities P, B, and R.

PROBLEMS

Use the hints you have learned to identify these three quantities in the following problems, then solve each problem.

(a) What percent of 25 is 30?

(b) 12 is $66\frac{2}{3}\%$ of what amount?

(c) Find 6% of 2400.

(d) **Office Services** Television City asked you to make a 15% down payment on your purchase of a new TV set. If the down payment was $78, what was the cost of the set?

(e) **Printing** A printer gives a 5% discount to preferred customers. If the normal cost of a certain job is $3400, what discount would a preferred customer receive?

(f) **Industrial Technology** Twelve computer chips were rejected out of a run of 400. What percent were rejected?

SOLUTIONS

(a) $\underbrace{\text{What percent}}_{R}$ $\underbrace{\text{of } 25}_{B}$ $\underbrace{\text{is } 30?}_{P}$ $\dfrac{30}{25} = \dfrac{R}{100}$

$25R = 3000$

$R = 120\%$

(b) $\underbrace{12}_{P}$ $\underbrace{\text{is}}$ $\underbrace{66\frac{2}{3}\%}_{R}$ $\underbrace{\text{of what?}}_{B}$ $\dfrac{12}{B} = \dfrac{66\frac{2}{3}}{100}$

$66\frac{2}{3} \cdot B = 1200$

$B = 18$

 12 $\boxed{\times}$ **100** $\boxed{\div}$ **66** $\boxed{a\frac{b}{c}}$ **2** $\boxed{a\frac{b}{c}}$ **3** $\boxed{=}$ → ▨▨▨▨ *18.*

(c) $\underbrace{\text{Find}}_{P}$ $\underbrace{6\%}_{R}$ $\underbrace{\text{of } \$2400.}_{B}$ $\dfrac{P}{2400} = \dfrac{6}{100}$

$100P = 6 \cdot 2400 = 14400$

$P = \$144$

(d) In this problem, 15% is the rate, so $78 must be either the part or the base. Since $78 is the down payment, a part of the whole price, $78 must be P.

$\dfrac{78}{B} = \dfrac{15}{100}$

$15B = 7800$

$B = \$520$ The total cost of the TV set was $520.

(e) 5% is the rate R, so $3400 must be either P or B. The discount is a part of the normal cost, so P is unknown and B is $3400.

$\dfrac{P}{3400} = \dfrac{5}{100}$

$100P = 5 \cdot 3400 = 17000$

$P = \$170$ The preferred customer receives a discount of $170.

(f) The percent R is not given. The 12 chips are a part (P) of the whole (B) run of 400.

$$\frac{12}{400} = \frac{R}{100}$$

$$400R = 1200$$

$$R = 3\% \qquad 3\% \text{ of the chips were rejected.}$$

Now go to Exercises 4-3 for more practice on the three basic kinds of percent problems.

Exercises 4-3

Percent Problems

A. Solve.

1. 4 is ___80___ % of 5.

2. What percent of 25 is 16?   *64*

3. 20% of what number is 3?

4. 8 is what percent of 8?   *100%*

5. 120% of 45 is ___54___.

6. 8 is ___12.5___ % of 64.

7. 3% of 5000 = ___150___.

8. 2.5% of what number is 2?   *80*

9. What percent of 54 is 36?   *66.7%*

10. 60 is ___500___ % of 12.

11. 17 is 17% of ___100___.   *289.*

12. 13 is what percent of 25?

$\frac{17}{x} = \frac{17\%}{100\%}$

13. 8½% of $250 is ___21.25___.

14. 12 is ___600___ % of 2.

15. 6% of 25 is ___1.5___.

16. 60% of what number is 14?   *23.3333*

17. What percent of 16 is 7?   *43.75 %*

18. 140 is ___75___ % of 105?

19. Find 24% of 10.   *2.4*

20. 45 is 12% of what number?   *375*

21. 30% of what number is equal to 12?   *40*

22. What is 65% of 5?   *3.25*

23. Find a number such that 15% of it is 750.   *5000*

24. 16% of 110 is what number?   *17.6*

B. Solve.

1. 75 is 33⅓% of ___225___.

2. What percent of 10 is 2.5?   *25%*

3. 6% of $3.29 is ___.1974___.

4. 63 is ___55.56___ % of 35.

5. 12.5% of what number is 20?   *160*

6. 33⅓% of $8.16 = ___2.72___.

7. 9.6 is what percent of 6.4?   *66.67%*

8. $0.75 is ___2___ % of $37.50.

9. 6¼% of 280 is ___17.5___.

10. 1.28 is ___400___ % of 0.32.

11. 42.7 is 10% of ___427___.

12. 260% of 8.5 is ___22.1___.

13. ½ is ___50___ % of 25.

14. 7¼% of 50 is ___3.625___.

15. 287.5% of 160 is ___460___.

16. 0.5% of ___1400___ is 7.

17. 112% of ___50___ is 56.

18. 2¼% of 110 is ___2.475___.   *2.475*

C. Word Problems

1. If you answered 37 problems correctly on a 42-question test, what percent score would you have?

2. **Office Services** The profits from Ed's Plumbing Co. increased by $14,460, or 30%, this year. What profit did it earn last year?

3. **Machine Technology**   A casting weighed 146 lb out of the mold. It weighed 138 lb after finishing. What percent of the weight was lost in finishing?

4. The Smiths bought a house for $484,500 and made a down payment of 20%. What was the actual amount of the down payment?

5. If your salary is $2760 per month and you get a 6% raise, what is the amount of your raise?

6. If state sales tax is 6%, what is the price before tax if the tax is $7.80?

7. A carpenter earns $28.50 per hour. If he receives a $7\frac{1}{2}\%$ pay raise, what is the amount of his hourly raise?

8. If 7.65% of your salary is withheld for Social Security and Medicare, what amount is withheld from monthly earnings of $1845.00?

9. **Printing**   If an 8-in. by 10-in. photocopy is enlarged to 120% of its size, what are its new dimensions?

10. **Police Science**   Year-end statistics for a community revealed that in 648 of the 965 residential burglaries, the burglar entered through an unlocked door or window. What percent of the entries were made in this way?

11. **Carpentry**   Preparing an estimate for a decking job, a carpenter assumes about a 15% waste factor. If the finished deck contains 1230 lineal feet of boards, how many lineal feet should the carpenter order? (*Hint:* If the waste is 15%, then the finished deck is 100% − 15% or 85% of the total order.)

12. **Printing**   Jiffy Print needs to produce 4000 error-free copies of a brochure. For this type of run there is usually a 2% spoilage rate. What is the minimum number of brochures that should be printed to assure 4000 good copies?

13. **Hotel Management**   The Bide-A-While Hotel manager is evaluated on the basis of the occupancy rate of the hotel, that is, the average daily percent of the available rooms that are actually occupied. During one particular week daily occupancy totals were 117, 122, 105, 96, 84, 107, and 114. If the hotel has 140 rooms available each night, what was the occupancy rate for that week?

14. **Restaurant Management**   The manager of the Broccoli Palace restaurant needs 25 lb of fresh broccoli for an evening's business. Because of damage and spoilage, he expects to use only about 80% of the broccoli delivered by his supplier. What is the minimum amount he should order?

15. **Construction**   The BiltWell Construction Company has submitted a bid on construction of a spa deck. If their $4320 bid includes $1630 for lumber, what percent of the bid was the lumber? (Round to the nearest percent.)

16. **Building Construction**   White oak flooring boards 3 in. wide are used to construct a hallway. If the flooring is laid at an angle, approximately 7200 lineal feet of boards are needed for the job. How many lineal feet should be ordered if an allowance of 18% is made for waste? (*Hint:* If there is 18% waste, then the amount needed is 100% − 18% = 82% of the amount that should be ordered. Round to the nearest foot.)

17. **Building Construction**   Quality Homes Construction Company uses only lumber that is at least 80% *clear*—free of defects, including knots and edge defects. What minimum amount of usable lumber would they expect to get in a lot containing 65,500 lineal feet?

176

When you have had the practice you need, check your answers in the Appendix, then continue with the study of some applications of percent to practical problems.

4-4 APPLICATIONS OF PERCENT CALCULATIONS

Now that you have seen and solved the three basic percent problems, you need some help in applying what you have learned to practical situations.

Discount
An important kind of percent problem, especially in business, involves the idea of *discount*. In order to stimulate sales, a merchant may offer to sell some item at less than its normal price. A discount is the amount of money by which the normal price is reduced. It is a percentage or part of the normal price. The *list price* is the normal or original price before the discount is subtracted. The *net price* is the new, reduced price. The net price is sometimes called the *discounted* price or the *sale* price. It is always less than the list price.

> Net price = list price − discount

The discount rate (R) is a percent number that enables you to calculate the discount (the percentage or part P) as a part of the list price (the base B). Therefore, the percent proportion

$$\frac{\text{part}}{\text{base}} = \frac{\text{rate}}{100}$$

becomes

$$\frac{\text{discount } D}{\text{list price } L} = \frac{\text{discount rate } R}{100}$$

EXAMPLE

The list price of a tool is \$18.50. On a special sale it is offered at 20% off. What is the net price?

$$\frac{\text{discount}}{\text{list price}} = \frac{\text{discount rate}}{100}$$

$$\frac{D}{18.50} = \frac{20}{100}$$

$$100D = 18.50 \times 20 = 370$$

$$D = \$3.70$$

Net price = list price − discount

$$= \$18.50 - \$3.70$$

$$= \$14.80$$

Think of it this way:

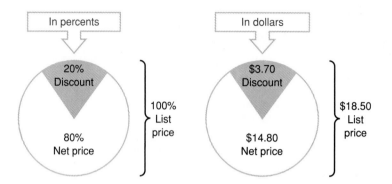

Ready for another problem?

After a 25% discount, the net price of a power sander is $144. What was its list price?

We can use a pie chart diagram to help set up the correct percent proportion.

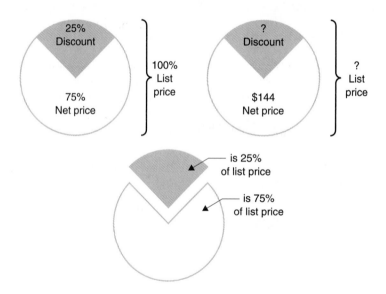

The net price, $144, is 75% of the list price. Therefore, the correct proportion is

$$\frac{\text{net price}}{\text{list price}} = \frac{\text{net price rate}}{100}$$

or $\quad \dfrac{144}{L} = \dfrac{75}{100}$

Solving for L, we have

$75L = 14400$

$\quad L = \$192 \qquad$ The list price is $192.

☑ $\dfrac{\text{discount}}{\text{list price}} = \dfrac{\text{discount rate}}{100}$

$$\dfrac{D}{192} = \dfrac{25}{100}$$

$$100D = 4800$$

$$D = \$48 \qquad \text{The discount is \$48.}$$

Net price = list price − discount

$$= \$192 - \$48$$

$$= \$144 \qquad \text{which is the correct list price as given in the original problem.}$$

PROBLEMS

Here are a few problems to test your understanding of the idea of discount.

(a) **Carpentry** A preinventory sale advertises all tools 20% off. What would be the net price of a hammer drill that cost $159.95 before the sale?

(b) **Painting** A compressor is on sale for $334.40 and is advertised as "12% off regular price." What was its regular price?

(c) **Office Services** A set of four 205/65/15 automobile tires is on sale for 15% off list price. What is the sale price if their list price is $79.50 each?

(d) **Plumbing** A kitchen sink that retails for $475 is offered to a plumbing contractor for $308.75. What discount rate is the contractor receiving?

SOLUTIONS

(a) $\dfrac{D}{L} = \dfrac{R}{100}$ becomes $\dfrac{D}{159.95} = \dfrac{20}{100}$

$$100D = 159.95 \times 20 = 3199$$

$$D = \$31.99$$

Net price = list price − discount

$$= \$159.95 - \$31.99$$

$$= \$127.96$$

A pie chart provides a quicker way to do this problem.

In percents

If the discount is 20% of the list price, then the net price is 80% of the list price.

$$\dfrac{\text{net price}}{\text{list price}} = \dfrac{80\%}{100\%}$$

or $\dfrac{N}{159.95} = \dfrac{80}{100}$

$$100N = 12796$$

$$N = \$127.96$$

(b) Pie charts will help you set up the correct proportions.

Use the net price proportion.

$$\frac{\text{net price}}{\text{list price}} = \frac{\text{net price rate}}{100}$$

$$\frac{334.40}{L} = \frac{88}{100}$$

$$88L = 33440$$

$$L = \$380.00$$

(c) Use the discount proportion.

$$\frac{D}{L} = \frac{R}{100} \quad \text{becomes} \quad \frac{D}{4 \times 79.50} = \frac{15}{100}$$

Four tires at $79.50 each

$$\frac{D}{318} = \frac{15}{100}$$

$$100D = 4770$$

$$D = \$47.70$$

Net (sale) price = list price − discount

$$= \$318 - \$47.70$$

$$= \$270.30$$

Using a calculator, first find the discount and store it in memory.

 4 $\times$ **79.50** $\times$ **15** $\div$ **100** $=$ STO $\rightarrow$ ▨ 47.70

Then subtract it from the list price.

318 $-$ RCL $=$ $\rightarrow$ ▨ 270.30   (318 = 4 × 79.50)

(d) **First,** subtract to find the discount.

Discount = list price − net price

$$= \$475 - \$308.75$$

$$= \$166.25$$

Then, set up the discount proportion.

$$\frac{D}{L} = \frac{R}{100} \quad \text{becomes} \quad \frac{166.25}{475} = \frac{R}{100}$$

$$475R = 16625$$

$$R = 35\% \qquad \text{The contractor receives a 35\%}$$
discount.

Using the proportion method for solving percent problems makes it easier to distinguish among the different types of problems. However, many practical applications involve finding the percentage, and there is a quicker way to do this, especially using a calculator. Simply change the percent to a decimal number and multiply by the base.

EXAMPLE

Find 20% of $640.

Step 1 Change the percent to a decimal number.

$$20\% = 20.\% = 0.20 = 0.2$$

Step 2 Multiply this decimal equivalent by the base.

$$0.2 \times \$640 = \$128$$

This is the same answer you would get using a proportion.

$$\frac{P}{640} = \frac{20}{100} \qquad \text{or} \qquad P = \frac{640 \times 20}{100} = 640 \times 0.2$$

YOUR TURN

Use this quick method to solve the following problems.

(a) Find 35% of 80.

(b) What is 4.5% of 650?

(c) What is 125% of 18?

(d) Find 0.3% of 5000.

SOLUTIONS

(a) $35\% = 35.\% = 0.35$
     $0.35 \times 80 = 28$

(b) $4.5\% = 04.5\% = 0.045$
     $0.045 \times 650 = 29.25$

(c) $125\% = 125.\% = 1.25$
     $1.25 \times 18 = 22.5$

(d) $0.3\% = 00.3\% = 0.003$
     $0.003 \times 5000 = 15$

Sales Tax *Taxes* are almost always calculated as a percent of some total amount. *Property taxes* are written as some fraction of the value of the property involved. *Income taxes* are most often calculated from complex formulas that depend on many factors. We cannot consider either income or property taxes here.

A *sales tax* is an amount calculated from the actual price of a purchase and added to the buyer's cost. Retail sales tax rates are set by the individual states in the United States and vary from 0 to 8% of the sales price. A sales tax of 6% is often stated as "6 cents on the dollar," since 6% of $1.00 equals 6 cents.

EXAMPLE

If the retail sales tax rate is 7.75% in Santa Barbara, how much would you pay for a pair of shoes costing $76.50?

Using a proportion equation, we have

$$\frac{\text{sales tax}}{\text{cost}} = \frac{\text{tax rate}}{100}$$

or $\text{Sales tax} = \dfrac{\text{tax rate}}{100} \times \text{cost}$

$$= \dfrac{7.75}{100} \times 76.50$$

$$= \$5.93 \text{ rounded}$$

Using the equivalent quick method, we have

Sales tax = 7.75% of \$76.50   7.75% = 07.75 = 0.0775
$\qquad\qquad$ = 0.0775 × \$76.50
$\qquad\qquad$ = \$5.92875 = \$5.93 rounded

Actual cost = list price + sales tax
$\qquad\qquad$ = \$76.50 + \$5.93
$\qquad\qquad$ = \$82.43

.0775 $\boxed{\times}$ 76.5 $\boxed{=}$ → 5.92875 $\boxed{+}$ 76.5 $\boxed{=}$ → 82.42875

Many stores provide their salesclerks with computerized cash registers that calculate sales tax automatically. However, all consumers and most small business or shop owners still need to be able to do the sales tax calculations.

YOUR TURN

Here are a few problems in calculating sales tax.

Find the tax and total cost for each of the following if the tax rate is as shown.

(a) A roll of plastic electrical tape at \$2.79   (5%).

(b) A random orbit sander priced at \$67.50   (6%).

(c) A tape measure priced at \$6.95 $(3\frac{1}{2}\%)$.

(d) A new car priced at \$23,550   (4%).

(e) A band saw priced at \$346.50 $(6\frac{1}{2}\%)$.

(f) A toggle switch priced at \$4.99 $(7\frac{3}{4}\%)$.

ANSWERS

	Tax	Total Cost	Quick Calculation
(a)	14 cents	\$2.93	0.05 × \$2.79 = 13.95 cents ≈ 14 cents
(b)	\$4.05	\$71.55	0.06 × \$67.50 = \$4.05
(c)	\$0.24	\$7.19	0.035 × \$6.95 = \$0.24325 ≈ \$0.24
(d)	\$942.00	\$24,492	0.04 × \$23,550 = \$942.00
(e)	\$22.52	\$369.02	0.065 × \$346.50 = \$22.5225 ≈ \$22.52
(f)	\$0.39	\$5.38	0.0775 × \$4.99 = \$0.386725 ≈ \$0.39

Interest In modern society we have set up complex ways to enable you to use someone else's money. A *lender*, with money beyond his or her needs, supplies cash to a *borrower*, whose needs exceed his money. The money is called a *loan*.

Interest is the amount the lender is paid for the use of his money. Interest is the money you pay to use someone else's money. The more you use and the longer you use it, the more interest you must pay. *Principal* is the amount of money lent or borrowed.

When you purchase a house with a bank loan, a car or a refrigerator on an install-ment loan, or gasoline on a credit card, you are using someone else's money, and you pay interest for that use. If you are on the other end of the money game, you may earn interest for money you invest in a savings account or in shares of a business.

Interest can be calculated from the following formulas.

$$\frac{\text{annual interest}}{\text{principal}} = \frac{\text{annual interest rate}}{100}$$

$$\text{Total interest} = \text{annual interest} \times \text{time in years}$$

EXAMPLE

Suppose that you have $500 in savings earning $5\frac{1}{2}\%$ interest annually in your local bank. How much interest do you receive in a year?

$$\frac{\text{annual interest } I}{\$500} = \frac{5\frac{1}{2}}{100}$$

$$100I = 2750$$

$$I = \$27.50$$

The total interest for one year is $27.50.

Using the quick calculation method to calculate $5\frac{1}{2}\%$ of $500,

$$\text{Annual interest } I = \$500 \times 5\frac{1}{2}\%$$

$$= 500 \times 0.055$$

$$= \$27.50$$

By placing your $500 in a bank savings account, you allow the bank to use it in various profitable ways, and you are paid $27.50 per year for its use.

YOUR TURN

Most of us play the money game from the other side of the counter. Suppose that you find yourself in need of cash and arrange to obtain a loan from a bank. You bor-row $600 at 14% per year for 3 months. How much interest must you pay?

Try to set up and solve this problem exactly as we did in the previous problem.

SOLUTION

$$\frac{\text{annual interest } I}{\$600} = \frac{14}{100}$$

$$I = \frac{600 \times 14}{100} = \$84$$

The time of the loan is 3 months, so the time in years is

$$\frac{3 \text{ months}}{12 \text{ months}} = \frac{3}{12} \text{ yr}$$

$$= 0.25 \text{ yr}$$

Total interest = annual interest × time in years

$$= \$84 \times 0.25$$

$$= \$21$$

By the quick calculation method,

$$\text{Total interest} = \$600 \times 14\% \times \frac{3}{12} \text{ yr}$$

$$= 600 \times 0.14 \times 0.25$$

$$= \$21$$

 600 $\times$ **.14** $\times$ **3** $\div$ **12** $=$ → *21.*

Depending on how you and the bank decide to arrange it, you may be required to pay the total principal ($600) plus interest ($21) all at once, at the end of three months, or according to some sort of regular payment plan—for example, pay $207 each month for three months.

HOW DOES A CREDIT CARD WORK?

Many forms of small loans, such as credit card loans, charge interest by the month. These are known as "revolving credit" or "charge account" plans. (If you borrow very much money this way, *you* do the revolving and may run in circles for years trying to pay it back!) Essentially, you buy now and pay later. Generally, if you repay the full amount borrowed within 25 or 30 days, there is no charge for the loan.

After the first pay period of 25 or 30 days, you pay a percent of the unpaid balance each month, usually between $\frac{1}{2}$ and 2%. In addition, you must pay a yearly fee and some minimum amount each month, usually $10 or 10% of the unpaid balance, whichever is larger. Often, you are also charged a small monthly amount for insurance premiums. (The credit card company insures itself against your defaulting on the loan or disappearing, and you pay for their insurance.)

Let's see how it works. Suppose that you go on a short vacation and pay for gasoline, lodging, and meals with your Handy Dandy credit card. A few weeks later you receive a bill for $1000. You can pay it within 30 days and owe no interest or you can pay over a longer period of time. Suppose you agree to pay $100 per month. The calculations look like this:

Month 1 $1\frac{1}{2}\%$ of $1000 = $1000 × 0.015 = $15.00   owe $1015.00
pay $100 and carry $915.00 over to next month

Month 2 $1\frac{1}{2}\%$ of $915 = 13.73   owe $928.73
pay $100 and carry $828.73 over to next month

. . . and so on.

Eleven months later you will have repaid the $1000 loan and all interest.

The $1\frac{1}{2}\%$ per month interest rate seems small, but it is equivalent to between 15 and 18%. You pay at a high rate for the convenience of using the credit card and the no-questions-asked ease of getting the loan.

Commission The simplest practical use of percent is in the calculation of a part or percentage of some total. For example, salespersons are often paid on the basis of their success at selling and receive a *commission* or share of the sales receipts. Commission is usually described as a percent of sales income.

$$\frac{\text{commission}}{\text{sales}} = \frac{\text{rate}}{100}$$

EXAMPLE

Suppose that your job as a lumber salesman pays 12% commission on all sales. How much do you earn from a sale of $400?

$$\frac{\text{commission } C}{400} = \frac{12}{100}$$

$$C = \frac{12 \times 400}{100}$$

$$C = \$48 \qquad \text{You earn \$48 commission.}$$

By the quick calculation method,

Commission = $400 \times 12\% = \$400 \times 0.12 = \$48$

YOUR TURN

Try this one yourself.

At the Happy Bandit Used Car Company each salesperson receives a 6% commission on sales. What would a salesperson earn if she sold a 1960 Airedale for $1299.95?

SOLUTION

$$\frac{C}{\$1299.95} = \frac{6}{100}$$

$$C = \frac{6 \times 1299.95}{100}$$

$$C = \$77.9970 \qquad \text{or} \qquad \$78 \text{ rounded}$$

or

Commission = $1299.95 \times 6\% = \$1299.95 \times 0.06$
$$\approx \$78$$

Some situations require that you determine the total sales or the rate of commission.

PROBLEMS

In the following two problems, concentrate on identifying each quantity in the commission proportion.

(a) **Office Services** A pharmaceutical sales rep generated $360,000 in sales last year and was paid a salary of $27,000. If she were to switch to straight commission compensation, what rate of commission would be needed for her commission to match her salary?

(b) **Office Services** A computer salesperson is paid a salary of $800 per month plus a 4% commission on sales. What amount of sales must he generate in order to earn $40,000 per year in total compensation?

SOLUTIONS

(a) Sales = $360,000
Commission = $27,000
Rate is unknown.

$$\frac{\$27,000}{\$360,000} = \frac{R}{100}$$

$$R = \frac{27,000 \times 100}{360000}$$

$R = 7.5\%$ She needs to be paid a 7.5% commission in order to match her salary.

(b) Salary = $800 per month $\times$ 12 months
       = $9600 per year

Commission = total compensation − salary
            = $40,000 − $9600 = $30,400

The commission needed is $30,400. Write the commission proportion. The sales, S, is not known.

$$\frac{\$30,400}{S} = \frac{4}{100}$$

$$S = \frac{30400 \times 100}{4}$$
$$S = \$760,000$$

He must sell $760,000 worth of computers to earn a total of $40,000 per year.

 40000 ⊟ **800** ⊠ **12** ⊜ → ▨ *30400.* ▨ ⊠ **100** ⊡ **4** ⊜ → ▨ *760000.* ▨

SMALL LOANS

Sooner or later everyone finds it necessary to borrow money. When you do, you will want to know beforehand how the loan process works. Suppose that you borrow $200 and the loan company specifies that you repay it at $25 per month plus interest at 3% per month on the unpaid balance. What interest do you actually pay?

Month 1   3% of $200 = $ 6.00   you pay $25 + $6.00 = $ 31.00
Month 2   3% of $175 = $ 5.25   you pay $25 + $5.25 = $ 30.25
Month 3   3% of $150 = $ 4.50   you pay $25 + $4.50 = $ 29.50
Month 4   3% of $125 = $ 3.75   you pay $25 + $3.75 = $ 28.75
Month 5   3% of $100 = $ 3.00   you pay $25 + $3.00 = $ 28.00
Month 6   3% of $ 75 = $ 2.25   you pay $25 + $2.25 = $ 27.25

(continued)

Month 7	3% of \$ 50 = \$ 1.50	you pay	\$25 + \$1.50 = \$ 26.50
Month 8	3% of \$ 25 = \$ 0.75	you pay	\$25 + \$0.75 = \$ 25.75
	\$27.00		\$227.00

Total interest is \$27.00 ⟶

Total of eight loan payments ⟶

They might also set it up as eight equal payments of \$227.00 ÷ 8 = \$28.375 or \$28.38 per month.

The 3% monthly interest rate seems small, but it amounts to about 20% per year.

A bank loan for \$200 at 12% for 8 months would cost you

$$(12\% \text{ of } \$200) \times \frac{8}{12}$$

or $\qquad \$24 \times \dfrac{8}{12}$

or $\qquad$ \$16.00 $\qquad$ Quite a difference.

The loan company demands that you pay more, and in return they are less worried about your ability to meet the payments. For a bigger risk, they want a higher rate of interest.

Efficiency When energy is converted from one form to another in any machine or conversion process, it is useful to talk about the *efficiency* of the machine or the process. The efficiency of a process is a fraction comparing the energy or power output to the energy or power input. It is usually expressed as a percent.

$$\frac{\text{output}}{\text{input}} = \frac{\text{efficiency}}{100}$$

EXAMPLE

An auto engine is rated at 175 hp and is found to deliver only 140 hp to the transmission. What is the efficiency of the process?

$$\frac{\text{output}}{\text{input}} = \frac{\text{efficiency } E}{100}$$

$$\frac{140}{175} = \frac{E}{100}$$

$$E = \frac{140 \times 100}{175}$$

$$E = 80\%$$

YOUR TURN

Now you try one. A gasoline shop engine rated at 65 hp is found to deliver 56 hp through a belt drive to a pump. What is the efficiency of the drive system?

output = 56 hp,   input = 65 hp

$$\frac{56}{65} = \frac{E}{100}$$

$$E = \frac{56 \times 100}{65} = 86\% \text{ rounded}$$

56 $\times$ **100** $\div$ **65** $=$ →   **86.15384615**

The efficiency of any practical system will always be less than 100%—you can't produce an output greater than the input!

Tolerances   On technical drawings or other specifications, measurements are usually given with a *tolerance* showing the allowed error. For example, the fitting shown in this drawing has a length of

1.370 ± 0.015 in.

This means that the dimension shown must be between

1.370 + 0.015 in.   or   1.385 in.

and

1.370 − 0.015 in.   or   1.355 in.

The *tolerance limits* are 1.385 in. and 1.355 in.

Very often the tolerance is written as a percent.

$$\frac{\text{tolerance}}{\text{measurement}} = \frac{\text{percent tolerance}}{100}$$

In this case,

$$\frac{0.015}{1.370} = \frac{\text{percent tolerance } R}{100}$$

$$R = \frac{0.015 \times 100}{1.370}$$

$$R = 1.0948\ldots \quad \text{or} \quad 1.1\% \text{ rounded}$$

The dimension would be written 1.370 in. ± 1.1%.

Rewrite the following dimension with a percent tolerance:

1.426 ± 0.010 in. = _____

$$\frac{0.010}{1.426} = \frac{\text{percent tolerance } R}{100}$$

$$R = \frac{0.010 \times 100}{1.426}$$

$$R = 0.70126\ldots \quad \text{or} \quad 0.70\% \text{ rounded}$$

$$1.426 \pm 0.010 \text{ in.} = 1.426 \text{ in.} \pm 0.70\%$$

If the dimension is given in **metric units,** the procedure is exactly the same.

EXAMPLE

The dimension

315 ± 0.25 mm

can be converted to percent tolerance.

$$\frac{0.25}{315} = \frac{R}{100}$$

$$R = \frac{0.25 \times 100}{315}$$

$$R = 0.07936\ldots \quad \text{or} \quad 0.08\% \text{ rounded}$$

The dimension is 315 mm $\pm 0.08\%$.

 .25 $\boxed{\times}$ **100** $\boxed{\div}$ **315** $\boxed{=}$ → ⬛ *0.079365079*

If the tolerance is given as a **percent,** it is easy to use the proportion to calculate the tolerance as a dimension value.

$$\frac{\text{tolerance}}{\text{measurement}} = \frac{\text{percent tolerance}}{100}$$

EXAMPLE

2.450 in. $\pm 0.2\%$ can be converted to a dimension value as follows:

$$\frac{\text{tolerance } T}{2.450} = \frac{0.2}{100}$$

$$T = \frac{0.2 \times 2.45}{100}$$

$$T = 0.0049 \text{ in.} \quad \text{or} \quad 0.005 \text{ in. rounded}$$

Therefore,

2.450 in. $\pm 0.2\% = 2.450$ in. ± 0.005 in.

 .2 $\boxed{\times}$ **2.45** $\boxed{\div}$ **100** $\boxed{=}$ → ⬛ *0.0049*

PROBLEMS

Now, for some practice in working with tolerances, convert the measurement tolerances on page 190 to percents, and vice versa. Round the tolerances to three decimal places and round the percents to two decimal places.

Measurement	Tolerance	Percent Tolerance
2.345 in.	±0.001 in. ±0.002 in. ±0.005 in. ±0.010 in. ±0.125 in.	
274 mm	±0.10 mm ±0.20 mm ±0.50 mm	
3.475 in.		±0.10% ±0.20%
123 mm		±0.05% ±0.15%

Measurement	Tolerance	Percent Tolerance
2.345 in.	±0.001 in. ±0.002 in. ±0.005 in. ±0.010 in. ±0.125 in.	±0.04% ±0.09% ±0.21% ±0.43% ±5.33%
274 mm	±0.10 mm ±0.20 mm ±0.50 mm	±0.04% ±0.07% ±0.18%
3.475 in.	±0.003 in. ±0.007 in.	±0.10% ±0.20%
123 mm	±0.062 mm ±0.185 mm	±0.05% ±0.15%

Percent Change In many situations in technical work and the trades, you may need to find a *percent increase* or *percent decrease* in a given quantity.

EXAMPLE

Suppose that the output of a certain electrical circuit is 20 amperes (A), and the output increases by 10%. What is the new value of the output?

$$\frac{\text{amount of increase}}{\text{original amount}} = \frac{\text{percent increase}}{100}$$

$$\frac{\text{amount of increase } i}{20} = \frac{10}{100}$$

$$100i = 200$$

$$i = 2\,\text{A}$$

New value = original amount + increase

$$= 20\,A + 2A = 22\,A$$

A quicker way to solve the problem is to add 100% to the percent increase and use this total for the rate R. The revised proportion becomes

$$\frac{\text{new value } V}{\text{original amount}} = \frac{100\% + \text{percent increase}}{100}$$

$$\frac{V}{20} = \frac{100\% + 10\%}{100}$$

$$\frac{V}{20} = \frac{110}{100}$$

$$V = \frac{20 \times 110}{100}$$

$$V = 22\,A$$

Percent decrease can be calculated in a very similar way.

EXAMPLE

Suppose that the output of a certain machine at the Acme Gidget Co. is 600 gidgets per day. If the output decreases by 8%, what is the new output?

Percent *decrease* involves subtracting, and, as with percent increase, we can subtract either at the beginning or at the end of the calculation. Subtracting at the end, we have

$$\frac{\text{amount of decrease}}{\text{original amount}} = \frac{\text{percent decrease}}{100}$$

$$\frac{\text{amount of decrease } d}{600} = \frac{8}{100}$$

$$d = \frac{8 \times 600}{100}$$

$$d = 48 \text{ gidgets per day}$$

Now subtract to find the new value.

New value = original amount − amount of decrease

$$= 600 - 48 = 552 \text{ gidgets per day}$$

We can also solve the problem by subtracting at the beginning of the problem.

$$\frac{\text{new value}}{\text{original amount}} = \frac{100\% - \text{percent decrease}}{100}$$

$$\frac{\text{new value } V}{600} = \frac{100\% - 8\%}{100}$$

$$\frac{V}{600} = \frac{92}{100}$$

$$100V = 55200$$

$$V = 552 \text{ gidgets per day}$$

YOUR TURN

Try these problems involving percent change. In each, add or subtract the percents at the beginning of the problem.

(a) What is your new pay rate if your old rate of $9.72 per hour is increased 15%?

(b) **Electrician** Normal line voltage, 115 volts, drops 3.5% during a system malfunction. What is the reduced voltage?

(c) **Construction Technology** Because of changes in job specifications, the cost of a small construction job is increased 25% from the original cost of $2475. What is the cost of the job now?

(d) **Welding** By using automatic welding equipment, the time for a given job can be reduced by 40% from its original 30 hours. How long will the job take using the automatic equipment?

SOLUTIONS

(a) $\dfrac{\text{new pay rate}}{\$9.72} = \dfrac{115}{100}$ ← $\boxed{100\% + 15\%}$

New pay rate $= \dfrac{9.72 \times 115}{100} = \$11.178$ or $11.18 rounded

(b) $\dfrac{\text{reduced voltage}}{115 \text{ volts}} = \dfrac{96.5}{100}$ ← $\boxed{100\% - 3.5\%}$

Reduced voltage $= \dfrac{115 \times 96.5}{100} = 110.975$ or   111 volts rounded

115 ✕ **96.5** ÷ **100** = →   *110.975*

(c) $\dfrac{\text{new cost}}{\$2475} = \dfrac{125}{100}$

New cost $= \dfrac{2475 \times 125}{100} = \$3093.75$

(d) $\dfrac{\text{new time}}{30} = \dfrac{60}{100}$

New time $= \dfrac{30 \times 60}{100} = 18 \text{ hours}$

In trade and technical work, changes in a measured quantity are often specified as a percent increase or decrease.

EXAMPLE

If the reading on a pressure valve increases from 30 psi to 36 psi, the percent increase can be found by using the following proportion:

$\begin{array}{c} \text{The change} \to \\ \text{Original value} \longrightarrow \end{array} \dfrac{36 \text{ psi} - 30 \text{ psi}}{30 \text{ psi}} = \dfrac{\text{percent increase}}{100}$

$$\frac{6}{30} = \frac{R}{100}$$

$$R = \frac{100 \times 6}{30} = 20\%$$

The reading on the pressure valve increased by 20%.

100 ✕ (**36** − **30**) ÷ **30** = →   *20.*

The length of a heating duct is reduced from 110 in. to 96 in. because of design changes. The percent decrease is calculated as follows:

$$\frac{\text{The change} \rightarrow 110 \text{ in.} - 96 \text{ in.}}{\text{Original value} \longrightarrow 110 \text{ in.}} = \frac{\text{percent decrease}}{100}$$

$$\frac{14}{110} = \frac{R}{100}$$

$$R = \frac{100 \times 14}{110} = 12.7272\ldots \qquad \text{or} \approx 13\%$$

The length of the duct was reduced by about 13%.

Careful ▶ Students often are confused about which number to use for the base in percent change problems. The difficulty usually arises when they think in terms of which is the bigger or smaller value. Always use the *original* value as the base. In percent increase problems, the original or base number will be the smaller number, and in percent decrease problems, the original or base number will be the larger number. ◀

YOUR TURN

Now try these problems to sharpen your ability to work with percent changes.

(a) **Carpentry** If 8 in. is cut from a 12-ft board, what is the percent decrease in length?

(b) **Electrician** What is the percent increase in voltage when the voltage increases from 65 volts to 70 volts?

(c) **Industrial Technology** The measured value of the power output of a motor is 2.7 hp. If the motor is rated at 3 hp, what is the percent difference between the measured value and the expected value?

SOLUTIONS

(a) Use 12 ft = 144 in.

$$\frac{\text{The change} \longrightarrow 8 \text{ in.}}{\text{Original value} \rightarrow 144 \text{ in.}} = \frac{\text{percent decrease}}{100}$$

$$R = \frac{100 \times 8}{144} = 5.555\ldots \quad \text{or} \quad 5.6\% \text{ rounded}$$

The length of the board decreased by about 5.6%.

(b) To find the amount of increase, subtract 65 volts from 70 volts.

$$\frac{70 \text{ volts} - 65 \text{ volts}}{\text{Original value} \longrightarrow 65 \text{ volts}} = \frac{\text{percent increase}}{100}$$

$$\frac{5}{65} = \frac{R}{100}$$

$$R = \frac{100 \times 5}{65} = 7.6923\ldots \quad \text{or } 7.7\% \text{ rounded}$$

The voltage increased by about 7.7%.

(c) To find the amount of decrease, subtract 2.7 hp from 3 hp.

$$\frac{3\,\text{hp} - 2.7\,\text{hp}}{3\,\text{hp}} = \frac{\text{percent difference}}{100}$$

The expected value is used as the base.

$$\frac{0.3}{3} = \frac{R}{100}$$

$$R = \frac{100 \times 0.3}{3} = 10\%$$

There is a 10% difference between the measured value and the expected value.

Now, turn to Exercises 4-4 for a set of problems involving percent calculations.

Exercises 4-4

Applications of Percent Calculations

1. **Electronics** A CB radio is rated at 7.5 watts, and actual measurements show that it delivers 4.8 watts to its antenna. What is its efficiency?

2. **Electrical Technology** An electric motor uses 6 kilowatts at an efficiency of 63%. How much power does it deliver?

3. **Electrical Technology** An engine supplies 110 hp to an electric generator, and the generator delivers 70 hp of electrical power. What is the efficiency of the generator?

4. **Electronics** Electrical resistors are rated in ohms and color coded to show both their resistance and percent tolerance. For each of the following resistors, find its tolerance limits and actual tolerance.

Resistance (ohms)	Limits (ohms)	Tolerance (ohms)
5300 ± 5%	_____ to _____	_____
2750 ± 2%	_____ to _____	_____
6800 ± 10%	_____ to _____	_____
5670 ± 20%	_____ to _____	_____

5. **Auto Mechanics** A 120-hp automobile engine delivers only 81 hp to the driving wheels of the car. What is the efficiency of the transmission and drive mechanism?

6. **Electronics** An electrical resistor is rated at 500 ohms ±10%. What is the highest value its resistance could have within this tolerance range?

7. **Roofing** On a roofing job 21 of 416 shingles had to be rejected for minor defects. What percent is this?

8. If you earn 12% commission on sales of $4200, what actual amount do you earn?

9. **Metalworking** On a cutting operation 2 sq ft of sheet steel is wasted for every 16 sq ft used. What is the percent waste?

10. **Metalworking** Four pounds of a certain bronze alloy is one-sixth tin, 0.02 zinc, and the rest copper. Express the portion of each metal in (a) percents and (b) pounds.

11. **Metalworking** An iron casting is made in a mold with a hot length of 16.40 in. After cooling, the casting is 16.25 in. long. What is the shrinkage in percent?

12. **Manufacturing** Because of friction, a pulley block system is found to be only 83% efficient. What actual load can be raised if the theoretical load is 2000 lb?

13. **Manufacturing** A small gasoline shop engine develops 65 hp at 2000 rpm. At 2400 rpm its power output is increased by 25%. What actual horsepower does it produce at 2400 rpm?

14. **Building Construction** On the basis of past experience, a contractor expects to find 4.5% broken bricks in every truck load. If he orders 2000 bricks, will he have enough to complete a job requiring 1900 bricks?

15. **Machine Technology** Specifications call for a hole in a machined part to be 2.315 in. in diameter. If the hole is measured to be 2.318 in., what is the machinist's percent error?

16. **Welding** A welding shop charges the customer 185% of its labor and materials costs to cover overhead and profit. If a bill totals $219.78, what is the cost of labor and materials?

17. Complete the following table.

Measurement	Tolerance	Percent Tolerance
3.425 in.	±0.001 in.	(a)
3.425 in.	±0.015 in.	(b)
3.425 in.	(c)	±0.20%

18. If you receive a pay increase of 12%, what is your new pay rate, assuming that your old rate was $14.65 per hour?

19. **Manufacturing** The pressure in a hydraulic line increases from 40 psi to 55 psi. What is the percent increase in pressure?

20. **Painting and Decorating** The cost of the paint used in a redecorating job is $96.80. This is a reduction of 20% from its initial cost. What was the original cost?

21. What sales tax, at a rate of $6\frac{1}{2}\%$, must you pay on the purchase of a computer hard disk drive costing $256.75?

22. **Interior Design** An interior designer working for a department store is paid a weekly salary of $350 plus a 20% commission on total sales. What would be his weekly pay if his total sales were $1875.00 for the week?

23. **Electronics** Murph's electronic supply shop offered a 40-megahertz digital oscilloscope on sale at 25% off. If the original price was $739, what was the sale price?

24. **Upholstery** Chris paid for the initial expenses in setting up her upholstery shop with a bank loan for $17,000 for 5 years at $12\frac{1}{2}\%$ annual interest.
    (a) What total interest will she pay?
    (b) What will be her monthly payments, assuming equal monthly installments?

25. **Woodworking** A right-angle grinder is on sale for $65.80 and is marked as being 30% off. What was the original list price?

26. Driving into a 20-mph breeze cuts gas mileage by about 12%. What will be your gas mileage in a 20-mph wind if it is 24 miles/gallon with no wind?

27. **Auto Mechanics**   Supplier A offers a mechanic a part at 35% off the retail price of $68.40. Supplier B offers him the same part at 20% over the whole-sale cost of $35.60. Which is the better deal, and by how much?

28. **Auto Mechanics**   An auto mechanic purchases the parts necessary for a repair for $126.40 and sells them to the customer at a 35% markup. How much does the mechanic charge the customer?

29. **Auto Mechanics**   A new tire had a tread depth of $\frac{3}{8}$ in. After one year of driving the tire had $\frac{9}{32}$ in. of tread remaining. What percent of the tread was worn?

30. **Printing**   A printer agrees to give a nonprofit organization a 10% discount on a printing job. The normal price to the customer is $1020, and the printer's cost is $850. What is the printer's percent profit over cost after the discount is subtracted?

31. **Wastewater Technology**   A treatment plant with a capacity of 20 MGD (million gallons per day) has a normal daily flow equal to 60% of capacity. When it rains the flow increases by 30% of the normal flow. What is the flow on a rainy day?

32. **Police Science**   Last year a community had 23 homicides. This year they had 32 homicides. What was the percent of increase?

33. **Fire Science**   The pressure reading coming out of a pump is 180 lb. Every 50-ft section reduces the pressure by about 3%. What will the nozzle pressure be at the end of six 50-ft sections? (*Hint:* One reduction of 18% is *not* equivalent to six reductions of 3% each. You must do this the long way.)

34. **Machine Technology**   Complete the following table of tolerances for some machine parts.

Measurement	Tolerance	Maximum	Minimum	Percent Tolerance
1.58 in.	±0.002 in.			
	±0.005 in.			
				±0.15%
0.647 in.	±0.004 in.			
	±0.001 in.			
				±0.20%
165.00 mm	±0.15 mm			
	±0.50 mm			
				±0.05%
35.40 mm	±0.01 mm			
	±0.07 mm			
				±0.05%
				±0.08%
				±0.10%

35. **Electrical Technology**   An electrician purchases some outdoor lighting for a customer at a wholesale price of $928.80. The customer would have paid the retail price of $1290. What was the wholesale discount?

36. In problem 35, what will the total retail price be after 7.75% tax is added?

37. **Building Construction**   A contractor takes a draw of $46,500 from a home-owner to start a remodeling job. He uses $22,450 to purchase materials and deposits the remaining money in a three-month CD paying 1.75% annual interest. How much interest does he earn on the remaining funds during the three-month term?

38. **Food Services**   A wholesale food sales rep is paid a base salary of $1200 per month plus a $3\frac{1}{2}\%$ commission on sales. How much sales must she generate in order to earn $35,000 per year in total compensation?

39. **Roofing**   A sales rep for roofing materials figures he can generate about $1,800,000 in business annually. What rate of commission does he need in order to earn $50,000?

40. **Carpentry**   A power planer originally selling for $139.50 is on sale for $111.60. By what percent was it discounted?

41. **Auto Mechanics**   Last year, the owner of the Fixum Auto Repair Shop paid a premium of $9.92 per $100 of gross wages for workmen's compensation insurance. Because his shop had a superior safety record over the last 12 months, his premium rate is being discounted by 22% this year. If he has 20 employees averaging $22 per hour, 40 hours per week for 50 weeks, what will be his total premium for this year?

42. **Electrician**   Because they have a history of paying their bills on time, the Radius Electric Company receives an 8.25% discount on all purchases of electrical supplies from their supplier. What would they pay for cable and fixtures originally costing $4760?

43. **Heating and Air Conditioning**   The following table shows the recommended cooling capacity (in Btu per hour) of a room air conditioner based on the area (in square feet) of the room. If more than two people regularly occupy the room, add 600 Btu/hr per additional person; if the room is particularly sunny, add 10%; and if the room gets very little sun, subtract 10% from the listed capacity.

Area (sq ft)	Capacity (Btu/hr)
250–300	7,000
300–350	8,000
350–400	9,000
400–450	10,000

Determine the recommended cooling capacity of a room air conditioner for:

(a) A sunny 320-square-foot room normally occupied by one or two people.
(b) A shady 440-square-foot room normally occupied by four people.

When you have completed these exercises, check your answers in the Appendix, then turn to Section 4-5 for a more advanced look at ratio and proportion.

4-5   APPLICATIONS OF RATIO AND PROPORTION

Earlier in this chapter we used the concepts of ratio and proportion to solve percent problems and in some simple technical applications. In this section we will learn about additional applications of ratio and proportion in the trades and technical areas.

Scale Drawings   Proportion equations are found in a wide variety of practical situations. For example, when a drafter makes a drawing of a machine part, building layout, or other large structure, he or she must *scale it down*. The drawing must represent the object accurately, but it must be small enough to fit on the paper. The draftsperson reduces every dimension by some fixed ratio.

Drawings that are larger than life involve an expanded scale.

For the automobile shown, the ratio of the actual length to the scale-drawing length is equal to the ratio of the actual width to the scale-drawing width.

$$\frac{\text{actual length}}{\text{drawing length}} = \frac{\text{actual width}}{\text{drawing width}}$$

$$\frac{16\text{ ft }6\text{ in.}}{4\frac{1}{8}\text{ in.}} = \frac{6\text{ ft}}{1\frac{1}{2}\text{ in.}}$$ ⟵ This equation is a proportion

↑ Ratio of lengths   ↑ Ratio of widths

Rewrite all quantities in the same units:

$$\frac{198\text{ in.}}{4\frac{1}{8}\text{ in.}} = \frac{72\text{ in.}}{1\frac{1}{2}\text{ in.}}$$

You should notice first of all that each side of this equation is a ratio. Each side is a ratio of *like* quantities: lengths on the left and widths on the right.

Second, notice that the ratio $\dfrac{198\text{ in.}}{4\frac{1}{8}\text{ in.}}$ is equal to $\dfrac{48}{1}$.

Divide it out: $198 \div 4\frac{1}{8} = 198 \div \dfrac{33}{8}$

$$= 198 \times \frac{8}{33}$$

$$= 48$$

198 ⊠ **8** ⊟ **33** ⊜ → [　　　　　*48.*　　　　　]

Notice also that the ratio $\dfrac{72\text{ in.}}{1\frac{1}{2}\text{ in.}}$ is equal to $\dfrac{48}{1}$.

The common ratio $\dfrac{48}{1}$ is called the *scale* factor of the drawing.

In the following problem one of the dimensions is unknown:

Suppose that a rectangular room has a length of 18 ft and a width of 12 ft. An architectural scale drawing of this room is made so that on the drawing the length of the room is 4.5 in. What will be the width of the room on the drawing?

First, set up a ratio of lengths and a ratio of widths.

$$\text{Length ratio} = \frac{\text{actual length}}{\text{drawing length}}$$

$$= \frac{18 \text{ ft}}{4.5 \text{ in.}} = \frac{216 \text{ in.}}{4.5 \text{ in.}} \longleftarrow \text{Convert 18 ft to inches so that the top and the bottom of the fraction have the same units.}$$

$$\text{Width ratio} = \frac{\text{actual width}}{\text{drawing width}}$$

$$= \frac{12 \text{ ft}}{x \text{ in.}} = \frac{144 \text{ in.}}{x \text{ in.}} \longleftarrow \text{Change 12 ft to 144 in.}$$

Second, write a proportion equation.

$$\frac{216}{4.5} = \frac{144}{x}$$

Third, solve this proportion. Cross-multiply to get

$$216x = 4.5 \cdot 144$$

$$216x = 648$$

$$x = 3 \text{ in.}$$

For the room drawing shown, the scale factor is

$$\text{Scale factor} = \frac{18 \text{ ft}}{4.5 \text{ in}} = \frac{216 \text{ in.}}{4.5 \text{ in.}}$$

$$= 48 \qquad \text{or} \qquad 48 \text{ to } 1$$

One inch on the drawing corresponds to 48 in. or 4 ft on the actual object. A draftsperson would write this as $\frac{1}{4}$ in. = 1 ft.

The drawing of the triangular plate shown has a height of 18 in. and a base length of 14 in.

(a) What will be the corresponding height in a reduced copy of this drawing if the base length in the copy is $2\frac{3}{16}$ in.?

(b) Find the scale factor of the reduction.

SOLUTIONS

(a) $\dfrac{18 \text{ in.}}{x} = \dfrac{14 \text{ in.}}{2\frac{3}{16} \text{ in.}}$

Cross-multiply $\quad (18)\left(2\frac{3}{16}\right) = 14x$

$$14x = 39\frac{3}{8}$$

$$x = 2\frac{13}{16} \text{ in.}$$

2 $\boxed{a\frac{b}{c}}$ **3** $\boxed{a\frac{b}{c}}$ **16** $\boxed{\times}$ **18** $\boxed{=}$ $\boxed{\div}$ **14** $\boxed{=}$ → ▨ *2⌐13⌐16.* or ▨ *2.8125*

(b) Scale factor $= \dfrac{18 \text{ in.}}{2\frac{13}{16} \text{ in.}}$

$$= 6\frac{2}{5} \quad \text{or} \quad 6.4 \text{ to } 1$$

18 $\boxed{\div}$ **2** $\boxed{a\frac{b}{c}}$ **13** $\boxed{a\frac{b}{c}}$ **16** $\boxed{=}$ → ▨ *6⌐2⌐5.* or ▨ *6.4*

Similar Figures

In general, two geometric figures that have the same shape but are not the same size are said to be *similar* figures. The blueprint drawing and the actual object are a pair of similar figures. An enlarged photograph and the smaller original are similar.

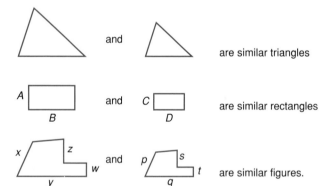

and are similar triangles

A and C are similar rectangles

and are similar figures.

In any similar figures, all parts of corresponding dimensions have the same scale ratio. For example, in the rectangles above

$$\frac{A}{C} = \frac{B}{D}$$

In the irregular figure above,

$$\frac{x}{p} = \frac{y}{q} = \frac{z}{s} = \frac{w}{t} \text{ and so on}$$

The triangles shown here are *not* similar:

Find the missing dimension in each of the following pairs of similar figures.

(a)

(b)

(c)

(d)

SOLUTIONS

(a) $\dfrac{A}{144} = \dfrac{4}{120}$ Convert ft to in.

$A = \dfrac{4(144)}{120} = 4.8 \text{ in.}$

(b) $\dfrac{B}{4\frac{1}{8}} = \dfrac{5\frac{1}{2}}{11}$

$B = \dfrac{5\frac{1}{2}\left(4\frac{1}{8}\right)}{11} = 2\frac{1}{16} \text{ in.}$

(c) $\dfrac{c}{1.25} = \dfrac{6.5}{2.6}$

$c = \dfrac{6.5(1.25)}{2.6} = 3.125 \text{ cm}$

(d) $\dfrac{D}{3\frac{1}{4}} = \dfrac{6.3}{1\frac{1}{2}}$

$D = \dfrac{6.3\left(3\frac{1}{4}\right)}{1\frac{1}{2}} = 13.65 \text{ in.}$

Work problem (d) on a calculator this way:

  3 $\boxed{a\frac{b}{c}}$ 1 $\boxed{a\frac{b}{c}}$ 4 $\boxed{\times}$ 6.3 $\boxed{\div}$ 1.5 $\boxed{=}$ → $\quad$ *13.65*

Direct and Inverse Proportion

Many trade problems can be solved by setting up a proportion involving four related quantities. But it is important that you recognize that there are *two* kinds of proportions—direct and inverse. Two quantities are said to be *directly proportional* if an increase in one quantity leads to a proportional increase in the other quantity, or if a decrease in one leads to a decrease in the other.

> **Direct proportion:**   increase → increase
> or   decrease → decrease

For example, the electrical resistance of a wire is directly proportional to its length—the longer the wire, the greater the resistance. If 1 ft of Nichrome heater element wire has a resistance of 1.65 ohms, what length of wire is needed to provide a resistance of 19.8 ohms?

Resistance = 1.65 ohms
1 ft

Resistance = 19.8 ohms
L

First, recognize that this problem involves a *direct* proportion. As the length of wire increases, the resistance increases proportionally.

As L increases . . . R increases

$$\frac{L}{1 \text{ ft}} = \frac{R}{1.65 \text{ ohms}}$$

Both ratios increase in size when L increases.

Second, set up a direct proportion and solve.

$$\frac{L}{1 \text{ ft}} = \frac{19.8 \text{ ohms}}{1.65 \text{ ohms}}$$

$$L = 12 \text{ ft}$$

YOUR TURN

Solve each of the following problems by setting up a direct proportion.

(a) If a widget machine produces 88 widgets in 2 hours, how many will it produce in $3\frac{1}{2}$ hours?

(b) If one gallon of paint covers 825 sq ft, how much paint is needed to cover 2640 sq ft?

(c) What is the cost of six filters if eight filters cost $39.92?

(d) A diesel truck was driven 273 miles on 42 gallons of fuel. How much fuel is needed for a trip of 600 miles? (Round to the nearest gallon.)

(e) A cylindrical oil tank holds 450 gallons when it is filled to its full height of 8 ft. When it contains oil to a height of 2 ft 4 in., how many gallons of oil are in the tank?

SOLUTIONS

(a) A direct proportion—the more time spent, the more widgets produced.

$$\frac{88}{x} = \frac{2 \text{ hours}}{3\frac{1}{2} \text{ hours}}$$

Cross-multiply:

$$2x = 308$$
$$x = 154 \text{ widgets}$$

88 ⊗ **3.5** ⊖ **2** ⊜ → *154.*

(b) A direct proportion—the more paint, the greater the area that can be covered:

$$\frac{1 \text{ gal}}{x \text{ gal}} = \frac{825 \text{ sq ft}}{2640 \text{ sq ft}}$$

$$x = 3.2 \text{ gallons}$$

(c) A direct proportion—the more you pay, the more you get:

$$\frac{6 \text{ filters}}{8 \text{ filters}} = \frac{x}{\$39.92}$$

$$x = \$29.94$$

(d) A direct proportion—the more miles you drive, the more fuel it takes:

$$\frac{273 \text{ mi}}{600 \text{ mi}} = \frac{42 \text{ gal}}{x \text{ gal}}$$

$$x \approx 92 \text{ gallons}$$

(e) A direct proportion—the volume is directly proportional to the height:

$$\frac{450 \text{ gal}}{x \text{ gal}} = \frac{8 \text{ ft}}{2\frac{1}{3} \text{ ft}} \quad \Longleftarrow \boxed{2'\ 4'' = 2\frac{4}{12}' = 2\frac{1}{3}'}$$

$$x = 131\frac{1}{4} \text{ gallons}$$

Two quantities are said to be *inversely proportional* if an increase in one quantity leads to a proportional decrease in the other quantity, or if a decrease in one leads to an increase in the other.

> **Inverse Proportion:** increase → decrease
> or   decrease → increase

For example, the time required for a trip of a certain length is *inversely* proportional to the speed of travel. If a certain trip takes 2 hours at 50 mph, how long will it take at 60 mph?

The correct proportion equation is

By inverting the time ratio, we have set it up so that both sides of the equation are in balance—both ratios decrease as speed increases.

Before attempting to solve the problem, make an estimate of the answer. We expect that the time to make the trip at 60 mph will be *less* than the time at 50 mph. The correct answer should be less than 2 hours.

Now solve it by cross-multiplying:

$$60x = 2 \cdot 50$$

$$x = 1\tfrac{2}{3} \text{ hours}$$

Learning Help ▶ Remember, in a *direct* proportion

$$\frac{A}{B} = \frac{C}{D}$$ These terms go together. If B and D are fixed, then an increase in A goes with an increase in C.

In an *inverse* proportion

$$\frac{X}{Y} = \frac{P}{Q}$$ These terms go together. If P and Y are fixed, then an increase in X goes with a decrease in Q. ◀

YOUR TURN

Try this problem. In an automobile cylinder, the pressure is inversely proportional to the volume if the temperature does not change. If the volume of gas in the cylinder is 300 cu cm when the pressure is 20 psi, what is the volume when the pressure is increased to 80 psi?

Set this up as an inverse proportion and solve.

SOLUTION

Pressure is inversely proportional to volume. If the pressure increases, we expect the volume to decrease.

The answer should be less than 300 cu cm.

$$\frac{P_1}{P_2} = \frac{V_2}{V_1}$$

$$\frac{20 \text{ psi}}{80 \text{ psi}} = \frac{V}{300 \text{ cu cm}}$$

Cross-multiply:

$$80V = 20 \cdot 300$$

$$V = 75 \text{ cu cm}$$

Gears and Pulleys A particularly useful kind of inverse proportion involves the relationship between the size of a gear or pulley and the speed with which it rotates.

In the following diagram, A is the driving gear.

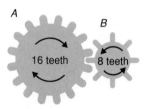

Because gear A has twice as many teeth as gear B, when A turns one turn, B will make two turns. If gear A turns at 10 turns per second, gear B will turn at 20

turns per second. The speed of the gear is inversely proportional to the number of teeth.

$$\frac{\text{speed of gear } A}{\text{speed of gear } B} = \frac{\text{teeth in gear } B}{\text{teeth in gear } A}$$

| This ratio has the A term on top | This ratio has the B term on top |

In this proportion, gear speed is measured in revolutions per minute, abbreviated rpm.

EXAMPLE

For the gear assembly shown, if gear A is turned by a shaft at 40 rpm, what will be the speed of gear B?

Because the relation is an inverse proportion, the smaller gear moves with the greater speed. We expect the speed of gear B to be faster than 40 rpm.

$$\frac{40 \text{ rpm}}{B} = \frac{8 \text{ teeth}}{16 \text{ teeth}}$$

$$B = 80 \text{ rpm}$$

On an automobile the speed of the drive shaft is converted to rear axle motion by the ring and pinion gear system.

$$\frac{\text{drive shaft speed}}{\text{rear axle speed}} = \frac{\text{teeth in ring gear on axle}}{\text{teeth in pinion gear on drive shaft}}$$

| This ratio has the drive shaft term on top | This ratio has the drive shaft term on the bottom |

Again, gear speed is inversely proportional to the number of teeth on the gear.

YOUR TURN

If the pinion gear has 9 teeth and the ring gear has 40 teeth, what is the rear axle speed when the drive shaft turns at 1200 rpm?

SOLUTION

$$\frac{1200 \text{ rpm}}{R} = \frac{40 \text{ teeth}}{9 \text{ teeth}}$$

Cross-multiply: $40R = 9 \cdot 1200$

$$R = 270 \text{ rpm}$$

Pulleys transfer power in much the same way as gears. For the pulley system shown, the speed of a pulley is inversely proportional to its diameter.

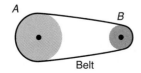

Belt

If pulley A has a diameter twice that of pulley B, when pulley A makes one turn, pulley B will make two turns, assuming of course that there is no slippage of the belt.

$$\frac{\text{speed of pulley } A}{\text{speed of pulley } B} = \frac{\text{diameter of pulley } B}{\text{diameter of pulley } A}$$

| This ratio has the A term on top | This ratio has the A term on the bottom |

YOUR TURN

If pulley B is 16 in. in diameter and is rotating at 240 rpm, what is the speed of pulley A if its diameter is 20 in.?

SOLUTION

$$\frac{A}{240 \text{ rpm}} = \frac{16 \text{ in.}}{20 \text{ in.}}$$

Cross-multiply: $20A = 16 \cdot 240$

$$A = 192 \text{ rpm}$$

PROBLEMS

Solve each of the following problems by setting up an inverse proportion.

(a) A 9-in. pulley on a drill press rotates at 960 rpm. It is belted to a 5-in. pulley on an electric motor. Find the speed of the motor shaft.

(b) A 12-tooth gear mounted on a motor shaft drives a larger gear. The motor shaft rotates at 1450 rpm. If the speed of the large gear is to be 425 rpm, how many teeth must be on the large gear?

(c) For gases, pressure is inversely proportional to volume if the temperature does not change. If 30 cu ft of air at 15 psi is compressed to 6 cu ft, what is the new pressure?

(d) If five assembly machines can complete a given job in 3 hours, how many hours will it take for two assembly machines to do the same job?

(e) The forces and lever arm distances for a lever obey an inverse proportion.

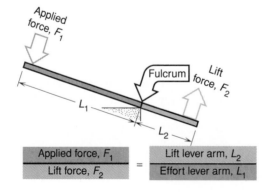

| Applied force, F_1 | | Lift lever arm, L_2 |
| Lift force, F_2 | $=$ | Effort lever arm, L_1 |

If a 100-lb force is applied to a 22-in. crowbar pivoted 2 in. from the end, what lift force is exerted?

SOLUTIONS

(a) $\dfrac{9 \text{ in.}}{5 \text{ in.}} = \dfrac{x}{960 \text{ rpm}}$ An inverse proportion: the larger pulley turns more slowly.

$$5x = 9 \cdot 960$$

$$x = 1728$$

(b) $\dfrac{12 \text{ teeth}}{x \text{ teeth}} = \dfrac{425 \text{ rpm}}{1450 \text{ rpm}}$ The larger gear turns more slowly.

$$425x = 12 \cdot 1450$$

$$x = 40.941 \ldots \quad \text{or 41 teeth, rounding to the nearest whole number. We can't have a part of a gear tooth!}$$

$$x \approx 41$$

(c) $\dfrac{30 \text{ cu ft}}{6 \text{ cu ft}} = \dfrac{P}{15 \text{ psi}}$ An inverse proportion: the higher the pressure, the smaller the volume.

$$6P = 30 \cdot 15$$

$$P = 75 \text{ psi}$$

(d) Careful on this one! An inverse proportion should be used. The *more* machines used, the *fewer* the hours needed to do the job.

$$\dfrac{5 \text{ machines}}{2 \text{ machines}} = \dfrac{x \text{ hours}}{3 \text{ hours}}$$

$$2x = 15$$

$$x = 7\tfrac{1}{2} \text{ hours} \quad \text{Two machines will take much longer to do the job than will five machines.}$$

(e) $\dfrac{100 \text{ lb}}{F} = \dfrac{2 \text{ in.}}{20 \text{ in.}}$ If the entire bar is 22 in. long, and $L_2 = 2$ in., then $L_1 = 20$ in.

$$2F = 20 \cdot 100$$

$$F = 1000 \text{ lb}$$

Now turn to Exercises 4-5 for a set of practice problems on applications of ratio and proportion.

Exercises 4-5   Applications of Ratio and Proportion

A. Complete the following tables.

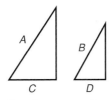
Similar triangles

1.

	A	B	C	D
(a)	$5\tfrac{1}{2}$ in.	$1\tfrac{1}{4}$ in.	$2\tfrac{3}{4}$ in.	
(b)		23.4 cm	20.8 cm	15.6 cm
(c)	12 ft		9 ft	6 ft
(d)	4.5 m	3.6 m		2.4 m

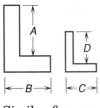

Similar figures

2.

	A	B	C	D
(a)	4 in.	5 in.	$3\frac{1}{2}$ in.	
(b)	5 ft		5 ft	2 ft
(c)	6.4 m	5.6 m		1.6 m
(d)		36 m	12 cm	14 cm

3.

	Number of Teeth on Gear 1	Number of Teeth on Gear 2	RPM of Gear 1	RPM of Gear 2
(a)	20	48	240	~~1,002~~ 100
(b)	25	70	150	420
(c)	64	40	160	100
(d)	32	40	960	1200

4.

	Diameter of Pulley 1	Diameter of Pulley 2	RPM of Pulley 1	RPM of Pulley 2
(a)	18 in.	24 in.	200	266.6667
(b)	12 in.		300	240
(c)		5 in.	400	640
(d)	14 in.	6 in.	300	

B. Practical Problems

1. **Metalworking**   A line shaft rotating at 250 rpm is connected to a grinding wheel by the pulley assembly shown. If the grinder shaft must turn at 1200 rpm, what size pulley should be attached to the line shaft?

Problem 1

2. **Architectural Technology**   The architectural drawing of an outside deck is $3\frac{1}{2}$ in. wide by $10\frac{7}{8}$ in. long. If the deck will actually be 14 ft wide, calculate the following:
   (a) The actual length of the deck
   (b) The scale factor

3. **Auto Mechanics**   Horsepower developed by an engine varies directly as its displacement. How many horsepower will be developed by an engine with a displacement of 240 cu in. if a 380-cu in. engine of the same kind develops 220 hp?

4. **Transportation Technology** The distance necessary to stop a subway train at a given speed is inversely proportional to its deceleration. If a train traveling at 30 mph requires 180 ft to stop when decelerating at 0.18 g, what is the stopping distance at the same speed when it is decelerating at 0.15 g?

5. A crowbar 28 in. long is pivoted 6 in. from the end. What force must be applied at the long end in order to lift a 400-lb object at the short end?

6. If 60 gal of oil flow through a certain pipe in 16 minutes, how long will it take to fill a 450-gal tank using this pipe?

7. **Auto Mechanics** If the alternator-to-engine drive ratio is 2.45 to 1, what rpm will the alternator have when the engine is idling at 400 rpm? (*Hint:* Use a direct proportion.)

8. The length of a wrench is inversely proportional to the amount of force needed to loosen a bolt. A wrench 6 in. long requires a force of 240 lb to loosen a rusty bolt. How much force would be required to loosen the same bolt using a 10-in. wrench?

9. **Metalworking** A bearing alloy is made up of 74% copper, 16% lead, and 10% tin. If 17 lb of copper is used to obtain a batch of this alloy, what weight of alloy is obtained? (Round to the nearest pound.)

10. **Machine Technology** A pair of belted pulleys have diameters of 20 in. and 16 in., respectively. If the larger pulley turns at 2000 rpm, how fast will the smaller pulley turn?

11. **Manufacturing** A 15-tooth gear on a motor shaft drives a larger gear having 36 teeth. If the motor shaft rotates at 1200 rpm, what is the speed of the larger gear?

12. **Electronics** The power gain of an amplifier circuit is defined as

$$\text{Power gain} = \frac{\text{output power}}{\text{input power}}$$

If the audio power amplifier circuit has an input power of 0.72 watt and a power gain of 30, what output power will be available at the speaker?

13. **Industrial Technology** If 12 assemblers can complete a certain job in 4 hours, how long will the same job take if the number of assemblers is cut back to 8?

14. **Electronics** A 115-volt power transformer has 320 turns on the primary. If it delivers a secondary voltage of 12 volts, how many turns are on the secondary? (*Hint:* Use a direct proportion.)

15. **Auto Mechanics** The headlights of a car are mounted at a height of 3.5 ft. If the light beam drops 1 in. per 35 ft, how far ahead in the road will the headlights illuminate?

16. A truck driver covers a certain stretch of the interstate in $4\frac{1}{2}$ hours traveling at the posted speed limit of 55 mph. If the speed limit is raised to 65 mph, how much time will the same trip require?

17. **Machine Technology** It is known that a cable with a cross-sectional area of 0.60 sq in. has a capacity to hold 2500 lb. If the capacity of the cable is proportional to its cross-sectional area, what size cable is needed to hold 4000 lb?

18. **Masonry** Cement, sand, and gravel are mixed to a proportion of 1:3:6 for a particular batch of concrete. How many cubic yards of each should be used to mix 125 cubic yards of concrete?

Problem 19

19. **Plumbing**   A cylindrical vent 6 in. in diameter must be cut at an angle to fit on a gable roof with a $\frac{2}{3}$ pitch. This means that for the vent itself the ratio of rise to run will be 2:3. Find the height x of the cut that must be made on the cylinder to make it fit the slope of the roof. (See the figure.)

20. **Architectural Technology**   A wrought-iron gate for a new house will be 18 ft 9 in. long and 9 ft 6 in. high. An architect makes a drawing of the gate using a scale factor of $\frac{1}{4}$ in. = 1 ft. What will be the dimensions of the gate on the drawing?

21. **Auto Mechanics**   When a tire is inflated, the air pressure is inversely proportional to the volume of the air. If the pressure of a certain tire is 28 psi when the volume is 120 cu in., what is the pressure when the volume is 150 cu in.?

22. **Electrical Technology**   The electrical resistance of a given length of wire is inversely proportional to the square of the diameter of the wire:

$$\frac{\text{resistance of wire } A}{\text{resistance of wire } B} = \frac{(\text{diameter of } B)^2}{(\text{diameter of } A)^2}$$

If a certain length of wire with a diameter of 34.852 mils has a resistance of 8.125 ohms, what is the resistance of the same length of the same composition wire with a diameter of 45.507 mils?

23. If you are paid $238.74 for $21\frac{1}{2}$ hours of work, what amount should you be paid for 34 hours of work at this same rate of pay?

24. **Sheet Metal Technology**   If the triangular plate shown is cut into eight pieces along equally spaced dashed lines, find the height of each cut. (Round to two decimal digits.)

$A =$ _____ $B =$ _____

$C =$ _____ $D =$ _____

$E =$ _____ $F =$ _____

$G =$ _____

Problem 24

25. **Wastewater Technology**   The time required for an outlet pipe to empty a tank is inversely proportional to the cross-sectional area of the pipe. A pipe with a cross-sectional area of 113.0 sq in. requires 6.4 hours to empty a certain tank. If the pipe was replaced with one with a cross-sectional area of 50.25 sq in., how long would it take this pipe to empty the same tank?

26. A gas has a volume of 2480 cu cm at a pressure of 63.5 psi. What is the pressure when the gas is compressed to 1830 cu cm? (Round to the nearest tenth.)

27. **Machine Technology**   The base of the flange shown in the drawing is actually 8 in. Its drawing width is 3 in. (a) What will be the actual dimension A if A is 2.25 in. on the drawing? (b) Find the scale factor of the drawing.

Problem 27

28. **Manufacturing** The drawing shows the actual length and width of a display panel. (a) What will be the corresponding drawing length of the panel if the drawing width is $2\frac{1}{4}$ in.? (b) Find the scale factor for this drawing.

Problem 28

29. **Plumbing** If a 45-gallon hot water tank holds 375 lb of water, what weight of water will a 55-gallon tank hold? (Round to the nearest pound.)

When you have completed these exercises, check your answers in the Appendix, then turn to Problem Set 4 for a set of practice problems on ratio, proportion, and percent.

4

Ratio, Proportion, and Percent

Answers are given in the Appendix.

A. Complete the following tables.

1.

	Diameter of Pulley A	Diameter of Pulley B	Pulley Ratio, $\dfrac{A}{B}$
(a)	12 in.		2 to 5
(b)	95 cm	38 cm	
(c)		15 in.	5 to 3

2.

	Teeth on Driving Gear A	Teeth on Driven Gear B	Gear Ratio, $\dfrac{A}{B}$
(a)	20	60	
(b)		10	6.5
(c)	56		4 to 3

B. Solve the following proportions.

1. $\dfrac{5}{6} = \dfrac{x}{42}$

2. $\dfrac{8}{15} = \dfrac{12}{x}$

3. $\dfrac{x}{12} = \dfrac{15}{9}$

4. $\dfrac{3\frac{1}{2}}{x} = \dfrac{5\frac{1}{4}}{18}$

5. $\dfrac{1.6}{5.2} = \dfrac{4.4}{x}$

6. $\dfrac{x}{12.4} = \dfrac{4 \text{ ft } 6 \text{ in.}}{6 \text{ ft } 3 \text{ in.}}$

C. Write each number as a percent.

1. 0.72   2. 0.06   3. 0.6   4. 0.358   5. 1.3

6. 3.03   7. 4   8. $\dfrac{7}{10}$ 9. $\dfrac{1}{6}$ 10. $2\dfrac{3}{5}$

D. Write each percent as a decimal number.

1. 4%   2. 37%   3. 11%   4. 94%

5. $1\frac{1}{4}$%   6. 0.09%   7. $\frac{1}{5}$%   8. 1.7%

9. $3\frac{7}{8}$%   10. 8.02%   11. 115%   12. 210%

E. Write each of the following percents as a fraction in lowest terms.

1. 28%   2. 375%   3. 81%   4. 4%

5. 0.5%   6. 70%   7. 14%   8. $41\frac{2}{3}$%

Name _____

Date _____

Course/Section _____

F. Solve.

1. 3 is _____ % of 5.
2. 5% of $120 is _____.
3. 25% of what number is 1.4?
4. 16 is what percent of 8?
5. 105% of 40 is _____.
6. 1.38 is _____ % of 1.15?
7. $7\frac{1}{4}$% of _____ is $2.10.
8. 250% of 50 is _____.
9. 0.05% of _____ is 4.
10. $8\frac{1}{4}$% of 1.2 is _____.

G. Solve.

1. **Metalworking** Extruded steel rods shrink 12% in cooling from furnace temperature to room temperature. If a standard tie rod is 34 in. exactly when it is formed, how long will it be after cooling?

2. **Metalworking** Cast iron contains up to 4.5% carbon, and wrought iron contains up to 0.08% carbon. How much carbon is in a 20-lb bar of each metal?

3. (a) A real estate saleswoman sells a house for $364,500. Her company pays her 70% of the 1.5% commission on the sale. How much does she earn on the sale?

   (b) All salespeople in the Ace Junk Store receive $360 per week plus a 2% commission. If you sold $1975 worth of junk in a week what would be your income?

   (c) A salesman at the Wasteland TV Company sold five HDTV sets last week and earned $509.70 in commissions. If his commission is 6%, what does an HDTV set cost?

4. **Carpentry** What is the net price of a table saw with a list price of $165.95 if it is on sale at a 35% discount?

5. If the retail sales tax in your state is 6%, what would be the total cost of each of the following items?
   (a) A $9.98 pair of pliers.
   (b) A $14.95 adjustable wrench.
   (c) 69 cents worth of washers.
   (d) A $63.60 textbook.
   (e) A $329.95 shop cabinet rollaway.

6. A computer printer sells for $376 after a 12% discount. What was its original or list price?

7. **Building Construction** How many running feet of matched 1-in. by 6-in. boards will be required to lay a subfloor in a house that is 28 ft by 26 ft? Add 20% to the area to allow for waste and matching.

8. **Sheet Metal Technology** In the Easy Does It Metal Shop, one sheet of metal is wasted for every 25 purchased. What percent of the sheets are wasted?

9. **Machine Technology** Complete the following table.

Measurement	Tolerance	Percent Tolerance
1.775 in.	± 0.001 in.	(a)
1.775 in.	± 0.05 in.	(b)
1.775 in.	(c)	± 0.50%
310 mm	± 0.1 mm	(d)
310 mm	(e)	± 0.20%

10. **Electronics** If 16 in. is cut from a 6-ft cable, what is the percent decrease in length?

11. **Manufacturing**   A production job is bid at $6275, but cost overruns amount to 15%. What is the actual job cost?

12. **Metalworking**   When heated, a metal rod expands 3.5%, to 15.23 cm. What is its cold length?

13. **Electrical Technology**   An electrical resistor is rated at 4500 ohms ± 3%. Express this tolerance in ohms and state the actual range of resistance.

14. **Auto Mechanics**   A 140-hp automobile engine delivers only 126 hp to the driving wheels of the car. What is the efficiency of the transmission and drive mechanism?

15. **Metalworking**   A steel casting has a hot length of 26.500 in. After cooling, the length is 26.255 in. What is the shrinkage expressed as a percent? Round to one decimal place.

16. **Auto Mechanics**   The parts manager for an automobile dealership can buy a part at a 25% discount off the retail price. If the retail price is $46.75, how much does he pay?

17. **Manufacturing**   An electric shop motor rated at 2.5 hp is found to deliver 1.8 hp to a vacuum pump when it is connected through a belt drive system. What is the efficiency of the drive system?

18. **Welding**   In order to purchase a truck for his mobile welding service, Jerry arranges a 36-month loan of $45,000 at 4.75% annual interest.
    (a)   What total interest will he pay?
    (b)   What monthly payment will he make? (Assume equal monthly payments on interest and principal.)

19. **Refrigeration**   The motor to run a refrigeration system is rated at 12 hp, and the system has an efficiency of 76%. What is the effective cooling power output of the system?

20. **Interior Design**   An interior designer is able to purchase a sofa from a design center for $1922. Her client would have paid $2480 for the same sofa at a furniture store. What rate of discount was the designer receiving off the retail price?

21. **Manufacturing**   The commission rate paid to a manufacturer's rep varies according to the type of equipment sold. If his monthly statement showed $124,600 in sales, and his total commission was $4438, what was his average rate of commission to the nearest tenth of a percent?

22. **Masonry**   Six square feet of a certain kind of brick wall contains 78 bricks. How many bricks are needed for 150 square feet?

23. If 60 miles per hour is equivalent to 88 feet per second, express
    (a)   45 miles per hour in feet per second.
    (b)   22 feet per second in miles per hour.

24. **Auto Mechanics**   The headlights on a car are mounted at the height of 28 in. The light beam must illuminate the road for 400 ft. That is, the beam must hit the ground 400 ft ahead of the car.
    (a)   What should be the drop ratio of the light beam in inches per foot?
    (b)   What is this ratio in inches per 25 ft?

25. **Printing**   A photograph measuring 8 in. wide by 12 in. long must be reduced to a width of $3\frac{1}{2}$ in. in order to fit on a printed page. What will be the corresponding length of the reduced photograph?

26. **Machine Technology**   If 250 ft of wire weighs 22 lb, what will be the weight of 100 ft of the same wire?

27. **Agricultural Technology**   A sprayer discharges 600 cc of herbicide in 45 sec. For what period of time should the sprayer be discharged in order to apply 2000 cc of herbicide?

28. **Auto Mechanics**   An auto mechanic currently rents a 1460-sq ft shop for $1650 per month. How much will his monthly rent be if he moves to a comparably priced 1920-sq ft shop? (Round to the nearest dollar.)

29. **Machine Technology**   Six steel parts weigh 1.8 lb. How many of these parts are in a box weighing 142 lb if the box itself weighs 7 lb?

30. **Manufacturing** Two belted pulleys have diameters of 24 in. and 10 in.
    (a) Find the pulley ratio.
    (b) If the larger pulley turns at 1500 rpm, how fast will the smaller pulley turn?

31. **Printing** If six printing presses can do a certain job in $2\frac{1}{2}$ hours, how long will it take four presses to do the same job? (*Hint:* Use an inverse proportion.)

32. **Drafting** Drafters usually use a scale of $\frac{1}{4}$ in. = 1 ft on their architectural drawings. Find the actual length of each of the following items if their blueprint length is given.
    (a) A printing press $2\frac{1}{4}$ in. long.   (b) A building $7\frac{1}{2}$ in. long.
    (c) A rafter $\frac{11}{16}$ in. long.   (d) A car $1\frac{7}{8}$ in. long.

33. **Construction Technology** A certain concrete mix requires one sack of cement for every 650 lb of concrete. How many sacks of cement are needed for 3785 lb of concrete? (*Remember:* You cannot buy a fraction of a sack.)

34. **Auto Mechanics** The rear axle of a car turns at 320 rpm when the drive shaft turns at 1600 rpm. If the pinion gear has 12 teeth, how many teeth are on the ring gear?

35. If a 120-lb force is applied to a 36-in. crowbar pivoted 4 in. from the end, what lift force is exerted?

36. **Machine Technology** How many turns will a pinion gear having 16 teeth make if a ring gear having 48 teeth makes 120 turns?

37. **Electrical Technology** The current in a certain electrical circuit is inversely proportional to the line voltage. If a current of 0.40 ampere is delivered at 440 volts, what is the current of a similar system operating at 120 volts? (Round to the nearest 0.01 ampere.)

38. **Printing** A printer needs 12,000 good copies of a flyer for a particular job. Previous experience has shown that she can expect no more than a 5% spoilage rate on this type of job. How many should she print to assure 12,000 clean copies?

39. **Auto Mechanics** If a mechanic's helper is paid $456.28 for $38\frac{1}{4}$ hours of work, how much should he receive for $56\frac{1}{2}$ hours at the same rate?

40. **Manufacturing** A pair of belted pulleys have diameters of $8\frac{3}{4}$ in. and $5\frac{5}{8}$ in. If the smaller pulley turns at 850 rpm, how fast will the larger one turn?

41. **Drafting** Each production employee in a plant requires an average of 120 sq ft of work area. How many employees will be able to work in an area that measures $8\frac{1}{16}$ in. by $32\frac{5}{8}$ in. on a blueprint if the scale of the drawing is $\frac{1}{4}$ in. = 1 ft?

42. Find the missing dimension in the following pair of similar figures.

43. **Machine Technology** In a closed container, the pressure is inversely proportional to the volume when the temperature is held constant. Find the pressure of a gas compressed to 0.386 cu ft if the pressure is 12.86 psi at 2.52 cu ft.

44. **Sports Technology** Use the following five steps to determine the NFL passing rating for quarterback Bubba Grassback.
    (a) Bubba completed 176 passes in 308 attempts. Calculate his pass completion rate as a percent. Subtract 30 from this number and multiply the result by 0.05.
    (b) He has 11 touchdown passes in 308 attempts. Calculate the percent of his attempts that resulted in touchdowns. Multiply that number by 0.2.
    (c) He has thrown 13 interceptions. Calculate the percent of his 308 attempts that have resulted in interceptions. Multiply this percent by 0.25 and then subtract the result from 2.375.

(d) Bubba passed for 2186 yards. Calculate the average number of yards gained per passing attempt. Subtract 3 from this number and then multiply the result by 0.25.

(e) Add the four numbers obtained in parts (a), (b), (c), and (d).

(f) Divide the total in part (e) by six and then multiply by 100. This is Bubba's NFL quarterback passing rating.

(*Note:* Ratings for starting quarterbacks usually range from a low of about 50 to a high of about 110.)

Measurement

Objective	Sample Problems		For Help Go to Page

When you finish this unit you will be able to:

1. Determine the precision and accuracy of measurement numbers.

Determine the precision of each of the following measurement numbers. Which one is most precise?
(a) 4.27 psi   (b) 6758 psi   (c) 350 psi

Determine the number of significant digits in each of the following measurement numbers.
(d) 9.6 kg   (e) 458 kg   (f) 6000 kg

(a) _____   223

(b) _____

(c) _____

(d) _____   222

(e) _____

(f) _____

2. Add and subtract measurement numbers. (Round to the correct precision.)

(a)   38.26 sec + 5.9 sec =

(b)   8.37 lb − 2.435 lb =

(a) _____   225

(b) _____

3. Multiply and divide measurement numbers. (Round to the proper accuracy.)

(a)   16.2 ft × 5.8 ft =

(b)   480 mi ÷ 32.5 mph =

(c)   240 lb/sq in. × 52.5 sq in. =

(a) _____   227

(b) _____

(c) _____

4. Convert units within the English system. (Round to the correct number of significant digits.)

(a)   65.0 lb = _____ oz

(b)   8200 yd = _____ mi

(c)   3.50 sq ft = _____ sq in.

236

_____
Name

_____
Date

_____
Course/Section

5. Think metric.

(a) Estimate the weight in pounds of a casting weighing 45 kg.

(a) _____ 251

(b) Estimate the volume in quarts of 80 liters.

(b) _____

6. Convert units within the metric system. (Round to the correct number of significant digits.)

(a) 6500 mg = _____ g

252

(b) 0.45 km = _____ m

(c) 0.285 ha = _____ sq m

7. Convert between English and metric system units.

(a) 48.0 liters = _____ gal

254

(b) 45 mi/hr = _____ km/h

(c) 42°C = _____ °F

8. Use common technical measuring instruments.

Correctly read a length rule, micrometer, and vernier caliper.

269

(Answers to these preview problems are given in the Appendix.)

If you are certain that you can work *all* these problems correctly, turn to page 293 for a set of practice problems. If you cannot work one or more of the preview problems, turn to the page indicated to the right of the problem. For those who wish to master this material with the greatest success, turn to Section 5-1 and begin work there.

5 Measurement

© Zits Partnership. Reprinted with special Permission of King Features Syndicate.

5-1 WORKING WITH MEASUREMENT NUMBERS

Most of the numbers used in practical or technical work either come from measurements or lead to measurements. A carpenter must measure a wall to be framed; a machinist may weigh materials; an electronic technician may measure circuit resistance or voltage output. At one time or another each of these technicians must make measurements, do calculations based on measurements, perhaps round measurement numbers, and worry about the units of measurement. Of course, they must also work with measurement numbers from drawings, blueprints, or other sources.

Units A measurement number usually has units associated with it, and the unit name for that measurement number *must* be included when you use that number, write it, or talk about it. For example, we know immediately that 20 is a number, 20 ft is a length, 20 lb is a weight or force, and 20 sec is a time interval.

Suppose that the length of a piece of pipe is measured to be 8 ft. Two bits of information are given in this measurement: the size of the measurement (8) and the units of measurement (ft). The measurement unit compares the size of the quantity being measured to some standard.

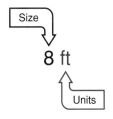

> **EXAMPLE**

If the piece of pipe is measured to be 8 ft in length, it is 8 times the standard 1-ft length.

Length = 8 ft = 8 × 1 ft

1 | 2 | 3 | 4 | 5 | 6 | 7 | 8

▬▬▬ = 1 ft

YOUR TURN

Complete the following.

(a) 6 in. = 6 × _____   (b) 10 lb = _____ × _____

(c) $5 = 5 × _____   (d) 14.7 psi = _____ × _____

ANSWERS

(a) 6 in. = 6 × 1 in.   (b) 10 lb = 10 × 1 lb

(c) $5 = 5 × $1   (d) 14.7 psi = 14.7 × 1 psi

Significant Digits   Every measuring instrument and every person making measurements has limitations. Measurement numbers—numbers used in technical work and in the trades—have a built-in uncertainty, and this uncertainty is not stated explicitly but is expressed by the number of digits used to communicate the measurement information.

In technical work, when we perform calculations with measurement numbers, we must round the result according to agreed-upon rules involving **significant digits.**

The significant digits in a number are those that represent an actual measurement. To determine the number of significant digits in a measurement number, follow these rules:

Rule 1   Digits other than zero are always significant.

Rule 2   A zero **is** significant when

    (a) it appears between two significant digits,

    (b) it is at the right end of a decimal number,

    (c) it is marked as significant with an overbar.

Rule 3   A zero **is not** significant when

    (a) it is at the right end of a whole number,

    (b) it is at the left end of a number.

- A tachometer reading of 84.2 rpm has three significant digits. All three digits in the number are nonzero. (Rule 1)
- A pressure reading of 3206 torr has four significant digits. The zero is significant because it appears between two nonzero digits. (Rule 2a)
- A voltage reading of 240 volts has two significant digits. The zero is not a significant digit because it is at the right end of a whole number. (Rule 3a)
- A voltage reading of 24$\overline{0}$ volts has three significant digits. The overbar tells us that the zero marked is significant. (Rule 2c) The person making the measurement had an instrument that could measure to the nearest unit.

- A voltage reading of 24.0 volts has three significant digits. The zero is significant because it appears at the right end of a decimal number. (Rule 2b)
- A current reading of 0.025 amp has two significant digits. The zeros are not significant because they are at the left end of the number. (Rule 3b)

YOUR TURN

Determine the number of significant digits in each of the following measurement numbers.

(a) 3.04 lb (b) 38,000 mi (c) 15.730 sec

(d) 0.005 volts (e) 18,4$\overline{0}$0 gal (f) 200.0 torr

ANSWERS

(a) Three significant digits. The zero is significant because it is between two non-zero digits.

(b) Two significant digits. The zeros are not significant because they are at the right end of a whole number.

(c) Five significant digits. The zero is significant because it is at the right end of a decimal number.

(d) One significant digit. The zeros are not significant because they are at the left end of the number.

(e) Four significant digits. The zero with the overbar is significant, but the right-end zero is not significant.

(f) Four significant digits. The right-end digit is significant because it is at the right end of a decimal number. The middle two zeros are significant because they are in between two significant digits.

Precision and Accuracy Because measurement numbers are never exact we need to be careful that the significant digits we specify in the number actually reflect the situation.

Look at this line:

If we measured the length of this line using a rough ruler, we would write that its length is

Length = 2 in.

To the nearest inch

But with a better measuring device, we might be able to say with confidence that its length is

Length = 2.1 in.

To the nearest tenth of an inch

Using a caliper, we might find that the length is

Length = 2.134 in.

To the nearest thousandth of an inch

We could go on using better and better measuring devices until we found ourselves looking through a microscope trying to decide where the ink line begins. If someone told us that the length of that line was

Length = 2.1304756 in.

To the nearest 0.0000001 inch

we would know immediately that the length had been measured with incredible precision—they couldn't get a number like that by using a carpenter's rule!

In order to work with measurement numbers, we define precision and accuracy* as follows:

> The **precision** of a measurement number is indicated by the place value of its right-most significant digit.
>
> The **accuracy** of a measurement number refers to the number of significant digits that it contains.

EXAMPLE

State the precision and accuracy of each of the following measurement numbers using the rules given.

(a) 4.01 sec   (b) 40 sec   (c) 420.1 sec

Measurement (a) is precise to the nearest hundredth of a second. It is accurate to three significant digits.

In (b) the zero is not significant. Therefore the number is precise to the nearest 10 seconds. It is accurate to one significant digit.

Measurement (c) is precise to the nearest tenth of a second. It is accurate to four significant digits.

Notice that measurement (a) is considered the most precise because it has the most decimal digits. However, measurement (c) is considered the most accurate because it contains the most significant digits.

YOUR TURN

Determine the precision and accuracy of each of the following measurement numbers. Then indicate which is the most precise and which is the most accurate.

(a) 34.1 lb   (b) 0.568 lb   (c) 2735 lb   (d) 5.09 lb   (e) 400 lb

*In engineering, *accuracy* is the degree of closeness between the measured value of some quantity and its true, or nominal, value. It has to do with the ability of an instrument to make measurements with small uncertainty. *Precision* is the degree of consistency and agreement among independent measurements of a quantity under the same conditions. For practical purposes we define them as shown in the box.

Measurement	Precise to the nearest . . .	Accurate to this number of significant digits
(a)	tenth of a pound	3
(b)	thousandth of a pound	3
(c)	pound	4
(d)	hundredth of a pound	3
(e)	hundred pounds	1

Measurement (b) is the most precise and measurement (c) is the most accurate.

TOLERANCE

Very often the technical measurements and numbers shown in drawings and specifications have a *tolerance* attached so that you will know the precision needed. For example, if a dimension on a machine tool is given as 2.835 ± 0.004 in., this means that the dimension should be 2.835 in. to within 4 thousandths of an inch. In other words, the dimension may be no more than 2.839 in. and no less than 2.831 in. These are the limits of size that are allowed or tolerated.

Addition and Subtraction of Measurement Numbers

When measurement numbers are added or subtracted, the resulting answer cannot be more precise than the least precise measurement number in the calculation. Therefore, we must always round a sum or difference to agree in precision with the least precise measurement number being added or subtracted.

EXAMPLE

To add

$$4.1 \text{ sec} + 3.5 \text{ sec} + 1.27 \text{ sec} = \underline{\quad ? \quad}$$

First, check to be certain that all numbers have the same units. In the given problem all numbers have units of seconds. If one or more of the numbers are given in other units, you will need to convert so that all have the same units.

Second, add the numerical parts.

$$\begin{array}{r} 4.1 \\ 3.5 \\ + 1.27 \\ \hline 8.87 \end{array}$$ Notice that we line up the decimal points and add as we would with any decimal numbers.

 4.1 ⊕ **3.5** ⊕ **1.27** ⊟ → ▓▓▓ 8.87

Third, attach the common units to the sum.

8.87 sec

Fourth, round the sum to agree in precision with the least precise number in the calculation. We must round 8.87 sec to the nearest tenth, or 8.9 seconds. Our answer agrees in precision with 4.1 sec and 3.5 sec, the least precise measurements in the calculation.

YOUR TURN

Perform the following subtraction and round your answer to the correct precision.

7.425 in. − 3.5 in. = ____?____

SOLUTION

First, check to be certain that both numbers have the same units. In this problem both numbers have inch units. If one of the numbers was given in other units, we would convert so that they have the same units. (You'll learn more about converting measurement units later in this chapter.)

Second, subtract the numerical parts.

$$\begin{array}{r} 7.425 \\ -\ 3.500 \\ \hline 3.925 \end{array}$$ ⟵ Attach zeros if necessary

 7.425 ⊟ **3.5** ⊜ → 　　*3.925*

Third, attach the common units to the difference.　　3.925 in.

Fourth, round the answer to agree in precision with the least precise number in the original problem. The least precise number is 3.5 in., which is precise to the nearest tenth. We should round our answer, 3.925 in., to the nearest tenth, or 3.9 in.

EXAMPLE

Here are a few more examples to help fix the concept in your mind.

(a)　4.6 gal + 2.145 gal = 6.745 gal ≈ 6.7 gal

Round to the nearest tenth to agree with the least precise measurement number, 4.6 gal, which is stated to the nearest tenth.

(b)　0.24 mi − 0.028 mi = 0.212 mi ≈ 0.21 mi

Round the answer to the nearest hundredth to agree with the least precise number, 0.24 mi.

(c)　2473 lb + 321.2 lb = 2794.2 lb ≈ 2794 lb

Round to the nearest whole number to agree with the least precise measurement, 2473 lb.

PROBLEMS

Try these problems for practice in adding and subtracting measurement numbers. Be careful to round your answers correctly.

(a)　10.5 in. + 8.72 in.　　　　　　　　(b)　2.46 cu ft + 3.517 cu ft + 4.8 cu ft

(c) 47.5 sec − 21.844 sec   (d) 56 lb − 4.8 lb

(e) 325 sq ft + 11.8 sq ft + 86.06 sq ft   (f) 0.027 volts + 0.18 volts + 0.009 volts

ANSWERS

(a) 10.5 in. + 8.72 in. = 19.22 in. ≈ 19.2 in.

(b) 2.46 cu ft + 3.517 cu ft + 4.8 cu ft = 10.777 cu ft ≈ 10.8 cu ft

(c) 47.5 sec − 21.844 sec = 25.656 sec ≈ 25.7 sec

(d) 56 lb − 4.8 lb = 51.2 lb ≈ 51 lb

(e) 325 sq ft + 11.8 sq ft + 86.06 sq ft = 422.86 sq ft ≈ 423 sq ft

(f) 0.027 volts + 0.18 volts + 0.009 volts = 0.216 volts ≈ 0.22 volts

---

Note ▶ If some of the numbers to be added are in fraction form, you should convert them to decimal form before adding. For example, the sum $3.2 \text{ lb} + 1\frac{3}{4} \text{ lb} + 2.4 \text{ lb}$ becomes

$$
\begin{array}{r}
3.2 \ \text{ lb} \\
1.75 \text{ lb} \\
+2.4 \ \text{ lb} \\
\hline
7.35 \text{ lb}
\end{array}
$$

If we do not know the true precision of the fractional measurement, the usual procedure is to round to the precision of the original *decimal* measurements. The least precise of these, 3.2 lb and 2.4 lb, are both given to the nearest tenth of a pound. Our answer, 7.35 lb, should be rounded to 7.4 lb. ◀

Multiplication and Division of Measurement Numbers

When multiplying or dividing measurement numbers, we need to pay special attention to both the proper units and the proper rounding of our final answer. First, let's consider the units.

With both addition and subtraction of measurement numbers, the answer to the arithmetic calculation has the same units as the numbers being added or subtracted. This is not true when we multiply or divide measurement numbers. For example, to multiply

$3 \text{ ft} \times 2 \text{ ft} =$

First, multiply the numerical parts: $3 \times 2 = 6$

Second, multiply the units: $1 \text{ ft} \times 1 \text{ ft} = 1 \text{ ft}^2$ or 1 sq ft

It may help if you remember that this multiplication is actually

$$
\begin{aligned}
3 \text{ ft} \times 2 \text{ ft} &= (3 \times 1 \text{ ft}) \times (2 \times 1 \text{ ft}) \\
&= (3 \times 2) \times (1 \text{ ft} \times 1 \text{ ft}) \\
&= 6 \times 1 \text{ sq ft} = 6 \text{ sq ft}
\end{aligned}
$$

The rules of arithmetic say that we can do the multiplications in any order we wish, so we multiply the units part separately from the number part.

Notice that the product $1 \text{ ft} \times 1 \text{ ft}$ can be written either as 1 ft^2 or as 1 sq ft.

EXAMPLE

Multiply

2 in. $\times$ 3 in. $\times$ 5 in.

First, multiply the numbers: $2 \times 3 \times 5 = 30$

Second, multiply the units: 1 in. $\times$ 1 in. $\times$ 1 in. $= 1$ in.^{3} or 1 cu in.

The answer is 30 cu in.

VISUALIZING UNITS

It is helpful to have a visual understanding of measurement units and the "dimension" of a measurement.

The length unit **inch,** in., specifies a **one**-dimensional or linear measurement. It gives the length of a straight line.

The area unit **square inch,** sq in., specifies a **two**-dimensional measurement. Think of a square inch as giving the area of a square whose sides are one inch in length.

The volume unit **cubic inch,** cu in., specifies a **three**-dimensional measurement. Think of a cubic inch as the volume of a cube whose edges are each one inch in length.

Division with measurement numbers requires that you take the same care with the units that is needed in multiplication.

EXAMPLE

To divide

24 mi $\div$ 1.5 hr = _____?_____

First, divide the numerical parts as usual.

$24 \div 1.5 = 16$

Second, divide the units.

$$1 \text{ mi} \div 1 \text{ hr} = \frac{1 \text{ mi}}{1 \text{ hr}} \quad \text{or} \quad 1 \text{ mph} \quad \text{or} \quad 1 \text{ mi/hr}$$

In this case the unit $\dfrac{1 \text{ mi}}{1 \text{ hr}}$ is usually written mi/hr or mph—miles per hour—and the answer to the problem is 16 mph. Remember that

$$24 \text{ mi} \div 1.5 \text{ hr} = \frac{24 \times 1 \text{ mi}}{1.5 \times 1 \text{ hr}} = \left(\frac{24}{1.5}\right) \times \left(\frac{1 \text{ mi}}{1 \text{ hr}}\right)$$

$$= 16 \times 1 \text{ mph} = 16 \text{ mph}$$

Units such as sq ft or mph, which are made up of a combination of simpler units, are called *compound* units.

Note ▶ Both h and hr are acceptable abbreviations for the unit *hour*. The standard metric abbreviation is h, and hr is in common usage in the United States. ◀

YOUR TURN

36 lb ÷ 1.5 sq in. = _____

SOLUTION

First, divide the numbers: 36 ÷ 1.5 = 24

Second, divide the units 1 lb ÷ 1 sq in. = $\dfrac{1\,\text{lb}}{1\,\text{sq in.}}$ = 1 lb/sq in. = 1 psi

$\left(\text{The unit } psi \text{ means "pounds per square inch" or } \dfrac{\text{lb}}{\text{sq in.}}\right)$

The answer is 24 psi.

When multiplying or dividing quantities having compound units, it is sometimes possible to "cancel" or divide out common units. This procedure is identical to the cancellation of common numerical factors when multiplying fractions.

EXAMPLE

(a) To multiply

45 mph × 4 hr

set it up this way:

$45\dfrac{\text{mi}}{\text{hr}} \times 4\,\text{hr} = (45 \times 4) \times \left(\dfrac{\text{mi}}{\cancel{\text{hr}}} \times \cancel{\text{hr}}\right)$

$= 180\,\text{mi}$

Notice that the "hr" units cancelled just as common numerical factors do.

(b) To divide

42 sq yd ÷ 6 yd

work with the numbers and the units separately as follows.

$\left(\dfrac{42}{6}\right) \times \left(\dfrac{\text{sq yd}}{\text{yd}}\right) = \left(\dfrac{42}{6}\right) \times \left(\dfrac{\text{yd} \times \cancel{\text{yd}}}{\cancel{\text{yd}}}\right) = 7\,\text{yd}$

PROBLEMS

Before we discuss rounding in multiplication and division, try the following practice problems. Be careful with units.

(a) 5 ft × 4 ft   (b) 4 ft ÷ 2 sec

(c) 2 yd × 6 yd × 0.5 yd   (d) 24 sq in. ÷ 3 in.

(e) 30 mph × 2 hr   (f) 70 ft ÷ 35 ft/sec

(a) 20 sq ft   (b) 2 ft/sec

(c) 6 cu yd   (d) 8 in.

(e) 60 mi   (f) 2 sec

Did you have difficulty with (d), (e), or (f)? Do them this way:

(d) $24 \text{ sq in.} \div 3 \text{ in.} = \left(\dfrac{24}{3}\right) \times \left(\dfrac{\text{sq in.}}{\text{in.}}\right)$

$$= 8 \times \left(\dfrac{\text{in.}^2}{\text{in.}}\right) = 8 \times \left(\dfrac{\text{in.} \times \cancel{\text{in.}}}{\cancel{\text{in.}}}\right) = 8 \text{ in.}$$

(e) $30 \text{ mph} \times 2 \text{ hr} = 30 \dfrac{\text{mi}}{\text{hr}} \times 2 \text{ hr}$

$$= (30 \times 2) \times \left(\dfrac{\text{mi}}{\cancel{\text{hr}}} \cdot \cancel{\text{hr}}\right)$$

$$= 60 \text{ mi}$$

(f) $70 \text{ ft} \div 35 \text{ ft/sec} = \left(\dfrac{70}{35}\right) \times (\text{ft} \div \text{ft/sec})$

$$= 2 \times \left(\cancel{\text{ft}} \times \dfrac{\text{sec}}{\cancel{\text{ft}}}\right)$$

$$= 2 \text{ sec}$$

Rounding Products and Quotients

When measurement numbers are multiplied or divided, the resulting answer cannot be expressed with greater accuracy than the least accurate measurement number in the calculation. The result of a calculation should not be written to reflect greater accuracy than any number used in that calculation.

Always round a product or quotient to the same number of significant digits as the least accurate number used in the calculation.

Multiply: 4.35 ft × 3.6 ft.

First, multiply the numerical factors: $4.35 \times 3.6 = 15.66$

Second, multiply the units: $1 \text{ ft} \times 1 \text{ ft} = 1 \text{ ft}^2$ or 1 sq ft

Third, round the answer to agree in accuracy with the least accurate number in the calculation.

   4.35 ft is accurate to three significant digits.

   3.6 ft is accurate to two significant digits.

The least accurate factor has two significant digits, so the answer must be rounded to two significant digits.

$15.66 \text{ sq ft} \approx 16 \text{ sq ft}$

Divide: 375 mi ÷ 65 mph.

First, divide the numerical parts: $375 \div 65 = 5.7692 \ldots$

Second, divide the units:

$$\text{mi} \div \frac{\text{mi}}{\text{hr}} = \cancel{\text{mi}} \times \frac{\text{hr}}{\cancel{\text{mi}}} = \text{hr}$$

Third, round the answer to two significant digits to agree in accuracy with 65, the least accurate measurement number in the calculation.

$$375 \text{ mi} \div 65 \text{ mph} \approx 5.8 \text{ hr}$$

Note ▶ Numbers that result from simple counting should be considered to be exact. They do not affect the accuracy of the calculation. For example, if a flywheel turns 48 rotations in 1.25 minutes, the speed of the flywheel is calculated as $48 \div 1.25 = 38.4$ rpm. We should assume that the 48 is an exact number—rotations are counted directly—so we express the answer as 38.4 with three significant digits to agree with 1.25, the least accurate measurement number in the calculation. ◀

PROBLEMS

Perform the following calculations. Round each answer to the proper accuracy.

(a) 7.2 in. × 0.48 in.
(b) 0.18 ft × 3.15 ft × 20 ft
(c) 82.45 ft ÷ 3.65 sec
(d) 28 sq yd ÷ 1.25 yd
(e) 2450 ton/sq ft × 3178 sq ft
(f) 148 mi ÷ 15 mph

ANSWERS

(a) 7.2 in. × 0.48 in. = 3.456 sq in. ≈ 3.5 sq in.

(b) 0.18 ft × 3.15 ft × 20 ft = 11.34 cu ft ≈ 10 cu ft

(c) 82.45 ft ÷ 3.65 sec = 22.58904 . . . ft/sec ≈ 22.6 ft/sec

(d) 28 sq yd ÷ 1.25 yd = 22.4 yd ≈ 22 yd

(e) 2450 ton/sq ft × 3178 sq ft = 7786100 ton ≈ 7,790,000 ton

(f) 148 mi ÷ 15 mph = 9.8666 . . . hr ≈ 9.9 hr

Careful ▶ When dividing by hand it is important that you carry the arithmetic to one digit more than will be retained after rounding. ◀

Follow these rounding rules for all calculations shown in this book unless specific instructions for rounding are given.

Decimal Equivalents Some of the measurements made by technical workers in a shop are made in fractions. But on many shop drawings and specifications, the dimensions may be given in decimal form.

The steel rule used in the shop or in other technical work may be marked in 8ths, 16ths, 32nds, or even 64ths of an inch. A common problem is to rewrite the decimal number to the nearest 32nd or 64th of an inch and to determine how much error is involved in using the fraction number rather than the decimal number.

EXAMPLE

In the drawing shown, what is the fraction, to the nearest 32nd of an inch, equivalent to 0.462 in.? The rule is to multiply the decimal number by the fraction $\frac{32}{32}$.

$$0.462 \text{ in.} = 0.462 \text{ in.} \times \frac{32}{32}$$

$$= \frac{0.462 \text{ in.} \times 32}{32}$$

$$= \frac{14.784}{32} \text{ in.} \quad \text{Round the top number to the nearest unit.}$$

$$\approx \frac{15}{32} \text{ in.} \quad \text{rounded to the nearest 32nd of an inch}$$

The error involved is the difference between

$$\frac{15}{32} \text{ in.} \quad \text{and} \quad \frac{14.784}{32} \text{ in.,} \quad \text{or} \quad \frac{15 - 14.784}{32} \text{ in.} = \frac{0.216}{32} \text{ in.}$$

$$= 0.00675 \text{ in.}$$

$$\approx 0.0068 \text{ in.} \quad \text{rounded}$$

The error involved in using $\frac{15}{32}$ in. instead of 0.462 in. is about 68 ten-thousandths of an inch. In problems of this kind we would usually round the error to the nearest ten-thousandth.

Fraction: **.462** $\times$ **32** $=$ STO $\rightarrow$ | 14.784 | Round to $\frac{15}{32}$.

Error: **15** $-$ RCL $=$ $\div$ **32** $=$ $\rightarrow$ | 0.00675 | about 0.0068 in.

YOUR TURN

Change 1.392 in. to a fraction expressed in 64ths of an inch, and find the error.

SOLUTION

$$1.392 \text{ in.} = 1.392 \text{ in.} \times \frac{64}{64}$$

$$= \frac{1.392 \times 64}{64} \text{ in.}$$

$$= \frac{89.088}{64} \text{ in.}$$

$$\approx \frac{89}{64} \text{ in.} \quad \text{rounded to the nearest unit}$$

$$\approx 1\frac{25}{64} \text{ in.}$$

The error in using the fraction rather than the decimal is

$$\frac{89.088}{64} \text{ in.} - \frac{89}{64} \text{ in.} = \frac{0.088}{64} \text{ in.}$$

$$\approx 0.0014 \text{ in.} \qquad \text{or about fourteen ten-thousandths of an inch.}$$

Fraction: **1.392** ⊗ **64** ⊜ → 89.088   Round to $\frac{89}{64}$.

Error: ⊖ **89** ⊜ ⊘ **64** ⊜ → 0.001375   about 0.0014 in.

PROBLEMS

(a) Express to the nearest 16th of an inch.

   (1)   0.438 in.   (2)   0.30 in.

   (3)   0.18 in.   (4)   2.70 in.

(b) Express to the nearest 32nd of an inch.

   (1)   1.650 in.   (2)   0.400 in.

   (3)   0.720 in.   (4)   2.285 in.

(c) Express to the nearest 64th of an inch and find the error, to the nearest ten-thousandth, involved in using the fraction in place of the decimal number.

   (1)   0.047 in.   (2)   2.106 in.

   (3)   1.640 in.   (4)   0.517 in.

ANSWERS

(a) (1) $\frac{7}{16}$ in.   (2) $\frac{5}{16}$ in.   (3) $\frac{3}{16}$ in.   (4) $2\frac{11}{16}$ in.

(b) (1) $1\frac{21}{32}$ in.   (2) $\frac{13}{32}$ in.   (3) $\frac{23}{32}$ in.   (4) $2\frac{9}{32}$ in.

(c) (1) $\frac{3}{64}$ in. The error is about 0.0001 in.

     (2) $2\frac{7}{64}$ in. The error is about 0.0034 in.

     (3) $1\frac{41}{64}$ in. The error is about 0.0006 in.

     (4) $\frac{33}{64}$ in. The error is about 0.0014 in.

A DIFFERENT WAY TO ROUND NUMBERS

Some engineers and technicians use the following rounding rule rather than the ones we presented in Chapter 3.

Examples

1. Determine the place to which the number is to be rounded. Mark it with a $_\wedge$.

   Round 4.786 to three digits.
            4.78$_\wedge$6

2. When the digit to the right of the mark is greater than 5, increase the digit to the left by 1.

   4.78$_\wedge$6   becomes 4.79
   1.701$_\wedge$72 becomes 1.702

3. When the digit to the right of the mark is less than 5, drop the digits to the right or replace them with zeros.

   3.81$_\wedge$2   becomes 3.81
   14$_\wedge$39   becomes 1400

(continued)

4. When the digit to the right of the mark is equal to 5, and there are no other nonzero digits to the right of the 5, round the left digit to the nearest *even* value.	$1.41_{\wedge}5$ $37.00_{\wedge}5$ $53_{\wedge}50$	becomes 1.42 becomes 37.00 becomes 5400
5. In step 4, if there are any nonzero digits after the 5, always round up.	$2.62_{\wedge}52$ $487_{\wedge}56$	becomes 2.63 becomes 48800

For a set of practice problems over this section of Chapter 5, turn to Exercises 5-1.

Exercises 5-1

Working with Measurement Numbers

A. State the precision and the accuracy of each of the following measurement numbers.

1. 4.7 gal
2. 0.008 ft
3. 562 psi
4. 3400 yd
5. 7.024 lb
6. 360 volts
7. $19,\overline{0}00$ tons
8. 19.65 oz
9. 10.0 in.
10. 0.20 sec

B. Add or subtract as shown. Round your answer to the correct precision.

1. 7 in. + 14 in.
2. 628 lb + 400 lb
3. 3.25 in. + 1.7 in. + 4.6 in.
4. 16.2 psi + 12.64 psi
5. 0.28 gal + 1.625 gal
6. 64.5 mi − 12.72 mi
7. 1.565 oz − 0.38 oz
8. 12.372 sec − 2.1563 sec
9. 28.3 psi + 16 psi + 5.9 psi
10. 6.4 tons + 8.26 tons − 5.0 tons

C. Multiply or divide as shown. Round to the appropriate number of significant digits.

1. 6 ft × 7 ft
2. 3.1 in. × 1.7 in.
3. 3.0 ft × 2.407 ft
4. 215 mph × 1.2 hr
5. $1\frac{1}{4}$ ft × 1.5 ft
6. 48 in. × 8.0 in.
7. 2.1 ft × 1.7 ft × 1.3 ft
8. 18 sq ft ÷ 2.1 ft
9. 458 mi ÷ 7.3 hr
10. \$1.37 ÷ 3.2 lb
11. $40\frac{1}{2}$ mi ÷ 25 mph
12. 2.0 sq ft ÷ 1.073 ft

D. Convert as shown.

1. Write each of the following fractions as a decimal number rounded to two decimal places.

    (a) $1\frac{7}{8}$ in. (b) $4\frac{3}{64}$ in. (c) $3\frac{1}{8}$ sec (d) $\frac{1}{16}$ in.

    (e) $\frac{3}{32}$ in. (f) $1\frac{3}{16}$ lb (g) $2\frac{17}{64}$ in. (h) $\frac{19}{32}$ in.

    (i) $\frac{3}{8}$ lb

2. Write each of the following decimal numbers as a fraction rounded to the nearest 16th of an inch.

    (a) 0.921 in. (b) 2.55 in. (c) 1.80 in. (d) 3.69 in.

(e)　0.802 in.　　(f)　0.306 in.　　(g)　1.95 in.　　(h)　1.571 in.

(i)　0.825 in.

3.　Write each of the following decimal numbers as a fraction rounded to the nearest 32nd of an inch.

(a)　1.90 in.　　(b)　0.85 in.　　(c)　2.350 in.　　(d)　0.666 in.

(e)　2.091 in.　　(f)　0.285 in.　　(g)　0.600 in.　　(h)　0.685 in.

(i)　1.525 in.

4.　Write each of the following decimal numbers as a fraction rounded to the nearest 64th of an inch, and find the error, to the nearest ten-thousandth, involved in using the fraction in place of the decimal number.

(a)　0.235 in.　　(b)　0.515 in.　　(c)　1.80 in.

(d)　2.420 in.　　(e)　3.175 in.　　(f)　2.860 in.

(g)　1.935 in.　　(h)　0.645 in.　　(i)　0.480 in.

E.　Practical Problems

1.　**Machine Technology**　Specifications call for drilling a hole 0.637 ± 0.005 in. in diameter. Will a $\frac{41}{64}$-in. hole be within the required tolerance?

2.　**Metalworking**　According to standard American wire size tables, 3/0 wire is 0.4096 in. in diameter. Write this size to the nearest 32nd of an inch. What error is involved in using the fraction rather than the decimal?

3.　**Auto Mechanics**　An auto mechanic converts a metric part to 0.473 in. The part comes only in fractional sizes given to the nearest 64th of an inch. What would be the closest size?

4.　**Manufacturing**　What size wrench, to the nearest 32nd of an inch, will fit a bolt with a 0.748-in. head?

5.　**Sheet Metal Technology**　Find the missing distance x in the drawing.

→| x |← 6.42" →|← 7.05" →|

|← 17 $\frac{3"}{4}$ →|

Problem 5

6.　**Carpentry**　A truckload of knotty pine paneling weighs 180 lb. If this kind of wood typically has a density of 38 lb/cu ft, find the volume of wood in the load. (*Hint:* Divide the weight by the density.) Round to the nearest 0.1 cu ft.

7.　**Auto Mechanics**　The wear on a cylinder has increased its bore to 3.334 in. The most likely original bore was the nearest 32nd-in. size less than this. What was the most likely original bore?

8.　**Interior Design**　If the rectangular floor of a room has an area of 455 sq ft and a length of 26 ft, what is its width?

9.　A truck, moving continuously, averages 52.5 mph for 6.75 hr. How far does it travel during this time?

10.　**Industrial Technology**　The volume of a cylindrical tank can be calculated as its base area multiplied by its height. If the tank has a capacity of 2760 cu ft of liquid and has a base area of 120 sq ft, what is its height?

11.　**Electronic Technology**　If an electromagnetic signal moves at the speed of light, about 186,282 miles per second, how long will it take this signal to travel from New Jersey to California, approximately 2500 miles?

When you have completed these problems, check your answers in the Appendix, then turn to Section 5-2 for some information on units and unit conversion.

When you talk about a 4-ft by 8-ft wall panel, a $1\frac{1}{4}$-in. bolt, 3 lb of solder, or a gallon of paint, you are comparing an object with a *standard*. The units or standards used in science, technology, or the trades have two important characteristics: they are convenient in size and they are standardized. Common units such as the foot, pound, gallon, or hour were originally chosen because they were related either to natural quantities or to body measurements.

A pinch of salt, a drop of water, a handful of sand, or a tank of gas are all given in *natural* units. The width of a finger, the length of a foot, or a stride length are all distances given in *body-related* units. In either case, there is a need to define and standardize these units before they are useful for business, science, or technology.

In ancient times, the *digit* was a length unit defined as the width of a person's finger. Four digits was called a *hand,* a unit still used to measure horses. Four hands was a *cubit,* the distance from fingertip to elbow, about 18 in. One *foot* was defined as the length of a man's foot. A *yard* was the distance from nose to tip of outstretched arm. Of course all these lengths depend on whose finger, hand, arm, or foot is used.

Standardization in the English system of units began in the fifteenth century, and one yard was defined as the length of a standard iron bar kept in a London vault. One foot was defined as exactly one-third the length of the bar, and one inch was one-twelfth of a foot. (The word *inch* comes from Latin and Anglo-Saxon words meaning "one-twelfth.") Today the inch is legally defined worldwide in terms of metric or SI units.

We will consider the familiar English system of units in this section and metric units in Section 5-3.

Because there are a great many units available for writing the same quantity, it is necessary that you be able to convert a given measurement from one unit to another. For example, a length of exactly one mile can be written as

1 mile = 1760 yd, for the traffic engineer
   = 5280 ft, for a landscape architect
   = 63,360 in., in a science problem
   = 320 rods, for a surveyor
   = 8 furlongs, for a horse racing fan
   = 1609.344 meters, in metric units

Here are some basic conversion factors for length, weight, and liquid capacity in the English system of units.

LENGTH UNITS

1 foot (ft) = 12 inches (in.)
1 yard (yd) = 3 feet = 36 in.
1 mile (mi) = 5280 ft = 1760 yd

WEIGHT UNITS

1 pound (lb) = 16 ounces (oz)
1 ton = 2000 lb

LIQUID CAPACITY UNITS

1 quart (qt) = 2 pints (pt)
1 gallon (gal) = 4 quarts
1 barrel (bbl) = 31.5 gallons

In this section we will show you a quick and mistake-proof way to convert measurements from one unit to another. Whatever the units given, if a new unit is defined, you should be able to convert the measurement to the new units.

EXAMPLE

Consider the following simple unit conversion. Convert 4 yards to feet.

4 yd = _____ ft

From the table we know that 1 yard is defined as exactly 3 feet; therefore,

4 yd = 4 × 3 ft = 12 ft

Easy.

YOUR TURN

Try another, more difficult, conversion. Surveyors use a unit of length called a *rod,* defined as exactly $16\frac{1}{2}$ ft.

Convert: 11 yd = _____ rods.

ANSWERS

The correct answer is 2 rods.

You might have thought to first convert 11 yd to 33 ft and then noticed that 33 ft divided by $16\frac{1}{2}$ equals 2. There are 2 rods in a 33-ft length.

Unity Fractions   Most people find it difficult to reason through a problem in this way. To help you, we have devised a method of solving *any* unit conversion problem quickly with no chance of error. This is the *unity fraction* method.

The first step in converting a number from one unit to another is to set up a unity fraction from the definition linking the two units.

EXAMPLE

To convert the distance 4 yd to foot units, take the equation relating yards to feet:

1 yd = 3 ft

and form the fractions $\dfrac{1\ \text{yd}}{3\ \text{ft}}$ and $\dfrac{3\ \text{ft}}{1\ \text{yd}}$.

These fractions are called unity fractions because they are both equal to 1. Any fraction whose top and bottom are equal has the value 1.

YOUR TURN

Practice this step by forming pairs of unity fractions from each of the following definitions.

(a)   12 in. = 1 ft   (b)   1 lb = 16 oz   (c)   4 quarts = 1 gallon

(a)   12 in. = 1 ft. The unity fractions are $\dfrac{12\text{ in.}}{1\text{ ft}}$ and $\dfrac{1\text{ ft}}{12\text{ in.}}$.

(b)   1 lb = 16 oz. The unity fractions are $\dfrac{1\text{ lb}}{16\text{ oz}}$ and $\dfrac{16\text{ oz}}{1\text{ lb}}$.

(c)   4 quarts = 1 gallon. The unity fractions are $\dfrac{4\text{ quarts}}{1\text{ gallon}}$ and $\dfrac{1\text{ gallon}}{4\text{ quarts}}$.

The second step in converting units is to multiply the original number by one of the unity fractions. Always choose the unity fraction that allows you to cancel out the units you do not want in your answer. The units you do not want will be in the denominator of the appropriate unity fraction.

EXAMPLE

To convert the distance 4 yd to foot units, multiply this way:

$$4\text{ yd} = (4 \times 1\ \cancel{\text{yd}}) \times \left(\frac{3\text{ ft}}{1\ \cancel{\text{yd}}}\right)$$

$$= 4 \times 3\text{ ft}$$

$$= 12\text{ ft}$$

You must choose one of the two fractions. Multiply by the fraction that allows you to cancel out the *yard* units you do not want and to keep the *feet* units you do want.

Did multiplying by a fraction give you any trouble? Remember that any number A can be written as $\dfrac{A}{1}$, so that we have

$$A \times \frac{B}{C} = \frac{A}{1} \times \frac{B}{C} \text{ or } \frac{A \times B}{C}$$

This   means   this

EXAMPLE

$$6 \times \frac{3}{4} = \frac{\overset{3}{\cancel{6}}}{1} \times \frac{3}{\underset{2}{\cancel{4}}}$$

$$= \frac{9}{2} \text{ or } 4\frac{1}{2}$$

(If you need to review the multiplication of fractions, turn back to page 70.)

Now try this problem. Use unity fractions to convert 44 oz to pounds. Remember that 1 lb = 16 oz.

▶ SOLUTION

Use the equation 1 lb = 16 oz to set up the unity fractions $\dfrac{1\,\text{lb}}{16\,\text{oz}}$ and $\dfrac{16\,\text{oz}}{1\,\text{lb}}$.

Multiply by the first fraction to convert ounces to pounds.

$$44\,\text{oz} = (44\,\cancel{\text{oz}}) \times \left(\frac{1\,\text{lb}}{16\,\cancel{\text{oz}}}\right)$$

$$= \frac{44 \times 1\,\text{lb}}{16}$$

$$= \frac{44}{16}\,\text{lb or } 2\frac{3}{4}\,\text{lb}$$

Note ▶ When measurement units are converted by division, the remainder often is expressed as the number of original units. For example, instead of giving this last answer as $2\frac{3}{4}$ lb, we can express it as 2 lb 12 oz. ◀

It may happen that several unity fractions are needed in the same problem.

▶ EXAMPLE

If you know that

$$1\,\text{rod} = 16\tfrac{1}{2}\,\text{ft} \quad \text{and} \quad 1\,\text{yd} = 3\,\text{ft}$$

then to convert 11 yd to rods multiply as follows:

$$11\,\text{yd} = (11\,\cancel{\text{yd}}) \times \left(\frac{3\,\cancel{\text{ft}}}{1\,\cancel{\text{yd}}}\right) \times \left(\frac{1\,\text{rod}}{16\frac{1}{2}\,\cancel{\text{ft}}}\right)$$

$$= \frac{11 \times 3 \times 1\,\text{rod}}{16\frac{1}{2}}$$

$$= \frac{33}{16\frac{1}{2}}\,\text{rod}$$

11 yd = 2 rods

11 ⊗ **3** ÷ **16.5** ⊜ → ▨▨▨▨▨▨▨▨▨ *2.*

Remember: To divide by a fraction, multiply by its reciprocal.

$$\frac{33}{16\frac{1}{2}} = 33 \div 16\tfrac{1}{2} = 33 \div \tfrac{33}{2} = 33 \times \tfrac{2}{33} = 2$$

Multiply by
the reciprocal

Be careful to set up the unity fraction so that the unwanted units cancel. At first it helps to write both fractions and then choose the one that fits. As you become an expert at unit conversion, you will be able to write down the correct fraction at first try.

Conversion factors are usually definitions and therefore are considered to be **exact** numbers rather than approximate measurement numbers. When converting units, round all answers to the same number of significant digits as the original measurement number. ◀

YOUR TURN

Try these practice problems. Use unity fractions to convert. Be sure to round your answers properly.

(a) 6.25 ft = _____ in.

(b) $5\frac{1}{4}$ yd = _____ ft

(c) 2.1 mi = _____ yd

(d) 13 gallons = _____ barrel or bbl

(e) 12.5 bbl = _____ pints

(f) 32 psi = _____ atm (1 atm = 14.7 psi)

(g) 87 in. = _____ ft _____ in.

(h) 9 lb 7 oz = _____ oz (Round to the nearest ounce.)

SOLUTIONS

(a) $6.25 \text{ ft} = (6.25 \text{ ft}) \times \left(\dfrac{12 \text{ in.}}{1 \text{ ft}} \right)$

$= 6.25 \times 12 \text{ in.}$

$= 75 \text{ in.}$

(b) $5\frac{1}{4} \text{ yd} = (5\frac{1}{4} \text{ yd}) \times \left(\dfrac{3 \text{ ft}}{1 \text{ yd}} \right)$

$= 5\frac{1}{4} \times 3 \text{ ft}$

$= 15\frac{3}{4} \text{ ft}$

(c) $2.1 \text{ mi} = (2.1 \text{ mi}) \times \left(\dfrac{1760 \text{ yd}}{1 \text{ mi}} \right)$

$= 2.1 \times 1760 \text{ yd}$

$= 3696 \text{ yd}$

$\approx 3700 \text{ yd rounded to 2}$
significant digits

(d) $13 \text{ gal} = (13 \text{ gal}) \times \left(\dfrac{1 \text{ bbl}}{31.5 \text{ gal}} \right)$

$= \dfrac{13}{31.5} \text{ bbl}$

$= 0.41269 \ldots \text{ bbl}$

$\approx 0.41 \text{ bbl rounded to 2}$
significant digits

(e) $12.5 \text{ bbl} = (12.5 \text{ bbl}) \times \left(\dfrac{31.5 \text{ gal}}{1 \text{ bbl}} \right) \times \left(\dfrac{4 \text{ qt}}{1 \text{ gal}} \right) \times \left(\dfrac{2 \text{ pt}}{1 \text{ qt}} \right)$

$= 12.5 \times 31.5 \times 4 \times 2 \text{ pt}$

$= 3150 \text{ pints}$

 12.5 ⊗ **31.5** ⊗ **4** ⊗ **2** ⊜ → ▨▨▨ *3150.*

Our answer has 3 significant digits so it agrees in accuracy with the original measurement number.

(f) $32 \text{ psi} = (32 \text{ psi}) \times \left(\dfrac{1 \text{ atm}}{14.7 \text{ psi}} \right)$

$= \dfrac{32}{14.7} \text{ atm}$

$\approx 2.176870748 \text{ atm}$

$\approx 2.2 \text{ atm} \quad \text{rounded to 2 significant digits}$

(g) $87 \text{ in.} = (87 \text{ in.}) \times \left(\dfrac{1\,\text{ft}}{12\,\text{in.}} \right)$

$$= \dfrac{87}{12}\,\text{ft}$$

$$= 7\text{R}3\,\text{ft} = 7\,\text{ft}\,3\,\text{in.}$$

(h) **First,** convert the pounds to ounces:

$9\,\text{lb} = (9\,\text{lb}) \times \left(\dfrac{16\,\text{oz}}{1\,\text{lb}} \right)$

$$= 9 \times 16\,\text{oz}$$

$$= 144\,\text{oz}$$

Then, add the ounces portion of the original measurement:

$9\,\text{lb}\,7\,\text{oz} = 144\,\text{oz} + 7\,\text{oz}$
$$= 151\,\text{oz}$$

Compound Units   In the English system of units commonly used in the United States in technical work, the basic units for length, weight, and time are named with a single word or abbreviation: ft, lb, and sec. We often define units for other similar quantities in terms of these. For example, 1 mile = 5280 ft, 1 ton = 2000 lb, 1 hr = 3600 sec, and so on. We may also define units for more complex quantities using these basic units. For example, the common unit of speed is miles per hour or mph, and it involves both distance (miles) and time (hour) units. Units named using the product or quotient of two or more simpler units are called *compound units*.

▮ EXAMPLE ▮

If a car travels 75 miles at a constant rate in 1.5 hours, it is moving at a speed of

$\dfrac{75\,\text{miles}}{1.5\,\text{hours}} = \dfrac{75 \times 1\,\text{mi}}{1.5 \times 1\,\text{hr}}$

$$= \dfrac{75}{1.5} \times \boxed{\dfrac{1\,\text{mi}}{1\,\text{hr}}}$$

$$= 50\,\text{mi/hr}$$

We write this speed as 50 mi/hr or 50 mph, and read it as "50 miles per hour." Ratios and rates are often written with compound units.

▮ YOUR TURN ▮

Which of the following are expressed in compound units?

(a)   Diameter, 4.65 in.

(b)   Density, 12 lb/cu ft

(c)   Gas usage, mi/gal or mpg

(d)   Current, 4.6 amperes

(e)   Rotation rate, revolution/min or rpm

(f)   Pressure, lb/sq in. or psi

(g)   Cost ratio, cents/lb

(h)   Length, 5 yd 2 ft

(i)   Area, sq ft

(j)   Volume, cu in.

Compound units are used in quantities (b), (c), (e), (f), (g), (i), and (j). Two different units are used in (h), but they have not been combined by multiplying or dividing.

In Chapters 8 and 9, we study the geometry needed to calculate the area and volume of the various plane and solid figures used in technical work. Here we look at the many different units used to measure area and volume.

Area

When the area of a surface is found, the measurement or calculation is given in "square units." If the lengths are measured in inches, the area is given in square inches or sq in.; if the lengths are measured in feet, the area is given in square feet or sq ft.

Area = 1 ft × 1 ft
       = 1 sq ft

To see the relation between sq ft area units and sq in. area units, rewrite the above area in inch units.

Area = 12 in. × 12 in.
       = 144 × 1 in. × 1 in.
       = 144 sq in.

Therefore,

1 sq ft = 144 sq in.

Note ▶ The area units "sq ft," "sq in.," and so on are often expressed as ft^2, in.^{2}, and so on. The reason for this will become clear when we study exponents in Chapter 6. ◀

Use this information to do the following conversion:

15 sq ft = _____ sq in.

$$15 \text{ sq ft} = (15 \times 1 \text{ sq ft}) \times \left(\frac{144 \text{ sq in.}}{1 \text{ sq ft}} \right)$$

$$= 15 \times 144 \text{ sq in.}$$

$$= 2160 \text{ sq in.} \quad \text{or} \quad 2200 \text{ sq in.} \quad \text{rounded}$$

 15 ⊗ **144** ⊜ → 2160.

If you do not remember the conversion number, 144, try it this way:

$$15 \text{ sq ft} = (15 \times 1 \text{ ft} \times 1 \text{ ft}) \times \left(\frac{12 \text{ in.}}{1 \text{ ft}} \right) \times \left(\frac{12 \text{ in.}}{1 \text{ ft}} \right)$$

$$= 15 \times 12 \times 12 \times 1 \text{ in.} \times 1 \text{ in.}$$

$$= 2160 \text{ sq in.}$$

$$= 2200 \text{ sq in.} \quad \text{rounded}$$

A number of other convenient area units have been defined.

AREA UNITS

1 square foot (sq ft or ft^2) $\quad= 144$ square inches (sq in. or in.^{2})

1 square yard (sq yd or yd^2) $= 9$ sq ft $= 1296$ sq in.

1 square rod (sq rod) $\qquad = 30.25$ sq yd

1 acre $= 160$ sq rod $\qquad = 4840$ sq yd

1 sq mile (sq mi or mi^2) $\quad = 640$ acres

PROBLEMS

Solve the following conversion problems.

(a) 4800 acres = _____ sq mi

(b) 2.50 sq yd = _____ sq in.

(c) 1210 sq yd = _____ acre

(d) Carpeting used to be sold strictly by the square yard, but now many companies are pricing it by the square foot. Suppose Carpet City offered a certain carpet at $42.95 per square yard, and a competitor, Carpeterium, offered a comparable carpet at $4.95 per square foot. Which is the better buy?

SOLUTIONS

(a) $4800 \text{ acres} = (4800 \times 1 \text{ acre}) \times \left(\dfrac{1 \text{ sq mi}}{640 \text{ acres}} \right)$

$\qquad = \dfrac{4800}{640} \text{ sq mi}$

$\qquad = 7.5 \text{ sq mi}$

(b) $2.50 \text{ sq yd} = (2.50 \times 1 \text{ sq yd}) \times \left(\dfrac{1296 \text{ sq in.}}{1 \text{ sq yd}} \right)$

$\qquad = 3240 \text{ sq in.}$

(c) $1210 \text{ sq yd} = (1210 \times 1 \text{ sq yd}) \times \left(\dfrac{1 \text{ acre}}{4840 \text{ sq yd}} \right)$

$\qquad = \dfrac{1210}{4840} \text{ acre}$

$\qquad = 0.25 \text{ acre}$

(d) To compare prices, we will convert Carpeterium's price to cost per square yard.

$\$4.95 \text{ per sq ft} = \left(\dfrac{\$4.95}{1 \text{ sq ft}} \right) \times \left(\dfrac{9 \text{ sq ft}}{1 \text{ sq yd}} \right) = \dfrac{\$44.55}{1 \text{ sq yd}} = \$44.55 \text{ per sq yd}$

Carpetarium's cost is $44.55 per square yard compared to Carpet City's cost of $42.95 per square yard. Carpet City has the better deal. Notice that we could also have converted Carpet City's price to cost per square foot and reached the same conclusion.

Volume

1 ft
1 ft
1 ft

Volume units are given in "cubic units." If the lengths are measured in inches, the volume is given in cubic inches or cu in.; if the lengths are measured in feet, the volume is given in cubic feet or cu ft.

$$\text{Volume} = 1 \text{ ft} \times 1 \text{ ft} \times 1 \text{ ft}$$
$$= 1 \text{ cu ft}$$

In inch units, this same volume is 12 in. × 12 in. × 12 in. = 1728 cu in., so that 1 cu ft = 1728 cu in.

Note ▶   Volume units such as cu ft and cu in. can also be written as ft^3 and in.^3 ◀

A very large number of special volume units have been developed for various uses.

VOLUME UNITS

1 cubic foot (cu ft or ft^3)   = 1728 cubic inches (cu in. or in.^3)

1 cubic yard (cu yd or yd^3) = 27 cu ft = 46,656 cu in.

1 gallon (gal)   = 231 cu in.   or   1 cu ft ≈ 7.48 gal

1 bushel (bu)   = 2150.42 cu in.

1 fluid ounce (fl oz)   = 1.805 cu in.

1 pint (pt)   = 28.875 cu in. (liquid measure)

Dozens of other volume units are in common use: cup, quart, peck, teaspoonful, and others. The cubic inch, cubic foot, and gallon are most used in trades work.

YOUR TURN

Convert the following volume measurements for practice. Round to the proper number of significant digits.

(a) 94 cu ft   = _____ cu yd

(b) 12.0 gal   = _____ cu in.

(c) 10,500 cu in. = _____ bu

SOLUTIONS

(a) $94 \text{ cu ft} = (94 \times 1 \text{ cu ft}) \times \left(\dfrac{1 \text{ cu yd}}{27 \text{ cu ft}} \right)$

$= \left(\dfrac{94}{27} \right) \text{cu yd}$

$= 3.481\ldots \text{ cu yd} \approx 3.5 \text{ cu yd}$

(b) $12.0 \text{ gal} = (12.0 \times 1 \text{ gal}) \times \left(\dfrac{231 \text{ cu in.}}{1 \text{ gal}} \right)$

$= 12.0 \times 231 \text{ cu in.}$

$= 2772 \text{ cu in.} \approx 2770 \text{ cu in.}$

(c) $\quad 10{,}500 \text{ cu in.} = (10{,}500 \times 1 \text{ cu in.}) \times \dfrac{1 \text{ bu}}{2150.42 \text{ cu in.}}$

$$= \dfrac{10{,}500}{2150.42} \text{ bu}$$

$$= 4.882 \ldots \text{bu} \approx 4.88 \text{ bu}$$

LUMBER MEASURE

Carpenters and workers in the construction trades use a special unit to measure the amount of lumber. The volume of lumber is measured in *board feet*. One board foot (bf) of lumber is a piece having an area of 1 sq ft and a thickness of 1 in. or less.

Number of board feet = thickness in inches × width in feet × length in feet

To find the number of board feet in a piece of lumber, multiply the length in feet by the width in feet by the thickness in inches. A thickness of less than 1 inch should be counted as 1 inch. If the lumber is dressed or finished, use the full size or rough stock dimension to calculate board feet.

For example, a 2 by 4 used in framing a house would actually measure $1\frac{1}{2}$ in. by $3\frac{1}{2}$ in., but we would use the dimensions 2 in. by 4 in. in calculating board feet. A 12-ft length of 2 in. by 4 in. would have a volume of

$$12 \text{ ft} \times \tfrac{4}{12} \text{ ft} \times 2 \text{ in.} = 8 \text{ board ft or 8 bf}$$

In lumber measure the phrase "per foot" means "per board foot." The phrase "per running foot" means "per foot of length."

Problems:

1. Find the number of board feet in each of the following pieces of lumber. Be sure to convert the middle dimension, width, to feet.

    (a) $\frac{3}{4}$ in. × 6 in. × 4 ft   (b)   2 in. × 6 in. × 16 ft

    (c) $\frac{1}{2}$ in. × 8 in. × 12 ft   (d)   4 in. × 4 in. × 8 ft

    (e)   1 in. × 6 in. × 14 ft   (f) $\frac{7}{8}$ in. × 12 in. × 3 ft

(continued)

2. How many board feet are there in a shipment containing 80 boards, 2 in. by 6 in., each 16 ft long?

3. To floor a small building requires 245 boards, each $1\frac{1}{2}$ in. by 12 in. by 12 ft long. How many board feet should be ordered? (*Hint:* Use 2 in. for the thickness to calculate board feet.)

The answers are in the Appendix.

Speed Speed is usually specified by a compound unit that consists of a distance unit divided by a time unit. For example, a speed given as 60 mph can be written as

$$60 \text{ mi/hr or } 60 \times \frac{1 \text{ mi}}{1 \text{ hr}}.$$

To convert speed units within the English system of units, use the length units from p. 236 and the following time conversion factors.

Time Units

1 minute (min) = 60 seconds (sec)

1 hour (hr or h) = 60 min = 3600 sec

1 day = 24 hr or 24 h

EXAMPLE

To convert 55 mph to ft/sec, use the conversion factors 1 mi = 5280 ft and 1 hr = 3600 sec. Set up the unity fractions like this:

> We want feet as our new numerator.

$$55 \text{ mph} = \frac{55 \text{ mi}}{1 \text{ hr}} \times \frac{5280 \text{ ft}}{1 \text{ mi}} \times \frac{1 \text{ hr}}{3600 \text{ sec}}$$

> We want seconds as our new denominator.

$$= \frac{55 \times 5280}{3600} \text{ ft/sec}$$

$$= 80.666 \ldots \text{ ft/sec} \approx 81 \text{ ft/sec}$$

YOUR TURN

Convert the following speed measurements.

(a) 3.8 in./sec = _____ ft/min

(b) 98.5 ft/sec = _____ mi/hr

(c) 160 rpm   = _____ rev/sec

SOLUTIONS

(a) $3.8 \text{ in./sec} = \dfrac{3.8 \text{ in.}}{1 \text{ sec}} \times \dfrac{1 \text{ ft}}{12 \text{ in.}} \times \dfrac{60 \text{ sec}}{1 \text{ min}}$

$= 19 \text{ ft/min}$

(b)　$98.5 \text{ ft/sec} = \dfrac{98.5 \text{ ft}}{1 \text{ sec}} \times \dfrac{1 \text{ mi}}{5280 \text{ ft}} \times \dfrac{3600 \text{ sec}}{1 \text{ hr}}$

$= 67.159\ldots \text{ mi/hr} \approx 67.2 \text{ mi/hr}$

98.5 ⊡ **5280** ⊠ **3600** ⊟ → ▓ *67.15909091*

(c)　$160 \text{ rpm} = 160 \text{ rev/min} = \dfrac{160 \text{ rev}}{1 \text{ min}} \times \dfrac{1 \text{ min}}{60 \text{ sec}}$

$= 2.666\ldots \text{ rev/sec} \approx 2.7 \text{ rev/sec}$

The following table summarizes all English conversion factors given in this section.

Length	Weight
1 ft = 12 in. 1 yd = 3 ft = 36 in. 1 mi = 5280 ft = 1760 yd	1 lb = 16 oz 1 ton = 2000 lb
Liquid Capacity	**Area**
1 qt = 2 pt 1 gal = 4 qt 1 bbl = 31.5 gal	1 sq ft = 144 sq in. 1 sq yd = 9 sq ft = 1296 sq in. 1 sq rod = 30.25 sq yd 1 acre = 160 sq rod = 4840 sq yd 1 sq mi = 640 acres
Volume	**Time**
1 cu ft = 1728 cu in. 1 gal = 231 cu in. 1 bu = 2150.42 cu in. 1 pt = 28.875 cu in. 1 cu yd = 27 cu ft = 46,656 cu in. 1 cu ft ≈ 7.48 gal 1 fl oz = 1.805 cu in.	1 min = 60 sec 1 hr or 1 h = 60 min = 3600 sec 1 day = 24 hr

In the next section we study metric units, and you will find only a few basic units connected by simple conversion numbers rather than the confusion of having many different length, area, and volume units.

Now, turn to Exercises 5-2 for a set of problems involving unit conversion in the English system.

Exercises 5-2　English Units and Unit Conversion

A.　Convert the units as shown. (Round your answer to the same number of significant digits as the given measurement.)

1.　4.25 ft　　= _____ in.　　　　2.　33 yd　　= _____ ft

3.　3.4 mi　　= _____ ft　　　　4.　17.3 lb　　= _____ oz

5. 126 gal   = _____ bbl   6. 8.5 mi   = _____ yd

7. 46 psi   = _____ atm   8. 7.5 lb   = _____ oz

9. 18.50 bbl   = _____ gal   10. 32 in.   = _____ ft

11. 4640 ft   = _____ mi   12. 19 yd   = _____ rods

13. 26 ft   = _____ yd   14. $23\overline{0}$ oz   = _____ lb

15. 5.10 gal   = _____ cu in.   16. 114 gal   = _____ cu in.

17. 1400 yd   = _____ mile   18. 0.85 cu ft   = _____ gal

19. 6 ft 11 in.   = _____ in.   20. 13.5 lb   = _____ oz

21. 326 oz = _____ lb _____ oz   22. 29 ft = _____ yd _____ ft

B. Convert the units as shown. (Round as needed.)

1. 6.0 sq ft   = _____ sq in.   2. 2.50 acre   = _____ sq ft

3. 17.25 bbl   = _____ cu ft   4. 385 sq ft   = _____ sq yd

5. 972 cu in.   = _____ cu ft   6. 325 rpm   = _____ rev/sec

7. $60 \dfrac{\text{lb}}{\text{cu ft}}$ = _____ $\dfrac{\text{lb}}{\text{cu in.}}$ 8. $3\overline{0}$ cu ft   = _____ bbl

9. 14,520 sq ft   = _____ acre   10. $2500 \dfrac{\text{ft}}{\text{min}}$ = _____ $\dfrac{\text{in.}}{\text{sec}}$

11. 20.0 sq mi   = _____ acre   12. 65 mph   = _____ ft/sec

13. 2.1 sq yd   = _____ sq in.   14. 42 sq yd   = _____ sq ft

15. 14,000 cu in. = _____ cu yd   16. 8.3 cu yd   = _____ cu ft

C. Solve:

1. **Wastewater Technology** Hydrological engineers often measure very large amounts of water in acre-feet. 1 acre-ft of water fills a rectangular volume of base 1 acre and height 1 ft. Convert 1 acre-ft to

(a) _____ cu ft   (b) _____ gal   (c) _____ bbl

(d) _____ cu yd

2. **Carpentry** The board foot defined on page 245 is a volume unit. Use the definition given to complete the following.

(a) 1 bd ft = _____ cu ft   (b) 1 bd ft = _____ cu in.

3. Martian greeb workers use the gronk, smersh, and pflug as length units. There are 3 gronks in a smersh, and 2 pflugs equal one smersh. Translate a length of 6 gronks to pflugs.

4. **Painting and Decorating** The area covered by a given volume of paint is determined by the consistency of the paint and the nature of the surface. The measure of covering power is called the *spreading rate* of the paint and is expressed in units of square feet per gallon. The following table gives some typical spreading rates.

Paint	Surface	Spreading Rate	
		One Coat	Two Coats
Oil-based paint	Wood, smooth	580	320
	Wood, rough	360	200
	Plaster, smooth	300	200
	Plaster, rough	250	160
Latex paint	Wood	500	350
Varnish	Wood	400	260

Use this table to answer the following questions.

(a) How many gallons of oil-based paint are needed to cover $25\overline{0}0$ sq ft of rough wood with two coats?

(b) At \$17.98 per gallon, how much would it cost to paint a room of $85\overline{0}$ sq ft area with latex paint? The surfaces are finished wood.

(c) How many square feet of rough plaster will 8 gallons of oil-based paint cover with one coat?

(d) How many gallons of varnish are needed to cover $6\overline{0}0$ sq ft of cabinets with two coats?

(e) At \$21.98 per gallon, how much will it cost to paint $75\overline{0}$ sq ft of smooth plaster with two coats of oil-based paint?

5. The earliest known unit of length to be used in a major construction project is the "megalithic yard" used by the builders of Stonehenge in southwestern Britain about 2600 B.C. If 1 megalithic yard $= 2.72 \pm 0.05$ ft, convert this length to inches. (Round to the nearest tenth of an inch.)

6. In the Bible (Genesis, Chapter 7), Noah built an ark 300 cubits long, 50 cubits wide, and 30 cubits high. In I Samuel, Chapter 17, it is reported that the giant Goliath was "six cubits and a span" in height. If 1 cubit = 18 in. and 1 span = 9 in.:

(a) What were the dimensions of the ark in English units?

(b) How tall was Goliath?

7. **Interior Decorating** A homeowner needing carpeting has calculated that 1265 sq ft of carpet must be ordered. However, carpet is sold in square yards. Make the necessary conversion for her. (Round up to the nearest whole number.)

8. **Metalworking** Cast aluminum has a density of 160.0 lb/cu ft. Convert this density to units of lb/cu in. and round to one decimal place.

9. **Machine Technology** A $\frac{5}{16}$-in. twist drill with a periphery speed of 50.0 ft/min has a cutting speed of 611 rpm (rev/min). Convert this speed to rps (rev/sec) and round to one decimal place.

10. **Welding** How many pieces $8\frac{3}{8}$ in. long can be cut from a piece of 20-ft stock? Allow for $\frac{3}{16}$ in. kerf (waste due to the width of a cut).

11. **Wastewater Technology**   A sewer line has a flow rate of 1.25 MGD (million gallons per day). Convert this to cfs (cubic feet per second).

12. **Wastewater Technology**   A reservoir has a capacity of 9000 cu ft. How long will it take to fill the reservoir at the rate of 250 gallons per minute?

13. **Sheet Metal Technology**   A machinist must cut 12 strips of metal that are each 3 ft $2\frac{7}{8}$ in. long. What is the total length needed in inches?

14. **Marine Technology**   A scientist wishes to treat a 430,000-barrel (bbl) oil spill with a bacterial culture. If the culture must be applied at the rate of 1 oz of culture per 100 cu ft of oil, how many ounces of culture will the scientist need? (Round to two significant digits.)

15. **Painting**   A painter must thin some paint for use in a sprayer. If the recommended rate is $\frac{1}{2}$ pint of water per gallon of paint, how many total gallons will there be after thinning 8 gallons of paint?

16. **Interior Design**   A designer is pricing a certain wool carpet for a new house. She finds it listed for $86.50 per square yard at Carpet City and $9.40 per square foot at Carpetorium. (a) Which is the better buy? (b) How much will she save on 166 square yards?

17. **Architecture**   Calculate the height of the church tower shown.

SANTA · BARBARA

WOOD RAILING

CHAPEL

18'5"

23'3"

13'7"

10'5"

13'1"

29'6"

Check your answers in the Appendix, then continue in Section 5-3 with the study of metric units.

5-3 METRIC UNITS

For all technical and many trades workers, the ability to use metric units is important. In many ways, working with meters, kilograms, and liters is easier and more logical than using our traditional units of feet, pounds, or quarts. In this section we define and explain the most useful metric units, show how they are related to the corresponding English units, explain how to convert from one unit to another, and provide practice problems designed to help you use the metric system and to "think metric."

The most important common units in the metric system are those for length or distance, speed, weight, volume, area, and temperature. Time units—year, day, hour, minute, second—are the same in the metric as in the English system.

Length

The basic unit of length in the metric system is the *meter,* pronounced *meet-ur* and abbreviated *m*. (The word is sometimes spelled *metre* but pronounced exactly the same.) One meter is roughly equal to one yard.

The meter is the appropriate unit to use in measuring your height, the width of a room, length of lumber, or the height of a building.

1 foot

1 yard

1 meter

YOUR TURN

Estimate the following lengths in meters:

(a) The length of the room in which you are now sitting = _____ m.

(b) Your height = _____ m.

(c) The length of a Ping-Pong table = _____ m.

(d) The length of a football field = _____ m.

Guess as closely as you can before looking at the solution.

ANSWERS

(a) A small room might be 3 or 4 m long, and a large one might be 6 or 8 m long.

(b) Your height is probably between $1\frac{1}{2}$ and 2 m. Note that 2 m is roughly 6 ft 7 in.

(c) A Ping-Pong table is a little less than 3 m in length.

(d) A football field is about 100 m long.

All other length units used in the metric system are defined in terms of the meter, and these units differ from one another by multiples of ten. For example, the *centimeter,* pronounced *cent-a-meter* and abbreviated *cm,* is defined as exactly one-hundredth of a meter. The *kilometer,* pronounced *kill-AH-meter* and abbreviated *km,* is defined as exactly 1000 meters. The *millimeter,* abbreviated *mm,* is defined as exactly $\frac{1}{1000}$ meter.

Because metric units increase or decrease in multiples of ten, they may be named using prefixes attached to a basic unit. For length units we have:

Metric Length Unit	Prefix	Multiplier	Money Analogy	Memory Aid
kilometer	kilo-	1000×1 meter	$1000	*kilo*watt
hectometer	hecto-	100×1 meter	$ 100	
decameter	deca-	10×1 meter	$ 10	*dec*ade = 10 years
meter	—	1 meter	$ 1	
decimeter	deci-	0.1×1 meter	10¢	*deci*mal $= \frac{1}{10}$
centimeter	centi-	0.01×1 meter	1¢	*cent*ury = 100 years
millimeter	milli-	0.001×1 meter	$\frac{1}{10}$¢	*mill*ennium = 1000 years

Note ▶ The shaded prefixes, hecto-, deca-, and deci-, are rarely used in trades work. Other prefixes are sometimes encountered in modern technology. These include, for example, nanosecond (billionth of a second, nsec), microsecond (millionth of a second, μsec), megabyte (million bytes, MB), gigabyte (billion bytes, GB), and terabyte (trillion bytes, TB). ◀

The following length conversion factors are especially useful for workers in the trades.

METRIC LENGTH UNITS

1 centimeter (cm) = 10 millimeters (mm)

1 meter (m) = 100 cm = 1000 mm

1 kilometer (km) = 1000 m

The millimeter (mm) unit of length is very often used on technical drawings, and the centimeter (cm) is handy for shop measurements.

1 cm is roughly the width of a large paper clip.

1 mm is roughly the thickness of the wire in a paper clip.

1 meter is roughly 10% more than a yard.

Metric to Metric Conversion One very great advantage of the metric system is that we can easily convert from one metric unit to another. There are no hard-to-remember conversion factors: 36 in. in a yard, 5280 ft in a mile, 220 yards in a furlong, or whatever.

For example, with English units, if we convert a length of 137 in. to feet or yards, we find

$$137 \text{ in.} = 11\tfrac{5}{12} \text{ ft} = 3\tfrac{29}{36} \text{ yard}$$

Here we must divide by 12 and then by 3 to convert the units. But in the metric system we simply shift the decimal point.

EXAMPLE

Convert 348 cm to meters like this:

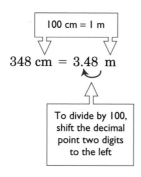

$$348 \text{ cm} = 3.48 \text{ m}$$

To divide by 100, shift the decimal point two digits to the left

Of course we may also use unity fractions:

$$348 \text{ cm} = 348 \text{ cm} \times \left(\frac{1 \text{ m}}{100 \text{ cm}}\right)$$

$$= \frac{348}{100} \text{ m}$$

$$= 3.48 \text{ m}$$

YOUR TURN

Try it. Complete the following unit conversions.

(a)   147 cm = _____ m

(b)   3.1 m = _____ cm

(c)   2.1 km = _____ m

(d)   307 m = _____ km

(e)   7 cm = _____ mm

(f)   20.5 mm = _____ cm

SOLUTIONS

(a) $147 \text{ cm} = 147 \text{ cm} \times \left(\frac{1 \text{ m}}{100 \text{ cm}}\right)$

$$= \frac{147}{100} \text{ m}$$

$$= 1.47 \text{ m} \qquad \text{The decimal point shifts two places to the left.}$$

(b) $3.1 \text{ m} = 3.1 \text{ m} \times \left(\frac{100 \text{ cm}}{1 \text{ m}}\right)$

$$= 3.1 \times 100 \text{ cm}$$

$$= 310 \text{ cm} \qquad \text{The decimal point shifts two places right.}$$

(c) $2.1 \text{ km} = 2.1 \text{ km} \times \left(\frac{1000 \text{ m}}{1 \text{ km}}\right)$

$$= 2100 \text{ m}$$

(d) $307 \text{ m} = 307 \text{ m} \times \left(\frac{1 \text{ km}}{1000 \text{ m}}\right)$

$$= 0.307 \text{ km}$$

(e) $7 \text{ cm} = 7 \cancel{\text{cm}} \times \left(\dfrac{10 \text{ mm}}{1 \cancel{\text{cm}}} \right)$

$\qquad = 70 \text{ mm}$

(f) $20.5 \text{ mm} = 20.5 \cancel{\text{mm}} \times \left(\dfrac{1 \text{ cm}}{10 \cancel{\text{mm}}} \right)$

$\qquad = 2.05 \text{ cm}$

Metric–English Conversion

Because metric units are the only international units, all English units are defined in terms of the metric system. Use the following table for metric–English length conversions.

1 inch

1 centimeter

> **METRIC–ENGLISH LENGTH CONVERSIONS**
>
> 1 in. = 2.54 cm ⟵ This is the exact legal definition of the inch.
>
> 1 ft = 30.48 cm
>
> 1 yd = 0.9144 m
>
> 1 mi ≈ 1.6093 km

To shift from English to metric units or from metric to English units, use either unity fractions or the conversion factors given in the table.

EXAMPLE

Convert 4.3 yd to meters.

From the conversion table, 1 yd = 0.9144 m.

Using unity fractions,

$$4.3 \cancel{\text{yd}} \times \frac{0.9144 \text{ m}}{1 \cancel{\text{yd}}} = 3.93192 \text{ m} \approx 3.9 \text{ m rounded to two significant digits.}$$

EXAMPLE

Sometimes more than one unity fraction is needed.

Convert 2.50 in. to millimeters.

The conversion table does not list an inch–millimeter conversion factor.

Therefore we must use two factors: 1 in. = 2.54 cm and 1 cm = 10 mm

$$2.50 \text{ in.} = 2.5 \cancel{\text{in.}} \times \frac{2.54 \cancel{\text{cm}}}{1 \cancel{\text{in.}}} \times \frac{10 \text{ mm}}{1 \cancel{\text{cm}}} = 63.5 \text{ mm}$$

The answer agrees in accuracy with the three significant digits of the original measurement.

PROBLEMS

Convert the following measurements as shown. Round answers to agree in accuracy with the original measurement.

(a) 16.0 in. = _____ cm   (b) 6.0 km = _____ mi

(c) 1.2 m = _____ ft   (d) 37.0 cm = _____ in.

(e) 4.2 m = _____ yd   (f) 140 cm = _____ ft

(g) 18 ft 6 in. = _____ m   (h) 2.25 in. = _____ mm

ANSWERS

(a) 40.6 cm   (b) 3.7 mi

(c) 3.9 ft   (d) 14.6 in.

(e) 4.6 yd   (f) 4.6 ft

(g) 5.6 m   (h) 57.2 mm

In problem (c), $1.2\, \cancel{m} \times \dfrac{1\, \cancel{yd}}{0.9144\, \cancel{m}} \times \dfrac{3\, ft}{1\, \cancel{yd}} = 3.937\, ft \ldots \approx 3.9\, ft.$

DUAL DIMENSIONING

Some companies involved in international trade use "dual dimensioning" on their technical drawings and specifications. With dual dimensioning both inch and metric dimensions are given. For example, a part might be labeled like this:

Notice that the metric measurement is written first or on top of the fraction bar. Diameter dimensions are marked with the symbol φ.

Weight   The weight of an object is the gravitational pull of the earth exerted on that object. The mass of an object is related to how it behaves when pushed or pulled. For scientists the difference between mass and weight can be important. For all practical purposes they are the same.

The basic unit of mass or metric weight in the International Metric System (SI) is the *kilogram,* pronounced *kill-o-gram* and abbreviated *kg.* The kilogram is defined as the mass of a standard platinum–iridium cylinder kept at the International Bureau of Weights and Measures near Paris. By law in the United States, the pound is defined as exactly equal to the weight of 0.45359237 kg, so that 1 kg weighs about 2.2046 lb.

One kilogram weighs about 10% more than 2 lb.

The kilogram was originally designed so that it was almost exactly equal to the weight of one liter of water.

Two smaller metric units of mass are also defined. The *gram,* abbreviated *g,* is equal to 0.001 kg and the *milligram,* abbreviated *mg,* is equal to 0.001 g.

The following conversion factors will make it easy for you to convert weight units within the metric system.

METRIC WEIGHT/MASS UNITS

1 gram (g) = 1000 milligrams (mg)

1 kilogram (kg) = 1000 g

1 metric ton = 1000 kg

The gram is a very small unit of weight, equal to roughly $\frac{1}{500}$ of a pound, or the weight of a paper clip. The milligram is often used by druggists and nurses to measure very small amounts of medication or other chemicals, and it is also used in scientific measurements.

Metric to Metric Conversion To convert weights within the metric system, use the conversion factors in the table to form unity fractions.

EXAMPLE

To convert 4500 mg to grams, use the conversion factor 1 g = 1000 mg.

$$4500 \text{ mg} = 4500 \text{ mg} \times \frac{1 \text{ g}}{1000 \text{ mg}} = 4.5 \text{ g}$$

YOUR TURN

Convert 0.0062 kg to milligrams.

SOLUTION

Here we need to use two unity fractions.

$$0.0062 \text{ kg} = 0.0062 \text{ kg} \times \frac{1000 \text{ g}}{1 \text{ kg}} \times \frac{1000 \text{ mg}}{1 \text{ g}} = 62{,}000 \text{ mg}$$

PROBLEMS

Convert the following measurements as shown.

(a)   21.4 g = _____ mg   (b)   156,000 mg = _____ kg

(c)   0.56 kg = _____ g   (d)   467 g = _____ kg

ANSWERS

(a)   21,400 mg   (b)   0.156 kg   (c)   560 g   (d)   0.467 kg

Metric–English Conversion

To convert weights from metric to English or English to metric units, use the conversion factors in the following table.

METRIC–ENGLISH WEIGHT CONVERSIONS

1 ounce (oz) ≈ 28.35 grams (g)

1 pound (lb) ≈ 0.4536 kilograms (kg)

1 ton (T) ≈ 907.2 kilograms (kg)

EXAMPLE

To convert 75 g to ounces, use the conversion factor 1 ounce = 28.35 grams to form a unity fraction.

$$75 \text{ g} = 75 \text{ g} \times \frac{1 \text{ oz}}{28.35 \text{ g}} = 2.6455 \ldots \approx 2.6 \text{ oz, rounded}$$

EXAMPLE

Culinary Arts Cactus sweetener sells for $7.20 per pound. What is its cost in cents per gram?

SOLUTION

$$\text{Cost} = \frac{\$7.20}{1 \text{ lb}} = \frac{720 \text{ ¢}}{1 \text{ lb}} \times \frac{1 \text{ lb}}{0.4536 \text{ kg}} \times \frac{1 \text{ kg}}{1000 \text{ g}}$$

$$= \frac{720}{0.4536 \times 1000} \text{ ¢/g} = 1.584 \text{ ¢/g} \approx 1.6 \text{ ¢/g}$$

YOUR TURN

Convert (a) 0.075 pounds to grams.

(b) 1.40 T to kg

SOLUTION

(a) Using two unity fractions we have

$$0.075 \text{ lb} = 0.075 \text{ lb} \times \frac{0.4536 \text{ kg}}{1 \text{ lb}} \times \frac{1000 \text{ g}}{1 \text{ kg}}$$

$$= 34.02 \text{ g} \approx 34 \text{ g, rounded}$$

 .075 ⊗ **.4536** ⊗ **1000** ⊜ → $\boxed{34.02}$

(b) $1.40 \text{ T} = 1.40 \text{ T} \times \dfrac{907.2 \text{ kg}}{1 \text{ T}}$

$$= 1.40 \times 907.2 \text{ kg}$$

$$\approx 1270 \text{ kg}$$

Convert the following measurements as shown. Round to the same number of significant digits.

(a) 150 lb = _____ kg   (b) 6.0 oz = _____ g

(c) 2.50 T = _____ kg   (d) $10\bar{0}$ kg = _____ lb

(e) 1.50 kg = _____ oz   (f) 35,600 kg = _____ T

(g) 150 g = _____ lb   (h) 4 lb 5 oz = _____ kg

ANSWERS

(a) 68 kg   (b) 170 g

(c) 2270 kg   (d) 220 lb

(e) 52.9 oz   (f) 39.2 tons

(g) 0.33 lb   (h) 2.0 kg

In problem (e) $1.50 \text{ kg} \times \dfrac{1 \text{ lb}}{0.4536 \text{ kg}} \times \dfrac{16 \text{ oz}}{1 \text{ lb}}$.

In problem (h) the answer 1.96 is 2.0 when rounded to two significant digits.

Area   Because an area is essentially the product of two lengths, it is specified by a compound unit. When the area of a surface is calculated, the units will be given in "square length units." If the lengths are measured in feet, the area will be given in square feet; if the lengths are measured in meters, the area will be given in square meters.

For example, the area of a carpet 2 m wide and 3 m long would be

2 m × 3 m = 6 square meters
           = 6 sq m

Areas roughly the size of carpets and room flooring are usually measured in square meters. Larger areas are usually measured in *hectares,* a metric surveyor's unit. One hectare (ha) is the area of a square 100 m on each side. The hectare is roughly $2\frac{1}{2}$ acres, or about double the area of a football field.

METRIC AREA UNITS

1 hectare (ha) = 100 m × 100 m

                  = 10,000 sq m (or m^2)

EXAMPLE

Convert 456,000 sq m to hectares.

Using unity fractions,

$$456,000 \text{ sq m} = 456,000 \text{ sq m} \times \frac{1 \text{ ha}}{10,000 \text{ sq m}} = 45.6 \text{ ha}$$

YOUR TURN

Convert 0.85 hectares to square meters.

SOLUTION

$$0.85 \text{ ha} = 0.85 \text{ ha} \times \frac{10{,}000 \text{ sq m}}{1 \text{ ha}} = 8500 \text{ sq m}$$

Area conversions within the metric system, such as sq m to sq km, are seldom used in the trades and we will not cover them here. To convert area units from metric to English or English to metric units, use the following conversion factors.

METRIC–ENGLISH AREA CONVERSIONS

1 square inch (sq in. or in.^{2}) $\approx$ 6.452 sq cm (or cm^2)

1 square foot (sq ft or ft^2) $\approx$ 0.0929 sq m (or m^2)

1 square yard (sq yd or yd^2) $\approx$ 0.836 sq m

PROBLEMS

Use this information to solve the following problems. Round carefully.

(a) 15 sq ft = _____ sq m

(b) 0.21 sq m = _____ sq ft

(c) $1\overline{0}$ sq cm = _____ sq in.

(d) 78 sq yd = _____ sq m

ANSWERS

(a) 1.4 sq m

(b) 2.3 sq ft

(c) 1.6 sq in.

(d) 65 sq m

Volume When the volume of an object is calculated or measured, the units will normally be given in "cubic length units." If the lengths are measured in feet, the volume will be given in cubic feet; if the lengths are measured in meters, the volume will be given in cubic meters.

For example, the volume of a box 2 m high by 4 m wide by 5 m long would be

$$2 \text{ m} \times 4 \text{ m} \times 5 \text{ m} = 40 \text{ cu m or } 40 \text{ m}^3$$

The cubic meter is an appropriate unit for measuring the volume of large amounts of water, sand, or gravel, or the volume of a room. However, a more useful metric volume unit for most practical purposes is the *liter*, pronounced *leet-ur*. You may find this volume unit spelled either *liter* in the United States or *litre* elsewhere in the world.

1 cm

1 cm

1 cm

1 ml = 1 cu cm

One liter is defined as 1000 cu cm, the volume of a cube 10 cm on each edge.

Because 1000 cubic centimeters are needed to make up 1 liter, a cubic centimeter is $\frac{1}{1000}$ of a liter or a *milliliter*.

Volumes smaller than one liter are usually measured in units of cubic centimeters or milliliters. The following table summarizes the most useful metric volume relationships.

METRIC VOLUME UNITS

1 cubic centimeter (cu cm or cc or cm^3) = 1 milliliter (ml)

1 liter = 1000 cu cm

YOUR TURN

Perform the following unit conversions.

(a)   0.24 liters = _____ cu cm   (b)   5780 milliliters = _____ liters

ANSWERS

(a)   240 cu cm   (b)   5.78 liters

Metric–English Conversion

In the United States, one quart is legally defined as 0.94635295 liters. This means that a volume of one liter is slightly larger than one quart.

One liter is roughly 6% more than one quart.

The following table summarizes the most useful metric–English volume conversion factors.

METRIC–ENGLISH VOLUME CONVERSIONS

1 cubic inch (cu in. or in.^{3}) $\approx$ 16.387 cubic centimeters (cu cm or cm^3)

1 cubic yard (cu yd or yd^3) $\approx$ 0.765 cubic meters (cu m or m^3)

1 fluid ounce $\approx$ 29.574 cu cm

1 quart (qt) $\approx$ 0.946 liters

1 gallon (gal) $\approx$ 3.785 liters

PROBLEMS

Convert the following measurements as shown. Round the answer to the same number of significant digits as in the original number.

(a) 13.4 gal = _____ liter

(b) 5.5 gal = _____ liter

(c) 9.1 fl oz = _____ cu cm

(d) 16.0 cu yd = _____ cu m

(e) 5.1 liters = _____ qt

(f) 140 cu cm = _____ cu in.

(g) 24.0 cu m = _____ cu yd

(h) 64.0 liters = _____ gal

(i) 12.4 qt = _____ liter

(j) 4.2 cu in. = _____ cu cm

ANSWERS

(a) 50.7 liters

(b) 19 liters

(c) 270 cu cm

(d) 12.2 cu m

(e) 5.4 qt

(f) 8.5 cu in.

(g) 31.4 cu yd

(h) 16.9 gal

(i) 11.7 liters

(j) 69 cu cm

Speed In the metric system, ordinary highway speeds are measured in kilometers per hour, abbreviated km/h. To "think metric" while you drive, remember that

100 km/h is approximately equal to 62 mph.

To convert speed units, either within the metric system or between metric and English units, use the appropriate combination of time and distance unity fractions.

EXAMPLE

To convert 2.4 ft/sec to m/min, use the following conversion factors:

1 min = 60 sec, 1 ft = 30.48 cm, and 1 m = 100 cm

Using unity fractions,

$$2.4 \text{ ft/sec} = \frac{2.4 \text{ ft}}{1 \text{ sec}} \times \frac{1 \text{ m}}{100 \text{ cm}} \times \frac{30.48 \text{ cm}}{1 \text{ ft}} \times \frac{60 \text{ sec}}{1 \text{ min}}$$

$$= (2.4 \times 30.48 \times 60) \div 100 \text{ m/min}$$

$$= 43.8912 \text{ m/min} \approx 44 \text{ m/min}$$

Convert 27.5 km/h to cm/sec.

First, determine the necessary conversion factors:

1 h = 3600 sec, 1 km = 1000 m, and 1 m = 100 cm

Then, set up the appropriate unity fractions:

$$27.5 \text{ km/h} = \frac{27.5 \text{ km}}{1 \text{ h}} \times \frac{1000 \text{ m}}{1 \text{ km}} \times \frac{100 \text{ cm}}{1 \text{ m}} \times \frac{1 \text{ h}}{3600 \text{ sec}}$$

$$= \frac{27.5 \times 1000 \times 100}{3600} \text{ cm/sec}$$

$$= 763.88 \ldots \text{ cm/sec} \approx 764 \text{ cm/sec}$$

27.5 ⊠ **1000** ⊠ **100** ÷ **3600** = → `763.8888889`

(a) 35 mph = _____ km/h   (b) 30.0 m/sec = _____ km/h

(c) 120 cm/sec = _____ in./sec   (d) 18 in./sec = _____ cm/sec

(e) 28$\overline{0}$ km/h = _____ mph   (f) 775 m/min = _____ cm/sec

(g) **Machine Technology**   The cutting speed for a soft steel part in a lathe is
     165 ft/min. Express this in cm/sec.

(a) 56 km/h   (b) 108 km/h   (c) 47 in./sec

(d) 46 cm/sec   (e) 174 mph   (f) 1290 cm/sec   (g) 83.8 cm/sec

Temperature   The Fahrenheit temperature scale is commonly used in the United States for
weather reports, cooking, and other practical work. On this scale, water boils at
212°F and freezes at 32°F. The metric or *Celsius* temperature scale is a simpler
scale originally designed for scientific work but now used worldwide for most temperature measurements. On the Celsius scale, water boils at 100°C and freezes
at 0°C.

Notice that the range from freezing to boiling is covered by 180 degrees on the
Fahrenheit scale and 100 degrees on the Celsius scale. The size and meaning of a
temperature degree is different for the two scales. A temperature of zero does not
mean "no temperature"—it is simply another point on the scale. Temperatures less
than zero are possible, and they are labeled with negative numbers.

Because zero does not correspond to the same temperature on both scales, converting Celsius to Fahrenheit or Fahrenheit to Celsius is not as easy as converting length or weight units. The simplest way to convert from one scale to the other is to use the chart on this page.

TEMPERATURE CONVERSION CHART
F = Fahrenheit   C = Celsius

EXAMPLE

On this chart equal temperatures are placed side by side. To convert 50°F to Celsius:

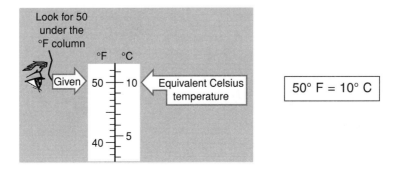

50° F = 10° C

Use this temperature conversion chart to convert

(a) 140°F to Celsius

(b) 20°C to Fahrenheit

ANSWERS

(a) On the second chart from the left, locate 140° under the °F column. Notice that 60 in the °C column is exactly aligned with 140°F. Therefore, 140°F = 60°C.

(b) On the third chart from the left, locate 20°C under the °C column. Notice that 68 in the °F column is approximately aligned with 20°C, to the nearest degree. Therefore, 20°C ≈ 68°F.

PROBLEMS

Now use this Temperature Conversion Chart to determine the following temperatures.

(a) 375°F = _____ °C   (b) −20°F = _____ °C

(c) 14°C = _____ °F   (d) 80°C = _____ °F

(e) 37°C = _____ °F   (f) 68°F = _____ °C

(g) −40°C = _____ °F   (h) 525°F = _____ °C

(i) 80°F = _____ °C   (j) −14°C = _____ °F

ANSWERS

(a) 191°C   (b) −29°C

(c) 57.5°F   (d) 176°F

(e) 98.6°F   (f) 20°C

(g) −40°F   (h) 274°C

(i) 27°C   (j) 7°F

If you must convert a temperature with more accuracy than this chart allows, or if you are converting a temperature not on the chart, algebraic conversion formulas should be used. These temperature conversion formulas will be explained in Chapter 7 when we study basic algebra.

This table summarizes conversions involving metric units.

Metric–Metric	Metric–English
Length	
1 cm = 10 mm	1 in. = 2.54 cm
1 m = 100 cm = 1000 mm	1 ft = 30.48 cm
1 km = 1000 m	1 yd = 0.9144 m
	1 mi ≈ 1.6093 km

(continued)

Weight/Mass	
1 g = 1000 mg 1 kg = 1000 g 1 metric ton = 1000 kg	1 oz = 28.35 g 1 lb = 0.4536 kg 1 T = 907.2 kg

Area	
1 ha = 10,000 sq m (m^2)	1 sq in. ($in.^2$) $\approx$ 6.452 sq cm (cm^2) 1 sq ft (ft^2) $\approx$ 0.0929 sq m (m^2) 1 sq yd (yd^2) $\approx$ 0.836 sq m

Volume	
1 cu cm (cm^3) = 1 ml 1 liter = 1000 cu cm	1 cu in. ($in.^3$) $\approx$ 16.387 cu cm (cm^3) 1 cu yd (yd^3) $\approx$ 0.765 cu m (m^3) 1 fluid ounce $\approx$ 29.574 cu cm 1 qt $\approx$ 0.946 liters 1 gal $\approx$ 3.785 liters

Now turn to Exercises 5-3 for a set of practice problems designed to help you use the metric system and to "think metric."

Exercises 5-3   Metric Units

A.   **Think Metric.** For each problem, circle the measurement closest to the first one given. No calculations are needed.

Remember: (1) A meter is a little more (about 10%) than a yard.
                 (2) A kilogram is a little more (about 10%) than 2 pounds.
                 (3) A liter is a little more (about 6%) than a quart.

		(a)		(b)		(c)	
1.	30 cm	(a)	30 in.	(b)	75 in.	(c)	1 ft
2.	5 ft	(a)	1500 cm	(b)	1.5 m	(c)	2 m
3.	1 yd	(a)	90 cm	(b)	110 cm	(c)	100 cm
4.	2 m	(a)	6 ft 6 in.	(b)	6 ft	(c)	2 yd
5.	3 km	(a)	3 mi	(b)	2 mi	(c)	1 mi
6.	200 km	(a)	20 mi	(b)	100 mi	(c)	120 mi
7.	50 km/h	(a)	30 mph	(b)	50 mph	(c)	60 mph
8.	55 mph	(a)	30 km/h	(b)	60 km/h	(c)	90 km/h
9.	100 m	(a)	100 yd	(b)	100 ft	(c)	1000 in.
10.	400 lb	(a)	800 kg	(b)	180 kg	(c)	250 kg
11.	6 oz	(a)	1.7 g	(b)	17 g	(c)	170 g
12.	5 kg	(a)	2 lb	(b)	5 lb	(c)	10 lb
13.	50 liters	(a)	12 gal	(b)	120 gal	(c)	50 gal
14.	6 qt	(a)	2 liters	(b)	3 liters	(c)	6 liters
15.	14 sq ft	(a)	1.3 sq m	(b)	13 sq m	(c)	130 sq m
16.	8 sq in.	(a)	50 sq cm	(b)	5 sq cm	(c)	500 sq cm

17.	100°F	(a) 38°C	(b) 212°C	(c) 32°C
18.	60°C	(a) 20°F	(b) 140°F	(c) 100°F
19.	212°F	(a) 100°C	(b) 400°C	(c) 50°C
20.	0°C	(a) 100°F	(b) 32°F	(c) −30°F

B. **Think Metric.** Choose the closest estimate.

1.	Diameter of a penny	(a) 3 cm	(b) 1.5 cm	(c) 5 cm
2.	Length of a man's foot	(a) 3 m	(b) 3 cm	(c) 30 cm
3.	Tank of gasoline	(a) 50 liters	(b) 5 liters	(c) 500 liters
4.	Volume of a wastebasket	(a) 10 liters	(b) 100 liters	(c) 10 cu cm
5.	One-half gallon of milk	(a) 1 liter	(b) 2 liters	(c) $\frac{1}{2}$ liter
6.	Hot day in Phoenix, Arizona	(a) 100°C	(b) 30°C	(c) 45°C
7.	Cold day in Minnesota	(a) −80°C	(b) −10°C	(c) 20°C
8.	200-lb barbell	(a) 400 kg	(b) 100 kg	(c) 40 kg
9.	Length of paper clip	(a) 10 cm	(b) 35 mm	(c) 3 mm
10.	Your height	(a) 17 m	(b) 17 cm	(c) 1.7 m
11.	Your weight	(a) 80 kg	(b) 800 kg	(c) 8 kg
12.	Boiling water	(a) 100°C	(b) 212°C	(c) 32°C

C. Perform the following metric–metric conversions. Round to the appropriate number of significant digits.

1.	5.6 cm = _____ mm	2.	0.08 g = _____ mg
3.	0.25 ha = _____ sq m	4.	64,600 cu cm = _____ liters
5.	0.045 kg = _____ g	6.	12.5 km/h = _____ cm/sec
7.	1.25 liters = _____ cu cm	8.	720,000 g = _____ kg
9.	4.2 m = _____ cm	10.	1026 mm = _____ cm
11.	0.16 m/sec = _____ cm/sec	12.	598 cm = _____ m
13.	9.62 km = _____ m	14.	785 mm = _____ m
15.	56,500 m = _____ km	16.	78,000 sq m = _____ ha
17.	9,500 mg = _____ g	18.	870 m/min = _____ km/h
19.	580 ml = _____ liters	20.	42 cm/sec = _____ m/min

D. Convert to the units shown. (Round to the appropriate number of significant digits.)

1.	3.0 in. = _____ cm	2.	4.2 ft = _____ cm
3.	2.0 mi = _____ km	4.	20 ft 6 in. = _____ m (three sig. digits)
5.	$9\frac{1}{4}$ in. = _____ cm (two sig. digits)	6.	31 kg = _____ lb

7. 152 lb = _____ kg

8. 3 lb 4 oz = _____ kg
(two sig. digits)

9. 8.5 kg = _____ lb

10. 10.4 oz = _____ g

11. 3.15 gal = _____ liters

12. 2.5 liters = _____ qt

13. 6.5 qt = _____ liters

14. 61 mph = _____ km/h

15. 8.0 km/h = _____ mph

16. 230°F = _____ °C

17. 10°C = _____ °F

18. 35 in./sec = _____ cm/sec

19. 63 ft/sec = _____ m/sec

20. 2.0 m = _____ ft

21. 285 sq ft = _____ sq m

22. 512 cu in. = _____ cu cm

23. 2.75 m^3 = _____ cu yd

24. 46.2 sq cm = _____ sq in.

E. Practical Problems. (Round as indicated.)

1. **Welding** Welding electrode sizes are presently given in inches. Convert the following set of electrode sizes to millimeters. (1 in. = 25.4 mm)

in.	mm
0.030	
0.035	
0.040	
0.045	

in.	mm
$\frac{1}{16}$	
$\frac{5}{64}$	
$\frac{3}{32}$	
$\frac{1}{8}$	

in.	mm
$\frac{5}{32}$	
$\frac{3}{16}$	
$\frac{3}{8}$	
$\frac{11}{64}$	

(Round to the nearest one-thousandth of a millimeter.)

2. One cubic foot of water weighs 62.4 lb. (a) What is the weight in pounds of 1 liter of water? (b) What is the weight in kilograms of 1 cubic foot of water? (c) What is the weight in kilograms of 1 liter of water? (Round to one decimal place.)

3. At the Munich Olympic Games in 1972 Frank Shorter of the United States won the 26.2-mile marathon race in 2 hr 12 min 19.8 sec. What was his average speed in mph and km/h? (Round to the nearest tenth.)

4. The metric unit of pressure is the *pascal,* where

1 lb/sq in. (psi) ≈ 6894 pascal (Pa), and 1 lb/sq ft (psf) ≈ 47.88 Pa

For very large pressures, the kilopascal (kPa) is the appropriate unit.

1 kPa = 1000 Pa and therefore

$$1\frac{\text{lb}}{\text{sq in.}} \approx 6.894\,\text{kPa} \quad \text{and} \quad 1\frac{\text{lb}}{\text{sq ft}} \approx 0.04788\,\text{kPa}$$

Convert the following pressures to metric units.

(a) 15.0 psi = _____ Pa

(b) 10$\overline{0}$ psi = _____ kPa

(c) 40 psf = _____ Pa

(d) 2500 psf = _____ kPa

5. Find the difference between

    (a) 1 mile and 1 kilometer   _____ km

    (b) 3 miles and 5000 meters   _____ ft

    (c) 120 yards and 110 meters   _____ ft

    (d) 1 quart and 1 liter   _____ qt

    (e) 10 pounds and 5 kilograms   _____ lb

6. The Minolta Maxxum 5 camera uses 35 mm film and produces 24-mm by 36-mm negatives. How would this negative size be given in inches? (Round to the nearest hundredth.)

7. Which is performing more efficiently, a car getting 12 km per liter of gas or one getting 25 miles per gallon of gas?

8. **Painting and Decorating** Industrial paint sells for $15.75 per gallon. What would 1 liter cost?

9. **Roofing** In roofing, the unit "one square" is sometimes used to mean 100 sq ft. Convert this unit to the metric system. (Round to one decimal place.)

    1 square = _____ sq m

10. A *cord* of wood is a volume of wood 8 ft long, 4 ft wide, and 4 ft high. (a) What are the equivalent metric dimensions for a "metric cord" of wood? (b) What is the volume of a cord in cubic meters? (Round to the nearest tenth.) (*Hint:* volume = length × width × height.)

11. Translate these well-known phrases into metric units.

    (a) A miss is as good as _____ km. (1 mile = _____ km)

    (b) A _____ cm TV screen. (30 in. = _____ cm)

    (c) An _____ cm by _____ cm sheet of paper. ($8\frac{1}{2}$ in. = _____ cm; 11 in. = _____ cm)

    (d) He was beaten within _____ cm of his life. (1 in. = _____ cm)

    (e) Race in the Indy _____ km. (500 miles = _____ km)

    (f) Take it with _____ grams of salt. (*Hint:* 437.5 grains = 1 oz)

    (g) _____ grams of prevention is worth _____ grams of cure. (1 oz = _____ g; 1 lb = _____ g)

    (h) See the world in _____ grams of sand. (1 grain = _____ g)

    (i) Peter Piper picked _____ liters of pickled peppers. (*Hint:* 1 peck = 8 quarts)

    (j) A cowboy in a _____ liter hat. (10 gal = _____ liter)

    (k) He's in _____ ml of trouble. (1 peck = _____ ml)

12. **Auto Mechanics** In an automobile, cylinder displacement is measured in either cubic inches or liters. What is the cylinder displacement in liters of a 400-cu in. engine? (Round to the nearest liter.)

13. **Electronics** Convert the following quantities as shown.

    (a) 18,000 volts = _____ kilovolts

    (b) 435 millivolts = _____ volts

14. **Culinary Arts** Fried crickets, served as an appetizer, sell for 4200 Cambodian riels per kg. If 3800 riels are equal to one dollar, how much will 5 lb of crickets cost in U.S. dollars?

When you have completed these exercises, check your answers in the Appendix, then turn to Section 5-4 to learn about direct measurement.

5-4 DIRECT MEASUREMENTS

The simplest kind of measuring device to use is one in which the output is a digital display. For example, in a digital clock or digital voltmeter, the measured quantity is translated into electrical signals, and each digit of the measurement number is displayed.

A digital display is easy to read, but most technical work involves direct measurement where the numerical reading is displayed on some sort of scale. Reading the scale correctly requires skill and the ability to estimate or *interpolate* accurately between scale divisions.

Length Measurements

Length measurement involves many different instruments that produce either fraction or decimal number readouts. Carpenters, electricians, plumbers, roofers, and machinists all use rulers, yardsticks, or other devices to make length measurements. Each instrument is made with a different scale that determines the precision of the measurement that can be made. The more precise the measurement has to be, the smaller the scale divisions needed.

For example, on the carpenter's rule shown, scale A is graduated in eighths of an inch. Scale B is graduated in sixteenths of an inch.

A carpenter or roofer may need to measure no closer than $\frac{1}{8}$ or $\frac{1}{16}$ in., but a machinist will often need a scale graduated in 32nds or 64ths of an inch. The finely engraved steel rule shown is a machinist's rule.

Some machinists' rules are marked in tenths or hundredths of an inch on one side and may be marked in metric units on the other side.

Notice that on all these rules only a few scale markings are actually labeled with numbers. It would be easy to read a scale marked like this:

Any useful scale has divisions so small that this kind of marking is impossible. An expert in any technical field must be able to identify the graduations on the scale correctly even though the scale is *not* marked. To use any scale, first determine the size of the smallest division by counting the number of spaces in a 1-in. interval.

EXAMPLE

There are four spaces in the first 1-in. interval. The smallest division is therefore $\frac{1}{4}$ in., and we can label the scale like this:

YOUR TURN

Try it. Label all marks on the following scales. The units are inches.

(a)

Smallest division = _____

(b)

Smallest division = _____

(c)

Smallest division = _____

(d)

Smallest division = _____

ANSWERS

(a) Smallest division = $\frac{1}{8}$ in.

(b) Smallest division = $\frac{1}{5}$ in.

(c) Smallest division = $\frac{1}{10}$ in.

(d) Smallest division = $\frac{1}{16}$ in.

To use the rule once you have mentally labeled all scale markings, first count inches, then count the number of smallest divisions from the last whole-inch mark.

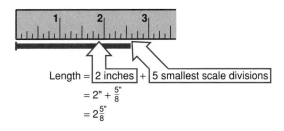

Length = $\boxed{2 \text{ inches}}$ + $\boxed{5 \text{ smallest scale divisions}}$

= 2" + $\frac{5"}{8}$

= $2\frac{5"}{8}$

If the length of the object being measured falls between scale divisions, record the nearest scale division.

EXAMPLE

In this measurement:

the endpoint of the object falls between $1\frac{3}{16}$ and $1\frac{4}{16}$ in.

To the nearest scale division, the length is $1\frac{3}{16}$ in. When a scale with $\frac{1}{16}$-in. divisions is used, we usually write the measurement to the nearest $\frac{1}{16}$ in., but we may estimate it to the nearest half of a scale division or $\frac{1}{32}$ in.

YOUR TURN

For practice in using length rulers, find the dimensions given on the following inch rules. Express all fractions in lowest terms.

(a)

(b)

(c) Express answers in decimal form.

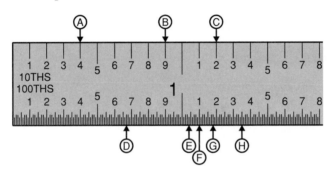

ANSWERS

(a) A. $\frac{3}{4}$ in. B. $2\frac{3}{8}$ in. C. $3\frac{1}{8}$ in. D. $3\frac{5}{8}$ in.

    E. $\frac{13}{16}$ in. F. $1\frac{9}{16}$ in. G. $2\frac{1}{4}$ in. H. $2\frac{15}{16}$ in.

(b) A. $\frac{7}{32}$ in. B. $1\frac{3}{32}$ in. C. $1\frac{15}{32}$ in.

    D. $\frac{28}{64}$ in. $= \frac{7}{16}$ in. E. $\frac{29}{32}$ in. F. $1\frac{37}{64}$ in.

(c) A. $\frac{4}{10}$ in. $= 0.4$ in. B. $\frac{9}{10}$ in. $= 0.9$ in. C. $1\frac{2}{10}$ in. $= 1.2$ in.

    D. $\frac{67}{100}$ in. $= 0.67$ in. E. $1\frac{4}{100}$ in. $= 1.04$ in. F. $1\frac{10}{100}$ in. $= 1.10$ in.

    G. $1\frac{18}{100}$ in. $= 1.18$ in. H. $1\frac{35}{100}$ in. $= 1.35$ in.

Micrometers A steel rule can be used to measure lengths of $\pm\frac{1}{32}$ in. or $\pm\frac{1}{64}$ in., or as fine as $\pm\frac{1}{100}$ in. To measure lengths with greater accuracy than this, more specialized instruments are needed. A *micrometer* can be used to measure lengths to ± 0.001 in., one one-thousandth of an inch.

A modern *digital micrometer* is accurate to 0.0001 inch, and the measurement is given as a digital readout in inches or millimeters.

Two scales are used on a micrometer. The first scale, marked on the sleeve, records the movement of a screw machined accurately to 40 threads per inch. The smallest divisions on this scale record movements of the spindle of one-fortieth of an inch or 0.025 in.

The second scale is marked on the rotating spindle. One complete turn of the thimble advances the screw one turn or $\frac{1}{40}$ in. The thimble scale has 25 divisions, so that each mark on the thimble scale represents a spindle movement of $\frac{1}{25} \times \frac{1}{40}$ in. $= \frac{1}{1000}$ in. or 0.001 in.

To read a micrometer, follow these steps.

Step 1 Read the largest numeral visible on the sleeve and multiply this number by 0.100 in.

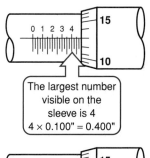

The largest number visible on the sleeve is 4
$4 \times 0.100" = 0.400"$

Step 2 Read the number of additional scale spaces visible on the sleeve. (This will be 0, 1, 2, or 3.) Multiply this number by 0.025 in.

Three additional spaces are visible
$3 \times 0.025" = 0.075"$

Step 3 Read the number on the thimble scale opposite the horizontal line on the sleeve and multiply this number by 0.001 in.

The thimble scale reads 13
13 × 0.001" = 0.013"

Step 4 Add these three products to find the measurement.

$$\begin{array}{r} 0.400 \text{ in.} \\ 0.075 \text{ in.} \\ + \ 0.013 \text{ in.} \\ \hline 0.488 \text{ in.} \end{array}$$

Careful ▶ To count as a "space" in step 2, the right end marker must be visible. ◀

Here are more examples.

EXAMPLE

Read first
Read third
Read second

Step 1 The largest number completely visible on the sleeve is 2.

2 × 0.100 in. = 0.200 in.

Step 2 Three additional spaces are visible on the sleeve.

3 × 0.025 in. = 0.075 in.

Step 3 The thimble scale reads 19.

19 × 0.001 in. = 0.019 in.

Step 4 Add:
$$\begin{array}{r} 0.200 \text{ in.} \\ 0.075 \text{ in.} \\ 0.019 \text{ in.} \\ \hline \text{Length} = 0.294 \text{ in.} \end{array}$$

EXAMPLE

Read first
Read third
Read second

Step 1 5 × 0.100 in. = 0.500 in.

Step 2 No additional spaces are visible on the sleeve.

0 × 0.025 in. = 0.000 in.

Step 3 6 × 0.001 in. = 0.006 in.

Step 4 Add:
$$\begin{array}{r} 0.500 \text{ in.} \\ 0.000 \text{ in.} \\ 0.006 \text{ in.} \\ \hline \text{Length} = 0.506 \text{ in.} \end{array}$$

YOUR TURN

For practice, read the following micrometers.

(a)   (b)

Metric micrometers are easier to read. Each small division on the sleeve represents 0.5 mm. The thimble scale is divided into 50 spaces; therefore, each division is $\frac{1}{50} \times 0.5$ mm $= \frac{1}{100}$ mm or 0.01 mm.

Turning the thimble scale of a metric micrometer by one scale division advances the spindle one one-hundredth of a millimeter, or 0.001 cm.

To read a metric micrometer, follow these steps.

Step 1   Read the largest mark visible on the sleeve and multiply this number by 1 mm.

Step 2 Read the number of additional half spaces visible on the sleeve. (This number will be either 0 or 1.) Multiply this number by 0.5 mm.

1 half space

1×0.5 mm = 0.5 mm

Step 3 Read the number on the thimble scale opposite the horizontal line on the sleeve and multiply this number by 0.01 mm.

The thimble scale reads 37.
37×0.01 mm = 0.37 mm

Step 4 Add these three products to find the measurement value.

$$\begin{array}{r} 8.00 \text{ mm} \\ 0.50 \text{ mm} \\ \underline{0.37 \text{ mm}} \\ 8.87 \text{ mm} \end{array}$$

EXAMPLE

$$\begin{aligned} 17 \times 1 \text{ mm} &= 17 \text{ mm} \\ 1 \times 0.5 \text{ mm} &= 0.5 \text{ mm} \\ 21 \times 0.01 \text{ mm} &= 0.21 \text{ mm} \\ \text{Length} &= 17 \text{ mm} + 0.5 \text{ mm} + 0.21 \text{ mm} \\ &= 17.71 \text{ mm} \end{aligned}$$

YOUR TURN

Read these metric micrometers.

(a)

(b)

(c)

(d)

ANSWERS

(a) 11.16 mm

(b) 15.21 mm

(c) 7.96 mm

(d) 21.93 mm

Vernier Micrometers When more accurate length measurements are needed, the *vernier* micrometer can be used to provide an accuracy of $\pm$ 0.0001 in.

On the vernier micrometer, a third or vernier scale is added to the sleeve. This new scale appears as a series of ten lines parallel to the axis of the sleeve above the usual scale.

The ten vernier marks cover a distance equal to nine thimble scale divisions—as shown on this spread-out diagram of the three scales.

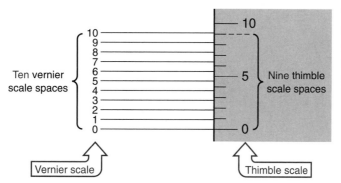

The difference between a vernier scale division and a thimble scale division is one-tenth of a thimble scale division, or $\frac{1}{10}$ of 0.001 in. = 0.0001 in.

To read the vernier micrometer, follow the first three steps given for the ordinary micrometer, then:

Step 4   Find the number of the line on the vernier scale that exactly lines up with any line on the thimble scale. Multiply this number by 0.0001 in. and add this amount to the other three distances.

EXAMPLE

For the vernier micrometer shown:

Sleeve scale: 3×0.100 in. = 0.300 in.   When adding these four lengths,
                 2×0.025 in. = 0.050 in.   be careful to align the decimal
Thimble scale: 8×0.001 in. = 0.008 in.   digits correctly.
Vernier scale: 3×0.0001 in. = 0.0003 in.
                 Length   = 0.3583 in.

The vernier scale provides a way of measuring accurately the distance to an additional tenth of the finest scale division.

Now for some practice in using the vernier scale, read each of the following.

(a)

(b)

(c)

(d)

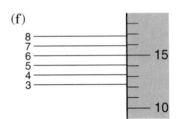

(e)

(f)

(a) 7   (b) 3   (c) 9

(d) 1   (e) 3   (f) 6

Now read the following vernier micrometers.

(a)

(b)

(c)

(d)

(e)

(f)

ANSWERS

(a) 0.3568 in.	(b) 0.4387 in.	(c) 0.7330 in.
(d) 0.1259 in.	(e) 0.6939 in.	(f) 0.2971 in.

Vernier Calipers   The vernier caliper is another length measuring instrument that uses the vernier principle and can measure lengths to ±0.001 in.

Modern direct *digital calipers,* battery or solar-powered, give measurements accurate to 0.001 in. They convert automatically between English and metric units.

Notice that on the manual instrument each inch of the main scale is divided into 40 parts, so that each smallest division on the scale is 0.025 in. Every fourth division or tenth of an inch is numbered. The numbered marks represent 0.100 in., 0.200 in., 0.300 in., and so on.

The sliding vernier scale is divided into 25 equal parts numbered by 5s. These 25 divisions on the vernier scale cover the same length as 24 divisions on the main scale.

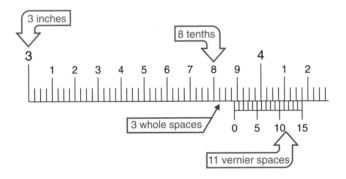

To read the vernier caliper above, follow these steps.

Step 1 Read the number of whole-inch divisions on the main scale to the left of the vernier zero. This gives the number of whole inches.

3 in. = 3.000 in.

Step 2 Read the number of tenths on the main scale to the left of the vernier zero. Multiply this number by 0.100 in.

8 tenths
8 × 0.100 in. = 0.800 in.

Step 3 Read the number of additional whole spaces on the main scale to the left of the vernier zero. Multiply this number by 0.025 in.

3 spaces
3 × 0.025 in. = 0.075 in.

Step 4 On the sliding vernier scale, find the number of the line that exactly lines up with any line on the main scale. Multiply this number by 0.001 in.

11 lines up exactly.
11 × 0.001 in. = 0.011 in.

Step 5 Add the four numbers.

$$\begin{array}{r} 3.000 \text{ in.} \\ 0.800 \text{ in.} \\ 0.075 \text{ in.} \\ + \, 0.011 \text{ in.} \\ \hline 3.886 \text{ in.} \end{array}$$

EXAMPLE

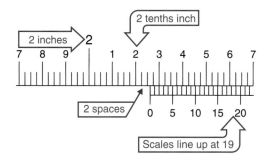

2 in.	= 2.000 in.
2 × 0.100 in.	= 0.200 in.
2 × 0.025 in.	= 0.050 in.
19 × 0.001 in.	= 0.019 in.
	2.269 in.

YOUR TURN

Read the following vernier calipers.

(a)

(b)

(c)

(d)

(e)

(f)

ANSWERS

(a) 2.478 in.   (b) 0.763 in.   (c) 1.084 in.

(d) 2.931 in.   (e) 3.110 in.   (f) 1.470 in.

What does "vernier" mean—as in vernier scale or vernier caliper?

The vernier scale was invented about 350 years ago by Pierre Vernier, a French mathematician.

Protractors A *protractor* is used to measure and draw angles. The simplest kind of protractor, shown here, is graduated in degrees with every tenth degree labeled with a number.

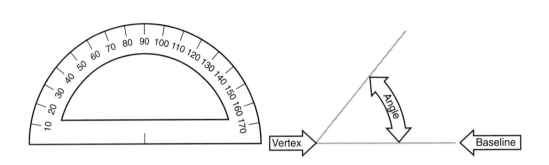

In measuring or drawing angles with a protractor, be certain that the vertex of the angle is exactly at the center of the protractor base. One side of the angle should be placed along the 0°–180° baseline.

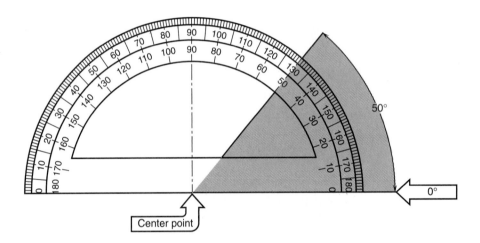

Notice that the protractor can be read from the 0° point on either the right or left side. Measuring from the right side, the angle is read as 50° on the inner scale. Measuring from the left side, the angle is 130° on the outer scale.

The *bevel protractor* is very useful in shopwork because it can be adjusted with an accuracy of ±0.5°. The angle being measured is read directly from the movable scale.

For very accurate angle measurements, the *vernier protractor* can be used with an accuracy of $\pm \frac{1}{60}$ of a degree or 1 minute, since 60 minutes = 1 degree.

To read the vernier protractor shown, follow these steps.

Step 1 Read the number of whole degrees on the main scale between the zero on the main scale and the zero on the vernier scale.

The mark on the vernier scale just to the right of 0° is 42°.

Step 2 Find the vernier scale line on the left of the vernier zero that lines up exactly with a main scale marking. The vernier protractor in the example reads 42°30′.

The vernier scale lines up with a mark on the main scale at 30. This gives the fraction of a degree in minutes.

YOUR TURN

Read the following vernier protractors.

(a)

(b)

(c)

(d)

ANSWERS

(a) 73°46′   (b) 11°15′   (c) 61°5′   (d) 27°25′

Meters Many measuring instruments convert the measurement into an electrical signal and then display that signal by means of a pointer and a decimal scale. Instruments of this kind, called *meters,* are used to measure electrical quantities such as voltage, current, or power, and other quantities such as air pressure, flow rates, or speed. As with length scales, not every division on the meter scale is labeled, and practice is required in order to read meters quickly and correctly.

The *range* of a meter is its full-scale or maximum reading. The *main* divisions of the scale are the numbered divisions. The *small* divisions are the smallest marked portions of a main division.

EXAMPLE

On the following meter scale, the *range* is 10 volts;

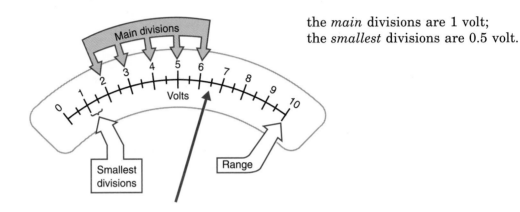

the *main* divisions are 1 volt;
the *smallest* divisions are 0.5 volt.

A *dual-scale meter* has an upper and lower scale and allows the user to switch ranges as needed.

YOUR TURN

Find the range, main divisions, and smallest divisions for each of the following meter scales.

(a) Range = _____
     main divisions = _____
     smallest divisions = _____

(b) Range = _____
     main divisions = _____
     smallest divisions = _____

(c) Range = _____
     main divisions = _____
     smallest divisions = _____

(d) Range = _____
main divisions = _____
smallest divisions = _____

ANSWERS

(a) Range = 2.0 V
main divisions = 1.0 V
smallest divisions = 0.1 V

(b) Range = 25 A
main divisions = 5 A
smallest divisions = 1 A

(c) Range = 1.0 mV
main divisions = 0.5 mV
smallest divisions = 0.1 mV

(d) Range = 500 psi
main divisions = 50 psi
smallest divisions = 10 psi

To read a meter scale, either we can choose the small scale division nearest to the pointer, or we can estimate between scale markers.

EXAMPLE

This meter

reads exactly 4.0 volts.

But this meter

reads 6.5 volts to the nearest scale marker, or about 6.7 volts if we estimate between scale markers. The smallest division is 0.5 volt.

On the following meter scale

the smallest scale division is $\frac{1}{5}$ of 0.1 ampere or 0.02 ampere. The reading at A is therefore 0.2 plus 4 small divisions or 0.28 ampere. The reading at B is 0.4 plus $3\frac{1}{2}$ small divisions or 0.47 ampere.

Read the following meters. Estimate between scale divisions where necessary.

(a)

(b)

(c)

(d)

(e)

(f)

ANSWERS

Note: Your estimated answers may differ slightly from ours.

(a) A. 5	B. 25	C. 43	D. 78 mph
(b) A. 2	B. 10.6	C. 15.8	D. 23 amperes
(c) A. 0.42	B. 0.7	C. 0.8	D. 0.93 volt
(d) A. 82	B. 207	C. 346	D. 458 psi
(e) A. 0.1	B. 0.36	C. 0.64	D. 0.96 millivolt
(f) A. 2.8	B. 7.2	C. 10.2	D. 15.7 volts

Notice in meter (f) that the smallest division is $\frac{1}{5}$ of 2 volts or 0.4 volt. If your answers differ slightly from ours, the difference is probably due to the difficulty of estimating the position of the arrows on the diagrams.

Gauge Blocks *Gauge blocks* are extremely precise rectangular steel blocks used for making, checking, or inspecting tools and other devices. Opposite faces of each block are ground and lapped to a high degree of flatness, so that when two blocks are fitted together they will cling tightly. Gauge blocks are sold in standard sets of from 36 to 81 pieces in precisely measured widths. These pieces may be combined to give any decimal dimension within the capacity of the set.

A standard 36-piece set would consist of blocks of the following sizes (in inches):

0.0501	0.051	0.050	0.110	0.100	1.000
0.0502	0.052	0.060	0.120	0.200	2.000
0.0503	0.053	0.070	0.130	0.300	
0.0504	0.054	0.080	0.140	0.400	
0.0505	0.055	0.090	0.150	0.500	
0.0506	0.056				
0.0507	0.057				
0.0508	0.058				
0.0509	0.059				

To combine gauge blocks to produce any given dimension, work from digit to digit, right to left. This procedure will produce the desired length with a minimum number of blocks.

EXAMPLE

Combine gauge blocks from the set given to equal 1.8324 in.

Step 1 To get the 4 in 1.8324 in., select block 0.0504 in. It is the only block in the set with a 4 in the fourth decimal place.

$$\begin{array}{r} 1.8324 \text{ in.} \\ -0.0504 \text{ in.} \\ \hline 1.7820 \text{ in.} \end{array}$$

The target length
The first block
The remainder

Step 2 To get the 2 in 1.782 in., select block 0.052 in. It is the only block in the set with a 2 in the third decimal place.

$$\begin{array}{r} 1.782 \text{ in.} \\ -0.052 \text{ in.} \\ \hline 1.730 \text{ in.} \end{array}$$

The new target length
The second block
The remainder

Step 3 To get the 3 in 1.73 in., select block 0.130 in.

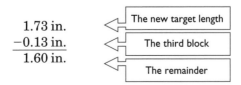

$$\begin{array}{r} 1.73 \text{ in.} \\ -0.13 \text{ in.} \\ \hline 1.60 \text{ in.} \end{array}$$

The new target length
The third block
The remainder

Step 4 To get 1.6, select blocks 1.000, 0.500, and 0.100 in. The six blocks can be combined or "stacked" to total the required 1.8324-in. length.

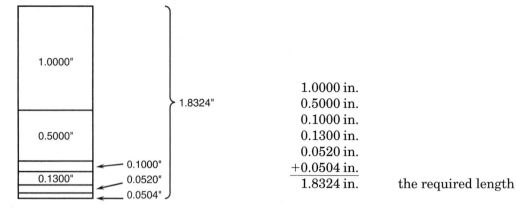

1.0000 in.
0.5000 in.
0.1000 in.
0.1300 in.
0.0520 in.
+0.0504 in.
1.8324 in.   the required length

YOUR TURN

Stack sets of gauge blocks that total each of the following.

(a) 2.0976 in.   (b) 1.8092 in.   (c) 3.6462 in.

(d) 0.7553 in.   (e) 1.8625 in.   (f) 0.7474 in.

ANSWERS

(a) 1.000 in. + 0.500 in. + 0.400 in. + 0.090 in. + 0.057 in. + 0.0506 in.

(b) 1.000 in. + 0.500 in. + 0.200 in. + 0.059 in. + 0.0502 in.

(c) 1.000 in. + 2.000 in. + 0.400 in. + 0.140 in. + 0.056 in. + 0.0502 in.

(d) 0.500 in. + 0.100 in. + 0.050 in. + 0.055 in. + 0.0503 in.

(e) 1.000 in. + 0.500 in. + 0.200 in. + 0.060 in. + 0.052 in. + 0.0505 in.

(f) 0.500 in. + 0.140 in. + 0.057 in. + 0.0504 in.

Now turn to Exercises 5-4 for a set of practice problems designed to help you become more expert in making direct measurements.

Exercises 5-4

Direct Measurements

A. Find the lengths marked on the following rules.

1.

2.

3.

B. Read the following micrometers.

1.   2.

3.   4.

5.   6.

7.   8.

9.   10.

11.

12.

13. **14.**

15. **16.**

17. **18.**

C. Read the following vernier calipers.

1. **2.**

3. **4.**

5. **6.**

7. **8.**

9. **10.**

D. Read the following vernier protractors.

1.

2.

3.

4.

5.

6.

7.

8.

E. Find combinations of gauge blocks that produce the following lengths. (Use the 36-block set given on page 287.)

1.	2.3573 in.	2.	1.9467 in.
3.	0.4232 in.	4.	3.8587 in.
5.	1.6789 in.	6.	2.7588 in.
7.	0.1539 in.	8.	3.0276 in.
9.	2.1905 in.	10.	1.5066 in.

Check your answers in the Appendix, then turn to Problem Set 5 for practice on working with measurement numbers.

Measurement

Answers are given in the Appendix.

A. **State the accuracy and precision of each of the following measurement numbers.**

1. 8.3 lb
2. 960 ft
3. 3.50 in.
4. 8775 gal
5. 0.04 kg
6. 17 psi
7. 4000 sq yd
8. 1.375 liters

B. **Perform the following calculations with measurement numbers.** Round to the correct precision or accuracy.

1. 5.75 sec − 2.3 sec
2. 3.1 ft × 2.2 ft
3. 4.56 sq in. ÷ 6.1 in.
4. 3.2 cm × 1.26 cm
5. $7\frac{1}{2}$ in. + $6\frac{1}{4}$ in. + $3\frac{1}{8}$ in.
6. 9 lb 6 oz − 5 lb 8 oz
7. 765 mi ÷ 36.2 gal
8. 1.5 cm + 2.38 cm + 5.8 cm
9. 34.38 psi − 16.732 psi
10. $7\frac{1}{4}$ ft − 4.1 ft
11. 7.064 in. − 3.19 in.
12. 22.0 psi × 18.35 sq in.
13. 8 ft 7 in. + 11 ft 11 in. + 9 in.
14. 2.6 in. × 3.85 in. × 14 in.

C. **Convert the following measurement numbers to new units as shown.** Round to the same number of significant digits as the original measurement.

1. 6.0 in. = _____ cm
2. 54 mph = _____ ft/sec
3. 22°C = _____ °F
4. 2.6 liters = _____ qt
5. 0.82 kg = _____ g
6. 1.62 cm = _____ mm
7. $26\frac{1}{2}$ cu ft = _____ cu in.
8. 41°F = _____ °C
9. 14.3 kg = _____ lb
10. 123.7 psi = _____ atm
11. 24.0 km/h = _____ cm/sec
12. 127,000 sq m _____ ha
13. $14\frac{1}{2}$ mi = _____ km
14. 350 cu in. = _____ cu cm
15. 14.3 bbl = _____ cu ft
16. 160 sq ft = _____ sq m
17. 1400 ml = _____ liters
18. 5650 m/min = _____ km/h

Name _____

Date _____

Course/Section _____

19. 24,680 cu in. = _____ bu

20. 6.9 liters = _____ gal

21. 6.8 sq ft = _____ sq in.

22. $12\frac{1}{4}$ qt = _____ liters

23. 0.095 g = _____ mg

24. 596,000 m = _____ km

25. 3.40 ft = _____ cm

26. 45.0 in./sec = _____ cm/sec

27. 1.85 liters = _____ cu cm

28. 4,000,000 mg = _____ g

29. 85 km/h = _____ mph

30. 28 lb = _____ kg

D. **Think metric.** Choose the closest estimate.

1. Weight of a pencil   **(a)** 4 kg   **(b)** 4 g   **(c)** 40 g

2. Size of a house   **(a)** 6 sq m   **(b)** 200 sq m   **(c)** 1200 sq m

3. Winter in New York   **(a)** −8°C   **(b)** −40°C   **(c)** 20°C

4. Volume of a 30-gal trash can   **(a)** 0.1 cu m   **(b)** 2.0 cu m   **(c)** 5.0 cu m

5. Height of an oak tree   **(a)** 500 mm   **(b)** 500 cm   **(c)** 50 km

6. Capacity of a 50-gal water heater   **(a)** 25 liters   **(b)** 100 liters   **(c)** 180 liters

E. **Read the following measuring devices.**

    1. Rulers

2. Micrometers. Problems (a) and (b) are English, (c) and (d) are metric, and (e) and (f) are English with a vernier scale.

(a)

(b)

(c)

(d)

(e)

(f)

3. Vernier calipers.

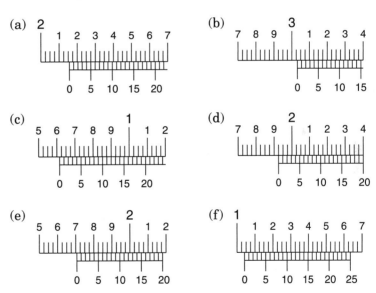

(a)

(b)

(c)

(d)

(e)

(f)

F. Practical Problems. Round appropriately.

1. **Machine Technology**   A piece $6\frac{1}{16}$ in. long is cut from a steel bar $28\frac{5}{8}$ in. long. How much is left?
2. An American traveling in Italy notices a road sign saying 85 km/h. What is this speed in miles per hour?
3. **Auto Mechanics**   An automotive technician testing cars finds that a certain bumper will prevent damage up to 9.0 ft/sec. Convert this to miles per hour.
4. **Machine Technology**   The gap in the piece of steel shown in the figure should be 2.4375 in. wide. Which of the gauge blocks in the set on page 287 should be used for testing this gap?
5. What size wrench, to the nearest 32nd of an inch, will fit a 0.455-in. bolt head?
6. **Metalworking**   The melting point of a casting alloy is 260°C. What temperature is this on the Fahrenheit scale?

Problem 4

7. **Sheet Metal Technology**   What area of sheet metal in square inches is needed to construct a vent with a surface area of 6.50 sq ft?

8. How many gallons of gas are needed to drive a car exactly 350 miles if the car normally averages 18.5 mi/gal?

9. **Metalworking**   A certain alloy is heated to a temperature of 560°F. What Celsius temperature does this correspond to? (Round to the nearest degree.)

10. **Auto Mechanics**   A Nissan sports car has a 2680-cc engine. What would this be in cubic inches? (Round to the nearest cubic inch.)

11. **Metalworking**   Find the total length of the piece of steel shown in the figure.

12. **Plumbing**   The Uniform Plumbing Code is being converted to the metric system. Convert the following measurements.
    (a)   The amount of gas used by a furnace (136 cu ft/hr) to cubic meters per hour.
    (b)   The length of a gas line (25.0 ft) to meters.
    (c)   The diameter of a water pipe (2.50 in.) to centimeters.

13. **Metalworking**   Wrought iron has a density of 480 lb/cu ft. Convert this to lb/cu in.

14. A cubic foot of water contains 7.48 gallons. What volume is this in liters?

15. **Carpentry**   What are the dimensions of a metric "two by four" (a piece of wood $1\frac{1}{2}$ in. by $3\frac{1}{2}$ in.) in centimeters?

16. **Architectural Technology**   A common scale used by architects and drafters in making blueprints is $\frac{1}{4}$ in. = 1 ft. What actual length in meters would correspond to a drawing dimension of $1\frac{1}{4}$ in.?

17. **Forestry**   The world's largest living organism is a fungus known as the Honey Mushroom, growing in the Malheur National Forest in Oregon. The fungus covers approximately 2200 acres. How many square miles is this?

2.759"

4.328"

0.536"

Problem 11

Pre-Algebra

Objective	Sample Problems	For Help Go to Page

When you finish this unit you will be able to:

1. Understand the meaning of signed numbers.

    (a) Which is smaller, -12 or -14?   _____   299

    (b) Mark the following points on a number line: $-5\frac{1}{4}$, $+1.0$, -0.75, $+3\frac{1}{2}$.   _____

    (c) Represent a loss of \$250 as a signed number.   _____

2. Add signed numbers.

    (a) $(-3) + (-5)$ _____   301

    (b) $18\frac{1}{2} + (-6\frac{3}{4})$ _____

    (c) $-21.7 + 14.2$ _____

    (d) $(-3) + 15 + 8 + (-20) + 6 + (-9)$ _____

3. Subtract signed numbers.

    (a) $4 - 7$ _____   308

    (b) $-11 - (-3)$ _____

    (c) $6.2 - (-4.7)$ _____

    (d) $-\frac{5}{8} - 1\frac{1}{4}$ _____

4. Multiply and divide signed numbers.

    (a) $(-12) \times 6$ _____   313

    (b) $(-54) \div (-9)$ _____

    (c) $(-4.73)(-6.5)$ (two significant digits)   _____

    (d) $\frac{3}{8} \div (-2\frac{1}{2})$ _____

    (e) $60 \div (-0.4) \times (-10)$ _____

_____

Name

_____

Date

_____

Course/Section

5. Work with exponents.

    (a) 4^3 _____   319

    (b) $(2.05)^2$ _____

6. Use the order of operations.

    (a) $2 + 3^2$ _____   321

    (b) $(4 \times 5)^2 + 4 \times 5^2$ _____

    (c) $6 + 8 \times 2^3 \div 4$ _____

7. Find square roots.

    (a) $\sqrt{169}$ _____   322

    (b) $\sqrt{14.5}$ (to two decimal places)   _____

(Answers to these preview problems are given in the Appendix.)

If you are certain that you can work *all* these problems correctly, turn to page 329 for a set of practice problems. If you cannot work one or more of the preview problems, turn to the page indicated to the right of the problem. Those who wish to master this material should turn to Section 6-1 and begin work there.

6 Pre-Algebra

© DRABBLE reprinted by permission of UFS, Inc.

6-1 ADDITION OF SIGNED NUMBERS

The Meaning of Signed Numbers

What kind of number would you use to name each of the following: a golf score two strokes below par, a $2 loss in a poker game, a debt of $2, a loss of 2 yards on a football play, a temperature 2 degrees below zero?

The answer is that they are all *negative numbers,* all equal to -2.

We can better understand the meaning of negative numbers if we can picture them on a number line. First we will display the familiar whole numbers as points along a number line:

Notice that the numbers become greater as you move to the right along the number line, and they become lesser as you move to the left. The numbers to the *right* of zero, *greater* than zero, are considered to be *positive numbers*. The negative numbers are all the numbers *less than* zero, and we must extend the number line to the left in order to picture them:

Think of the negative part of the number line as a reflection of the positive part through an imaginary mirror placed through zero. This means that ordering the negative numbers is the opposite of ordering the positive numbers. For example, 4 is greater than 1, but -4 is less than -1.

YOUR TURN

Circle the greater in each pair of numbers.

(a)　−5 or −3　　　　(b)　9 or 6　　　　(c)　−15 or −22

(d)　12 or 31　　　　(e)　−8 or 2　　　　(f)　43 or −6

ANSWERS

(a)　−3　　　　(b)　9　　　　(c)　−15

(d)　31　　　　(e)　2　　　　(f)　43

To help in solving the preceding problems, you may extend the number line in both directions. This is perfectly legitimate, since the number line is infinitely long. Notice we have marked only the positions of the positive and negative whole numbers. These numbers are called the *integers*. The positions of fractions and decimals, both positive and negative, can be located between the integers.

YOUR TURN

Try marking the following points on a number line.

(a)　$-3\frac{1}{2}$　　　　(b)　$\frac{1}{4}$　　　　(c)　−2.4　　　　(d)　1.75　　　　(e)　−0.6

ANSWERS

When we work with both positive and negative numbers and need to use names to distinguish them from each other, we often refer to them as *signed numbers*. Signed numbers are a bookkeeping concept. They were used in bookkeeping by the ancient Greeks, Chinese, and Hindus more than 2000 years ago. Chinese merchants wrote positive numbers in black and negative numbers in red in their account books. (Being "in the red" still means losing money or having a negative income.) The Hindus used a dot or circle to show that a number was negative. We use a minus sign (−) to show that a number is negative.

For emphasis, the "+" symbol is sometimes used to denote a positive number, although the absence of any symbol automatically implies that a number is positive. Because the + and − symbols are used also for addition and subtraction, we will use parentheses when writing problems involving addition and subtraction of signed numbers.

PROBLEMS

Before we attempt these operations, test your understanding of signed numbers by representing each of the following situations with the appropriate signed number.

(a)　Two seconds before liftoff

(b)　A profit of eight dollars

(c) An elevator going up six floors

(d) Diving 26 feet below the surface of the ocean

(e) A debt of $50

(f) Twelve degrees above zero

ANSWERS

(a) −2 sec   (b) +$8   (c) +6

(d) −26 ft   (e) −$50   (f) +12°

Addition of Signed Numbers

Addition of signed numbers may be pictured using the number line. A positive number may be represented on the line by an arrow to the right or positive direction. Think of a positive arrow as a "gain" of some quantity such as money. The sum of two numbers is the net "gain."

EXAMPLE

The sum of 4 + 3 or (+4) + (+3) is

The numbers +4 and +3 are represented by arrows directed to the right. The +3 arrow begins where the +4 arrow ends. The sum 4 + 3 is the arrow that begins at the start of the +4 arrow and ends at the tip of the +3 arrow. In terms of money, a "gain" of $4 followed by a "gain" of $3 produces a net gain of $7.

A negative number may be represented on the number line by an arrow to the left or negative direction. Think of a negative arrow as a loss of some quantity.

EXAMPLE

The sum (−4) + (−3) is −7.

The numbers −4 and −3 are represented by arrows directed to the left. The −3 arrow begins at the tip of the −4 arrow. Their sum, −7, is the arrow that begins at the start of the −4 arrow and ends at the tip of the −3 arrow. A loss of $4 followed by another loss of $3 makes a total loss of $7.

Try setting up number line diagrams for the following two sums.

(a)   4 + (−3)   (b)   (−4) + 3

(a)   4 + (−3) = 1

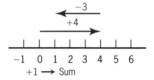

The number 4 is represented by an arrow directed to the right. The number −3 is represented by an arrow starting at the tip of +4 and directed to the left. A gain of $4 followed by a loss of $3 produces a net gain of $1.

(b)   (−4) + 3 = −1

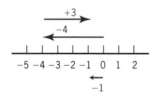

The number −4 is represented by an arrow directed to the left. The number 3 is represented by an arrow starting at the tip of −4 and directed to the right. A loss of $4 followed by a gain of $3 produces a net loss of $1.

ANOTHER WAY TO PICTURE ADDITION OF SIGNED NUMBERS

If you are having trouble with the number line, you might find it easier to picture addition of signed numbers using ordinary poker chips. Suppose that we designate white chips to represent positive integers and blue chips to represent negative integers. The sum of (+4) + (+3) can be found by taking four white chips and combining them with three more white chips. The result is seven white chips or +7:

OOOO + OOO = OOOOOOO

| 4 white chips | plus | 3 white chips | equals | 7 white chips |

(+4)   +   (+3)   =   +7

Similarly, $(-4) + (-3)$ can be illustrated by taking four *blue* chips and combining them with three additional *blue* chips. We end up with seven blue chips or -7:

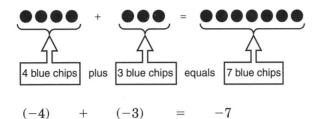

$$(-4) \quad + \quad (-3) \quad = \quad -7$$

To illustrate $4 + (-3)$, take four white chips and three blue chips. Realizing that a white and a blue will "cancel" each other out $((+1) + (-1) = 0)$, we rearrange them as shown below, and the result is one white chip or $+1$:

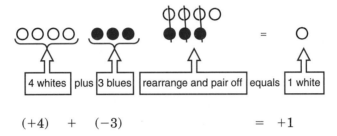

$$(+4) \quad + \quad (-3) \qquad\qquad = \quad +1$$

Finally, to represent $(-4) + 3$, picture the following:

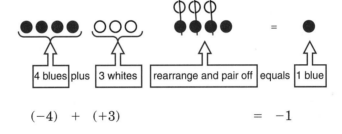

$$(-4) \quad + \quad (+3) \qquad\qquad = \quad -1$$

Number lines and poker chips provide a good way to see what is happening when we add signed numbers, but we need a nice easy rule that will allow us to do the addition more quickly.

Absolute Value   To add or subtract signed numbers you should work with their absolute value. The **absolute value** of a number is the distance along the number line between that number and the zero point. We indicate the absolute value by enclosing the number with vertical lines. The absolute value of any number is always a nonnegative number. For example,

$|4| = 4$ The distance between 0 and 4 is 4 units, so the absolute value of 4 is 4.

$|-5| = 5$ The distance between 0 and -5 is 5 units, so the absolute value of -5 is 5.

To add two signed numbers, follow these steps.

Step 1 Find the absolute value of each number.

Step 2 If the two numbers have the **same sign,** add their absolute values and give the sum the sign of the original numbers.

(a) $9 + 3 = +12$

Absolute value = 9
Absolute value = 3
Add absolute values: $9 + 3 = 12$
Sign is positive since both numbers are positive

(b) $(-6) + (-9) = -15$

Absolute value = 6
Absolute value = 9
Add absolute values: $6 + 9 = 15$
Sign is negative since both numbers are negative

If the two numbers have **opposite signs,** subtract the absolute values and give the difference the sign of the number with the larger absolute value.

(c) $7 + (-12) = -5$

Absolute value = 7
Absolute value = 12
Subtract absolute values: $12 - 7 = 5$
Same sign as the number with the larger absolute value (-12)

(d) $22 + (-14) = +8$

Absolute value = 22
Absolute value = 14
Subtract absolute values: $22 - 14 = 8$
Same sign as the number with the larger absolute value $(+22)$

Try it. Use this method to do the following calculations.

(a) $7 + (-2)$ (b) $-6 + (-14)$ (c) $(-15) + (-8)$

(d) $23 + 13$ (e) $-21.6 + 9.8$ (f) $-6\frac{3}{8} + 14\frac{3}{4}$

(g) $-5 + 8 + 11 + (-21) + (-9) + 12$

(a) $7 + (-2) = +5$ Signs are opposite; subtract absolute values. Answer has the sign of the number with the larger absolute value (7).

(b) $-6 + (-14) = -20$ Signs are the same; add absolute values. Answer has the sign of the original numbers.

(c) $(-15) + (-8) = -23$ Signs are the same; add absolute values. Answer has the sign of the original numbers.

(d) $23 + 13 = +36$ Signs are the same; add absolute values. Answer has the sign of the original numbers.

(e) $-21.6 + 9.8 = -11.8$ Signs are opposite; subtract absolute values. Answer has the sign of the number with the larger absolute value (-21.6).

(f) $-6\frac{3}{8} + 14\frac{3}{4} = +8\frac{3}{8}$ Signs are opposite; subtract absolute values. Answer has the sign of the number with the larger absolute value $(+14\frac{3}{4})$.

(g) When more than two numbers are being added, rearrange them so that numbers with like signs are grouped together:

$$-5 + 8 + 11 + (-21) + (-9) + 12 = \underbrace{(8 + 11 + 12)}_{} + \underbrace{[(-5) + (-21) + (-9)]}_{}$$

Rearrange: ⬆ Positives   ⬆ Negatives

Find the positive total: $8 + 11 + 12 = +31$

Find the negative total: $(-5) + (-21) + (-9) = -35$

Add the positive total to the negative total: $31 + (-35) = -4$

USING A CALCULATOR

To work with signed numbers on a calculator, we need only one new key, the $\boxed{+/-}$ key. This key allows us to enter a negative number on our display. For example, to enter "-4," simply press 4, then $\boxed{+/-}$. You must press $\boxed{+/-}$ *after* entering the number. Try the following addition problems using a calculator.

(a) $(-2675) + 1437$ (b) $(-6.975) + (-5.2452)$

(c) $0.026 + (-0.0045)$ (d) $(-683) + (-594) + 438 + (-862)$

Check your answers below.

(a) **2675** $\boxed{+/-}$ $\boxed{+}$ **1437** $\boxed{=}$ $\rightarrow$ *-1238.*

(b) **6.975** $\boxed{+/-}$ $\boxed{+}$ **5.2452** $\boxed{+/-}$ $\boxed{=}$ $\rightarrow$ *-12.2202*

(c) **.026** $\boxed{+}$ **.0045** $\boxed{+/-}$ $\boxed{=}$ $\rightarrow$ *0.0215*

(d) **683** $\boxed{+/-}$ $\boxed{+}$ **594** $\boxed{+/-}$ $\boxed{+}$ **438** $\boxed{+}$ **862** $\boxed{+/-}$ $\boxed{=}$ $\rightarrow$ *-1701.*

For more practice in addition of signed numbers, turn to Exercises 6-1 for a set of problems.

Exercises 6-1   Addition of Signed Numbers

A. Circle the smaller number.

    1. -4 or 7 2. -17 or -14 3. 23 or 18

    4. 13 or -15 5. -9 or -10 6. -31 or 41

    7. -65 or -56 8. 47 or 23 9. -47 or -23

10. −3 or 0

11. 6 or 0

12. −1.4 or −1.39

13. 2.7 or −2.8

14. $-3\frac{1}{4}$ or $-3\frac{1}{8}$

15. −11.4 or $-11\frac{1}{4}$

16. −0.1 or −0.001

B. For each problem, make a separate number line and mark the points indicated.

1. $-8.7, -9.4, -8\frac{1}{2}, -9\frac{1}{2}$

2. $-1\frac{3}{4}, +1\frac{3}{4}, -\frac{1}{4}, +\frac{1}{4}$

3. $-0.2, -1\frac{1}{2}, -2.8, -3\frac{1}{8}$

4. $+4.1, -1.4, -3.9, +1\frac{1}{4}$

C. Represent each of the following situations with a signed number.

1. A temperature of six degrees below zero

2. A debt of three hundred dollars

3. An airplane rising to an altitude of 12,000 feet

4. The year A.D. 240

5. A golf score of nine strokes below par

6. A quarterback getting sacked for a seven-yard loss

7. Diving eighty feet below the surface of the ocean

8. Winning fifteen dollars at poker

9. Taking five steps backward

10. Ten seconds before takeoff

11. A profit of six thousand dollars

12. An elevation of thirty feet below sea level

13. A checking account overdrawn by $17.50

14. A 26 point drop in the NAZWAC stock index

D. Add.

1. $-1 + 6$

2. $-11 + 7$

3. $-13 + (-2)$

4. $-6 + (-19)$

5. $31 + (-14)$

6. $23 + (-28)$

7. $21 + 25$

8. $12 + 19$

9. $-8 + (-8)$

10. $-7 + (-9)$

11. $-34 + 43$

12. $-56 + 29$

13. $45 + (-16)$

14. $22 + (-87)$

15. $-18 + (-12)$

16. $-240 + 160$

17. $-3.6 + 2.7$

18. $5.7 + (-7.9)$

19. $-12.43 + (-15.66)$

20. $-90.15 + (-43.69)$

21. $2\frac{5}{8} + (-3\frac{7}{8})$

22. $-4\frac{1}{4} + 6\frac{3}{4}$

23. $-9\frac{1}{2} + (-4\frac{1}{2})$

24. $-6\frac{2}{3} + (-8\frac{2}{3})$

25. $-2\frac{1}{2} + 1\frac{3}{4}$

26. $-5\frac{5}{8} + 8\frac{3}{4}$

27. $9\frac{7}{16} + (-7\frac{7}{8})$

28. $-13\frac{3}{10} + (-8\frac{4}{5})$

29. $-6 + 4 + (-8)$

30. $22 + (-38) + 19$

31. $-54 + 16 + 22$

32. $11 + (-6) + (-5)$

33. $-7 + 12 + (-16) + 13$

34. $23 + (-19) + 14 + (-21)$

35. $-36 + 45 + 27 + (-18)$

36. $18 + (-51) + (-8) + 26$

37. $-250 + 340 + (-450) + (-170) + 320 + (-760) + 880 + 190 + (-330)$

38. $-1 + 3 + (-5) + 7 + (-9) + 11 + (-13) + (15) + (-17) + 19$

39. $23 + (-32) + (-14) + 55 + (-66) + 44 + 28 + (-31)$

40. $-12.64 + 22.56 + 16.44 + (-31.59) + (-27.13) + 44.06 + 11.99$

41. $(-2374) + (5973)$

42. $(6.875) + (-11.765)$

43. $(-46,720) + (-67,850) + 87,950$

44. $(0.0475) + (-0.0875) + (-0.0255)$

E. Word Problems

1. **Printing** The Jiffy Print Shop has a daily work quota that it tries to meet. During the first two weeks of last month its daily comparisons to this quota showed the following: +$30, −$27, −$15, +$8, −$6, +$12, −$22, +$56, −$61, +$47. (In each case "+" means they were above the quota and "−" means they were below.) How far below or above their quota was the two-week total?

2. During a series of plays in a football game, a running back carried the ball for the following yardages: 7, −3, 2, −1, 8, −5, and −2. Find his total yards and his average yards per carry.

3. On a cold day in Fairbanks the temperature stood at 12 degrees below zero at 2 P.M. Over the next 5 hours the temperature rose 4 degrees, dropped 2 degrees, rose 5 degrees, dropped 1 degree, and finally, dropped 3 degrees. What was the temperature at 7 P.M.?

4. Pedro's checking account contained $489 on June 1. He then made the following transactions: deposited $150, withdrew $225, withdrew $34, deposited $119, withdrew $365, and deposited $750. What was his new balance after these transactions?

5. A struggling young company reported the following results for the four quarters of the year: A loss of $257,000, a loss of $132,000, a profit of $87,000, a profit of $166,000. What was the net result for the year?

6. **Plumbing** In determining the fitting allowance for a pipe configuration, a plumber has negative allowances of $-2\frac{1}{2}$ in. and $-1\frac{7}{16}$ in. and two positive allowances of $\frac{1}{2}$ in. each. What is the net fitting allowance for the job?

When you have completed these exercises, check your answers in the Appendix, then turn to Section 6-2 to learn how to subtract signed numbers.

6-2 SUBTRACTION OF SIGNED NUMBERS

Opposites   When the number line is extended to the left to include the negative numbers, every positive number can be paired with a negative number whose absolute value, or distance from zero, is the same. These pairs of numbers are called *opposites*. For example, -3 is the opposite of $+3$, $+5$ is the opposite of -5, and so on.

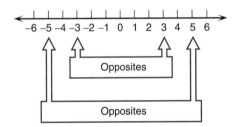

Note ▶   Zero has no opposite because it is neither positive nor negative. ◀

The concept of opposites will help us express a rule for subtraction of signed numbers.

Subtraction of Signed Numbers   We know from basic arithmetic that $5 - 2 = 3$. From the rules for addition of signed numbers, we know that $5 + (-2) = 3$ also. Therefore,

$$(+5) - (+2) = (+5) + (-2) = 3$$

This means that subtracting $(+2)$ is the same as adding its opposite, (-2). Our knowledge of arithmetic and signed numbers verifies that subtracting and adding the opposite are equivalent for all the following cases:

$$5 - 5 = 5 + (-5) = 0$$
$$5 - 4 = 5 + (-4) = 1$$
$$5 - 3 = 5 + (-3) = 2$$
$$5 - 2 = 5 + (-2) = 3$$
$$5 - 1 = 5 + (-1) = 4$$
$$5 - 0 \qquad\qquad = 5 \quad \text{(zero has no opposite)}$$

But what happens with a problem like $5 - 6 = ?$ If we rewrite this subtraction as addition of the opposite like the others, we have

$$5 - 6 = 5 + (-6) = -1$$

Notice that this result fits right into the pattern established by the previous results. The answers *decrease* as the numbers being subtracted *increase*: $5 - 3 = 2$, $5 - 4 = 1$, $5 - 5 = 0$, so it seems logical that $5 - 6 = -1$.

What happens if we subtract a negative? For example, how can we solve the problem $5 - (-1) = ?$ If we rewrite this subtraction as addition of the opposite, we have

$$5 - (-1) = 5 + (+1) = 6$$

| subtracting -1 | becomes | adding the opposite of -1 or $+1$ |

Notice that this too fits in with the pattern established by the previous results. The answers *increase* as the numbers being subtracted *decrease:* $5 - 2 = 3$, $5 - 1 = 4$, $5 - 0 = 5$, so again it seems logical that $5 - (-1) = 6$.

We can now state a rule for subtraction of signed numbers. To subtract two signed numbers, follow these steps.

Step 1 Rewrite the subtraction as an addition of the opposite.

Step 2 Use the rules for addition of signed numbers to add.

EXAMPLE

(a) $3 - 7 = (+3) - (+7) = 3 + (-7) = -4$

| subtracting $+7$ | adding -7 |
| is the same as |

(b) $-2 - (-6) = -2 + (+6) = +4$

| subtracting -6 | adding $+6$ |
| is the same as |

(c) $-8 - (-3) = -8 + (+3) = -5$

| subtracting -3 | adding $+3$ |
| is the same as |

(d) $-4 - 9 = -4 - (+9) = (-4) + (-9) = -13$

| subtracting $+9$ | adding -9 |
| is the same as |

Careful ▶ Never change the sign of the first number. You change the sign of the *second* number only *after* changing subtraction to addition. ◄

YOUR TURN

Try these:

(a) $-7 - 5$ (b) $8 - 19$

(c) $4 - (-4)$ (d) $-12 - (-8)$

(e) $\left(-\dfrac{3}{4}\right) - \left(\dfrac{5}{16}\right)$ (f) $-2 - (-3) + (-1)$

(g) $-3 - 9 - (-7) + (-8) + (+2)$ (h) $(-26.75) - (-44.38) - 16.96$

SOLUTIONS

(a) $-7 - 5 = -7 + (-5) = -12$

(b) $8 - 19 = (+8) + (-19) = -11$

(c) $4 - (-4) = (+4) + (+4) = +8$

(d) $-12 - (-8) = -12 + (+8) = -4$

(e) $\left(-\dfrac{3}{4}\right) - \left(\dfrac{5}{16}\right) = \left(-\dfrac{3}{4}\right) + \left(-\dfrac{5}{16}\right)$

$$= \left(-\dfrac{12}{16}\right) + \left(-\dfrac{5}{16}\right)$$

$$= -\dfrac{17}{16} = -1\tfrac{1}{16}$$

(f) $-2 - (-3) + (-1) = \underline{-2 + (+3)} + (-1)$

$$= (+1) + (-1)$$

$$= 0$$

(g) $-3 - 9 - (-7) + \underline{(-8) + (+2)} = -3 + (-9) + (+7) + \underline{(-8) + (+2)} = -11$

> These last two operations are additions, so they
> remain unchanged when the problem is rewritten.

 (h) **26.75** ⊞⁄⁻ ⊟ **44.38** ⊞⁄⁻ ⊟ **16.96** ⊜ →   *0.67*

SUBTRACTION WITH CHIPS

If you are having trouble visualizing subtraction of signed numbers, try the poker chip method that we used to show addition. Subtraction means "taking away" chips. First, a simple example: $7 - 3$ or $(+7) - (+3)$.

Begin with 7 white (positive) chips.   Take away 3 white chips.   Four white chips remain.

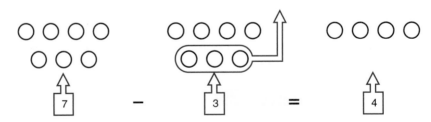

What about the problem: $(+3) - (+7) = ?$ How can you remove 7 white chips if you have only 3 white chips? Here we must be very clever. Remember that a positive chip and a negative chip "cancel" each other $[(+1) + (-1) = 0]$.

So if we add both a white (positive) and a blue (negative) chip to a collection, we do not really change its value. Do $(+3) - (+7)$ this way:

| Start with 3 white chips. | Add in 4 more whites and 4 more blues | Now we can remove 7 white chips. | Four blue chips remain. |

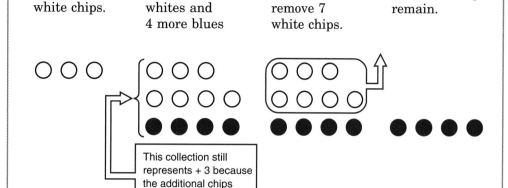

This collection still represents + 3 because the additional chips cancel each other out.

$$(+3) \quad - \quad (+7) \quad = \quad (-4)$$

Our rules for subtraction verify that this is correct.

Finally, let's try subtracting a negative. Consider the problem $-2 - (-5) = ?$

| Begin with -2: 2 blue chips. | To subtract -5 we must remove 5 blue chips. We have only 2 blues, so add 3 more blues and 3 more whites to keep the value the same. | Now remove 5 blues. | Three whites remain. |

$$(-2) \quad - \quad (-5) \quad = \quad (+3)$$

Our rules for subtraction verify that this is correct.

Now turn to Exercises 6-2 for a set of problems on subtraction of signed numbers.

Exercises 6-2

Subtraction of Signed Numbers

A. Find the value of the following.

1. $3 - 5$ -2

2. $7 - 20$ -14

3. $-4 - 12$ -16

4. $-9 - 15$ -24

5. $6 - (-3)$ 9

6. $14 - (-7)$ 21

7. $-11 - (-5)$ ~~-6~~   8. $-22 - (-10)$ ~~-11~~

9. $12 - 4$ ~~8~~   10. $23 - 9$ ~~14~~

11. $-5 - (-16)$ ~~11~~   12. $-14 - (-17)$ ~~3~~

13. $12 - 30$ ~~-18~~   14. $30 - 65$ ~~-35~~

15. $-23 - 8$ ~~-31~~   16. $-18 - 11$

17. $2 - (-6)$ ~~8~~   18. $15 - (-12)$ ~~27~~

19. $44 - 18$ ~~26~~   20. $70 - 25$ ~~45~~

21. $-34 - (-43)$ ~~9~~   22. $-27 - (-16) = -11$

23. $0 - 19$ ~~-19~~   24. $0 - 26$ ~~-26~~

25. $-17 - 0$ ~~-17~~   26. $-55 - 0$ ~~-55~~

27. $0 - (-8)$ ~~8~~   28. $0 - (-14)$ ~~14~~

29. $-56 - 31$ ~~-87~~   30. $-45 - 67$ ~~-112~~

31. $22 - 89$ ~~-67~~   32. $17 - 44$ ~~-27~~

33. $-12 - 16 + 18$ ~~-10~~   34. $23 - (-11) - 30$ ~~4~~

35. $19 + (-54) - 7$ ~~-42~~   36. $-21 + 6 - (-40)$ ~~25~~

37. $-2 - 5 - (-7) + 6$ 38. $5 - 12 + (-13) - (-7)$

39. $-3 - 4 - (-5) + (-6) + 7$ ~~-1~~   40. $2 - (-6) - 10 + 4 + (-11)$ ~~-9~~

            $8 - 10 + 4 - 11$

41. $\left(-\frac{3}{8}\right) - \frac{1}{4}$ ~~-\frac{1}{8}~~   42. $2\frac{1}{2} - \left(-3\frac{1}{4}\right)$ $12 - 21 =$

43. $6\frac{9}{16} - 13\frac{1}{16}$ 44. $\left(-\frac{5}{6}\right) - \left(-\frac{2}{3}\right)$

[margin: $-7 + 5 - 6 + 7$ -13]

45. $6.5 - (-2.7)$ 46. $(-0.275) - 1.375$

47. $(-86.4) - (-22.9)$ 48. $0.0025 - 0.075$

49. $22.75 - 36.45 - (-54.72)$ 50. $(-7.35) - (-4.87) - 3.66$

51. $3\frac{1}{2} - 6.6 - \left(-2\frac{1}{4}\right)$ 52. $(-5.8) - \left(-8\frac{1}{8}\right) - 3.9$

B. Word Problems

1. The height of Mt. Whitney is +14,410 ft. The lowest point in Death Valley is −288 ft. What is the difference in altitude between these two points?

2. The highest temperature ever recorded on Earth was +136°F at Libya in North Africa. The coldest was −127°F at Vostok in Antarctica. What is the difference in temperature between these two extremes?

3. At 8 P.M. the temperature was −2°F in Anchorage, Alaska. At 4 A.M. it was −37°F. By how much did the temperature drop in that time?

4. A vacation trip involved a drive from a spot in Death Valley 173 ft below sea level (−173 ft) to the San Bernardino Mountains 4250 ft above sea level (+4250 ft). What was the vertical ascent for this trip?

5. **Auto Mechanics** Dave's Auto Shop was $2465 in the "red" (−$2465) at the end of April—they owed this much to their creditors. At the end of December they were $648 in the red. How much money did they make in profit between April and December to achieve this turnaround?

6. A golfer had a total score of six strokes over par going into the last round of a tournament. He ended with a total score of three strokes under par. What was his score relative to par during the final round?

7. **Roofing** Pacific Roofing was $3765 in the "red" (see problem 5) at the end of June. At the end of September they were $8976 in the "red." Did they make or lose money between June and September? How much?

8. **Sheet Metal Technology** To establish the location of a hole relative to a fixed zero point, a machinist must make the following calculation:

$$y = 5 - (3.750 - 0.500) - 2.375$$

Find y.

9. **Electronics** The peak-to-peak voltage of an ac circuit is found by subtracting the negative peak voltage from the positive peak voltage. Calculate the peak-to-peak voltage V in the diagram shown.

Problem 9

Check your answers in the Appendix, then turn to Section 6-3 to learn about multiplication and division of signed numbers.

6-3 MULTIPLICATION AND DIVISION OF SIGNED NUMBERS

Multiplication of Signed Numbers As we learned earlier, multiplication is the shortcut for repeated addition. For example,

$$5 \times 3 = \underbrace{3 + 3 + 3 + 3 + 3}_{} = 15$$

| Five times three | means | three added five times |

Notice that the *product of two positive numbers* is always a *positive* number.

We can use repeated addition to understand multiplication involving negative numbers. Consider the problem $(+5) \times (-3)$.

$$(+5) \times (-3) = \underbrace{(-3) + (-3) + (-3) + (-3) + (-3)}_{} = -15$$

| Five times negative three | | negative three added five times |

Notice that the product of a *positive number* and a *negative number* is a *negative number*. The order in which you multiply two numbers does not matter, so a negative times a positive is also a negative. For example,

$$(-5) \times (+3) = (+3) \times (-5) = (-5) + (-5) + (-5) = -15$$

To understand the product of a negative times a negative, look at the pattern in the following sequence of problems:

$$(-5) \times (+3) = -15$$
$$(-5) \times (+2) = -10$$
$$(-5) \times (+1) = -5$$
$$(-5) \times 0 = 0$$
$$(-5) \times (-1) = ?$$

Each time we *decrease* the factor on the right by 1, we *increase* the product by 5. So to preserve this pattern, the product of $(-5) \times (-1)$ must be $+5$. In other words, *a negative times a negative* is a *positive*.

Rules for Multiplication

In summary, to multiply two signed numbers, follow these steps.

Step 1   Find the product of their absolute values.

Step 2   Attach a sign to the answer:

- If both numbers have the **same** sign, the answer is **positive.**

  $(+) \times (+) = +$ $(-) \times (-) = +$

- If the two numbers have **opposite** signs, the answer is **negative.**

  $(+) \times (-) = -$ $(-) \times (+) = -$

EXAMPLE

Multiply:

(a) $(+3) \times (-2)$ (b) $(-4) \times (-8)$ (c) $\left(-\frac{1}{4}\right) \times \left(+\frac{2}{3}\right)$

SOLUTIONS

(a)   **Step 1**   The product of their absolute values is $3 \times 2 = 6$

   Step 2   The original numbers have opposite signs, so the product is negative. $(+3) \times (-2) = -6$

(b)   **Step 1**   The product of their absolute values is $4 \times 8 = 32$

   Step 2   The original numbers have the same sign, so the product is positive. $(-4) \times (-8) = +32$

(c)   **Step 1**   The product of their absolute values is $\dfrac{1}{\underset{2}{4}} \times \dfrac{\overset{1}{2}}{3} = \dfrac{1}{6}$

   Step 2   The original fractions have opposite signs, so the product is negative. $\left(-\frac{1}{4}\right) \times \left(+\frac{2}{3}\right) = -\frac{1}{6}$

YOUR TURN

Use these rules on the following problems.

(a) $(-4) \times (+7)$ (b) $(-8) \times (-6)$ (c) $(+5) \times (+9)$

(d) $(+10) \times (-3)$ (e) $(-3) \times (-3)$ (f) $(-2) \times (-2) \times (-2)$

(g) $(-2) \times (-3) \times (+4) \times (-5) \times (-6)$ (h) $\left(-2\frac{1}{3}\right) \times \left(\frac{9}{14}\right)$

(i) $(-3.25) \times (-1.40)$

SOLUTIONS

(a) $(-4) \times (+7) = -28$

   opposite signs   answer is negative

(b) $(-8) \times (-6) = +48$

(c) $+45$ (same sign, answer is positive)

(d) -30 (opposite signs, answer is negative)

(e) $+9$

(f) -8

(g) $+720$

(h) $-1\frac{1}{2}$

   (i)   **3.25** [+/-] [×] **1.4** [+/-] [=] → ▨▨▨▨▨▨ *4.55*

Did you discover an easy way to do problems (f) and (g) where more than two signed numbers are being multiplied? You could go left to right, multiply two numbers, then multiply the answer by the next number. You would then need to apply the sign rules for each separate product. However, since every *two* negative factors results in a positive, an *even* number of negative factors gives a positive answer, and an *odd* number of negative factors gives a negative answer. In a problem with more than two factors, count the total number of negative signs to determine the sign of the answer.

In problem (f) there are three negative factors—an odd number—so the answer is negative. Multiply the absolute values as usual to get the numerical part of the answer, 8. In problem (g) there are four negative factors—an even number—so the answer is positive. The product of the absolute values of all five numbers gives us the numerical part of the answer, 720.

Division of Signed Numbers   Recall that division is the reverse of the process of multiplication. The division

$24 \div 4 = \square$ can be written as the multiplication $24 = 4 \times \square$

and in this case the number $\square$ is 6. Notice that *a positive divided by a positive equals a positive*. We can use this "reversing" process to verify that the rest of the sign rules for division are exactly the same as those for multiplication. To find $(-24) \div (+4)$, rewrite the problem as follows:

$(-24) \div (+4) = \square$ $-24 = (+4) \times \square$

From our knowledge of signed number multiplication, $\square = -6$. Therefore, *a negative divided by a positive equals a negative*. Similarly,

$(+24) \div (-4) = \square$ means $(+24) = (-4) \times \square$ so that $\square = -6$

A positive divided by a negative is also a negative. Finally, let's try a negative divided by a negative:

$(-24) \div (-4) = \square$ means $(-24) = (-4) \times \square$ so that $\square = +6$

A negative divided by a negative equals a positive.

When dividing two signed numbers, determine the sign of the answer the same way you do for multiplication:

• If both numbers have the **same** sign, the answer is **positive.**

   $(+) \div (+) = +$ $(-) \div (-) = +$

- If the two numbers have **opposite** signs, the answer is **negative.**

$$(+) \div (-) = - \qquad (-) \div (+) = -$$

Divide the absolute values of the two numbers to get the numerical value of the quotient.

EXAMPLE

Divide:

(a) $(+28) \div (-4)$ 　　　　(b) $(-45) \div (-5)$ 　　　　(c) $\left(-\frac{2}{5}\right) \div \left(+\frac{7}{10}\right)$

SOLUTIONS

(a) **Step 1**　The quotient of their absolute values is 　　$28 \div 4 = 7$

　　Step 2　The original numbers have opposite signs,
　　　　　　so the quotient is negative. 　　　　　　$(+28) \div (-4) = -7$

(b) **Step 1**　The quotient of their absolute values is 　　$45 \div 5 = 9$

　　Step 2　The original numbers have the same sign,
　　　　　　so the quotient is positive. 　　　　　　$(-45) \div (-5) = +9$

(c) **Step 1**　The quotient of their absolute values is 　　$\dfrac{2}{5} \div \dfrac{7}{10} = \dfrac{2}{\underset{1}{\cancel{5}}} \times \dfrac{\overset{2}{\cancel{10}}}{7} = \dfrac{4}{7}$

　　Step 2　The original fractions have opposite sign,
　　　　　　so the quotient is negative. 　　　　　　$\left(-\dfrac{2}{5}\right) \div \left(+\dfrac{7}{10}\right) = -\dfrac{4}{7}$

YOUR TURN

Try these division problems:

(a) $(-36) \div (+3)$ 　　　　(b) $(-72) \div (-9)$ 　　　　(c) $(+42) \div (-6)$

(d) $(-5.4) \div (-0.9)$ 　　　　(e) $\left(\frac{3}{8}\right) \div \left(-\frac{1}{4}\right)$ 　　　　(f) $(-3.75) \div (6.4)$
　　　　　　　　　　　　　　　　　　　　　　　　　(Round to two decimal places.)

ANSWERS

(a) -12 　　　(b) $+8$ 　　　(c) -7 　　　(d) $+6$ 　　　(e) $-1\frac{1}{2}$ 　　　(f) -0.59

 Enter (f) this way on a calculator:

3.75 $\boxed{+/-}$ $\boxed{\div}$ **6.4** $\boxed{=}$ $\rightarrow$ ▨ *−0.5859375*

For more practice in multiplication and division of signed numbers, turn to Exercises 6-3 for a set of problems.

Exercises 6-3

Multiplication and Division of Signed Numbers

A. Multiply or divide as indicated.

　　1. $(-7) \times (9)$ 　　　　2. $8 \times (-5)$ 　　　　3. $(-7) \times (-11)$

　　4. $(-4) \times (-9)$ 　　　　5. $28 \div (-7)$ 　　　　6. $(-64) \div 16$

7. $(-72) \div (-6)$ 8. $(-44) \div (-11)$ 9. $(-8.4) \div 2.1$

10. $(-10.8) \div (-1.2)$ 11. $\left(-4\frac{1}{2}\right) \times \left(-2\frac{3}{4}\right)$ 12. $\frac{2}{3} \times \left(-\frac{3}{4}\right)$

13. $(-3.4) \times (-1.5)$ 14. $(-4) \div (16)$ 15. $(-5) \div (-25)$

16. $(-13) \times (-4)$ 17. $15 \times (-6)$ 18. $(-18) \times 3$

19. $(-12) \times (-7)$ 20. $(-120) \div 6$ 21. $140 \div (-5)$

22. $(-200) \div (-10)$ 23. $(-350) \div (-70)$ 24. $2.3 \times (-1.5)$

25. $(-1.6) \times (5.0)$ 26. $\left(-6\frac{2}{5}\right) \times \left(-1\frac{1}{4}\right)$ 27. $\left(-\frac{5}{8}\right) \times \left(-\frac{4}{15}\right)$

28. $(-2.25) \div 0.15$ 29. $8\frac{1}{2} \div \left(-3\frac{1}{3}\right)$ 30. $\left(-\frac{5}{6}\right) \div \left(-\frac{3}{4}\right)$

31. $(-4.8) \div (-12)$ 32. $(-2) \times \left(-5\frac{7}{8}\right)$ 33. $(-3) \times \left(6\frac{1}{2}\right)$

34. $(-0.07) \times (-1.1)$ 35. $3.2 \times (-1.4)$ 36. $(-0.096) \div 1.6$

37. $(-5) \times (-6) \times (-3)$ 38. $(-8) \times 4 \times (-7)$

39. $(-24) \div (-6) \times (-5)$ 40. $35 \div (-0.5) \times (-10)$

41. $(-12) \times (-0.3) \div (-9)$ 42. $0.2 \times (-5) \div (-0.02)$

B. Multiply or divide using a calculator and round as indicated.

1. $(-3.87) \times (4.98)$ (three significant digits)

2. $(-5.8) \times (-9.75)$ (two significant digits)

3. $(-0.075) \times (-0.025)$ (two significant digits)

4. $(648) \times (-250)$ (two significant digits)

5. $(-4650) \div (1470)$ (three significant digits)

6. $(-14.5) \div (-3.75)$ (three significant digits)

7. $(-0.58) \div (-2.5)$ (two significant digits)

8. $0.0025 \div (-0.084)$ (two significant digits)

9. $(-4.75) \times 65.2 \div (-6.8)$ (two significant digits)

10. $78.4 \div (-8.25) \times (-22.6)$ (three significant digits)

11. $(-2400) \times (-450) \times (-65)$ (two significant digits)

12. $(-65,000) \div (2.75) \times 360$ (two significant digits)

C. Practical Problems

1. To convert 17°F to degrees Celsius, it is necessary to perform the calculation

$$°C = \frac{5}{9} \times (°F - 32) = \frac{5}{9} \times (17 - 32)$$

What is the Celsius equivalent of 17°F?

2. In the first four months of the year, a certain country had an average monthly trade deficit (a *negative* quantity) of $16.5 million. During the final eight months of the year, the country had an average monthly trade surplus (a *positive* quantity) of $5.3 million. What was the overall trade balance for the year? (Use the appropriate sign to indicate an overall deficit or surplus.)

3. Due to the relative sizes of two connected pulleys A and B, pulley A will turn $1\frac{1}{2}$ times more than pulley B and in the opposite direction. If positive numbers are used to indicate counterclockwise revolutions and negative numbers are used for clockwise revolutions, express as a signed number the result for:
   (a) Pulley A when pulley B turns 80 revolutions counterclockwise.
   (b) Pulley B when pulley A turns 60 revolutions clockwise.

4. **Machinist** A long metal rod changes in length by 0.4 mm for every 1°C change in temperature. If it expands when heated and contracts when cooled, express its change in length as a signed number as the temperature changes from 12°C to 4°C.

5. An airplane descends from 42,000 ft to 18,000 ft in 12 minutes. Express its rate of change in altitude in feet per minute as a signed number.

6. To convert −8°C to degrees Fahrenheit, you must do the following calculation:

$$°F = \frac{9 \times °C}{5} + 32 = \frac{9 \times (-8)}{5} + 32$$

What is the Fahrenheit equivalent of −8°C?

Check your answers in the Appendix, then turn to Section 6-4 to learn about exponents and square roots.

6-4 EXPONENTS AND SQUARE ROOTS

When the same number appears many times in a multiplication, writing the product may become monotonous, tiring, and even inaccurate. It is easy, for example, to miscount the twos in

$$32{,}768 = 2 \times 2 \times 2 \times 2 \times 2 \times 2 \times 2 \times 2 \times 2 \times 2 \times 2 \times 2 \times 2 \times 2 \times 2$$

or the tens in

$$100{,}000{,}000{,}000 = 10 \times 10 \times 10 \times 10 \times 10 \times 10 \times 10 \times 10 \times 10 \times 10 \times 10$$

Products of this sort are usually written in a shorthand form as 2^{15} and 10^{11}. In this *exponential* form the raised integer 15 shows the number of times that 2 is to be used as a factor in the multiplication.

EXAMPLE

$2 \times 2 = 2^2$ Product of two factors of 2

$2 \times 2 \times 2 = 2^3$ Product of three factors of 2

$2 \times 2 \times 2 \times 2 = 2^4$ Product of four factors of 2

Four 2s

Write $3 \times 3 \times 3 \times 3 \times 3$ in exponential form.

$3 \times 3 \times 3 \times 3 \times 3 =$ _____

Counting 3s,

$$\underbrace{3 \times 3 \times 3 \times 3 \times 3}_{\text{Five factors of 3}} = 3^5$$

Exponents In this expression, 3 is called the *base,* and the integer 5 is called the *exponent.* The exponent 5 tells you how many times the base 3 must be used as a factor in the multiplication.

EXAMPLE

Find the value of $4^3 =$ _____

The exponent 3 tells us to use 4 as a factor 3 times.

$4^3 = 4 \times 4 \times 4 = (4 \times 4) \times 4 = 16 \times 4 = 64$

It is important that you be able to read exponential forms correctly.

2^2 is read "two to the second power" or "two squared"

2^3 is read "two to the third power" or "two cubed"

2^4 is read "two to the fourth power"

2^5 is read "two to the fifth power" and so on

Students studying basic electronics, or those going on to another mathematics or science class, will find that it is important to understand exponential notation.

PROBLEMS

Do the following problems for practice in using exponents.

(a) Write in exponential form.

$5 \times 5 \times 5 \times 5$	= ____ base = ____	exponent = ____
7×7	= ____ base = ____	exponent = ____
$10 \times 10 \times 10 \times 10 \times 10$	= ____ base = ____	exponent = ____
$3 \times 3 \times 3 \times 3 \times 3 \times 3 \times 3$	= ____ base = ____	exponent = ____
$9 \times 9 \times 9$	= ____ base = ____	exponent = ____
$1 \times 1 \times 1 \times 1$	= ____ base = ____	exponent = ____

(b) Write as a product of factors and multiply out.

2^6	= _____ = ____ base = ____	exponent = ____
10^7	= _____ = ____ base = ____	exponent = ____

$(-3)^4 =$ _____ = ____   base = ____   exponent = ____

5^2 = _____ = ____   base = ____   exponent = ____

$(-6)^3 =$ _____ = ____   base = ____   exponent = ____

4^5 = _____ = ____   base = ____   exponent = ____

12^3 = _____ = ____   base = ____   exponent = ____

1^5 = _____ = ____   base = ____   exponent = ____

ANSWERS

(a) $5 \times 5 \times 5 \times 5$ $= \underline{5^4}$ base = $\underline{5}$ exponent = $\underline{4}$

     7×7 $= \underline{7^2}$ base = $\underline{7}$ exponent = $\underline{2}$

     $10 \times 10 \times 10 \times 10 \times 10$ $= \underline{10^5}$ base = $\underline{10}$ exponent = $\underline{5}$

     $3 \times 3 \times 3 \times 3 \times 3 \times 3 \times 3$ $= \underline{3^7}$ base = $\underline{3}$ exponent = $\underline{7}$

     $9 \times 9 \times 9$ $= \underline{9^3}$ base = $\underline{9}$ exponent = $\underline{3}$

     $1 \times 1 \times 1 \times 1$ $= \underline{1^4}$ base = $\underline{1}$ exponent = $\underline{4}$

(b) $2^6 = 2 \times 2 \times 2 \times 2 \times 2 \times 2$ $= \underline{64}$ base = $\underline{2}$ exponent = $\underline{6}$

     $10^7 = 10 \times 10 \times 10 \times 10 \times 10 \times 10 \times 10 = 10,000,000$

                             base = $\underline{10}$ exponent = $\underline{7}$

     $(-3)^4 = (-3) \times (-3) \times (-3) \times (-3)$ $= \underline{+81}$ base = $\underline{-3}$ exponent = $\underline{4}$

     $5^2 = 5 \times 5$ $= \underline{25}$ base = $\underline{5}$ exponent = $\underline{2}$

     $(-6)^3 = (-6) \times (-6) \times (-6)$ $= \underline{-216}$ base = $\underline{-6}$ exponent = $\underline{3}$

Note ▶ Notice that $(-3)^4 = +81$, but $(-6)^3 = -216$. The rules for multiplying signed numbers tell us that a negative number raised to an **even** power is positive, but a negative number raised to an **odd** power is negative. ◀

     $4^5 = 4 \times 4 \times 4 \times 4 \times 4$ $= \underline{1024}$ base = $\underline{4}$ exponent = $\underline{5}$

     $12^3 = 12 \times 12 \times 12$ $= \underline{1728}$ base = $\underline{12}$ exponent = $\underline{3}$

     $1^5 = 1 \times 1 \times 1 \times 1 \times 1$ $= \underline{1}$ base = $\underline{1}$ exponent = $\underline{5}$

Careful ▶ Be careful when there is a negative sign in front of the exponential expression. For example,

$3^2 = 3 \times 3 = 9$ and $(-3)^2 = (-3) \times (-3) = 9$

but

$-3^2 = -(3^2) = -(3 \times 3) = -9$

To calculate $-(-2)^3$ first calculate

$(-2)^3 = (-2) \times (-2) \times (-2) = -8$

Then

$-(-2)^3 = -(-8) = 8$. A useful way of remembering this: exponents act only on what they touch. ◀

Any power of 1 is equal to 1, of course.

$1^2 = 1 \times 1 = 1$
$1^3 = 1 \times 1 \times 1 = 1$
$1^4 = 1 \times 1 \times 1 \times 1 = 1$ and so on

Notice that when the base is ten, the product is easy to find.

$10^2 = 10 \times 10 = 100$
$10^3 = 10 \times 10 \times 10 = 1000$
$10^4 = 10000$
$10^5 = \underbrace{100000}$

5 zeros

The exponent number is always exactly equal to the number of zeros in the final product when the base is 10.

Continue the pattern to find the value of 10^0 and 10^1.

$10^5 = 100000$	$10^2 = 100$
$10^4 = 10000$	$10^1 = 10$
$10^3 = 1000$	$10^0 = 1$

For any base, powers of 1 or 0 are easy to find.

$2^1 = 2$	$2^0 = 1$
$3^1 = 3$	$3^0 = 1$
$4^1 = 4$	$4^0 = 1$ and so on

Order of Operations with Exponents

In Chapter 1 we discussed the order of operations when you add, subtract, multiply, or divide. To evaluate an arithmetic expression containing exponents, first calculate the value of the exponential factor, then perform the other operations using the order of operations already given. If the calculations are to be performed in any other order, parentheses will be used to show the order. The revised order of operations is:

1. Simplify all operations within **parentheses.**

2. Simplify all **exponents,** left to right.

3. Perform all **multiplications** and **divisions,** left to right.

4. Perform all **additions** and **subtractions,** left to right.

Learning Help ▶

If you think you will have trouble remembering this order of operations, try memorizing this phrase: **P**lease **E**xcuse **M**y **D**ear **A**unt **S**ally. The first letter of each word should remind you of **P**arentheses, **E**xponents, **M**ultiply/**D**ivide, **A**dd/**S**ubtract. ◀

Here are some examples showing how to apply this expanded order of operations.

EXAMPLE

(a) To calculate $12 + 54 \div 3^2$

First, simplify the exponent factor	$= 12 + 54 \div 9$
Second, divide	$= 12 + 6$
Finally, add	$= 18$

(b) To calculate $237 - (2 \times 3)^3$

First, work within parentheses	$= 237 - (6)^3$
Second, simplify the exponent term	$= 237 - 216$
Finally, subtract	$= 21$

YOUR TURN

Calculate.

(a) $2^4 \times 5^3$ (b) $6 + 3^2$ (c) $(6 + 3)^2$

(d) $(24 \div 6)^3$ (e) $2^2 + 4^2 \div 2^3$ (f) $4 + 3 \times 2^4 \div 6$

SOLUTIONS

(a) $2^4 \times 5^3 = 16 \times 125 = 2000$ (b) $6 + 3^2 = 6 + 9 = 15$

(c) $(6 + 3)^2 = (9)^2 = 81$ (d) $(24 \div 6)^3 = (4)^3 = 64$

(e) $2^2 + 4^2 \div 2^3 = 4 + 16 \div 8 = 4 + 2 = 6$ (Do the division before adding.)

(f) $4 + 3 \times 2^4 \div 6 = 4 + 3 \times 16 \div 6 = 4 + 48 \div 6 = 4 + 8 = 12$

To square a number using a calculator, enter the number, then press the $\boxed{x^2}$ key. For example, 38^2 can be found by entering

38 $\boxed{x^2}$ → `1444.`

Notice that it is not necessary to use the $\boxed{=}$ key.

To raise a number to any other power, enter the base number, press the $\boxed{y^x}$ or $\boxed{x^y}$ or $\boxed{\wedge}$ key, enter the exponent number, and finally press the $\boxed{=}$ key. For example, to calculate 3.8^5 enter **3.8** $\boxed{y^x}$ **5** $\boxed{=}$ → `792.35168`

YOUR TURN

A bit of practice will help. Calculate the following:

(a) 68^2 (b) 65^3 (c) 2.75^4 (Round to three significant digits.)

(d) 0.475^2 (Round to three significant digits.)

ANSWERS

(a) **68** $\boxed{x^2}$ → `4624.`

(b) **65** $\boxed{y^x}$ **3** $\boxed{=}$ → `274625.`

(c) **2.75** $\boxed{y^x}$ **4** $\boxed{=}$ → `57.19140625` or 57.2 rounded

(d) **.475** $\boxed{x^2}$ → `0.225625` or 0.226 rounded

Square Roots What is interesting about the numbers

1, 4, 9, 16, 25, 36, 49, 64, 81, 100, . . .?

Do you recognize them?

These numbers are the squares or second powers of the counting numbers,

$1^2 = 1$
$2^2 = 4$
$3^2 = 9$
$4^2 = 16$, and so on. 1, 4, 9, 16, 25, . . . are called *perfect squares*.

PERFECT SQUARES

$1^2 = 1$	$6^2 = 36$	$11^2 = 121$	$16^2 = 256$
$2^2 = 4$	$7^2 = 49$	$12^2 = 144$	$17^2 = 289$
$3^2 = 9$	$8^2 = 64$	$13^2 = 169$	$18^2 = 324$
$4^2 = 16$	$9^2 = 81$	$14^2 = 196$	$19^2 = 361$
$5^2 = 25$	$10^2 = 100$	$15^2 = 225$	$20^2 = 400$

If you have memorized the multiplication table for one-digit numbers, you will recognize them immediately. The number 3^2 is read "three squared." What is "square" about $3^2 = 9$? The name comes from an old Greek idea about the nature of numbers. Ancient Greek mathematicians called certain numbers "square numbers" or "perfect squares" because they could be represented by a square array of dots.

4   9   16   . . . and so on

The number of dots along the side of the square was called the "root" or origin of the square number. We call it the *square root*.

The square root of 16 is 4, since $4 \times 4 = 16$.

What is the square root of 64?

The square root of 64 is equal to 8 because $8 \times 8 = 64$.

The sign $\sqrt{}$ is used to indicate the square root.

$\sqrt{16} = 4$ read it "square root of 16"

$\sqrt{9} = 3$ read it "square root of 9"

$\sqrt{169} = 13$ read it "square root of 169"

YOUR TURN

Find $\sqrt{81} =$ _____ $\sqrt{361} =$ _____ $\sqrt{289} =$ _____

Try using the table of perfect squares if you do not recognize these.

ANSWERS

$\sqrt{81} = 9$ ☑ $9 \times 9 = 81$

$\sqrt{361} = 19$ ☑ $19 \times 19 = 361$

$\sqrt{289} = 17$ ☑ $17 \times 17 = 289$

Always check your answer as shown.

How do you find the square root of any number? Using a calculator is the easiest way to find the square or other power of a number, or to find a square root.

To calculate a square root enter the number and then press the $\boxed{\sqrt{}}$ key. There is no need to press the $\boxed{=}$ key. For example, to calculate $\sqrt{237}$ enter

237 $\boxed{\sqrt{}}$ → $\boxed{15.39480432}$

Note ▶ On some scientific calculators and on all graphing calculators, you must press the square root key *first*, then enter the number. ◀

YOUR TURN

Perform the following calculations.

(a) $\sqrt{760}$ (b) $\sqrt{5.74}$

ANSWERS

(a) **760** $\boxed{\sqrt{}}$ → $\boxed{27.5680975}$

(b) **5.74** $\boxed{\sqrt{}}$ → $\boxed{2.39582971}$

Careful ▶ Notice that you cannot take the square root of a negative number. There is no number $\square$ such that $\square \times \square = -81$.

If $\square$ is positive, then $\square \times \square$ is positive, and if $\square$ is negative, then $\square \times \square$ is also positive. You cannot multiply a number by itself and get a negative result. (In order to perform higher-level algebra, mathematicians have invented imaginary numbers, which are used to represent the square roots of negative numbers. Students who go on to higher-level courses will learn about these numbers later.) ◀

Square roots have the same place in the order of operations as exponents. Simplify square roots before performing multiplication, division, addition, or subtraction.

The expanded order of operations now reads:

1. Simplify all operations within **parentheses,**

2. Simplify all **exponents and roots,** left to right,

3. Perform all **multiplications and divisions,** left to right,

4. Perform all **additions and subtractions,** left to right.

EXAMPLE

$3\sqrt{49} = 3 \times \sqrt{49} = 3 \times 7 = 21$ Note that $3\sqrt{49}$ is a shorthand

$2 + \sqrt{64} = 2 + 8 = 10$ way of writing $3 \times \sqrt{49}$.

$\dfrac{\sqrt{81}}{3} = \dfrac{9}{3} = 3$

If an arithmetic operation appears under the square root symbol, perform this operation before taking the square root. For example,

$$\sqrt{9 + 16} = \sqrt{25} = 5$$

Careful ▶ Notice $\sqrt{9 + 16}$ is not the same as $\sqrt{9} + \sqrt{16} = 3 + 4 = 7$. ◀

Calculate:

(a) $100 - \sqrt{100}$
(b) $\dfrac{4^2 \times \sqrt{36}}{3}$
(c) $\sqrt{169 - 144}$
(d) $\sqrt{64} + \sqrt{121}$

(e) $6 + 4\sqrt{25}$
(f) $49 - 3^2\sqrt{4}$
(g) $\dfrac{\sqrt{225} - \sqrt{81}}{\sqrt{36}}$
(h) $\sqrt{\dfrac{225 - 81}{36}}$

SOLUTIONS

(a) $100 - \sqrt{100} = 100 - 10 = 90$

(b) $\dfrac{4^2 \times \sqrt{36}}{3} = \dfrac{16 \times (6)}{3} = \dfrac{96}{3} = 32$

(c) $\sqrt{169 - 144} = \sqrt{25} = 5$

(d) $\sqrt{64} + \sqrt{121} = 8 + 11 = 19$

(e) $6 + 4\sqrt{25} = 6 + 4 \times 5 = 6 + 20 = 26$

(f) $49 - 3^2\sqrt{4} = 49 - 9\sqrt{4} = 49 - 9 \times 2 = 49 - 18 = 31$

(g) $\dfrac{\sqrt{225} - \sqrt{81}}{\sqrt{36}} = \dfrac{15 - 9}{6} = \dfrac{6}{6} = 1$

(h) $\sqrt{\dfrac{225 - 81}{36}} = \sqrt{\dfrac{144}{36}} = \sqrt{4} = 2$

SIMPLIFYING SQUARE ROOTS

When the square root is not an exact whole number, a calculator or table can be used to find a decimal approximation of its value. For example, a plumber calculating the length of a connecting pipe might approximate $\sqrt{12}$ as about 3.5. But in more mathematical or theoretical work, $\sqrt{12}$ might be expressed in what is known as Simplified Radical Form. Follow these steps to write the simplified radical form of a square root.

Step 1 Find the largest perfect square that exactly divides the number.   The largest perfect square that exactly divides 12 is 4.

Step 2 Use this perfect square to factor the number. $\sqrt{12} = \sqrt{4 \times 3}$

Step 3 Since $\sqrt{a \times b} = \sqrt{a} \times \sqrt{b}$ for any two positive numbers a and b, we can rewrite **Step 2** as a product. $\sqrt{4 \times 3} = \sqrt{4} \times \sqrt{3}$

Step 4 Take the square root of the perfect square factor. $= 2 \times \sqrt{3}$

Therefore $2\sqrt{3}$ is the simplest radical form for $\sqrt{12}$.

Write the following square roots in simplest radical form.

(a) $\sqrt{18}$
(b) $\sqrt{75}$
(c) $\sqrt{72}$

(d) $\sqrt{48}$
(e) $\sqrt{500}$
(f) $\sqrt{162}$

(continued)

Check your work below.

(a) $\sqrt{18} = \sqrt{9 \times 2} = \sqrt{9} \times \sqrt{2} = 3\sqrt{2}$

(b) $\sqrt{75} = \sqrt{25 \times 3} = \sqrt{25} \times \sqrt{3} = 5\sqrt{3}$

(c) $\sqrt{72} = \sqrt{36 \times 2} = \sqrt{36} \times \sqrt{2} = 6\sqrt{2}$

(d) $\sqrt{48} = \sqrt{16 \times 3} = \sqrt{16} \times \sqrt{3} = 4\sqrt{3}$

(e) $\sqrt{500} = \sqrt{100 \times 5} = \sqrt{100} \times \sqrt{5} = 10\sqrt{5}$

(f) $\sqrt{162} = \sqrt{81 \times 2} = \sqrt{81} \times \sqrt{2} = 9\sqrt{2}$

In problems (c), (d), and (f) it is important to find the *largest* perfect square that divides each number.

Now go to Exercises 6-4 for a set of problems on exponents and square roots.

Exercises 6-4

Exponents and Square Roots

A. Find the value of these.

1. 2^4	2. 3^2	3. 4^3
4. 5^3	5. 10^3	6. 7^2
7. 2^8	8. 6^2	9. 8^3
10. 3^4	11. 5^4	12. 10^5
13. $(-2)^3$	14. $(-3)^5$	15. 9^3
16. 6^0	17. 5^1	18. 1^4
19. 2^5	20. 8^2	21. $(-14)^2$
22. $(-21)^2$	23. 15^3	24. 16^2
25. $2^2 \times 3^3$	26. $2^6 \times 3^2$	27. $3^2 \times 5^3$
28. $2^3 \times 7^3$	29. $6^2 \times 5^2 \times 3^1$	30. $2^{10} \times 3^2$
31. $2 + 4^2$	32. $14 - 2^3$	33. $(8 - 2)^3$
34. $(14 \div 2)^2$	35. $3^3 - 8^2 \div 2^3$	36. $12 - 5 \times 2^5 \div 2^4$
37. $5 \times 6^2 - 4^3$	38. $64 \div 4^2 + 18$	39. $(28 \div 7)^2 - 14$
40. $3(8^2 - 3^3)$	41. $26 + 40 \div 2^3 \times 3^2$	42. $18 + 4 \times 5^2 - 3^4$

B. Calculate. (Round to two decimal places if necessary.)

1. $\sqrt{81}$	2. $\sqrt{144}$	3. $\sqrt{36}$
4. $\sqrt{16}$	5. $\sqrt{25}$	6. $\sqrt{9}$
7. $\sqrt{256}$	8. $\sqrt{400}$	9. $\sqrt{225}$

10. $\sqrt{49}$ 11. $\sqrt{324}$ 12. $\sqrt{121}$

13. $\sqrt{4.5}$ 14. $\sqrt{500}$ 15. $\sqrt{12.4}$

16. $\sqrt{700}$ 17. $\sqrt{210}$ 18. $\sqrt{321}$

19. $\sqrt{810}$ 20. $\sqrt{92.5}$ 21. $\sqrt{1000}$

22. $\sqrt{2000}$ 23. $\sqrt{25000}$ 24. $\sqrt{2500}$

25. $\sqrt{150}$ 26. $\sqrt{300}$ 27. $\sqrt{3000}$

28. $\sqrt{30000}$ 29. $\sqrt{1.25}$ 30. $\sqrt{1.008}$

31. $3\sqrt{20}$ 32. $4 + 2\sqrt{5}$ 33. $55 - \sqrt{14}$

34. $\dfrac{3^2\sqrt{24}}{3 \times 2}$ 35. $\sqrt{2^5 - 3^2}$ 36. $5.2^2 - 3\sqrt{52 - 5^2}$

C. Practical Problems

1. Find the length of the side of a square whose area is 184.96 sq ft. (*Hint:* The area of a square is equal to the square of one of its sides. The side of a square equals the square root of its area.)

2. A square building covers an area of 1269 sq ft. What is the length of each side of the building? Round to the nearest tenth.

3. **Mechanical Engineering** The following calculation must be performed to find the stress (in psi) on a plunger. Find this stress to the nearest thousand psi.

$$\frac{3000\sqrt{34}}{0.196}$$

4. **Electronics** The resistance (in ohms) in a certain silicon diode is found using the expression

$$\sqrt{15 \times 10^6 \times 24}$$

Calculate this resistance to the nearest thousand ohms.

5. **Fire Protection** The velocity (in fps) of water discharged from a hose with a nozzle pressure of 62 psi is given by

$$12.14\sqrt{62}$$

Calculate and round to the nearest whole number.

6. **Police Science** Skid marks can be used to determine the maximum speed of a car involved in an accident. If a car leaves a skid mark of 175 ft on a road with a 35% coefficient of friction, its speed (in mph) can be estimated using

$$\sqrt{30 \times 0.35 \times 175}$$

Find this speed to the nearest whole number.

7. **Health Sciences** The Fit-4-U weight loss center calculates the Body Mass Index for each prospective client.

$$\text{BMI} = \frac{703 \times (\text{Weight in pounds})}{(\text{Height in inches})^2}$$

A person with a BMI greater than 30 is considered to be obese and a BMI over 25 is considered overweight. Calculate the BMI of a person 5 ft 6 in. tall weighing 165 lb.

8. **Roofing**   The length L of a rafter on a shed roof is calculated using the formula

$$L = \sqrt{(H_2 - H_1)^2 + W^2}$$

If H_2, the maximum roof height, is 13 ft, H_1, the minimum wall height, is 7.5 ft, and W, the width of the shed, is 12 ft, calculate the rafter length, not including allowance for overhang. (Round to the nearest tenth.)

9. **Construction Technology**   The Golden Gate bridge is about 1 mile long. On a warm day it expands about 2 ft in length. If there were no expansion joints to compensate for the expansion, how high a bulge would this produce? Use the following formula for your calculation, and round to the nearest foot.

$$\text{Height} \atop \text{(in feet)} = \sqrt{\left(\frac{5280}{2} + 1\right)^2 - \left(\frac{5280}{2}\right)^2}$$

Check your answers in the Appendix, then turn to Problem Set 6 for practice problems on Chapter 6.

Pre-Algebra

Answers are given in the Appendix.

A. Rewrite each group of numbers in order from smallest to largest.

1. $-13, 7, 4, -8, -2$

2. $-0.6, 0, 0.4, -0.55, -0.138, -0.96$

3. $-150, -140, -160, -180, -120, 0$

4. $-\dfrac{5}{8}, \dfrac{3}{4}, \dfrac{1}{2}, -\dfrac{1}{2}, -\dfrac{3}{4}, \dfrac{3}{8}$

B. Add or subtract as indicated.

1. $-16 + 6$
2. $-5 + (-21)$

3. $7 - 13$
4. $-4 - (-4)$

5. $23 + (-38)$
6. $-13 - 37$

7. $20 - (-5)$
8. $11 + (-3)$

9. $-2.6 + 4.9$
10. $2\frac{1}{4} - 4\frac{3}{4}$

11. $-\$5.26 - \$3.89$
12. $-2\frac{3}{8} + \left(-7\frac{1}{2}\right)$

13. $-9 - 6 + 4 + (-13) - (-8)$
14. $-12.4 + 6\frac{1}{2} - \left(-1\frac{3}{4}\right) + (-18) - 5.2$

C. Multiply or divide as indicated.

1. $6 \times (-10)$
2. $(-7) \times (-5)$

3. $-48 \div (-4)$
4. $(-320) \div 8$

5. $\left(-1\frac{1}{8}\right) \times \left(-6\frac{1}{4}\right)$
6. $\left(-5\frac{1}{4}\right) \div 7$

7. $6.4 \div (-16)$
8. $(-2.4) \times 15$

9. $(-4) \times (-5) \times (-6)$
10. $72 \div (-4) \times (-3)$

11. $(2.4735) \times (-6.4)$ (two significant digits)

12. $(-0.675) \div (-2.125)$ (three significant digits)

Name _____

Date _____

Course/Section _____

D. Find the numerical value of each expression.

1. $12.5 \times (-6.3 + 2.8)$
2. $(-3) \times 7 - 4 \times (-9)$
3. $\dfrac{14 - 46}{(-4) \times (-2)}$

4. $\left(-4\frac{1}{2} - 3\frac{3}{4}\right) \div 8$
5. 2^6
6. 3^5

7. 17^2
8. 23^3
9. 0.5^2

10. 1.2^3
11. 0.02^2
12. 0.03^3

13. 0.001^3
14. 2.01^2
15. 4.02^2

16. $(-3)^3$
17. $(-2)^4$
18. $\sqrt{1369}$

19. $\sqrt{784}$
20. $\sqrt{4.41}$
21. $\sqrt{0.16}$

22. $\sqrt{5.29}$
23. $5.4 + 3.3^2$
24. $2.1^2 + 3.1^3$

25. 4.5×5.2^2
26. $4.312 \div 1.4^2$
27. $(3 \times 6)^2 + 3 \times 6^2$

28. $28 - 4 \times 3^2 \div 6$
29. $2 + 2\sqrt{16}$
30. $5\sqrt{25} - 4.48$

31. $\sqrt{0.49} - \sqrt{0.04}$
32. $10 - 3\sqrt{1.21}$

Round to two decimal digits.

33. $\sqrt{80}$
34. $\sqrt{106}$
35. $\sqrt{310}$

36. $\sqrt{1.8}$
37. $\sqrt{4.20}$
38. $\sqrt{3.02}$

39. $\sqrt{1.09}$
40. $\sqrt{0.08}$
41. $7 - 3\sqrt{2}$

42. $\sqrt{6} - \sqrt{5}$
43. $3\sqrt{2} - 2\sqrt{3}$
44. $3\sqrt{2} + 2\sqrt{3}$

E. Practical Problems

1. **Electrician** The current in a circuit changes from -2 amperes to 5 amperes. What is the change in current?
2. Find the side of a square whose area is 225 sq ft. (The side of a square is the square root of its area.)
3. In Hibbing, Minnesota, the temperature dropped 8°F overnight. If the temperature read -14°F before the plunge, what did it read afterward?
4. **Plumbing** Morgan Plumbing had earnings of $3478 in the last week of January and expenses of $4326. What was the net profit or loss? (Indicate the answer with the appropriate sign.)
5. **Electrician** The current capacity (in amperes) of a service using a three-phase system for 124 kilowatts of power and a line voltage of 440 volts is given by the calculation

$$\frac{124{,}000}{440\sqrt{3}}$$

Determine this current to three significant digits.
6. **Fire Protection** The flow rate (in gallons per minute) of water from a $1\frac{1}{2}$-in. hose at 65 psi is calculated from the expression

$$29.7 \times (1.5)^2 \sqrt{65}$$

Find the flow rate to the nearest gallon per minute.

7. **Civil Engineer** The crushing load (in tons) for a square pillar 8 in. thick and 12 ft high is given by the formula

$$L = \frac{25 \times (8)^4}{12^2}$$

Find this load to the nearest hundred tons.

8. **Automotive Services** An auto shop was $14,560 in the "red" at the beginning of the year and $47,220 in the "black" at the end of the year. How much profit did the shop make during the year?

9. **Electronics** In determining the bandwidth of a high-fidelity amplifier, a technician would first perform the following calculation to find the rise time of the signal:

$$T = \sqrt{3200^2 - 22.5^2}$$

Calculate this rise time to the nearest hundred. The time will be in nanoseconds.

10. **Electronics** The following calculation is used to determine one of the voltages for a sweep generator:

$$V = 10 - 9.5 \div 0.03$$

Calculate this voltage. Round to one significant digit.

11. **Meteorology** Use the formula

$$°C = \frac{5}{9} \times (°F - 32)$$

to convert the following Fahrenheit temperatures to Celsius rounded to the nearest degree.

(a) −4°F   (b) 24°F

12. **Meteorology** Use the formula

$$°F = \frac{9 \times °C}{5} + 32$$

to convert the following Celsius temperatures to Fahrenheit.

(a) −21°C   (b) −12°C

PREVIEW

7

Basic Algebra

Objective	Sample Problems		For Help Go to Page

When you finish this unit you will be able to:

1. Evaluate formulas and literal expressions.

If $x = 2$, $y = 3$, $a = 5$, $b = 6$, find the value of

340

(a) $2x + y$ _____

(b) $\dfrac{1 + x^2 + 2a}{y}$ _____

(c) $A = x^2 y$ $A =$ _____

(d) $T = \dfrac{2(a + b + 1)}{3x}$ $T =$ _____

(e) $P = abx^2$ $P =$ _____

2. Perform the basic algebraic operations.

(a) $3ax^2 + 4ax^2 - ax^2$ $=$ _____ 348

(b) $5x - 3y - 8x + 2y$ $=$ _____

(c) $5x - (x + 2)$ $=$ _____

(d) $3(x^2 + 5x) - 4(2x - 3)$ $=$ _____

3. Solve linear equations in one unknown and solve formulas.

(a) $3x - 4 = 11$ $x =$ _____ 365

(b) $2x = 18$ $x =$ _____ 360

(c) $2x + 7 = 43 - x$ $x =$ _____ 377

(d) Solve the following formula

for N: $S = \dfrac{N}{2} + 26$ $N =$ _____ 382

(e) Solve the following formula

for A: $M = \dfrac{(A - B)L}{8}$ $A =$ _____

_____
Name

_____
Date

_____
Course/Section

Preview Chapter 7 Basic Algebra

333

4. Translate simple English phrases and sentences into algebraic expressions and equations.

Write each phrase as an algebraic expression or equation.

(a) Four times the area.

_____ 389

(b) Current squared times resistance.

_____

(c) Efficiency is equal to 100 times the output divided by the input.

_____

(d) Resistance is equal to 12 times the length divided by the diameter squared.

_____

(e) An electrician collected $1265.23 for a job that included $840 labor, which was not taxed, and the rest for parts, which were taxed, at 6%. Determine how much of the total was tax.

_____ 394

(f) A plumber's helper earns $18 per hour plus $27 per hour overtime. If he works 40 regular hours during a week, how much overtime does he need in order to earn $1000?

_____

5. Multiply and divide simple algebraic expressions.

(a) $2y \cdot 3y$

_____ 399

(b) $(6x^4y^2)(-2xy^2)$

_____

(c) $3x(y - 2x)$

_____

(d) $\dfrac{10x^7}{-2x^2}$

_____ 401

(e) $\dfrac{6a^2b^4}{18a^5b^2}$

_____

(f) $\dfrac{12m^4 - 9m^3 + 15m^2}{3m^2}$

_____

6. Use scientific notation.

Write in scientific notation.

(a) 0.000184

_____ 404

(b) 213,000

_____

Calculate.

(c) $(3.2 \times 10^{-6}) \times (4.5 \times 10^2)$

_____ 407

(d) $(1.56 \times 10^{-4}) \div (2.4 \times 10^3)$

_____

334

(Answers to these preview problems are given in the Appendix.)

If you are certain that you can work *all* these problems correctly, turn to page 413 for the set of practice problems. If you cannot work one or more of the preview problems, turn to the page indicated to the right of the problem. Those who wish to master this material should turn to Section 7-1 and begin work there.

7 Basic Algebra

I'M PRACTICING MY BRACKETS...

DID YOU KNOW THAT BRACKETS ARE ALWAYS USED IN PAIRS?

IF YOU EVER SEE A BRACKET BY ITSELF, YOU CAN BE SURE IT'S UP TO NO GOOD!

Peanuts reprinted by permission of UFS, Inc.

In this chapter you will study algebra, but not the very formal algebra that deals with theorems, proofs, sets, and abstract problems. Instead, we shall study practical or applied algebra as actually used by technical and trades workers. Let's begin with a look at the language of algebra and algebraic formulas.

7-1 ALGEBRAIC LANGUAGE AND FORMULAS

The most obvious difference between algebra and arithmetic is that in algebra letters are used to replace or to represent numbers. A mathematical statement in which letters are used to represent numbers is called a *literal* expression. Algebra is the arithmetic of literal expressions—a kind of symbolic arithmetic.

Any letters will do, but in practical algebra we use the normal lower- and upper-case letters of the English alphabet. It is helpful in practical problems to choose the letters to be used on the basis of their memory value: t for time, D for diameter, C for cost, A for area, and so on. The letter used reminds you of its meaning.

Multiplication Most of the usual arithmetic symbols have the same meaning in algebra that they have in arithmetic. For example, the addition $(+)$ and subtraction $(-)$ signs are used in exactly the same way. However, the multiplication sign $(\times)$ of arithmetic looks like the letter x and to avoid confusion we have other ways to show multiplication in algebra. The product of two algebraic quantities a and b, "a times b," may be written using

A raised dot	$a \cdot b$
Parentheses	$a(b)$ or $(a)b$ or $(a)(b)$
Nothing at all	ab

Obviously, this last way of showing multiplication won't do in arithmetic; we cannot write "two times four" as "24"—it looks like twenty-four. But it is a quick and easy way to write a multiplication in algebra.

Placing two quantities side by side to show multiplication is not new and it is not only an algebra gimmick; we use it every time we write 20 cents or 4 feet.

20 cents = 20 × 1 cent
4 feet   = 4 × 1 foot

YOUR TURN

Write the following multiplications using no multiplication symbols.

(a) 8 times a = _____

(b) m times p = _____

(c) 2 times s times t = _____

(d) 3 times x times x = _____

ANSWERS

(a) $8a$ (b) mp (c) $2st$ (d) $3x^2$

Did you notice in problem (d) that powers are written just as in arithmetic:

$x \cdot x = x^2$

$x \cdot x \cdot x = x^3$

$x \cdot x \cdot x \cdot x = x^4$ and so on.

Parentheses Parentheses ( ) are used in arithmetic to show that some complicated quantity is to be treated as a unit. For example,

$2 \cdot (13 + 14 - 6)$

means that the number 2 multiplies *all* of the quantity in the parentheses.

In exactly the same way in algebra, parentheses ( ), brackets [ ], or braces { } are used to show that whatever is enclosed in them should be treated as a single quantity. An expression such as

$(3x^2 - 4ax + 2by^2)^2$

should be thought of as (something)^2. The expression

$(2x + 3a - 4) - (x^2 - 2a)$

should be thought of as (first quantity) − (second quantity). Parentheses are the punctuation marks of algebra. Like the period, comma, or semicolon in regular sentences, they tell you how to read an expression and get its correct meaning.

Division In arithmetic we would write "48 divided by 2" as

$2\overline{)48}$ or $48 \div 2$ or $\frac{48}{2}$

But the first two ways of writing division are used very seldom in algebra. Division is usually written as a fraction.

"x is divided by y" is written $\dfrac{x}{y}$

"$(2n + 1)$ divided by $(n - 1)$" is written $\dfrac{(2n + 1)}{(n - 1)}$ or $\dfrac{2n + 1}{n - 1}$

Write the following using algebraic notation.

(a) 8 times $(2a + b)$ = _____

(b) $(a + b)$ times $(a - b)$ = _____

(c) x divided by y^2 = _____

(d) $(x + 2)$ divided by $(2x - 1)$ = _____

ANSWERS

(a) $8(2a + b)$

(b) $(a + b)(a - b)$

(c) $\dfrac{x}{y^2}$

(d) $\dfrac{x + 2}{2x - 1}$

Algebraic Expressions

The word "expression" is used very often in algebra. An *expression* is a general name for any collection of numbers and letters connected by arithmetic signs. For example,

$$x + y \qquad 2x^2 + 4 \qquad 3(x^2 - 2ab)$$

$$\frac{D}{T} \qquad \sqrt{x^2 + y^2} \qquad \text{and} \qquad (b - 1)^2$$

are all algebraic expressions.

If the algebraic expression has been formed by multiplying quantities, each multiplier is called a *factor* of the expression.

Expression	**Factors**
ab	a and b
$2x(x + 1)$	$2x$ and $(x + 1)$
$(R - 1)(2R + 1)$	$(R - 1)$ and $(2R + 1)$

The algebraic expression can also be a sum or difference of simpler quantities or *terms*.

Expression	**Terms**	
$x + 4y$	x and $4y$	⟵ The first term is x / The second term is $4y$
$2x^2 + xy$	$2x^2$ and xy	
$A - R$	A and R	

YOUR TURN

Now let's check to see if you understand the difference between terms and factors. In each algebraic expression below, tell whether the portion quoted is a term or a factor.

(a) $2x^2 - 3xy$ y is a _____

(b) $7x - 4$ $7x$ is a _____

(c) $4x(a + 2b)$ $4x$ is a _____

(d) $-2y(3y - 5)$ $(3y - 5)$ is a _____

(e) $3x^2y + 8y^2 - 9$ $3x^2y$ is a _____

Evaluating Formulas

One of the most useful algebraic skills for any technical or practical work involves finding the value of an algebraic expression when the letters are given numerical values. A *formula* is a rule for calculating the numerical value of one quantity from the values of the other quantities. The formula or rule is usually written in mathematical form because algebra gives a brief, convenient to use, and easy to remember form for the rule. Here are a few examples of rules and formulas used in the trades.

1. **Rule:** Ohm's law: The voltage V (in volts) across a simple resistor is equal to the product of the current i (in amperes) through the resistor and the value of its resistance R (in ohms).

   Formula: $V = iR$

2. **Rule:** The cost C of setting type is equal to the total ems set T in the job multiplied by the hourly wage H divided by the rate E at which the type is set, in ems per hour.

   Formula: $C = \dfrac{TH}{E}$

3. **Rule:** The number of standard bricks N needed to build a wall is about 21 times the volume of the wall.

   Formula: $N = 21\,LWH$

Evaluating a formula or algebra expression means to find its value by substituting numbers for the letters in the expression.

EXAMPLE

In retail stores the following formula is used:

$M = R - C$ where M is the markup on an item, R is the retail selling price, and C is the original cost

Find M if R = $25 and C = $21.

$M =$ _____

$M = \$25 - \$21 = \$4$

Easy? Of course. Simply substitute the numbers for the correct letters and then do the arithmetic. A formula is a recipe for a calculation.

YOUR TURN

Automotive engineers use the following formula to calculate the SAE rating (R) of an engine.

$R = \dfrac{D^2N}{2.5}$ where D is the diameter of a cylinder in inches, and N is the number of cylinders

Find R when $D = 3\frac{1}{2}$ in. and $N = 6$.

$$R = \frac{(3.5)^2(6)}{2.5}$$

3.5 $\boxed{x^2}$ $\boxed{\times}$ **6** $\boxed{\div}$ **2.5** $\boxed{=}$ → ▨▨▨▨ 29.4

$R = 29.4$ or roughly 29.

To be certain you do it correctly, follow this two-step process:

Step 1 Place the numbers being substituted in parentheses, and then substitute them in the formula.

Step 2 Do the arithmetic carefully *after* the numbers are substituted.

EXAMPLE

(a) Find $A = x + 2y$ for $x = 5$ ft, $y = 3$ ft.

        Step 1 $A = (5 \text{ ft}) + 2(3 \text{ ft})$ Put 5 ft and 3 ft in parentheses.

        Step 2 $A = 5 \text{ ft} + 6 \text{ ft}$ Do the arithmetic.

                 $A = 11 \text{ ft}$

(b) Find $B = x^2 - y$ for $x = 4$, $y = -3$.

        Step 1 $B = (4)^2 - (-3)$ Put the numbers in parentheses.

        Step 2 $B = 16 + 3$ Do the arithmetic.

                 $B = 19$

Using parentheses in this way may seem like extra work for you, but it is the key to avoiding mistakes when evaluating formulas.

YOUR TURN

Evaluate the following formulas.

(a) $D = 2R$ for $R = 3.45$ in.

(b) $W = T - C$ for $T = 1420$ lb, $C = 385$ lb

(c) $P = 0.433H$ for $H = 11.4$ in. Round to three significant digits.

(d) $A = bh - 2$ for $b = 1.75$ in., $h = 4.20$ in.

(e) $P = 2(8 + x)$ for $x = -3$

Use the two-step process and round your answer appropriately.

SOLUTIONS

(a) $D = 2(3.45) = 6.90$ in.   (b) $W = (1420) - (385)$
                                                                $= 1420 - 385$
                                                                $= 1035$ lb

(c) $P = 0.433(11.4)$
$\quad\ = 4.9362$ or 4.94 in., rounded

(d) $A = (1.75)(4.2) - 2$
$\quad\ = 7.35 - 2$
$\quad\ = 5.35$ sq in.

For problem (d),

1.75 ⊠ **4.2** ⊟ **2** ⊟ → [*5.35*]

(e) $P = 2(8 + (-3))$
$\quad\ P = 2(8 - 3)$
$\quad\ P = 2 \cdot 5$
$\quad\ P = 10$

The order of operations for arithmetic calculations should be used when evaluating formulas. Remember:

1. Do any operations **inside parentheses.**

2. Find all **powers** and **roots.**

3. Do all **multiplications** or **divisions** left to right.

4. Do all **additions** or **subtractions** left to right.

PROBLEMS

1. Evaluate the following formulas using the standard order of operations.

$a = 3, b = 4, c = 6$

(a) $3ab$ $\quad = $ _____

(b) $2a^2c$ $\quad = $ _____

(c) $2a^2 - b^2 = $ _____

(d) $3a^2bc - ab$ $\quad = $ _____

(e) $b + 2a$ $\quad = $ _____

(f) $a + 3(2b - c)$ $\quad = $ _____

(g) $2(a + b) = $ _____

(h) $(a^2 - 1) - (2 + b) = $ _____

2. To convert Fahrenheit F to Celsius C temperature, use the following formula:

$$°C = \frac{5(°F - 32)}{9}$$

To convert Celsius C to Fahrenheit F temperature, use this formula:

$$°F = \frac{9°C}{5} + 32$$

Use these formulas to find the following temperatures. Round to the nearest degree.

(a) $650°F = $ _____ °C

(b) $-160°C = $ _____ °F

(c) $398°C = $ _____ °F

(d) $2200°F = $ _____ °C

(e) $80°F = $ _____ °C

(f) $52.5°C = $ _____ °F

1. (a) $3(3)(4) = 36$

   (b) $2(3)^2(6) = 2 \cdot 9 \cdot 6 = 108$

   (c) $2(3)^2 - (4)^2 = 2 \cdot 9 - 16$
   $$= 18 - 16$$
   $$= 2$$

   (d) $3(3)^2(4)(6) - (3)(4) = (3 \cdot 9 \cdot 4 \cdot 6) - (3 \cdot 4)$
   $$= 648 - 12$$
   $$= 636$$

 3 $\boxed{\times}$ **3** $\boxed{x^2}$ $\boxed{\times}$ **4** $\boxed{\times}$ **6** $\boxed{-}$ **3** $\boxed{\times}$ **4** $\boxed{=}$ → _636._

   (e) $(4) + 2(3) = 4 + 2 \cdot 3$
   $$= 4 + 6$$
   $$= 10$$

   (f) $(3) + 3(2(4) - (6)) = 3 + 3(2 \cdot 4 - 6)$
   $$= 3 + 3(8 - 6)$$
   $$= 3 + 3 \cdot 2 = 3 + 6$$
   $$= 9$$

   (g) $2((3) + (4)) = 2 \cdot (3 + 4)$
   $$= 2 \cdot 7$$
   $$= 14$$

   (h) $((3)^2 - 1) - (2 + (4)) = (9 - 1) - (2 + 4)$
   $$= 8 - 6$$
   $$= 2$$

2. (a) $C = \dfrac{5 \times (650 - 32)}{9}$

   $$= \dfrac{5 \times (618)}{9}$$

   $$= \dfrac{3090}{9}$$

   $$\approx 343°C$$

   (b) $F = \dfrac{9 \times (-160)}{5} + 32$

   $$= \dfrac{-1440}{5} + 32$$

   $$= -288 + 32$$

   $$= -256°F$$

   (c) $F = \dfrac{9 \times 398}{5} + 32$

   $$= \dfrac{3582}{5} + 32$$

   $$= 716.4 + 32$$

   $$\approx 748°F$$

   (d) $C = \dfrac{5 \times (2200 - 32)}{9}$

   $$= \dfrac{5 \times 2168}{9}$$

   $$= \dfrac{10840}{9}$$

   $$\approx 1204°C$$

   (e) $C = \dfrac{5 \times (80 - 32)}{9}$

   $$= \dfrac{5 \times 48}{9}$$

   $$= \dfrac{240}{9}$$

   $$\approx 27°C$$

   (f) $F = \dfrac{9 \times 52.5}{5} + 32$

   $$= \dfrac{472.5}{5} + 32$$

   $$= 94.5 + 32$$

   $$\approx 127°F$$

For problem (e) using a calculator.

 5 ⊠ (80 ⊟ 32) ÷ 9 🟰 → 26.66666667

Careful ▶ Avoid the temptation to combine steps when you evaluate formulas. Take it slowly and carefully, follow the standard order of operations, and you will arrive at the correct answer. Rush through problems like these and you usually make mistakes. ◀

Now turn to Exercises 7-1 for a set of practice problems in evaluating formulas.

Exercises 7-1

Evaluating Formulas

A. Find the value of each of these formulas for $x = 2$, $y = 3$, $z = 4$, $R = 5$.

1. $A = 3x$	2. $D = 2R - y$
3. $T = x^2 + y^2$	4. $H = 2x + 3y - z$
5. $K = 3z - x^2$	6. $Q = 2xyz - 10$
7. $F = 2(x + y^2) - 3$	8. $W = 3(y - 1)$
9. $L = 3R - 2(y^2 - x)$	10. $A = R^2 - y^2 - xz$
11. $B = 3R - y + 1$	12. $F = 3(R - y + 1)$

B. Find the value of each of the following formulas. (Round to the nearest whole number if necessary.)

1. $A = 2(x + y) - 1$ for $x = 2, y = 4$

2. $V = (L + W)(2L + W)$ for $L = 7.5$ ft, $W = 5.0$ ft

3. $I = PRT$ for $P = 150, R = 0.05, T = 2$

4. $H = 2(a^2 + b^2)$ for $a = 2$ cm, $b = 1$ cm

5. $T = \dfrac{(A + B)H}{2}$ for $A = 3.26$ m, $B = 7.15$ m, $H = 4.4$ m

6. $V = \dfrac{\pi D^2 H}{4}$ for $\pi = 3.14, D = 6.25$ in., $H = 7.2$ in.

7. $P = \dfrac{NR(T + 273)}{V}$ for $N = 5, R = 0.08, T = 27, V = 3$

8. $W = D(AB - \pi R^2)H$ for $D = 9$ lb/in.^{3}, $A = 6.3$ in., $B = 2.7$ in., $\pi = 3.14, R = 2$ in., $H = 1.0$ in.

9. $V = LWH$ for $L = 16.25$ m, $W = 3.1$ m, $H = 2.4$ m

10. $V = \pi R^2 A$ for $\pi = 3.14, R = 3.2$ ft, $A = 0.425$ ft

C. Practical Problems

1. The perimeter of a rectangle is given by the formula $P = 2L + 2W$, where L is the length of the rectangle and W is its width. Find P when L is $8\frac{1}{2}$ in. and W is 11 in.

2. **Electronics**   The current in amperes in a simple electrical circuit is given by Ohm's law $i = V/R$, where V is the voltage and R is the resistance of the circuit. Find the current in a circuit whose resistance is 10 ohms and which is connected across a 120-volt power source.

3. **Electrical Technology** Find the power used in an electric light bulb, $P = i^2R$, if the current $i = 0.80$ ampere and the resistance $R = 150$ ohms. P will be in watts.

4. Find the surface area of a sphere, $A = 4\pi R^2$, when π equals roughly 3.14 and the radius $R = 10.0$ cm. (Round to the nearest 10 sq cm.)

5. The Fahrenheit temperature °F is related to the Celsius temperature °C by the formula $°F = \frac{9}{5}°C + 32$. Find the Fahrenheit temperature when °C = 40°.

6. **Machine Technology** The volume of a round steel bar depends on its length L and diameter D according to the formula $V = \pi D^2 L/4$. Find the volume of a bar 20.0 in. long and 3.0 in. in diameter. Use $\pi \approx 3.14$. (Round to the nearest 10 cu in.)

7. If D dollars is invested at p percent simple interest for t years, the amount A of the investment is

$$A = D\left(1 + \frac{pt}{100}\right)$$

Find A if $D = \$10,000$, $p = 5\%$, and $t = 5$ years.

8. **Electronics** The total resistance R of two resistances a and b connected in parallel is $R = ab/(a + b)$. What is the total resistance if $a = 200$ ohms and $b = 300$ ohms?

9. The area of a trapezoid is $T = \dfrac{(A + B)H}{2}$, where A and B are the lengths of its parallel sides and H is the height. Find the area of the trapezoid shown.

10. **Plumbing** Use the formula $T = \frac{1}{2}(D - d)$ to find the wall thickness T of tubing having the following dimensions.

D, outside diameter	d, inside diameter
(a) 2.125 in	1.500 in.
(b) 0.785 in.	0.548 in.
(c) 1.400 cm	0.875 cm
(d) $\frac{15}{16}$ in.	$\frac{5}{8}$ in.

11. **Sheet Metal Technology** To make a right-angle inside bend in sheet metal, the length of sheet used is given by the formula $L = x + y + \frac{1}{2}T$. Find L when $x = 6\frac{1}{4}$ in., $y = 11\frac{7}{8}$ in., $T = \frac{1}{4}$ in.

12. **Sheet Metal Technology** The length of a chord of a circle is given by the formula $L = 2RH - H^2$, where R is the radius of the circle and H is the height of the arc above the chord. If you have a portion of a circular disk for which $R = 4\frac{1}{4}$ in. and $H = 5\frac{1}{2}$ in., what is the length of the chord? (See the figure.)

13. **Building Technology** The rope capacity of a drum is given by the formula $L = ABC(A + D)$. How many feet of $\frac{1}{2}$-in. rope can be wound on a drum where $A = 6$ in., $C = 30$ in., $D = 24$ in., and $B = 1.05$ for $\frac{1}{2}$-in rope? (Round to the nearest hundred inches.)

14. **Manufacturing** A millwright uses the following formula as an approximation to find the required length of a pulley belt. (See the figure.)

$$L = 2C + 1.57(D + d) + \frac{(D + d)}{4C}.$$

A = 3.5 cm
H = 4.0 cm
B = 6.0 cm

Problem 9

T
d
D

Problem 10

x
y
Dimensioned surface
T

Problem 11

H
R
L

Problem 12

C
D
A

Problem 13

C
D
d

Problem 14

Find the length of belt needed if

$C = 36$ in. between pulley centers
$D = 24$-in. follower
$d = 4$-in. driver

Round to the nearest 10 in.

15. **Electrical Technology** An electrician uses a bridge circuit to locate a ground in an underground cable several miles long. The formula

$$\frac{R_1}{L - x} = \frac{R_2}{x} \qquad \text{or} \qquad x = \frac{R_2 L}{R_1 + R_2}$$

is used to find x, the distance to the ground. Find x if $R_1 = 750$ ohms, $R_2 = 250$ ohms, and $L = 4000$ ft.

16. **Machine Technology** The cutting speed of a lathe is the rate, in feet per minute, that the revolving workpiece travels past the cutting edge of the tool. Machinists use the following formula to calculate cutting speed:

$$\text{Cutting speed, } C = \frac{\pi D N}{12}$$

where $\pi \approx 3.1416$, D is the diameter in inches of the work, and N is the turning rate in rpm. Find the cutting speed if a steel shaft 3.25 in. in diameter is turned at 210 rpm. (Round to the nearest whole number.)

17. **Electronics** Find the power load P in kilowatts of an electrical circuit that takes a current I of 12 amperes at a voltage V of 220 volts if

$$P = \frac{VI}{1000}$$

18. **Aeronautical Mechanics** A jet engine developing T pounds of thrust and driving an airplane at V mph has a thrust horsepower, H, given approximately by the formula

$$H = \frac{TV}{375} \qquad \text{or} \qquad V = \frac{375H}{T}$$

Find the airspeed V if $H = 16,000$ hp and $T = 10,000$ lb.

19. **Electronics** The resistance R of a conductor is given by the formula

$$R = \frac{PL}{A} \qquad \text{or} \qquad L = \frac{AR}{P}$$

where P = coefficient of resistivity
L = length of conductor
A = cross-sectional area of the conductor

Find the length in cm of No. 16 Nichrome wire needed to obtain a resistance of 8 ohms. For this wire $P = 0.000113$ ohm-cm and $A = 0.013$ cm^2. (Round to the nearest centimeter.)

20. **Auto Mechanics** The pressure P and total force F exerted on a piston of diameter D are approximately related by the equation

$$P = \frac{4F}{\pi D^2} \qquad \text{or} \qquad F = \frac{\pi P D^2}{4}$$

Find the total force on a piston of diameter 3.25 in. if the pressure exerted on it is 150 lb/sq in. (Use $\pi \approx 3.14$ and round to the nearest 50 lb.)

21. **Printing** The formula for the number of type lines L set solid in a form is

$$L = \frac{12D}{T}$$

where D is the depth in picas and T is the point size of the type. How many lines of 10-point type can be set in a space 30 picas deep?

22. **Plumbing** To determine the number N of smaller pipes that provide the same total flow as a larger pipe, use the formula

$$N = \frac{D^2}{d^2}$$

where D is the diameter of the larger pipe and d is the diameter of each smaller pipe. How many $1\frac{1}{2}$-in. pipes will it take to produce the same flow as a $2\frac{1}{2}$-in. pipe?

23. **Auto Mechanics** Find the volume of a spherical ball bearing of radius $R = 1.0076$ in.:

$$\text{Volume } V = \frac{4\pi R^3}{3} \quad \text{Use 3.1416 for } \pi.$$

(Round to 0.0001 cu in.)

24. **Machine Technology** How many minutes will it take a lathe to make 17 cuts each 24.5 in. in length on a steel shaft if the tool feed F is 0.065 in. per revolution and the shaft turns at 163 rpm? Use the formula

$$T = \frac{LN}{FR}$$

where T is the cutting time (min), N is the number of cuts, L the length of cut (in.), F the tool feed rate (in./rev), and R the rpm rate of the workpiece. Round to 0.1 minute.

25. **Machine Technology** Find the area of each of the following circular holes using the formula

$$A = \frac{\pi D^2}{4}$$

(a) $D = 1.04$ in. (b) $D = \frac{7}{8}$ in. (c) $D = 4.13$ in. (d) $D = 2.06$ cm

Use 3.14 for π and round to three significant digits.

26. Suppose that a steel band was placed tightly around the earth at the equator. If the temperature of the steel is raised 1°F, the metal will expand 0.000006 in. each inch. How much space would there be between the earth and the steel band if the temperature was raised 1°F? Use the formula

$$D \text{ (in ft)} = \frac{(0.000006)(\text{diameter of the earth})(5280)}{\pi}$$

where diameter of the earth = 7917 miles. Use $\pi \approx 3.1415927$.

(Round to two decimal places.)

27. **Plumbing** When a cylindrical container is lying on its side, the following formula can be used for calculating the volume of liquid in the container:

$$V = \frac{4}{3}h^2L\sqrt{\frac{d}{h} - 0.608}$$

Use this formula to calculate the volume of water in such a tank 6.0 ft long (L), 2.0 ft in diameter (d), and filled to a height (h) of 0.75 ft. Round to two significant digits.

28. **Meteorology** The following formula is used to calculate the wind chill factor W in degrees Celsius:

$$W = 33 - \frac{(10.45 + 10\sqrt{V} - V)(33 - T)}{22.04}$$

where T is the air temperature in degrees Celsius, and V is the wind speed in meters per second. Determine the wind chill for the following situations:

(a) Air temperature 7°C and wind speed 20 meters per second.

(b) Air temperature 28°F and wind speed 22 mph. (*Hint:* Use the conversion factors from Section 5-3 and the temperature formulas given earlier in this chapter to convert to the required units. Convert your final answer to degrees Fahrenheit.)

$C_1 = 4\ \mu F$

$C_2 = 14\ \mu F$

$C_T = \dfrac{C_1\,C_2}{C_1 + C_2}$

Problem 29

29. **Electronics** Calculate the total capacitance C_T in the circuit shown. Round to the nearest 0.1 μF (microfarad).

When you have completed these exercises, check your answers in the Appendix, then turn to Section 7-2 to learn how to add and subtract algebraic expressions.

7-2 ADDING AND SUBTRACTING ALGEBRAIC EXPRESSIONS

In the preceding section we learned how to find the value of an algebraic expression after substituting numbers for the letters. There are other useful ways of using algebra where we must manipulate algebraic expressions *without* first substituting numbers for letters. In this section we learn how to simplify algebraic expressions by adding and subtracting terms.

Combining Like Terms Two algebraic terms are said to be *like terms* if they contain exactly the same literal part. For example, the terms

2x and 5x are like terms. The literal part, x, is the same for both terms.

| Numerical part | Literal part | | Numerical part | Literal part |

The number multiplying the letters is called the *numerical coefficient* of the term.

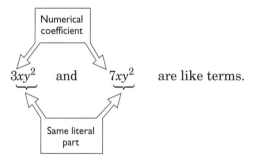

$3xy^2$ and $7xy^2$ are like terms.

Numerical coefficient

Same literal part

But

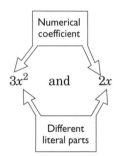

$3x^2$ and $2x$ are unlike terms. Their literal parts, x and x^2, are different.

Numerical coefficient

Different literal parts

You can add and subtract like terms but not unlike terms. To add or subtract like terms, add or subtract their numerical coefficients and keep the same literal part.

(a)

We are adding like quantities

(b)

$$8 - 3 = 5$$

$$8a^2x - 3a^2x = 5a^2x$$

Same literal part

Try these problems for practice.

(a) $12d^2 + 7d^2$ = _____   (b) $2ax - ax - 5ax =$ _____

(c) $3(y + 1) + 9(y + 1) =$ _____   (d) $8x^2 + 2xy - 2x^2 =$ _____

(e) $x - 6x + 2x$ = _____   (f) $4xy - xy + 3xy =$ _____

(a) $12d^2 + 7d^2 = 19d^2$

(b) $2ax - ax - 5ax = -4ax$ The term ax is equal to $1 \cdot ax$.

(c) $3(y + 1) + 9(y + 1) = 12(y + 1)$

(d) $8x^2 + 2xy - 2x^2 = 6x^2 + 2xy$

We cannot combine the x^2-term and the xy-term because the literal parts are not the same. They are unlike terms.

(e) $x - 6x + 2x = -3x$

(f) $4xy - xy + 3xy = 6xy$

In general, to simplify a series of terms being added or subtracted, first group together like terms, then add or subtract.

EXAMPLE

$3x + 4y - x + 2y + 2x - 8y$

Be careful not to lose the negative sign on 8y.

becomes

$(3x - x + 2x) + (4y + 2y - 8y)$ after grouping like terms,

Be careful not to lose the negative sign on x.

$(3x - x + 2x) + (4y + 2y - 8y) = 4x + (-2y)$ or $4x - 2y$

Adding a negative | subtracting a positive.

is the same as

It is simpler to write the final answer as a subtraction rather than as an addition.

YOUR TURN

Simplify the following expressions by adding and subtracting like terms.

(a) $5x + 4xy - 2x - 3xy$

(b) $3ab^2 + a^2b - ab^2 + 3a^2b - a^2b$

(c) $x + 2y - 3z - 2x - y + 5z - x + 2y - z$

(d) $17pq - 9ps - 6pq + ps - 6ps - pq$

(e) $4x^2 - x^2 + 2x + 2x^2 + x$

SOLUTIONS

(a) $(5x - 2x) + (4xy - 3xy) = 3x + xy$

(b) $(3ab^2 - ab^2) + (a^2b + 3a^2b - a^2b) = 2ab^2 + 3a^2b$

(c) $(x - 2x - x) + (2y - y + 2y) + (-3z + 5z - z) = -2x + 3y + z$

(d) $(17pq - 6pq - pq) + (-9ps + ps - 6ps) = 10pq - 14ps$

(e) $(4x^2 - x^2 + 2x^2) + (2x + x) = 5x^2 + 3x$

Expressions with Parentheses

Parentheses are used in algebra to group together terms that are to be treated as a unit. Adding and subtracting expressions usually involves working with parentheses.

EXAMPLE

To add

$(a + b) + (a + d)$

First, remove all parentheses: $(a + b) + (a + d)$

$$= a + b + a + d$$

Second, add like terms: $= \underline{a + a} + b + d$

$$= 2a + b + d$$

Removing parentheses can be a tricky business. Remember these two rules:

Rule 1 If the parenthesis has a plus sign in front, simply remove the parentheses.

EXAMPLE

(a) $1 + (3x + y) = 1 + 3x + y$

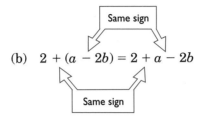

(b) $2 + (a - 2b) = 2 + a - 2b$

Rule 2 If the parenthesis has a negative sign in front, change the sign of each term inside, then remove the parentheses.

EXAMPLE

(a) $2 - (x + 2y) = 2 - x - 2y$

(b) $a - (2x - y) = a - 2x + y$

(c) $5 - (-2a + b) = 5 + 2a - b$

Note ▶ In using Rule 2 you are simply rewriting subtraction as addition of the opposite. However, you must add the opposite of *all* terms inside parentheses. ◀

Simplify the following expressions by using these two rules to remove parentheses.

(a) $x + (2y - a^2)$

(b) $4 - (x^2 - y^2)$

(c) $-(x + 1) + (y + a)$

(d) $ab - (a - b)$

(e) $(x + y) - (p - q)$

(f) $-(-x - 2y) - (a + 2b)$

(g) $3 + (-2p - q^2)$

(h) $-(x - y) + (-3x^2 + y^2)$

ANSWERS

(a) $x + 2y - a^2$

(b) $4 - x^2 + y^2$

(c) $-x - 1 + y + a$

(d) $ab - a + b$

(e) $x + y - p + q$

(f) $x + 2y - a - 2b$

(g) $3 - 2p - q^2$

(h) $-x + y - 3x^2 + y^2$

A third rule is needed when a multiplier is in front of the parentheses.

Rule 3   If the parenthesis has a multiplier in front, multiply each term inside the parentheses by the multiplier.

EXAMPLE

(a) $+2(a + b) = +2a + 2b$

Think of this as $(+2)(a + b) = (+2)a + (+2)b$
$$= +2a + 2b$$

Each term inside the parentheses is multiplied by $+2$.

(b) $-2(x + y) = -2x - 2y$

Think of this as $(-2)(x + y) = (-2)x + (-2)y$

Each term inside the parentheses is multiplied by -2.

(c) $-(x - y) = (-1)(x - y)$
$$= (-1)(x) + (-1)(-y) = -x + y$$

(d) $2 - 5(3a - 2b) = 2 + (-5)(3a - 2b) = 2 + (-5)(3a) + (-5)(-2b)$
$$= 2 + (-15a) + (+10b) \quad \text{or} \quad 2 - 15a + 10b$$

Careful ▶   In the last example it would be incorrect first to subtract the 5 from the 2. The order of operations requires that we multiply before we subtract. ◀

When you multiply negative numbers, you may need to review the arithmetic of negative numbers starting on page 299.

Notice that we must multiply *every* term inside the parentheses by the number outside the parentheses. Once the parentheses have been removed, you can add and subtract like terms, as explained previously.

Simplify the following expressions by removing parentheses. Use the three rules.

(a) $2(2x - 3y)$ (b) $1 - 4(x + 2y)$ (c) $a - 2(b - 2x)$

(d) $x^2 - 3(x - y)$ (e) $p - 2(-y - 2x)$ (f) $3(x - y) - 2(2x^2 + 3y^2)$

ANSWERS

(a) $4x - 6y$ (b) $1 - 4x - 8y$ (c) $a - 2b + 4x$

(d) $x^2 - 3x + 3y$ (e) $p + 2y + 4x$ (f) $3x - 3y - 4x^2 - 6y^2$

Once you can simplify expressions by removing parentheses, it is easy to add and subtract like terms.

EXAMPLE

$(3x - y) - 2(x - 2y) = 3x - y - 2x + 4y$ Simplify by removing parentheses.

$\qquad\qquad\qquad = \underbrace{3x - 2x}\ \underbrace{-\,y + 4y}$ Group like terms.

$\qquad\qquad\qquad = x + 3y$ Combine like terms.

PROBLEMS

Try these problems for practice.

(a) $(3y + 2) + 2(y + 1)$ (b) $(2x + 1) + 3(4 - x)$

(c) $(a + b) - (a - b)$ (d) $2(a + b) - 2(a - b)$

(e) $2(x - y) - 3(y - x)$ (f) $2(x^3 + 1) - 3(x^3 - 2)$

(g) $(x^2 - 2x) - 2(x - 2x^2)$ (h) $-2(3x - 5) - 4(x - 1)$

ANSWERS

(a) $5y + 4$ (b) $-x + 13$ (c) $2b$ (d) $4b$

(e) $5x - 5y$ (f) $-x^3 + 8$ (g) $5x^2 - 4x$ (h) $-10x + 14$

Some step-by-step solutions:

(a) $(3y + 2) + 2(y + 1) = 3y + 2 + 2y + 2$
$\qquad\qquad\qquad\quad = 3y + 2y + 2 + 2$
$\qquad\qquad\qquad\quad = 5y + 4$

(b) $(2x + 1) + 3(4 - x) = 2x + 1 + 12 - 3x$
$\qquad\qquad\qquad\quad = 2x - 3x + 1 + 12$
$\qquad\qquad\qquad\quad = -x + 13$

(c) $(a + b) - (a - b) = a + b - a - (-b)$
$\qquad\qquad\qquad\quad = a + b - a + b$
$\qquad\qquad\qquad\quad = a - a + b + b = 0 + 2b = 2b$

(h) $-2(3x - 5) - 4(x - 1) = -6x - 2(-5) - 4x - 4(-1)$
$$= -6x + 10 - 4x + 4$$
$$= -6x - 4x + 10 + 4 = -10x + 14$$

Now turn to Exercises 7-2 for additional practice on addition and subtraction of algebraic expressions.

Exercises 7-2

Adding and Subtracting Algebraic Expressions

A. Simplify by adding or subtracting like terms.

1. $3y + y + 5y$
2. $4x^2y + 5x^2y$
3. $E + 2E + 3E$
4. $ax - 5ax$
5. $9B - 2B$
6. $3m - 3m$
7. $3x^2 - 5x^2$
8. $4x + 7y + 6x + 9y$
9. $6R + 2R^2 - R$
10. $1.4A + 0.05A - 0.8A^2$
11. $x - \frac{1}{2}x - \frac{1}{4}x - \frac{1}{8}x$
12. $x + 2\frac{1}{2}x - 5\frac{1}{2}x$
13. $2 + W - 4.1W - \frac{1}{2}$
14. $q - p - 1\frac{1}{2}p$
15. $2xy + 3x + 4xy$
16. $ab + 5ab - 2ab$
17. $x^2 + x^2y + 4x^2 + 3x$
18. $1.5p + 0.3pq + 3.1p$

B. Simplify by removing parentheses and, if possible, combining like terms.

1. $3x^2 + (2x - 5)$
2. $6 + (-3a + 8b)$
3. $8m + (4m^2 + 2m)$
4. $9x + (2x - 5x^2)$
5. $2 - (x + 5y)$
6. $7x - (4 + 2y)$
7. $3a - (8 - 6b)$
8. $5 - (w - 6z)$
9. $4x - (10x^2 + 7x)$
10. $12m - (6n + 4m)$
11. $15 - (3x - 8)$
12. $-12 - (5y - 9)$
13. $-(x - 2y) + (2x + 6y)$
14. $-(3 + 5m) + (11 - 4m)$
15. $-(14 + 5w) - (2w - 3z)$
16. $-(16x - 8) - (-2x + 4y)$
17. $3(3x - 4y)$
18. $4(5a + 6b)$
19. $-8(7m + 6)$
20. $-2(6x - 3)$
21. $x - 5(3 + 2x)$
22. $4y - 2(8 + 3y)$
23. $9m - 7(-2m + 6)$
24. $w - 5(4w - 3)$
25. $3 - 4(2x + 3y)$
26. $6 - 2(3a - 5b)$
27. $12 - 2(3w - 8)$
28. $2 - 11(7 + 5a)$
29. $2(3x + 4y) - 4(6x^2 - 5y^2)$
30. $-4(5a - 6b) + 7(2ab - 4b^2)$
31. $(22x - 14y) + 3(8y - 6x)$
32. $8(3w - 5z) + (9z - 6w)$
33. $6(x + y) - 3(x - y)$
34. $-5(2x - 3x^2) + 2(9x^2 + 8x)$
35. $4(x^2 - 6x + 8) - 6(x^2 + 3x - 5)$
36. $2(3a - 5b + 6ab) - 8(2a + b - 4ab)$

When you have completed these exercises, check your answers in the Appendix, then turn to Section 7-3 to learn how to solve algebraic equations.

7-3 SOLVING SIMPLE EQUATIONS

An arithmetic equation such as $3 + 2 = 5$ means that the number named on the left $(3 + 2)$ is the same as the number named on the right (5).

An algebraic equation such as $x + 3 = 7$ is a statement that the sum of some number x and 3 is equal to 7. If we choose the correct value for x, the number $x + 3$ will be equal to 7.

x is a *variable,* a symbol that stands for a number in an equation, a blank space to be filled. Many numbers might be put in the space, but only one makes the equation a true statement.

YOUR TURN

Find the missing numbers in the following arithmetic equations.

(a) $37 +$ _____ $= 58$ (b) _____ $- 15 = 29$

(c) $4 \times$ _____ $= 52$ (d) $28 \div$ _____ $= 4$

ANSWERS

(a) $37 +$ 21 $= 58$ (b) 44 $- 15 = 29$

(c) $4 \times 13 = 52$ (d) $28 \div 7 = 4$

We could have written these equations as follows, with variables instead of blanks.

$$37 + A = 58 \qquad B - 15 = 29 \qquad 4C = 52 \qquad \frac{28}{D} = 4$$

Of course, any letters would do in place of A, B, C, and D in these algebraic equations.

How did you solve these equations? You probably "eyeballed" them—mentally juggled the other information in the equation until you found a number that made the equation true. Solving algebraic equations is very similar except that we can't "eyeball" it entirely. We need certain and systematic ways of solving the equation that will produce the correct answer quickly every time.

In this section you will learn first what a solution to an algebraic equation is—how to recognize it if you stumble over it in the dark—then how to solve linear equations.

Solution   Each value of the variable that makes an equation true is called a *solution* of the equation.

EXAMPLE

(a) The solution of $x + 3 = 7$ is $x = 4$.

☑ $(4) + 3 = 7$

(b) The solution of the equation $2x - 9 = 18 - 7x$ is $x = 3$.

☑ $2(3) - 9 = 18 - 7(3)$
$6 - 9 = 18 - 21$
$-3 = -3$

For certain equations more than one value of the variable may make the equation true.

The equation $x^2 + 6 = 5x$ is true for $x = 2$,

☑ $(2)^2 + 6 = 5(2)$
$4 + 6 = 5 \cdot 2$
$10 = 10$

and it is also true for $x = 3$.

☑ $(3)^2 + 6 = 5(3)$
$9 + 6 = 5 \cdot 3$
$15 = 15$

PROBLEMS

Determine whether each given value of the variable is a solution to the equation.

(a) For $4 - x = -3$ is $x = -7$ a solution?

(b) For $3x - 8 = 22$ is $x = 10$ a solution?

(c) For $-2x + 15 = 5 + 3x$ is $x = -2$ a solution?

(d) For $3(2x - 8) = 5 - (6 - 4x)$ is $x = 11.5$ a solution?

(e) For $3x^2 - 5x = -2$ is $x = \frac{2}{3}$ a solution?

ANSWERS

(a) No—the left side equals 11.

(b) Yes.   (c) No—the left side equals 19, the right side equals -1.

(d) Yes.   (e) Yes.

☑ Here is the check for (e): $3\left(\dfrac{2}{3}\right)^2 - 5\left(\dfrac{2}{3}\right) = -2$

$$3\left(\frac{4}{9}\right) - 5\left(\frac{2}{3}\right) = -2$$

$$\frac{4}{3} - \frac{10}{3} = -2$$

$$-\frac{6}{3} = -2$$

$$-2 = -2$$

Equivalent Equations Equations with the exact same solution are called *equivalent* equations. The equations $2x + 7 = 13$ and $3x = 9$ are equivalent because substituting the value 3 for x makes them both true.

We say that an equation with the variable x is *solved* if it can be put in the form

$x = \square$ where $\square$ is some number

For example, the solution to the equation

$2x - 1 = 7$ is $x = 4$

because $2(4) - 1 = 7$

or $8 - 1 = 7$

is a true statement.

Solving Equations Equations as simple as the previous one are easy to solve by guessing, but guessing is not a very dependable way to do mathematics. We need some sort of rule that will enable us to rewrite the equation to be solved ($2x - 1 = 7$, for example) as an equivalent solution equation ($x = 4$).

The general rule is to treat every equation as a balance of the two sides.

Any changes made in the equation must not disturb this balance.

Any operation performed on one side of the equation must also be performed on the other side.

Two kinds of balancing operations may be used.

1. Adding or subtracting a number on both sides of the equation does not change the balance

and

2. Multiplying or dividing both sides of the equation by a number (but not zero) does not change the balance.

> **EXAMPLE**

Let's work through an example.

Solve: $x - 4 = 2$.

Step 1 We want to change this equation to an equivalent equation with only x on the left, so we add 4 to each side of the equation.

$$x - 4 \boxed{+ 4} = 2 \boxed{+ 4}$$

Step 2 Combine terms.

$$x \underbrace{- 4 \boxed{+ 4}}_{0} = 2 \boxed{+ 4}$$

$$x = 6 \quad \text{Solution}$$

☑ $(6) - 4 = 2$
   $2 = 2$

Use these balancing operations to solve the equation

$8 + x = 14$

Solve: $8 + x = 14$

Step 1   We want to change this equation to an equivalent equation with only x on the left, so we subtract 8 from each side of the equation.

$8 + x \boxed{-8} = 14 \boxed{-8}$

Step 2   Combine terms.

$x + 8 \underbrace{\boxed{-8}}_{0} = 14 \boxed{-8}$ where $8 + x = x + 8$

$x = 6$ Solution

   $8 + (6) = 14$
$14 = 14$

Solve these in the same way.

(a) $x - 7 = 10$ (b) $12 + x = 27$ (c) $x + 6 = 2$

(d) $8.4 = 3.1 + x$ (e) $6.7 + x = 0$ (f) $\frac{1}{4} = x - \frac{1}{2}$

(g) $-11 = x + 5$ (h) $x - 5.2 = -3.7$

(a) **Solve:** $x - 7 = 10$

$x - 7 \boxed{+7} = 10 \boxed{+7}$ Add 7 to each side.

$x \underbrace{- 7 \boxed{+7}}_{0} = 10 \boxed{+7}$ Combine terms.

$x = 17$ Solution

$(17) - 7 = 10$
$10 = 10$

(b) **Solve:** $12 + x = 27$

$12 + x \boxed{-12} = 27 \boxed{-12}$ Subtract 12 from each side.

$x \underbrace{+ 12 \boxed{-12}}_{0} = 27 \boxed{-12}$ Combine terms.
(Note that $12 + x = x + 12$.)

$x = 15$ Solution

$12 + (15) = 27$
$27 = 27$

(c) **Solve:** $x + 6 = 2$

$x + 6 \;\boxed{-\,6} = 2 \;\boxed{-\,6}$ Subtract 6 from each side.

$x + \underbrace{\boxed{6 \;-\; 6}}_{0} = 2 \;\boxed{-\,6}$ Combine terms.

$x = -4$ Solution

The solution is a negative number. Remember, any number, positive or negative, may be the solution of an equation.

$\checkmark$ $(-4) + 6 = 2$
$2 = 2$

(d) **Solve:** $8.4 = 3.1 + x$

The variable x is on the right side of this equation.

$8.4 \;\boxed{-\,3.1} = 3.1 + x \;\boxed{-\,3.1}$ Subtract 3.1 from each side.

$8.4 - 3.1 = x$ Combine terms.

$5.3 = x$ Decimal numbers often appear in practical problems.

or $x = 5.3$ Solution $5.3 = x$ is the same as $x = 5.3$.

$\checkmark$ $8.4 = 3.1 + (5.3)$
$8.4 = 8.4$

(e) **Solve:** $6.7 + x = 0$

$6.7 + x \;\boxed{-\,6.7} = 0 \;\boxed{-\,6.7}$ Subtract 6.7 from each side.

$x = -6.7$ Solution A negative-number answer is reasonable.

$\checkmark$ $6.7 + (-6.7) = 0$
$6.7 - 6.7 = 0$
$0 = 0$

(f) **Solve:** $\dfrac{1}{4} = x - \dfrac{1}{2}$

$\underbrace{\dfrac{1}{4} + \boxed{\dfrac{1}{2}}} = x - \underbrace{\dfrac{1}{2} + \boxed{\dfrac{1}{2}}}_{0}$ Add $\dfrac{1}{2}$ to each side.

$\dfrac{1}{4} + \dfrac{1}{2} = \dfrac{3}{4}$

$\dfrac{3}{4} = x$ or $x = \dfrac{3}{4}$ Solution

$\checkmark$ $\dfrac{1}{4} = \left(\dfrac{3}{4}\right) - \dfrac{1}{2}$

$\dfrac{1}{4} = \dfrac{1}{4}$

(g) **Solve:** $-11 = x + 5$

$-11 \;\boxed{-\,5} = x + 5 \;\boxed{-\,5}$ Subtract 5 from each side.

$-16 = x$ or $x = -16$ Solution

$\checkmark$ $-11 = (-16) + 5$
$-11 = -11$

(h) **Solve:** $x - 5.2 = -3.7$

$$x - 5.2 \boxed{+ 5.2} = -3.7 \boxed{+ 5.2} \qquad \text{Add 5.2 to each side.}$$

$$x = 1.5 \quad \text{Solution}$$

☑ $(1.5) - 5.2 = -3.7$
$ -3.7 = -3.7$

Learning Hint ▶ Notice that to solve these equations, we always do the opposite of whatever operation is being performed on the variable. If a number is being added to x in the equation, we must subtract that number in order to solve. If a number is being subtracted from x in the equation, we must add that number in order to solve. Finally, whatever we do to one side of the equation, we must also do to the other side of the equation. ◀

In the equations solved so far, the variable has appeared as simply x. In some equations the variable may be multiplied or divided by a number, so that it appears as $2x$ or $x/3$, for example. To solve such an equation, do the operation opposite of the one being performed on the variable. If the variable appears as $2x$, divide by 2; if the variable appears as $x/3$, multiply by 3, and so on.

EXAMPLE

To solve the equation $27 = -3x$ follow these steps:

$27 = -3x$ Notice that the x-term is on the right side of the equation. It is x multiplied by -3. To change $-3x$ into $1x$ or x, we must perform the opposite operation, that is, divide by -3.

Step 1 $\dfrac{27}{-3} = \dfrac{-3x}{-3}$ Divide each side by -3.

Step 2 $-9 = \left(\dfrac{-3}{-3}\right)x$

$$-9 = x \qquad \text{or} \qquad x = -9 \qquad \text{Solution}$$

☑ $27 = -3(-9)$
$27 = 27$

EXAMPLE

Here is a slightly different problem to solve.

$\dfrac{x}{3} = 13$ In this equation the variable x is divided by 3.

Step 1 $\dfrac{x}{3} \cdot \boxed{3} = 13 \cdot \boxed{3}$ We want to change this equation to an equivalent equation with x alone on the left, so performing the opposite operation, we multiply both sides by 3.

$$\dfrac{3x}{3} = \left(\dfrac{3}{3}\right)x = 1x = x$$

Step 2 $x = 13 \cdot 3$
$x = 39$ Solution

☑ $\dfrac{(39)}{3} = 13$
$13 = 13$

To solve $-\dfrac{x}{4} = 5$

Step 1 $\left(-\dfrac{x}{4}\right)\,(-4) = 5\,(-4)$ Multiply both sides by -4.

Step 2 $\dfrac{(-x)(-4)}{4} = -20$ $-\dfrac{x}{4} = \dfrac{-x}{4}$

$\qquad\qquad \dfrac{4x}{4} = -20$ $(-x)(-4) = +4x$

$\qquad\qquad\quad x = -20$ $\dfrac{4x}{4} = \left(\dfrac{4}{4}\right)x = 1x = x$

☑ $-\dfrac{(-20)}{4} = 5$

$\quad -(-5) = 5$

$\qquad\quad 5 = 5$

Note ▶ When a negative sign precedes a fraction, we may move it either to the numerator or to the denominator, but not to both. In the previous problem, we placed the negative sign in the numerator in step 2. If we had moved it to the denominator instead, we would have the following:

$\left(\dfrac{x}{-4}\right)\,(-4) = -20$ or $\dfrac{-4x}{-4} = -20$ or $x = -20$

which is exactly the same solution. ◀

Try the following problems for more practice with one-step multiplication and division equations.

(a) $-5x = -25$ (b) $\dfrac{x}{6} = 6$ (c) $-16 = 2x$

(d) $7x = 35$ (e) $-\dfrac{x}{2} = 14$ (f) $\dfrac{2x}{3} = 6$

(g) $-4x = 22$ (h) $\dfrac{x+3}{2} = 4$ (i) $2.4x = 0.972$

(a) **Solve:** $-5x = -25$ The variable x appears multiplied by -5.

$\qquad$ **Step 1** $\dfrac{-5x}{-5} = \dfrac{-25}{-5}$ Divide both sides by -5.

$\qquad$ **Step 2** $\quad x = 5$ $\dfrac{-5x}{-5} = \left(\dfrac{-5}{-5}\right)x = 1x = x$

☑ $\quad -5(5) = -25$

$\qquad -25 = -25$

(b) **Solve:** $\dfrac{x}{6} = 6$ The variable x appears divided by 6.

Step 1 $\left(\dfrac{x}{6}\right)(6) = 6(6)$ Multiply both sides by 6.

Step 2 $x = 36$ Solution $\left(\dfrac{x}{6}\right)(6) = x\left(\dfrac{6}{6}\right) = x(1) = x$

☑ $\dfrac{(36)}{6} = 6$

   $6 = 6$

(c) **Solve:** $-16 = 2x$

Step 1 $\dfrac{-16}{2} = \dfrac{2x}{2}$ Divide each side by 2.

Step 2 $-8 = x$ $\dfrac{2x}{2} = \left(\dfrac{2}{2}\right)x = 1x = x$

   or $x = -8$ Solution

☑ $-16 = 2(-8)$
   $-16 = -16$

(d) **Solve:** $7x = 35$

Step 1 $\dfrac{7x}{7} = \dfrac{35}{7}$ Divide both sides by 7.

Step 2 $x = \dfrac{35}{7}$ $\dfrac{7x}{7} = \left(\dfrac{7}{7}\right)x = x$

   $x = 5$ Solution

☑ $7 \cdot (5) = 35$
   $35 = 35$

(e) **Solve:** $-\dfrac{x}{2} = 14$

Step 1 $\left(-\dfrac{x}{2}\right)\left(-2\right) = (14)\left(-2\right)$ Multiply both sides by -2.

Step 2 $x\left(-\dfrac{1}{2}\right)(-2) = 14(-2)$ $\left(-\dfrac{x}{2}\right) = -\dfrac{1}{2}x = x\left(-\dfrac{1}{2}\right)$

   $\underbrace{x\left(-\dfrac{1}{2}\right)(-2)}_{1} = -28$ $\left(-\dfrac{1}{2}\right)(-2) = \dfrac{2}{2} = 1$

   $x = -28$ Solution

☑ $-\dfrac{(-28)}{2} = 14$

   $\dfrac{28}{2} = 14$

   $14 = 14$

(f) **Solve:** $\dfrac{2x}{3} = 6$

Step 1 $\left(\dfrac{2x}{3}\right)3 = (6)\,3$ Multiply both sides by 3.

   $2x = 6 \cdot 3$ $\left(\dfrac{2x}{3}\right)3 = \dfrac{2 \cdot x \cdot 3}{3} = 2 \cdot x$

   $2x = 18$

Step 2 $\dfrac{2x}{2} = \dfrac{18}{2}$ $\qquad\qquad$ Divide both sides by 2. $\qquad \dfrac{2x}{2} = x$

$x = 9$ Solution

☑ $\dfrac{2 \cdot (9)}{3} = 6$

$\dfrac{18}{3} = 6$

$6 = 6$

(g) **Solve:** $-4x = 22$

$\dfrac{-4x}{-4} = \dfrac{22}{-4}$ $\qquad\qquad$ Divide both sides by -4.

$x = -5.5$ Solution

☑ $-4(-5.5) = 22$
$22 = 22$

(h) **Solve:** $\dfrac{x + 3}{2} = 4$ $\qquad\qquad$ Multiply by 2.

$x + 3 = 8$ $\qquad\qquad \left(\dfrac{x + 3}{2}\right) 2 = \dfrac{(x + 3) \cdot 2}{2} = x + 3$

$x + 3 \boxed{- 3} = 8 \boxed{- 3}$ $\qquad\qquad$ Subtract 3.

$x = 5$ Solution

☑ $\dfrac{(5) + 3}{2} = 4$

$\dfrac{8}{2} = 4$

(i) **Solve:** $2.4x = 0.972$ Divide by 2.4.
$x = 0.405$ Solution

 .972 ÷ **2.4** = → $\boxed{ 0.405 }$

PROBLEMS

We have now covered the four basic single-operation equations. Before moving on, try the following set of problems covering these four operations. For each problem, think carefully whether you must add, subtract, multiply, or divide in order to make the equation read $x = $ _____ or _____ $= x$.

(a) $14 = x - 8$ $\qquad$ (b) $\dfrac{y}{6} = 3$ $\qquad$ (c) $n + 11 = 4$

(d) $3x = 33$ $\qquad$ (e) $7 = -\dfrac{a}{5}$ $\qquad$ (f) $z - 2 = -10$

(g) $-8 = -16x$ $\qquad$ (h) $26 = y + 5$ $\qquad$ (i) $\dfrac{5x}{4} = -30$

SOLUTIONS

Each solution states the operation you should have performed on both sides and then gives the final answer.

(a) Add 8. $x = 22$. $\qquad\qquad\qquad$ (b) Multiply by 6. $y = 18$.

(c) Subtract 11. $n = -7$.   (d) Divide by 3. $x = 11$.

(e) Multiply by -5. $a = -35$.   (f) Add 2. $z = -8$.

(g) Divide by -16. $x = \dfrac{1}{2}$.   (h) Subtract 5. $y = 21$.

(i) Multiply by $\dfrac{4}{5}$ (or multiply by 4, divide by 5). $x = -24$.

Now turn to Exercises 7-3 for practice on solving simple equations.

Exercises 7-3

Solving Simple Equations

A. Solve the following equations.

1. $x + 4 = 13$ 2. $23 = A + 6$

3. $17 - x = 41$ 4. $z - 18 = 29$

5. $6 = a - 2\dfrac{1}{2}$ 6. $73 + x = 11$

7. $y - 16.01 = 8.65$ 8. $11.6 - R = 3.7$

9. $-39 = 3x$ 10. $-9y = 117$

11. $13a = 0.078$ 12. $\dfrac{x}{3} = 7$

13. $\dfrac{z}{1.3} = 0.5$ 14. $\dfrac{N}{2} = \dfrac{3}{8}$

15. $m + 18 = 6$ 16. $-34 = x - 7$

17. $-6y = -39$ 18. $\dfrac{a}{-4} = 9$

19. $22 - T = 40$ 20. $79.2 = 2.2y$

21. $-5.9 = -6.6 + Q$ 22. $\dfrac{5}{4} = \dfrac{Z}{8}$

23. $12 + x = 37$ 24. $66 - y = 42$

25. $\dfrac{K}{0.5} = 8.48$ 26. $-12z = 3.6$

B. Practical Problems

1. **Electrician**   Ohm's law is often written in the form

$$I = \dfrac{E}{R}$$

where I is the current in amperes, E is the voltage, and R is the resistance in ohms. What is the voltage necessary to push a 0.80-ampere current through a resistance of 450 ohms?

2. **Airline Pilot**   The formula

$$D = RT$$

is used to calculate the distance D traveled by an object moving at a constant average speed R during an elapsed time T. How long would it take a pilot to fly 1240 miles at an average speed of 220 mi/hr?

3. **Wastewater Technology** A wastewater treatment operator uses the formula

$$A = 8.34FC$$

to determine the amount A of chlorine in pounds to add to a basin. The flow F through the basin is in millions of gallons per day, and the desired concentration C of chlorine is in parts per million.

If 1800 pounds of chlorine was added to a basin with a flow rate of 7.5 million gallons per day, what would be the resulting concentration?

4. **Sheet Metal Technology** The water pressure P in pounds per square foot is related to the depth of water D in feet by the formula

$$P = 62.4D$$

where 62.4 is the weight of 1 cu ft of water. If the material forming the bottom of a tank is made to withstand 800 pounds per square foot of pressure, what is the maximum safe height for the tank?

5. **Music Technology** The frequency f (in waves per second) and the wavelength w (in meters) of a sound traveling in air are related by the equation

$$fw = 343$$

where 343 m/s is the speed of sound in air. Find the wavelength of a musical note with a frequency of 200 waves per second.

6. At a constant temperature, the pressure P (in psi) and the volume V (in cubic feet) of a particular gas are related by the equation

$$PV = 1080$$

If the volume is 60 cu ft, what is the pressure?

7. **Electrical Technology** For a particular transformer, the voltage E in the circuits is related to the number of windings W of wire around the core by the equation

$$E = 40W$$

How many windings will produce a voltage of 840 V?

8. **Metalworking** The surface speed S in fpm (feet per minute) of a rotating cylindrical object is

$$S = \frac{\pi dn}{12}$$

where d is the diameter of the object in inches and n is the speed of rotation in rpm (revolutions per minute). If an 8-in. grinder must have a surface speed of 6000 fpm, what should the speed of rotation be? Use $\pi = 3.14$, and round to the nearest hundred rpm.

When you have completed these exercises check your answers in the Appendix, then continue in Section 7-4.

7-4 SOLVING EQUATIONS INVOLVING TWO OPERATIONS

Solving many simple algebraic equations involves both kinds of operations: addition/subtraction and multiplication/division.

Solve: $2x + 6 = 14$

Step 1 We want to change this equation to an equivalent equation with only x or terms that include x on the left, so subtract 6 from both sides.

$$2x + 6 \boxed{- \; 6} = 14 \boxed{- \; 6}$$

$$2x + \underbrace{6 \boxed{- \; 6}}_{0} = 14 \boxed{- \; 6}$$ Combine terms. (Be careful. Some careless students will add $2x$ and 6 to get $8x$! You may add only *like* terms.)

$$2x = 8$$ Now change this to an equivalent equation with only x on the left.

Step 2 $$\dfrac{2x}{\boxed{2}} = \dfrac{8}{\boxed{2}}$$ Divide both sides of the equation by 2.

$$x = \dfrac{8}{2} \qquad \dfrac{2x}{2} = x$$

$$x = 4 \quad \text{Solution}$$

☑ $2 \cdot (4) + 6 = 14$
$$8 + 6 = 14$$
$$14 = 14$$

Try this one to test your understanding of the process.

Solve: $3x - 7 = 11$

Solve: $3x - 7 = 11$

Step 1 $3x - 7 \boxed{+ \; 7} = 11 \boxed{+ \; 7}$ Add 7 to each side.

$$3x - \underbrace{7 \boxed{+ \; 7}}_{0} = 11 \boxed{+ \; 7}$$

$$3x = 18$$

Step 2 $$\dfrac{3x}{\boxed{3}} = \dfrac{18}{\boxed{3}}$$ Divide both sides of the equation by 3.

$$x = 6 \quad \text{Solution}$$

☑ $3 \cdot (6) - 7 = 11$
$$18 - 7 = 11$$
$$11 = 11$$

Here is another two-operation equation.

Solve: $23 = 9 - \dfrac{y}{3}$

Solve: $23 = 9 - \dfrac{y}{3}$

Step 1 Notice that the variable y is on the right side of the equation. We must first eliminate the 9 by subtracting 9 from both sides. Be sure to keep the negative sign in front of the $\dfrac{y}{3}$ term.

$$23 - 9 = 9 - \dfrac{y}{3} - 9$$

$$\boxed{9 - 9 = 0}$$

$$14 = -\dfrac{y}{3}$$

Step 2 $(14)\,(-3) = \left(-\dfrac{y}{3}\right)(-3)$ Multiply both sides by -3.

$$-42 = (-3)\left(-\dfrac{1}{3}\right)y \qquad -\dfrac{y}{3} = -\dfrac{1}{3}y$$

$$-42 = 1y \qquad\qquad (-3)\left(-\dfrac{1}{3}\right) = 1$$

or $y = -42$ Solution

☑ $23 = 9 - \dfrac{(-42)}{3}$

$23 = 9 - (-14)$

$23 = 23$

Note ▶ In these two-step problems, we did the addition or subtraction in step 1 and the multiplication or division in step 2. We could have reversed this order and arrived at the correct solution, but the problem might have become more complicated to solve. Always add or subtract first, and when you multiply or divide, do so to *all* terms. ◀

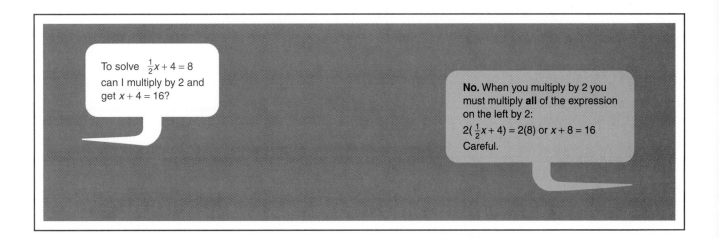

To solve $\frac{1}{2}x + 4 = 8$
can I multiply by 2 and
get $x + 4 = 16$?

No. When you multiply by 2 you must multiply **all** of the expression on the left by 2:
$2(\frac{1}{2}x + 4) = 2(8)$ or $x + 8 = 16$
Careful.

If more than one variable term appears on the same side of an equation, combine these like terms before performing any operation to both sides.

Solve: $2x + 5 + 4x = 17$

Step 1 $(2x + 4x) + 5 = 17$ Combine the x-terms on the left side.

$6x + 5 = 17$

Step 2 $6x + 5 \boxed{- 5} = 17 \boxed{- 5}$ Subtract 5 from each side.

$6x = 12$

Step 3 $\dfrac{6x}{6} = \dfrac{12}{6}$ Divide each side by 6.

$x = 2$ Solution

☑ $2(2) + 5 + 4(2) = 17$ Substitute 2 for x in each x-term.

$4 + 5 + 8 = 17$

$17 = 17$

YOUR TURN

Now you try one.

Solve: $32 = x - 12 - 5x$

SOLUTION

Step 1 $32 = (x - 5x) - 12$ Combine the x-terms on the right.

$32 = -4x - 12$

Step 2 $32 \boxed{+ 12} = -4x - 12 \boxed{+ 12}$ Add 12 to both sides.

$44 = -4x$

Step 3 $\dfrac{44}{-4} = \dfrac{-4x}{-4}$ Divide both sides by -4.

$-11 = x$ or $x = -11$ Solution

☑ $32 = (-11) - 12 - 5(-11)$

$32 = -11 - 12 + 55$

$32 = -23 + 55$

$32 = 32$

PROBLEMS

Here are a few more two-operation equations for practice.

(a) $7x + 2 = 51$ (b) $18 - 5x = 3$ (c) $15.3 = 4x - 1.5$

(d) $\dfrac{x}{5} - 4 = 6$ (e) $11 - x = 2$ (f) $5 = 7 - \dfrac{x}{4}$

(g) $2 + \dfrac{x}{2} = 3$ (h) $2x - 9.4 = 0$ (i) $2.75 = 14.25 - 0.20x$

(j) $3x + x = 18$ (k) $12 = 9x + 4 - 5x$ (l) $6 - 3x + x = 22$

(a) **Solve:** $7x + 2 = 51$ Change this equation to an equivalent equation with only an x-term on the left.

Step 1 $7x + 2 \boxed{-2} = 51 \boxed{-2}$ Subtract 2 from each side.

$7x \underbrace{+ 2 \boxed{-2}}_{0} = 51 \boxed{-2}$ Combine terms.

$7x = 49$

Step 2 $\dfrac{7x}{\boxed{7}} = \dfrac{49}{\boxed{7}}$ Divide both sides by 7.

$x = \dfrac{49}{7}$

$x = 7$ Solution

☑ $7 \cdot (7) + 2 = 51$
$49 + 2 = 51$
$51 = 51$

(b) **Solve:** $18 - 5x = 3$

Step 1 $18 - 5x \boxed{-18} = 3 \boxed{-18}$ Subtract 18 from each side.

$-5x \underbrace{+ 18 - 18}_{0} = 3 - 18$ Rearrange terms.

$-5x = -15$ Combine terms.

Step 2 $\dfrac{-5x}{-5} = \dfrac{-15}{-5}$ Divide both sides by -5.

$x = 3$ Solution

☑ $18 - 5 \cdot (3) = 3$
$18 - 15 = 3$
$3 = 3$

(c) **Solve:** $15.3 = 4x - 1.5$

Step 1 $15.3 \boxed{+ 1.5} = 4x \underbrace{- 1.5 \boxed{+ 1.5}}_{0}$ Add 1.5 to each side.

$16.8 = 4x$ Combine terms.

$4x = 16.8$

Step 2 $\dfrac{4x}{\boxed{4}} = \dfrac{16.8}{\boxed{4}}$ Divide both sides by 4.

$x = 4.2$ Solution   Decimal number solutions are common in practical problems.

 15.3 $\boxed{+}$ **1.5** $\boxed{=}$ $\boxed{\div}$ **4** $\boxed{=}$ →   *4.2*

☑ $15.3 = 4(4.2) - 1.5$
$15.3 = 16.8 - 1.5$
$15.3 = 15.3$

(d) **Solve:** $\frac{x}{5} - 4 = 6$

    Step 1 $\quad \frac{x}{5} - 4 \boxed{+4} = 6 \boxed{+4}$ Add 4 to both sides.

$$\frac{x}{5} = 10$$

    Step 2 $\quad \left(\frac{x}{5}\right) \cdot \boxed{5} = 10 \cdot \boxed{(5)}$ Multiply both sides by 5.

$$x = 50 \quad \text{Solution}$$

☑ $\quad \frac{(50)}{5} - 4 = 6$

$$10 - 4 = 6$$
$$6 = 6$$

(e) **Solve:** $11 - x = 2$

$$11 - x \boxed{-11} = 2 \boxed{-11} \qquad \text{Subtract 11 from each side.}$$

$$-x + 11 \underbrace{\boxed{-11}}_{0} = 2 \boxed{-11} \qquad \text{Combine terms.}$$

$$-x = -9 \qquad \text{Multiply each side by } -1.$$

$$x = 9 \quad \text{Solution}$$

☑ $\quad 11 - (9) = 2$

$$2 = 2$$

(f) **Solve:** $5 = 7 - \frac{x}{4}$

    Step 1 $\quad 5 \boxed{-7} = 7 - \frac{x}{4} \boxed{-7} \qquad$ Subtract 7 from both sides.

$$-2 = -\frac{x}{4} \qquad\qquad 7 - \frac{x}{4} - 7 = (7-7) - \frac{x}{4} = -\frac{x}{4}$$

    Step 2 $\quad -2 \boxed{(-4)} = \left(-\frac{x}{4}\right)\boxed{(-4)} \qquad$ Multiply both sides by -4.

$$8 = x \qquad\qquad \left(-\frac{x}{4}\right)(-4) = (-x)\left(\frac{-4}{4}\right) = (-x)(-1) = x$$

$$\text{or} \quad x = 8 \quad \text{Solution}$$

☑ $\quad 5 = 7 - \frac{(8)}{4}$

$$5 = 7 - 2$$
$$5 = 5$$

(g) **Solve:** $2 + \frac{x}{2} = 3$

$$\frac{x}{2} + 2 \boxed{-2} = 3 \boxed{-2} \qquad\qquad \text{Subtract 2 from each side.}$$

$$\frac{x}{2} = 1$$

$$\left(\frac{x}{2}\right) \cdot \boxed{2} = 1 \cdot \boxed{2} \qquad\qquad \text{Multiply each side by 2.}$$

$$x = 2 \quad \text{Solution}$$

$\checkmark$ $\quad 2 + \dfrac{(2)}{2} = 3$

$\qquad 2 + 1 = 3$

(h) **Solve:** $\;2x - 9.4 = 0$

$\qquad 2x - 9.4 \;\boxed{+\,9.4} = 0 \;\boxed{+\,9.4}$ — Add 9.4 to each side.

$\qquad\qquad 2x = 9.4$ — Divide each side by 2.

$\qquad\qquad x = 4.7 \quad$ Solution

$\checkmark$ $\quad 2(4.7) - 9.4 = 0$
$\qquad 9.4 - 9.4 = 0$

(i) **Solve:** $\;2.75 = 14.25 - 0.20x$

Step 1 $\quad 2.75 - 14.25 = 14.25 - 0.20x - 14.25 \quad$ Subtract 14.25 from both sides.

$\qquad\qquad\qquad -11.5 = -0.20x$

Step 2 $\qquad \dfrac{-11.5}{\boxed{-0.20}} = \dfrac{-0.20x}{\boxed{-0.20}} \qquad$ Divide both sides by -0.20.

$\qquad\qquad\qquad 57.5 = x \qquad$ or $\qquad x = 57.5$

2.75 $\boxminus$ **14.25** $\boxed{=}$ $\boxed{\div}$ **.2** $\boxed{+/-}$ $\boxed{=}$ $\rightarrow$ $\boxed{\;57.5\;}$

$\checkmark$ $\quad 2.75 = 14.25 - 0.20(57.5)$
$\qquad 2.75 = 14.25 - 11.5$
$\qquad 2.75 = 2.75$

(j) **Solve:** $\;3x + x = 18$

Step 1 $\qquad 4x = 18$ — Combine like terms on the left side.

Step 2 $\qquad \dfrac{4x}{\boxed{4}} = \dfrac{18}{\boxed{4}}$ — Divide each side by 4.

$\qquad\qquad x = 4.5$

$\checkmark$ $\quad 3(4.5) + (4.5) = 18$
$\qquad 13.5 + 4.5 = 18$
$\qquad\qquad 18 = 18$

(k) **Solve:** $\;12 = 9x + 4 - 5x$

Step 1 $\qquad 12 = 4x + 4$ — Combine like terms.

Step 2 $\quad 12 \;\boxed{-\,4} = 4x + 4 \;\boxed{-\,4}$ — Subtract 4 from each side.

$\qquad\qquad 8 = 4x$

Step 3 $\qquad \dfrac{8}{\boxed{4}} = \dfrac{4x}{\boxed{4}}$ — Divide each side by 4.

$\qquad\qquad 2 = x \qquad$ or $\qquad x = 2$

The check is left to you.

(l) **Solve:** $\;6 - 3x + x = 22$

Step 1 $\qquad 6 - 2x = 22$ — Combine like terms.

Careful: The $3x$-term is negative: $\qquad -3x + x = -2x$

Step 2 $\quad 6 - 2x \;\boxed{-\,6} = 22 \;\boxed{-\,6}$ — Subtract 6 from both sides.

$\qquad\qquad -2x = 16$ — (Keep the negative sign with the $2x$-term.)

Step 3 $\dfrac{-2x}{-2} = \dfrac{16}{-2}$ Divide both sides by -2.

$x = -8$ Solution

Be sure to check your solution.

Now turn to Exercises 7-4 for more practice on solving equations.

Exercises 7-4

Solving Two-Step Equations

A. Solve.

1. $2x - 3 = 17$ 2. $4x + 6 = 2$

3. $\dfrac{x}{5} = 7$ 4. $-8y + 12 = 32$

5. $4.4m - 1.2 = 9.8$ 6. $\dfrac{3}{4} = 3x + \dfrac{1}{4}$

7. $14 - 7n = -56$ 8. $38 = 58 - 4a$

9. $3z - 5z = 12$ 10. $17 = 7q - 5q - 3$

11. $23 - \dfrac{x}{4} = 11$ 12. $9m + 6 + 3m = -60$

13. $-15 = 12 - 2n + 5n$ 14. $2.6y - 19 - 1.8y = 1$

15. $\dfrac{1}{2} + 2x = 1$ 16. $3x + 16 = 46$

17. $-4a + 45 = 17$ 18. $\dfrac{x}{2} + 1 = 8$

19. $-3Z + \dfrac{1}{2} = 17$ 20. $2x + 6 = 0$

21. $1 = 3 - 5x$ 22. $23 = 17 - \dfrac{x}{4}$

23. $-5P + 18 = 3$ 24. $5x - 2x = 24$

25. $6x + 2x = 80$ 26. $x + 12 - 6x = -18$

27. $27 = 2x - 5 + 4x$ 28. $-13 = 22 - 3x + 8x$

B. Practical Problems

1. **Office Services** A repair service charges $45 for a house call and an additional $36 per hour of work. The formula

$T = 45 + 36H$

represents the total charge T for H hours of work. If the total bill for a customer was $225, how many hours of actual labor were there?

2. **Meteorology** The air temperature T (in degrees Fahrenheit) at an altitude h (in feet) above a particular area can be approximated by the formula

$T = -0.002h + G$

where G is the temperature on the ground directly below. If the ground temperature is 76°F, at what altitude will the air temperature drop to freezing, 32°F?

3. **Sports Technology**   Physical fitness experts sometimes use the following formula to approximate the maximum target heart rate R during exercise based on a person's age A:

$$R = -0.8A + 176$$

At what age should the heart rate during exercise not exceed 150 beats per minute?

4. The formula $A = p + prt$ is used to determine the total amount of money A in a bank account after an amount p is invested for t years at a rate of interest r. What rate of interest is needed for $8000 to grow to $12,000 after 5 years? (Be sure to convert your decimal answer to a percent and round to the nearest tenth of a percent.)

5. **Auto Mechanics**   The formula

$$P + 2T = C$$

gives the overall diameter C of the crankshaft gear for a known pitch diameter P of the small gear and the height T of teeth above the pitch diameter circle. If $P = 2.875$ in. and $C = 3.125$ in., find T.

6. **Sheet Metal Technology**   The allowance A for a Pittsburgh lock is given by

$$A = 2w + \frac{3}{16} \text{ in.}$$

where w is the width of the pocket. If the allowance for a Pittsburgh lock is $\frac{11}{16}$ in., what is the width of the pocket?

7. **Plumbing**   A plumber's total bill A can be calculated using the formula

$$A = RT + M$$

where R is his hourly rate, T is the total labor time in hours, and M is the cost of materials. A plumber bids a particular job at $2820. If materials amount to $980, and his hourly rate is $64 per hour, how many hours should the job take for the estimate to be accurate?

8. **Machinist**   The formula

$$L = 2d + 3.26(r + R)$$

can be used under certain conditions to approximate the length L of belt needed to connect two pulleys of radii r and R if their centers are a distance d apart. How far apart can two pulleys be if their radii are 8 in. and 6 in., and the total length of the belt connecting them is 82 in.? (Round to the nearest inch.)

When you have completed these exercises, check your answers in the Appendix, then continue in Section 7-5.

7-5 SOLVING MORE EQUATIONS AND FORMULAS

Parentheses in Equations   In Section 7-2 you learned how to deal with algebraic expressions involving parentheses. Now you will learn to solve equations containing parentheses by using these same skills.

Solve: $2(x + 4) = 27$

Step 1 Use Rule 3 on page 352. Multiply each term inside the parentheses by 2.

$$2x + 8 = 27$$

Step 2 Now solve this equation using the techniques of the previous section.

$$2x + 8 \boxed{- 8} = 27 \boxed{- 8}$$ Subtract 8 from both sides.

$$2x = 19$$

$$\frac{2x}{\boxed{2}} = \frac{19}{\boxed{2}}$$ Divide both sides by 2.

$$x = 9.5$$

☑ $2[(9.5) + 4] = 27$

      $2(13.5) = 27$

         $27 = 27$

YOUR TURN

Try this similar example.

$$-3(y - 4) = 36$$

SOLUTION

Solve: $-3(y - 4) = 36$

Step 1 $-3y + 12 = 36$ Multiply each term inside parentheses by -3.

                                         $\boxed{-3}(y - 4) = \boxed{-3}(y) + \boxed{(-3)}(-4)$

Step 2 $-3y + 12 \boxed{- 12} = 36 \boxed{- 12}$ Subtract 12 from each side.

           $-3y = 24$

Step 3 $\dfrac{-3y}{\boxed{-3}} = \dfrac{24}{\boxed{-3}}$ Divide each side by -3.

            $y = -8$

☑ $-3[(-8) - 4] = 36$

      $-3(-12) = 36$

         $36 = 36$

Note ▶ In each of the last two examples, you could have first divided both sides by the number in front of parentheses.

Here is how each solution would have looked.

                                                              Chap. 7 Basic Algebra

To solve this equation: $2x - 3 = 10$
Why can't I divide both sides by 2
to get $x - 3 = 5$ so that
$x = 8$? Isn't this ok?

No. When you divide both sides of an equation by a number, you must divide **all** of both sides. Divide **all** of the expression on the left, not just part of it. $\frac{2x-3}{2} = \frac{10}{2}$ or $x - \frac{3}{2} = 5$.

First Example	**Second Example**
$\dfrac{2(x + 4)}{2} = \dfrac{27}{2}$	$\dfrac{-3(y - 4)}{-3} = \dfrac{36}{-3}$
$x + 4 = 13.5$	$y - 4 = -12$
$x + 4 - 4 = 13.5 - 4$	$y - 4 + 4 = -12 + 4$
$x = 9.5$	$y = -8$

Some students may find this technique preferable, especially when the right side of the equation is exactly divisible by the number in front of the parentheses. ◀

Here is a more difficult equation involving parentheses.

EXAMPLE

Solve: $5x - (2x - 3) = 27$

Step 1 Use Rule 2 on page 351. Change the sign of each term inside parentheses and then remove them.

$5x - 2x + 3 = 27$ $\qquad\qquad -(2x - 3) = -2x + 3$

Step 2 $\qquad 3x + 3 = 27$ $\qquad\qquad$ Combine the like terms on the left.
$\qquad\qquad\qquad\qquad\qquad\qquad\qquad\qquad 5x - 2x = 3x$

Step 3 $\quad 3x + 3 \boxed{- 3} = 27 \boxed{- 3}$ $\qquad$ Subtract 3 from both sides.

$\qquad\qquad\qquad 3x = 24$

Step 4 $\qquad\quad \dfrac{3x}{\boxed{3}} = \dfrac{24}{\boxed{3}}$ $\qquad\qquad$ Divide both sides by 3.

$\qquad\qquad\qquad\quad x = 8$

☑ $\quad 5(8) - [2(8) - 3] = 27$

$\qquad\quad 40 - (16 - 3) = 27$

$\qquad\qquad\quad 40 - 13 = 27$

$\qquad\qquad\qquad\quad 27 = 27$

Try this problem.

Solve: $8 - 3(2 - 3x) = 34$

Step 1 Be very careful here. Some students are tempted to subtract the 3 from the 8. However, the order of operations rules on page 321 specify that multiplication must be performed before addition or subtraction. Therefore, your first step is to multiply the expression in parentheses by -3.

$$8 - 6 + 9x = 34 \qquad\qquad 8 - 3(2 - 3x) = 8 + (-3)(2) + (-3)(-3x)$$

Step 2 $\qquad 2 + 9x = 34 \qquad\qquad$ Combine the like terms on the left.

Step 3 $2 + 9x - 2 = 34 - 2 \qquad$ Subtract 2 from each side.

$$9x = 32$$

Step 4 $\qquad \dfrac{9x}{9} = \dfrac{32}{9}$

$$x = 3\tfrac{5}{9}$$

Be sure to check your answer.

Here are more equations with parentheses for you to solve.

(a) $4(x - 2) = 26$ $\qquad\qquad$ (b) $11 = -2(y + 5)$

(c) $23 = 6 - (3n + 4)$ $\qquad\quad$ (d) $4x - (6x - 9) = 41$

(e) $7 + 3(5x + 2) = 38$ $\qquad\;$ (f) $20 = 2 - 5(9 - a)$

(g) $(3m - 2) - (5m - 3) = 19 \qquad$ (h) $2(4x + 1) + 5(3x + 2) = 58$

(a) $x = 8.5$ $\qquad$ (b) $y = -10.5$ $\qquad$ (c) $n = -7$ $\qquad$ (d) $x = -16$

(e) $x = 1\tfrac{2}{3}$ $\qquad$ (f) $a = 12.6$ $\qquad$ (g) $m = -9$ $\qquad$ (h) $x = 2$

Here are worked solutions to (b), (d), (f), and (h). The checks are left to you.

(b) **Solve:** $11 = -2(y + 5)$

$\qquad$ **Step 1** $\qquad 11 = -2y - 10 \qquad\qquad$ Multiply each term inside parentheses by -2.

$\qquad$ **Step 2** $\quad 11 + 10 = -2y - 10 + 10 \qquad$ Add 10 to each side.

$\qquad\qquad\qquad\qquad 21 = -2y$

Step 3
$$\frac{21}{-2} = \frac{-2y}{-2}$$
Divide each side by -2.

$$-10.5 = y \quad \text{or} \quad y = -10.5$$

(d) **Solve:** $4x - (6x - 9) = 41$

Step 1 $4x - 6x + 9 = 41$ Change the sign of each term inside parentheses and then remove parentheses.

Step 2 $-2x + 9 = 41$ Combine like terms on the left side.

Step 3 $-2x + 9 - 9 = 41 - 9$ Subtract 9 from each side.

$$-2x = 32$$

Step 4 $\dfrac{-2x}{-2} = \dfrac{32}{-2}$ Divide each side by -2.

$$x = -16$$

(f) **Solve:** $20 = 2 - 5(9 - a)$

Step 1   Multiply both terms inside parentheses by -5 and then remove parentheses.

$$20 = 2 - 45 + 5a \qquad\qquad 2 - 5(9 - a)$$
$$= 2 + (-5)(9) + (-5)(-a)$$

Step 2 $20 = -43 + 5a$ Combine like terms on the right side.

Step 3 $20 + 43 = -43 + 5a + 43$ Add 43 to both sides.

$$63 = 5a$$

Step 4 $\dfrac{63}{5} = \dfrac{5a}{5}$ Divide both sides by 5.

$$12.6 = a \quad \text{or} \quad a = 12.6$$

(h) **Solve:** $2(4x + 1) + 5(3x + 2) = 58$

Step 1   Multiply each term inside the first parentheses by 2 and each term inside the second parentheses by 5.

$$8x + 2 + 15x + 10 = 58$$

Step 2 $23x + 12 = 58$ Combine both pairs of like terms on the left side.

Step 3 $23x + 12 - 12 = 58 - 12$ Subtract 12 from both sides.

$$23x = 46$$

Step 4 $\dfrac{23x}{23} = \dfrac{46}{23}$ Divide both sides by 23.

$$x = 2$$

Variable on Both Sides   In all the equations we have solved so far, the variable has been on only one side of the equation. Sometimes it is necessary to solve equations with variable terms on both sides.

To solve

$$3x - 4 = 8 - x$$

First, move all variable terms to the left side by adding x to both sides.

$3x - 4 + x = 8 - x + x$

$\qquad 4x - 4 = 8$

Next, proceed as before.

$4x - 4 \boxed{+ 4} = 8 \boxed{+ 4}$ $\qquad$ Add 4 to both sides.

$\qquad 4x = 12$

$\qquad \dfrac{4x}{4} = \dfrac{12}{4}$ $\qquad$ Divide both sides by 4.

$\qquad x = 3$

Finally, check your answer.

$3(3) - 4 = 8 - (3)$

$\quad 9 - 4 = 5$

$\qquad 5 = 5$

YOUR TURN

Ready to attempt one yourself? Give this one a try, then check it with our solution.

$5y - 21 = 8y$

SOLUTION

Did you move the variable terms to the left? Here you can save a step by moving them to the right instead of to the left.

Solve: $\qquad 5y - 21 = 8y$

Step 1 $\quad 5y - 21 \boxed{- 5y} = 8y \boxed{- 5y}$ $\qquad$ Subtract $5y$ from both sides.

$\qquad\qquad -21 = 3y$

Step 2 $\qquad\qquad \dfrac{-21}{3} = \dfrac{3y}{3}$ $\qquad$ Divide both sides by 3.

$\qquad\qquad -7 = y \quad$ or $\quad y = -7$

 $5(-7) - 21 = 8(-7)$

$\quad -35 - 21 = -56$

$\qquad\quad -56 = -56$

Learning Hint ▶ $\quad$ If a variable term is already by itself on one side of the equation, move all variable terms to this side. As in the last example, this will save a step. ◀

Now try these problems for practice in solving equations in which the variable appears on both sides.

(a) $x - 6 = 3x$ (b) $5(x - 2) = x + 4$

(c) $2(x - 1) = 3(x + 1)$ (d) $4x + 9 = 7x - 15$

(e) $4n + 3 = 18 - 2n$ (f) $2A = 12 - A$

(g) $6y = 4(2y + 7)$ (h) $8m - (2m - 3) = 3(m - 4)$

(i) $3 - (5x - 8) = 6x + 22$ (j) $-2(3x - 5) = 7 - 5(2x + 3)$

SOLUTIONS

(a) **Solve:** $x - 6 = 3x$

$x - 6\ \boxed{-\ x} = 3x\ \boxed{-\ x}$ Subtract x from each side, so that x-terms will appear only on the right.

$-6 = 2x$ Combine terms.

or $2x = -6$ Divide by 2.

$x = -3$ Solution

☑ $(-3) - 6 = 3(-3)$
  $-9 = -9$

(b) **Solve:** $5(x - 2) = x + 4$

$5x - 10 = x + 4$ Multiply each term inside the parentheses by 5.

$5x - 10\ \boxed{-\ x} = x + 4\ \boxed{-\ x}$ Subtract x from each side.

$4x - 10 = 4$

$4x - 10\ \boxed{+\ 10} = 4\ \boxed{+\ 10}$ Add 10 to each side.

$4x = 14$ Divide each side by 4.

$x = 3\frac{1}{2}$ Solution

☑ $5\left(3\frac{1}{2} - 2\right) = \left(3\frac{1}{2}\right) + 4$
  $5\left(1\frac{1}{2}\right) = 7\frac{1}{2}$
  $7\frac{1}{2} = 7\frac{1}{2}$

(c) **Solve:** $2(x - 1) = 3(x + 1)$ Remove parentheses by multiplying.

$2x - 2 = 3x + 3$

$2x - 2\ \boxed{-\ 3x} = 3x + 3\ \boxed{-\ 3x}$ Subtract $3x$ from each side.

$-2 - x = 3$ Combine terms.

$-2 - x\ \boxed{+\ 2} = 3\ \boxed{+\ 2}$ Add 2 to each side.

$-x = 5$

or $x = -5$ Solution   Solve for x, not $-x$.

Do you see that $-x = 5$ is the same as $x = -5$? Multiply both sides of the equation $-x = 5$ by -1 to get $x = -5$.

✓ $2(-5 - 1) = 3(-5 + 1)$
$2(-6) = 3(-4)$
$-12 = -12$

(d) **Solve:** $4x + 9 = 7x - 15$

$4x + 9 \;-\; 7x = 7x - 15 \;-\; 7x$ Subtract $7x$ from each side.

$-3x + 9 = -15$ Combine terms.

$-3x + 9 \;-\; 9 = -15 \;-\; 9$ Subtract 9 from each side.

$-3x = -24$ Divide each side by -3.

$x = 8$ Solution

✓ $4(8) + 9 = 7(8) - 15$
$32 + 9 = 56 - 15$
$41 = 41$

(e) **Solve:** $4n + 3 = 18 - 2n$

$4n + 3 \;+\; 2n = 18 - 2n \;+\; 2n$ Add $2n$ to both sides.

$6n + 3 = 18$ Now the n-terms appear only on the left side.

$6n + 3 \;-\; 3 = 18 \;-\; 3$ Subtract 3 from both sides.

$6n = 15$

$\dfrac{6n}{6} = \dfrac{15}{6}$ Divide by 6.

$n = 2.5$ Solution

✓ $4(2.5) + 3 = 18 - 2(2.5)$
$10 + 3 = 18 - 5$
$13 = 13$

(f) **Solve:** $2A = 12 - A$

$2A \;+\; A = 12 - A \;+\; A$ Add A to each side.

$3A = 12$ The A-terms now appear only on the left.

$\dfrac{3A}{3} = \dfrac{12}{3}$ Divide by 3.

$A = 4$ Solution

✓ $2(4) = 12 - (4)$
$8 = 8$

(g) **Solve:** $6y = 4(2y + 7)$

$6y = 8y + 28$ Multiply to remove parentheses.

$6y \;-\; 8y = 8y + 28 \;-\; 8y$ Subtract $8y$ from both sides.

$-2y = 28$

$\dfrac{-2y}{-2} = \dfrac{28}{-2}$ Divide by -2.

$y = -14$ Solution

Check the solution.

(h) **Solve:** $8m - (2m - 3) = 3(m - 4)$

$$8m - 2m + 3 = 3(m - 4)$$ Change signs to remove parentheses on the left.

$$8m - 2m + 3 = 3m - 12$$ Multiply by 3 to remove parentheses on the right.

$$6m + 3 = 3m - 12$$ Combine like terms.

$$6m + 3 \boxed{- 3m} = 3m - 12 \boxed{- 3m}$$ Subtract $3m$ from both sides.

$$3m + 3 = -12$$

$$3m + 3 \boxed{- 3} = -12 \boxed{- 3}$$ Subtract 3 from both sides.

$$3m = -15$$

$$\frac{3m}{\boxed{3}} = \frac{-15}{\boxed{3}}$$ Divide by 3.

$$m = -5 \quad \text{Solution}$$

Check the solution.

(i) **Solve:** $3 - (5x - 8) = 6x + 22$

$$3 - 5x + 8 = 6x + 22$$ Remove parentheses by changing the signs of both terms.

$$11 - 5x = 6x + 22$$ Combine terms.

$$11 - 5x \boxed{- 6x} = 6x + 22 \boxed{- 6x}$$ Subtract $6x$ from both sides.

$$11 - 11x = 22$$ Combine terms.

$$11 - 11x \boxed{- 11} = 22 \boxed{- 11}$$ Subtract 11 from both sides.

$$-11x = 11$$ Divide by -11.

$$x = -1 \quad \text{Solution}$$

Check it.

(j) **Solve:** $-2(3x - 5) = 7 - 5(2x + 3)$

$$-6x + 10 = 7 - 10x - 15$$ Remove parentheses. Multiply each term inside parentheses on the left by -2. Multiply each term inside parentheses on the right by -5 (leave the 7 alone!).

$$-6x + 10 = -8 - 10x$$ Combine terms.

$$-6x + 10 \boxed{+ 10x} = -8 - 10x \boxed{+ 10x}$$ Add $10x$ to both sides.

$$4x + 10 = -8$$ Combine terms.

$$4x + 10 \boxed{- 10} = -8 \boxed{- 10}$$ Subtract 10 from both sides.

$$4x = -18$$ Divide by 4.

$$x = -4.5 \quad \text{Solution}$$

Check it.

Remember:

1. Do only legal operations: add or subtract the same quantity from both sides of the equation; multiply or divide both sides of the equation by the same nonzero quantity.

2. Remove all parentheses carefully.

3. Combine like terms when they are on the same side of the equation.

4. Use legal operations to change the equation so that you have only x by itself on one side of the equation and a number on the other side of the equation.

5. Always check your answer.

Solving Formulas   To *solve a formula* for some letter means to rewrite the formula as an equivalent formula with that letter isolated on the left of the equals sign.

For example, the area of a triangle is given by the formula

$$A = \frac{BH}{2}$$ where A is the area, B is the length of the base, and H is the height.

Solving for the base B gives the equivalent formula

$$B = \frac{2A}{H}$$

Solving for the height H gives the equivalent formula

$$H = \frac{2A}{B}$$

Solving formulas is a very important practical application of algebra. Very often a formula is not written in the form that is most useful. To use it you may need to rewrite the formula, solving it for the letter whose value you need to calculate.

To solve a formula, use the same balancing operations that you used to solve equations. You may add or subtract the same quantity on both sides of the formula and you may multiply or divide both sides of the formula by the same nonzero quantity.

> **EXAMPLE**

To solve the formula

$$S = \frac{R + P}{2}$$ for R

First, multiply both sides of the equation by 2.

$$2 \cdot S = 2 \cdot \left(\frac{R + P}{2} \right)$$

$$2S = R + P$$

Second, subtract P from both sides of the equation.

$$2S \boxed{-P} = R + P \underbrace{\boxed{-P}}_{0}$$

$$2S - P = R$$

This formula can be reversed to read

$$R = 2S - P$$ We have solved the formula for R.

Solve the following formulas for the variable indicated.

(a) $V = \dfrac{3K}{T}$ for K

(b) $Q = 1 - R + T$ for R

(c) $V = \pi R^2 H - AB$ for H

(d) $P = \dfrac{T}{A - B}$ for A

SOLUTIONS

(a) $V = \dfrac{3K}{T}$

First, multiply both sides by T to get $\qquad VT = 3K$

Second, divide both sides by 3 to get $\qquad \dfrac{VT}{3} = K$

Solved for K, the formula is $\qquad K = \dfrac{VT}{3}$

(b) $Q = 1 - R + T$

First, subtract T from both sides to get $\qquad Q - T = 1 - R$

Second, subtract 1 from both sides to get $\qquad Q - T - 1 = -R$

This is equivalent to $\qquad -R = Q - T - 1$

or $\quad R = -Q + T + 1$

> We have multiplied *all terms by* -1.

or $\quad R = 1 - Q + T$

(c) $V = \pi R^2 H - AB$

First, add AB to both sides to get $\qquad V + AB = \pi R^2 H$

Second, divide both sides by πR^2 to get $\qquad \dfrac{V + AB}{\pi R^2} = H$

Notice that we divide *all* of the left side by πR^2.

Solved for H, the formula is $\qquad H = \dfrac{V + AB}{\pi R^2}$

(d) $P = \dfrac{T}{A - B}$

First, multiply each side by $A - B$ to get $\qquad P(A - B) = T$

Second, multiply to remove the parentheses. $\qquad PA - PB = T$

Next, add PB to each side. $\qquad PA = T + PB$

Finally, divide each side by P. $\qquad A = \dfrac{T + PB}{P}$

Solved for A, the formula is $\qquad A = \dfrac{T + PB}{P}$

Remember, when using the multiplication/division rule, you must multiply or divide *all* of both sides of the formula by the same quantity. ◀

PROBLEMS

Practice solving formulas with the following problems.

Solve:

(a) $P = 2A + 3B$ for A

(b) $E = MC^2$ for M

(c) $S = \dfrac{A - RT}{1 - R}$ for A

(d) $S = \dfrac{1}{2}gt^2$ for g

(e) $P = i^2R$ for R

(f) $I = \dfrac{V}{R + a}$ for R

(g) $A = \dfrac{2V - W}{R}$ for V

(h) $F = \dfrac{9C}{5} + 32$ for C

(i) $A = \dfrac{\pi R^2 S}{360}$ for S

(j) $P = \dfrac{t^2dN}{3.78}$ for d

(k) $C = \dfrac{AD}{A + 12}$ for D

(l) $V = \dfrac{\pi LT^2}{6} + 2$ for L

ANSWERS

(a) $A = \dfrac{P - 3B}{2}$

(b) $M = \dfrac{E}{C^2}$

(c) $A = S - SR + RT$

(d) $g = \dfrac{2S}{t^2}$

(e) $R = \dfrac{P}{i^2}$

(f) $R = \dfrac{V - aI}{I}$

(g) $V = \dfrac{AR + W}{2}$

(h) $C = \dfrac{5F - 160}{9}$

(i) $S = \dfrac{360A}{\pi R^2}$

(j) $d = \dfrac{3.78P}{t^2N}$

(k) $D = \dfrac{CA + 12C}{A}$

(l) $L = \dfrac{6V - 12}{\pi T^2}$

USING SQUARE ROOTS IN SOLVING EQUATIONS

The equations you learned to solve in this chapter are all *linear* equations. The variable appears only to the first power—no x^2 or x^3 terms appear in the equations. You will learn how to solve more difficult algebraic equations in Chapter 11, but equations that look like

$x^2 = a$ where a is some positive number

can be solved easily.

To solve such an equation, simply take the square root of each side of the equation. The solution can be either

$$x = \sqrt{a} \text{ or } x = -\sqrt{a}$$

Example: Solve $x^2 = 36$.

Taking square roots, we have $x = +6$ or $x = -6$.
There are two possible solutions, one negative and one positive. Be careful, one of them, usually the negative one, may not be a reasonable answer to a practical problem.

Example: If the cross-sectional area of a square heating duct is 75 sq in., what must be the width of the duct?

Solve $x^2 = 75$.

Taking the square root of each side, we obtain the positive solution

$$x = \sqrt{75}$$

 $x \approx 8.7$ in.   rounded

75 $\boxed{\vee}$ → 8.660254038

If you need to review the concept of square roots, return to Section 6-4 on page 323.

We will look at more problems like this in Chapter 11.

Now turn to Exercises 7-5 for a set of practice problems on solving equations and formulas.

Exercises 7-5

Solving More Equations and Formulas

A. Solve the following equations.

1. $5(x - 3) = 30$
2. $22 = -2(y + 6)$
3. $3(2n + 4) = 41$
4. $-6(3a - 7) = 21$
5. $2 - (x - 5) = 14$
6. $24 = 5 - (3 - 2m)$
7. $6 + 2(y - 4) = 13$
8. $7 - 11(2z + 3) = 18$
9. $8 = 5 - 3(3x - 4)$
10. $7 + 9(2w + 3) = 25$
11. $6c - (c - 4) = 29$
12. $9 = 4y - (y - 2)$
13. $5x - 3(2x - 8) = 31$
14. $6a + 2(a + 7) = 8$
15. $9t - 3 = 4t - 2$
16. $7y + 5 = 3y + 11$
17. $12x = 4x - 16$
18. $22n = 16n - 18$
19. $8y - 25 = 13 - 11y$
20. $6 - 2p = 14 - 4p$
21. $9x = 30 - 6x$
22. $12 - y = y$
23. $2(3t - 4) = 10t + 7$
24. $5A = 4(2 - A)$
25. $2 - (3x - 20) = 4(x - 2)$
26. $2(2x - 5) = 6x - (5 - x)$
27. $8 + 3(6 - 5x) = 11 - 10x$
28. $2(x - 5) - 3(2x - 8) = 16 - 6(4x - 3)$

B. Solve the following formulas for the variable shown.

1. $S = LW$ for L 2. $A = \dfrac{1}{2}BH$ for B

3. $V = IR$ for I 4. $H = \dfrac{D - R}{2}$ for D

5. $S = \dfrac{W}{2}(A + T)$ for T 6. $V = \pi R^2 H$ for H

7. $P = 2A + 2B$ for B 8. $H = \dfrac{R}{2} + 0.05$ for R

9. $T = \dfrac{RP}{R + 2}$ for P 10. $I = \dfrac{E + V}{R}$ for V

C. Practical Problems

Problem 1

1. **Sheet Metal Technology**   The length of arc of a sector of a circle is given by the formula

$$L = \frac{2\pi Ra}{360}$$

R is the radius of the circle and a is the central angle in degrees.

(a) Solve for a.   (b) Solve for R.

(c) Find L when $R = 10$ in. and $a = 30°$. Use $\pi \approx 3.14$.

2. **Sheet Metal Technology**   The area of the sector shown in problem 1 is $A = \pi R^2 a/360$.

(a) Solve this formula for a.

(b) Find A if $R = 12$ in., $a = 45°$, $\pi \approx 3.14$. Round to two significant digits.

3. **Electronics**   When two resistors R_1 and R_2 are put in series with a battery giving V volts, the current through the resistors is

$$i = \frac{V}{R_1 + R_2}$$

(a) Solve for R_1.

(b) Find R_2 if $V = 100$ volts, $i = 0.4$ ampere, $R_1 = 200$ ohms.

4. **Machine Technology**   Machinists use a formula known as Pomeroy's formula to determine roughly the power required by a metal punch machine.

$$P \approx \frac{t^2 dN}{3.78}$$

where P = power needed, in horsepower
            t = thickness of the metal being punched
            d = diameter of the hole being punched
           N = number of holes to be punched at one time

(a) Solve this formula for N.

(b) Find the power needed to punch six 2-in.-diameter holes in a sheet $\frac{1}{8}$ in. thick. Round to one significant digit.

5. **Marine Technology** When a gas is kept at constant temperature and the pressure on it is changed, its volume changes in accord with the pressure–volume relationship known as Boyle's law:

$$\frac{V_1}{V_2} = \frac{P_2}{P_1}$$

where P_1 and V_1 are the beginning volume and pressure, and P_2 and V_2 are the final volume and pressure.

(a) Solve for V_1. (b) Solve for V_2.

(c) Solve for P_1. (d) Solve for P_2.

(e) Find P_1 when $V_1 = 10$ cu ft, $V_2 = 25$ cu ft, and $P_2 = 120$ psi.

6. The volume of a football is roughly $V = \pi LT^2/6$, where L is its length and T is its thickness.

(a) Solve for L.

(b) Solve for T^2.

7. **Medical Technology** Nurses use a formula known as Young's rule to determine the amount of medicine to give a child under 12 years of age when the adult dosage is known.

$$C = \frac{AD}{A + 12}$$

C is the child's dose; A is the age of the child in years; D is the adult dose.

(a) Work backward and find the adult dose in terms of the child's dose. Solve for D.

(b) Find D if $C = 0.05$ gram and $A = 7$.

8. **Carpentry** The projection or width P of a protective overhang of a roof is determined by the height T of the window, the height H of the header above the window, and a factor F that depends on the latitude of the construction site.

$$P = \frac{T + H}{F} \qquad \text{Solve this equation for the header height } H.$$

9. **Electronics** For a current transformer, $\dfrac{i_L}{i_S} = \dfrac{T_P}{T_S}$

where i_L = line current
i_S = secondary current
T_P = number of turns of wire in the primary coil
T_S = number of turns of wire in the secondary coil

(a) Solve for i_L.

(b) Solve for i_S.

(c) Find i_L when $i_S = 1.5$ amperes, $T_P = 1500$, $T_S = 100$.

10. **Electronics** The electrical power P dissipated in a circuit is equal to the product of the current I and the voltage V, where P is in watts, I is in amperes, and V is in volts.

(a) Write an equation giving P in terms of I and V.

(b) Solve for I.

(c) Find V when $P = 15{,}750$ watts, $I = 42$ amperes.

11. **Electronics** The inductance L in microhenrys of a coil constructed by a ham radio operator is given by the formula

$$L = \frac{R^2N^2}{9R + 10D}$$

where $R =$ radius of the coil
       $D =$ length of the coil
       $N =$ number of turns of wire in the coil

Find L if $R = 3$ in., $D = 6$ in., $N = 200$.

12. **Sheet Metal Technology** A sheet metal technician uses the following formula for calculating bend allowance, BA, in inches:

$$BA = N(0.01743R + 0.0078T)$$

where $N =$ number of degrees in the bend
       $R =$ inside radius of the bend, in inches
       $T =$ thickness of the metal, in inches

Find BA for each of the following situations. (You'll want to use a calculator on this one.)

	N	R	T
(a)	$50°$	$1\frac{1}{4}$ in.	0.050 in.
(b)	$65°$	0.857 in.	0.035 in.
(c)	$40°$	1.025 in.	0.0856 in.

13. **Machinist** Suppose that, on the average, 3% of the parts produced by a particular machine have proven to be defective. Then the formula

$$N - 0.03N = P$$

will give the number of parts N that must be produced in order to manufacture a total of P nondefective ones. How many parts should be produced by this machine in order to end up with 7500 nondefective ones?

14. **Office Services** The formula

$$A = P(1 + rt)$$

is used to find the total amount A of money in an account when an original amount or principal P is invested at a rate of simple interest r for t years. How long would it take $8000 to grow to $10,000 at 8% simple interest?

15. **Civil Engineering** The formula

$$I = 0.000014L(T - t)$$

gives the expansion I of a particular highway of length L at a temperature of T degrees Fahrenheit. The variable t stands for the temperature at which the highway was built. If a 2-mile stretch of highway was built at an average temperature of 60°F, what is the maximum temperature it can withstand if expansion joints allow for 7.5 ft of expansion? (*Hint:* The units of L must be the same as the units of I.)

When you have completed these exercises check your answers in the Appendix, then turn to Section 7-6 to learn about word problems.

7-6 SOLVING WORD PROBLEMS

Translating English to Algebra

Algebra is a very useful tool for solving real problems. But in order to use it you may find it necessary to translate simple English sentences and phrases into mathematical expressions or equations. In technical work especially, the formulas to be used are often given in the form of English sentences, and they must be rewritten as algebraic formulas before they can be used.

EXAMPLE

The statement

> Horsepower required to overcome vehicle air resistance is equal to the cube of the vehicle speed in mph multiplied by the frontal area in square feet divided by 150,000

translates to the formula

$$\text{hp} = \frac{\text{mph}^3 \cdot \text{area}}{150,000}$$

or

$$P = \frac{v^3 A}{150,000} \qquad \text{in algebraic form, where } v \text{ is the vehicle speed}$$

In the next few pages of this chapter we show you how to translate English statements into algebraic formulas. To begin, try the following problem.

YOUR TURN

An automotive technician found the following statement in a manual:

> The pitch diameter of a cam gear is twice the diameter of the crank gear.

Translate this sentence into an algebraic equation.

ANSWER

The equation is $P = 2C$ where P is the pitch diameter of the cam gear, and C is the diameter of the crank gear

You may use any letters you wish of course, but we have chosen letters that remind you of the quantities they represent: P for *pitch* and C for *crank*.

Notice that the phrase "twice C" means "two times C" and is written as $2C$ in algebra.

Certain words and phrases appear again and again in statements to be translated. They are signals alerting you to the mathematical operations to be used. Here is a handy list of the *signal words* and their mathematical translations.

English Term	Math Translation	Example
Equals Is, is equal to, was, are, were The same as . . . What is left is . . . The result is . . . Gives, makes, leaves, having	=	$A = B$
Plus, sum of Increased by, more than	+	$A + B$
Minus B, subtract B Less B Decreased by B, take away B Reduced by B, diminished by B B less than A B subtracted from A Difference between A and B	−	$A - B$
Times, multiply, of Multiplied Product of	×	AB
Divide, divided by B Quotient of	÷	$A \div B$ or $\dfrac{A}{B}$
Twice, twice as much Double	×2	$2A$
Squared, square of A Cubed, cube of A		A^2 A^3

YOUR TURN

Translate the phrase "length plus 3 inches" into an algebraic expression. Try it, then check your answer with ours.

SOLUTION

First, make a word equation by using parentheses.

(length)   (plus)   (3 inches)

Second, substitute mathematical symbols.

(length)   (plus)   (3 inches)

   ↓   ↓   ↓

   L +   3   or $L + 3$

Notice that signal words, such as "plus," are translated directly into math symbols. Unknown quantities are represented by letters of the alphabet, chosen to remind you of their meaning.

Translate the following phrases into math expressions.

(a) Weight divided by 13.6   _____

(b) $6\frac{1}{4}$ in. more than the width   _____

(c) One-half of the original torque   _____

(d) The sum of the two lengths   _____

(e) The voltage decreased by 5   _____

(f) Five times the gear reduction   _____

(g) 8 in. less than twice the height   _____

SOLUTIONS

(a) (weight)   (divided by)   (13.6)

    ↓   ↓   ↓

    W ÷   13.6   or $\dfrac{W}{13.6}$

(b) ($6\frac{1}{4}$ in.)   (more than)   (the width)

    ↓   ↓   ↓

    $6\frac{1}{4}$ + W or $6\frac{1}{4} + W$

(c) (one-half)   (of)   (the original torque)

    ↓   ↓   ↓

    $\frac{1}{2}$ × T or $\frac{1}{2} T$ or $\dfrac{T}{2}$

(d) (the sum of)   (the two lengths)

            ↓ ↓

           $L_1 + L_2$

(e) (the voltage)   (decreased by)   (5)

     ↓   ↓   ↓

     V −   5   or $V - 5$

(f) (five)   (times)   (the gear reduction)

    ↓   ↓   ↓

    5 × G or $5G$

(g) 8 less than twice the height means
    (twice the height)   (less)   (8)

     ↘   ↙   ↓ ↓

    2   H − 8 or $2H - 8$

Of course, any letters could be used in place of the ones used above.

Note ▶ The phrase "less than" has a meaning very different from the word "less."

"8 *less* 5" means $8 - 5$

"8 *less than* 5" means $5 - 8$ ◀

Translating Sentences to Equations

So far we have translated only phrases, pieces of sentences, but complete sentences can also be translated. An English phrase translates into an algebraic expression, and an English sentence translates into an algebraic formula or equation.

EXAMPLE

The sentence

The height of the duct | is equal to | its width

translates to H $=$ W or $H = W$

Each word or phrase in the sentence becomes a mathematical term, letter, number, expression, or arithmetic operation sign.

YOUR TURN

Translate the following sentence into algebraic form as we did above.

The size of a drill for a tap is equal to the tap diameter minus the depth.

$S = T - D$

Follow these steps when you must translate an English sentence into an algebraic equation or formula.

Step 1 Cross out all unnecessary words.

~~The~~ size ~~of a drill for a tap~~ is equal to ~~the~~ tap diameter minus ~~the~~ depth.

Step 2 Make a word equation using parentheses.

(Size) (is equal to) (tap diameter) (minus) (depth)

↓   ↓   ↓   ↓   ↓

Step 3 Substitute a letter or an arithmetic symbol for each parentheses.

S $=$ T $-$ D

Step 4 Combine and simplify.

$S = T - D$

In most formulas the units for the quantities involved must be given. In the formula above, T and D are in inches.

Learning Help ▶

Translating English sentences or verbal rules into algebraic formulas requires that you read the sentences very differently from the way you read stories or newspaper articles. Very few people are able to write out the math formula after reading the problem only once. You should expect to read it several times, and you'll want to read it slowly. No speed reading here! ◀

The ideas in technical work and formulas are usually concentrated in a few key words, and you must find them. If you find a word you do not recognize, stop reading and look it up in a dictionary, textbook, or manual. It may be important.

Translating and working with formulas is one of the skills you must have if you are to succeed at any technical occupation.

EXAMPLE

Here is another example of translating a verbal rule into an algebraic formula:

The electrical resistance of a length of wire is equal to the resistivity of the metal times the length of the wire divided by the square of the wire diameter.

Step 1	Eliminate all but the key words.	"~~The electrical~~ resistance ~~of a length of wire~~ is equal to ~~the~~ resistivity ~~of the metal~~ times ~~the~~ length ~~of the wire~~ divided by ~~the~~ square of ~~the wire~~ diameter."
Step 2	Make a word equation.	(Resistance) (is equal to) (resistivity) (times) (length) (divided by) (square of diameter)
Step 3	Substitute letters and symbols.	$R = \dfrac{rL}{D^2}$

If the resistivity r has units of ohms times inches, L and D will be in inches.

PROBLEMS

The more translations you do, the easier it gets. Translate each of the following technical statements into algebraic formulas.

(a) A sheet metal worker measuring a duct cover finds that the width is $8\frac{1}{2}$ in. less than the height.

(b) One quarter of a job takes $3\frac{1}{2}$ days.

(c) One half of a coil of wire weighs $16\frac{2}{3}$ lb.

(d) The volume of an elliptical tank is approximately 0.7854 times the product of its height, length, and width.

(e) The engine speed is equal to 168 times the overall gear reduction multiplied by the speed in mph and divided by the rolling radius of the tire.

(f) The air resistance force in pounds acting against a moving vehicle is equal to 0.0025 times the square of the speed in mph times the frontal area of the vehicle.

(g) Two pieces of wire have a combined length of 24 in. The longer piece is five times the length of the shorter piece. (*Hint:* Write two separate equations.)

(h) Two shims are to have a combined thickness of 0.090 in. The larger shim must be 3.5 times as thick as the smaller shim. (*Hint:* Write two separate equations.)

ANSWERS

(a) $W = H - 8\dfrac{1}{2}$

(b) $\dfrac{1}{4}J = 3\dfrac{1}{2}$ or $\dfrac{J}{4} = 3\dfrac{1}{2}$

(c) $\dfrac{1}{2}C = 16\dfrac{2}{3}$ or $\dfrac{C}{2} = 16\dfrac{2}{3}$

(d) $V = 0.7854HLW$

(e) $S = \dfrac{168Gv}{R}$ where v is in mph.

(f) $R = 0.0025v^2A$ where v is in mph.

(g) $24 = L + S$ and $L = 5S$

(h) $S + L = 0.090$ and $L = 3.5S$

Notice in problems (g) and (h) that two equations can be written. These can be combined to form a single equation.

(g) $24 = L + S$ and $L = 5S$ give $24 = 5S + S$

(h) $S + L = 0.090$ and $L = 3.5S$ give $S + 3.5S = 0.090$

General Word Problems

Now that you can translate English phrases and sentences into algebraic expressions and equations, you should be able to solve many practical word problems.

The following six-step method will help you to solve any word problems.

1. **Read** the problem carefully, noting what information is given and what is unknown.

2. **Assign a variable** to represent the unknown. If there are additional unknowns, try to represent these with an expression involving the same variable.

3. **Write an equation** using the unknown(s) and some given fact about the unknown(s) in the problem statement.

4. **Solve** the equation.

5. **Answer** the original question, being careful to state the value of all unknown quantities asked for in the problem.

6. **Check** your solution.

EXAMPLE

Consider this problem:

A machinist needs to use two shims with a combined thickness of 0.084 in. One shim is to be three times as thick as the other. What are the thicknesses of the two shims?

Read the problem carefully, then note that there are two unknowns—a thinner shim and a thicker shim. We know their total thickness and we know that one is three times thicker than the other. We use this last fact to assign variable expressions to the unknowns.

Let x = thickness of the thinner shim

then $3x$ = thickness of the thicker shim

Use their combined thickness to write an equation.

$3x + x = 0.084$ in.   To solve the equation, first combine terms.

     $4x = 0.084$ in.   Then divide each side by 4.

       $x = 0.021$ in. $3x = 3(0.021$ in.$) = 0.063$ in.

Notice that two answers are required here.

The thin shim is 0.021 in. and the thicker one is 0.063 in.

 0.021 in. + 3(0.021 in.) = 0.084 in.
 0.021 in. + 0.063 in. = 0.084 in.   which is correct

YOUR TURN

Use your knowledge of algebra to translate the following problem to an algebraic equation and then solve it.

Two carpenters, Al and Bill, produced 42 assembly frames in one day. Al worked faster and produced 8 more than Bill. How many frames did each build?

SOLUTION

There are two unknowns, the number of frames built by Bill and the number of frames built by Al. We use the fact that Al produced 8 more than Bill to assign variable expressions to the unknowns:

Let B = number of frames built by Bill
then $B + 8$ = number of frames built by Al

Now use the fact that they produced a total of 42 to write an equation.

$B + (B + 8) = 42$	To solve, first combine terms, $B + B = 2B$.
$2B + 8 = 42$	Then subtract 8 from each side.
$2B = 34$	Finally, divide both sides by 2.
$B = 17$	Bill built 17 frames.
then $B + 8 = 25$	Al built 25 frames.

Notice that two answers are required.

 $17 + 25 = 42$

Some of the more difficult percent problems can be simplified using algebraic equations. One such problem, backing the tax out of a total, is commonly encountered by all those who own their own businesses.

EXAMPLE

Suppose that an auto mechanic has collected $1468.63 for parts, including 6% tax. For accounting purposes he must determine exactly how much of the total is sales tax. If we let x stand for the dollar amount of the parts before tax was added, then the tax is 6% of that or $0.06x$, and we have

$$x \quad + \quad 0.06x \quad = \quad \$1468.63$$

Cost of the parts	6% Tax on the parts	Total amount

Solving for x yields

$x + 0.06x = \$1468.63$	
$1.00x + 0.06x = \$1468.63$	$(x = 1x = 1.00x)$. Combine like terms.
$1.06x = \$1468.63$	Now divide by 1.06.
$x = \$1385.50$	x is the cost of the parts.
$0.06x = 0.06(\$1385.50) = \$83.13$	Multiply by 0.06 or
$\$1468.63 - \$1385.50 = \$83.13$	subtract from $1468.63 to get the tax.

Practice makes perfect. Work the following set of word problems.

(a) The area of a rectangular shop floor is 400 sq ft. If the width is 16 ft, what is the length? (*Hint:* Area = length · width.)

(b) A 14-ft-long steel rod is cut into two pieces. The longer piece is $2\frac{1}{2}$ times the length of the shorter piece. Find the length of each piece. (Ignore waste.)

(c) A carpenter wants to cut a 12-ft board into three pieces. The longest piece must have three times the length of the shortest piece, and the medium-sized piece is 2 ft longer than the shortest piece. Find the actual length of each piece. (Ignore waste in cutting.)

(d) Find the dimensions of a rectangular cover plate if its length is 6 in. longer than its width and if its perimeter is 68 in.
(*Hint:* Perimeter = 2 · length + 2 · width.)

(e) Mike and Jeff are partners in a small manufacturing firm. Because Mike provided more of the initial capital for the business, they have agreed that Mike's share of the profits should be $\frac{1}{4}$ greater than Jeff's. The total profit for the first quarter of this year was $17,550. How should they divide it?

(f) A plumber collected $2784.84 during the day, $1650 for labor, which is not taxed, and the rest for parts, which includes 5% sales tax. Determine the total amount of tax that was collected.

(g) A printer knows from past experience that about 2% of a particular run of posters will be spoiled. How many should she print in order to end up with 1500 usable posters?

(a) Width = 16 ft   Length = L
    Area = 400 sq ft = 16 ft · L
    $16L = 400$ Divide by 16.
    $L = 25$ ft

(b) x = shorter piece

   $(2\frac{1}{2})x$ = longer piece

   $x + (2\frac{1}{2})x = 14$ ft

   $(3\frac{1}{2})x = 14$ ft

or $\dfrac{7x}{2} = 14$ Multiply by 2.

   $7x = 28$ Divide by 7.
   $x = 4$ ft   Longer piece = $2\frac{1}{2}x = 2\frac{1}{2}(4) = 10$ ft.

   The shorter piece is 4 ft long and the longer piece is 10 ft long.

(c) Let x = the shortest piece
    then $3x$ = longest piece
    and $x + 2$ = medium-sized piece
    $x + (x + 2) + (3x) = 12$ ft   Combine terms.
    $5x + 2 = 12$ Subtract 2 from each side.
    $5x = 10$ Divide by 5.
    $x = 2$ ft   The shortest piece is 2 ft long.
    $3x = 6$ ft   The longest piece is 6 ft long.
    $x + 2 = 4$ ft   The third piece is 4 ft long.

(d) Let width = W
then length = $W + 6$
and perimeter = 68 in. = $2W + 2(W + 6)$ Remove parentheses,
                                            $2(W + 6) = 2W + 12$.

$$68 = 2W + 2W + 12$$ Combine terms.
or $4W + 12 = 68$ Subtract 12 from each side.
$$4W = 56$$ Divide by 4.
$$W = 14 \text{ in.}$$
$$\text{length} = W + 6 = 20 \text{ in.}$$

(e) Let Jeff's share = J
then Mike's share = $J + \frac{1}{4}J$

$$\$17{,}550 = J + \left(J + \tfrac{1}{4}J\right)$$ Combine terms; remember $J = 1 \cdot J$.

$$= 2\tfrac{1}{4}J$$ Write the mixed number as a fraction.

$$= \frac{9J}{4}$$

or $\dfrac{9J}{4} = 17{,}550$ Multiply by 4.

$$9J = 70{,}200$$ Divide by 9.

$$J = \$7800 \quad \text{Jeff's share}$$

$$\text{Mike's share} = \$7800 + \frac{\$7800}{4}$$

$$= \$7800 + \$1950$$

$$= \$9750$$

(f) Let x = cost of the parts
then $0.05x$ = amount of the tax

$$x + 0.05x = \$2784.84 - \$1650$$ Subtract out the labor to get the dollar amount of the parts.
$$1.05x = \$1134.84$$ Divide by 1.05.
$$x = \$1080.80$$ This represents the cost of the parts.
$$0.05(\$1080.80) = \$54.04$$ The total tax is $54.04.

(g) Let x = number of posters she needs to print
then $0.02x$ = number that will be spoiled

$$1.00x - 0.02x = 1500$$ $x = 1x = 1.00\,x$. Now subtract like terms.
$$0.98x = 1500$$ Divide by 0.98.
$$x = 1531 \text{ (rounded)}$$ She must print 1531 posters to end up with 1500 unspoiled ones.

Now turn to Exercises 7-6 for a set of word problems.

Exercises 7-6   Solving Word Problems

A. Translate the following into algebraic equations:

1. The height of the tank is 1.4 times its width.

2. The sum of the two weights is 167 lb.

3. The volume of a cylinder is equal to $\frac{1}{4}$ of its height times π times the square of its diameter.

4. The weight in kilograms is equal to 0.454 times the weight in pounds.

5. **Machine Technology** The volume of a solid bar is equal to the product of the cross-sectional area and the length of the bar.

6. The volume of a cone is $\frac{1}{3}$ times π times the height times the square of the radius of the base.

7. **Machine Technology** The pitch diameter D of a spur gear is equal to the number of teeth on the gear divided by the pitch.

8. **Machine Technology** The cutting time for a lathe operation is equal to the length of the cut divided by the product of the tool feed rate and the revolution rate of the workpiece.

9. **Machine Technology** The weight of a metal cylinder is approximately equal to 0.785 times the height of the cylinder times the density of the metal times the square of the diameter of the cylinder.

10. The voltage across a simple circuit is equal to the product of the resistance of the circuit and the current flowing in the circuit.

B. Set up equations and solve.

1. **Wastewater Technology** Two tanks must have a total capacity of 400 gallons. If one tank needs to be twice the size of the other, how many gallons should each tank hold?

2. **Machine Technology** Two metal castings weigh a total of 84 lb. One weighs 12 lb more than the other. How much does each one weigh?

3. **Plumbing** The plumber wants to cut a 24-ft length of pipe into three sections. The largest piece should be twice the length of the middle piece, and the middle piece should be twice the length of the smallest piece. How long should each piece be?

4. **Electrical Technology** The sum of three voltages in a circuit is 38 volts. The middle-sized one is 2 volts more than the smallest. The largest is 6 volts more than the smallest. What is the value of each voltage?

5. **Masonry** There are 156 concrete blocks available to make a retaining wall. The bottom three rows will all have the same number of blocks. The next six rows will each have two blocks fewer than the row below it. How many blocks are in each row?

6. **Construction Technology** A concrete mix is in volume proportions of 1 part cement, 2 parts water, 2 parts aggregate, and 3 parts sand. How many cubic feet of each ingredient are needed to make 54 cu ft of concrete?

7. **Office Services** Jo and Ellen are partners in a painting business. Because Jo is the office manager, she is to receive one-third more than Ellen when the profits are distributed. If their profit is $85,400, how much should each of them receive?

8. **Construction Technology** A total of $1,120,000 is budgeted for constructing a roadway. The rule of thumb for this type of project is that the pavement costs twice the amount of the base material, and the sidewalk costs one-fourth the amount of the pavement. Using these figures, how much should each item cost?

9. **Photography** A photographer has enough liquid toner for fifty 5-in. by 7-in. prints. How many 8-in. by 10-in. prints will this amount cover? (*Hint:* The toner covers the *area* of the prints: Area = length × width.)

10. **Construction Technology** A building foundation has a length of 82 ft and a perimeter of 292 ft. What is the width?
(*Hint:* Perimeter = 2 × length + 2 × width.)

11. **Roofing** A roofer earns $24 per hour plus $36 per hour overtime. If he puts in 40 hours of regular time during a certain week and he wishes to earn $1200, how many hours of overtime should he work?

12. **Agricultural Technology** An empty avocado crate weighs 4.2 kg. How many avocados weighing 0.3 kg each can be added before the total weight of the crate reaches 15.0 kg?

13. **Machine Technology** One-tenth of the parts tooled by a machine are rejects. How many parts must be tooled to assure 4500 acceptable ones?

14. **Office Services** An auto mechanic has total receipts of $1861.16 for a certain day. He determines that $1240 of this is labor. The remaining amount is for parts plus 6% sales tax on the parts only. Find the amount of the sales tax he collected.

15. **Office Services** A travel agent receives a 7% commission on the base fare of a customer—that is, the fare *before* sales tax is added. If the total fare charged to a customer comes to $753.84 *including* 5.5% sales tax, how much commission should the agent receive?

16. **Office Services** A newly established carpenter wishes to mail out letters advertising her services to the families of a small town. She has a choice of mailing the letters first class or obtaining a bulk-rate permit and mailing them at the cheaper bulk rate. The bulk-rate permit and mailing fee cost $300 total and each piece of bulk mail then costs 35.2 cents. If the first-class rate is 37 cents, how many pieces would she have to mail in order to make bulk rate the cheaper way to go?

Check your answers in the Appendix, then turn to Section 7-7 to learn more algebra.

7-7 MULTIPLYING AND DIVIDING ALGEBRAIC EXPRESSIONS

Multiplying Simple Factors Earlier we learned that only like algebraic terms, those with the same literal part, can be added and subtracted. However, any two terms, like or unlike, can be multiplied.

In order to multiply two terms such as $2x$ and $3xy$, first remember that $2x$ means 2 times x. Second, recall from arithmetic that the order in which you do multiplications does not make a difference. For example, in arithmetic

$$2 \cdot 3 \cdot 4 = (2 \cdot 4) \cdot 3 = (3 \cdot 4) \cdot 2$$

and in algebra

$$a \cdot b \cdot c = (a \cdot c) \cdot b = (c \cdot b) \cdot a \qquad \text{or} \qquad 2 \cdot x \cdot 3 \cdot x \cdot y = 2 \cdot 3 \cdot x \cdot x \cdot y$$

Remember that $x \cdot x = x^2, x \cdot x \cdot x = x^3$, and so on. Therefore,

$$2x \cdot 3xy = 2 \cdot 3 \cdot x \cdot x \cdot y = 6x^2y$$

> **EXAMPLE**

The following examples show how to multiply two terms.

(a) $a \cdot 2a = a \cdot 2 \cdot a$

$\qquad = 2 \cdot \underbrace{a \cdot a}$ Group like factors together.

$\qquad = 2 \cdot a^2$

$\qquad = 2a^2$

(b) $2x^2 \cdot 3xy = 2 \cdot x \cdot x \cdot 3 \cdot x \cdot y$

$\qquad = \underline{2 \cdot 3} \cdot \underline{x \cdot x \cdot x} \cdot y \qquad$ Group like factors together.

$\qquad = 6 \cdot x^3 \cdot y$

$\qquad = 6x^3y$

(c) $3x^2yz \cdot 2xy = 3 \cdot x^2 \cdot y \cdot z \cdot 2 \cdot x \cdot y$

$\qquad = \underline{3 \cdot 2} \cdot \underline{x^2 \cdot x} \cdot \underline{y \cdot y} \cdot z$

$\qquad = 6 \cdot x^3 \cdot y^2 \cdot z$

$\qquad = 6x^3y^2z$

Remember to group like factors together before multiplying.

If you need to review exponents, return to page 318.

YOUR TURN

Now try the following problems.

(a) $x \cdot y$ $\quad = $ _____ (b) $2x \cdot 3x$ $\quad = $ _____

(c) $2x \cdot 5xy$ $\quad = $ _____ (d) $4a^2b \cdot 2a$ $\quad = $ _____

(e) $3x^2y \cdot 4xy^2$ $\quad = $ _____ (f) $5xyz \cdot 2ax^2$ $\quad = $ _____

(g) $3x \cdot 2y^2 \cdot 2x^2y = $ _____ (h) $x^2y^2 \cdot 2x \cdot y^2$ $\quad = $ _____

(i) $2x^2(x + 3x^2)$ $\quad = $ _____ (j) $-2a^2b(a^2 - 3b^2) = $ _____

ANSWERS

(a) xy (b) $6x^2$

(c) $10x^2y$ (d) $8a^3b$

(e) $12x^3y^3$ (f) $10ax^3yz$

(g) $12x^3y^3$ (h) $2x^3y^4$

(i) $2x^3 + 6x^4$ (j) $-2a^4b + 6a^2b^3$

Were the last two problems tricky for you? Recall from Rule 3 on page 352 that when there is a multiplier in front of parentheses, every term inside must be multiplied by this factor. Try it this way:

(i) $2x^2(x + 3x^2) = (2x^2)(x) + (2x^2)(3x^2)$

$\qquad = 2 \cdot \underline{x^2 \cdot x} + \underline{2 \cdot 3} \cdot \underline{x^2 \cdot x^2}$

$\qquad = 2 \cdot x^3 + 6 \cdot x^4$

$\qquad = 2x^3 + 6x^4$

(j) $-2a^2b(a^2 - 3b^2) = (-2a^2b)(a^2) + (-2a^2b)(-3b^2)$

$\qquad = (-2) \cdot \underline{a^2 \cdot a^2} \cdot b + \underline{(-2) \cdot (-3)}\, a^2 \cdot \underline{b \cdot b^2}$

$\qquad = -2 \cdot a^4 \cdot b + 6 \cdot a^2 \cdot b^3$

$\qquad = -2a^4b + 6a^2b^3$

Rule for Multiplication

As you did the preceding problems, you may have noticed that when you multiply like factors together, you actually add their exponents. For example,

$$x^2 \cdot x^3 = x \cdot x \cdot x \cdot x \cdot x = x^5$$

In general, we can state the following rule:

Rule 1 To multiply numbers written in exponential form having the same base, add the exponents.

$$x^m \cdot x^n = x^{m+n}$$

m + n

Same base

EXAMPLE

(a) $x^2 \cdot x^3 = x^{2+3} = x^5$

(b) $3^5 \times 3^2 = 3^{5+2} = 3^7$

Careful ▶

1. Notice in the second example that the like bases remain unchanged when you multiply—that is, $3^5 \times 3^2 = 3^7$, not 9^7.

2. Exponents are added during multiplication, but numerical factors are multiplied as usual. For example,

Add exponents

$2x^2 \cdot 3x^3 = 6x^5$

Multiply factors ◀

Dividing Simple Factors

To divide a^5 by a^2, think of it this way:

$$\frac{a^5}{a^2} = \frac{a \cdot a \cdot a \cdot a \cdot a}{a \cdot a} = a^3$$

Cancel common factors

Notice that the final exponent, 3, is the difference between the original two exponents, 5 and 2. Remember that multiplication and division are reverse operations. It makes sense that if you add exponents when multiplying, you would subtract exponents when dividing. We can now state the following rule:

Rule 2 To divide numbers written in exponential form having the same base, subtract the exponents.

$$\frac{x^m}{x^n} = x^{m-n}$$

EXAMPLE

(a) $\dfrac{x^7}{x^3} = x^{7-3} = x^4$

(b) $\dfrac{4^5}{4^2} = 4^{5-2} = 4^3$

Careful ▶

1. As with multiplication, the like bases remain unchanged when dividing with exponents. This is especially important to keep in mind when the base is a number, as in the second preceding example.

2. When dividing expressions that contain both exponential expressions and numerical factors, *subtract* the exponents but *divide* the numerical factors. For example,

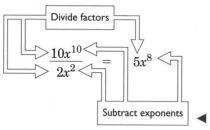

◀

YOUR TURN

Try out the division rule on these problems:

(a) $\dfrac{x^6}{x^3}$ (b) $\dfrac{6a^4}{2a}$ (c) $\dfrac{-12m^3n^5}{4m^2n^2}$ (d) $\dfrac{5^8}{5^2}$ (e) $\dfrac{4x^5 - 8x^4 + 6x^3}{2x^2}$

ANSWERS

(a) x^3 (b) $3a^3$ (c) $-3mn^3$ (d) 5^6 (e) $2x^3 - 4x^2 + 3x$

Did you have trouble with (e)? Unlike the other problems, which have only one term in the numerator (dividend), problem (e) has *three* terms in the numerator. Just imagine that the entire numerator is enclosed in parentheses and divide each of the three terms separately by the denominator (divisor). Write it out this way:

$$\frac{4x^5 - 8x^4 + 6x^3}{2x^2} = \frac{4x^5}{2x^2} - \frac{8x^4}{2x^2} + \frac{6x^3}{2x^2} = 2x^3 - 4x^2 + 3x$$

Note ▶

The rule for division confirms the fact stated in Section 6-4 that $a^0 = 1$. We know from arithmetic that any quantity divided by itself is equal to 1, so

$$\frac{a^n}{a^n} = 1 \qquad (\text{for } a \neq 0)$$

But according to the division rule,

$$\frac{a^n}{a^n} = a^{n-n} = a^0 \qquad \text{Therefore, } a^0 \text{ must be equal to 1.} \blacktriangleleft$$

Negative Exponents   If we apply the division rule to a problem in which the divisor has a larger exponent than the dividend, the answer will contain a negative exponent. For example,

$$\frac{x^4}{x^6} = x^{4-6} = x^{-2}$$

If we use the "cancellation method" to do this problem, we have

$$\frac{x^4}{x^6} = \frac{\cancel{x} \cdot \cancel{x} \cdot \cancel{x} \cdot \cancel{x}}{\cancel{x} \cdot \cancel{x} \cdot \cancel{x} \cdot \cancel{x} \cdot x \cdot x} = \frac{1}{x^2} \qquad \text{Therefore,} \quad x^{-2} = \frac{1}{x^2}.$$

In general, we define negative exponents as follows:

$$x^{-n} = \frac{1}{x^n}$$

⬤ **YOUR TURN**

Practice using the definition of negative exponents by giving two answers for each of the following problems—one with negative exponents, and one using fractions to eliminate the negative exponents.

(a) $\dfrac{a^2}{a^6}$ 　　(b) $\dfrac{8x^2}{-2x^3}$ 　　(c) $\dfrac{6^3}{6^5}$ 　　(d) $\dfrac{5x^4y^2}{15xy^7}$ 　　(e) $\dfrac{12x^6 - 9x^3 + 6x}{3x^2}$

⬤ **ANSWERS**

(a) a^{-4} or $\dfrac{1}{a^4}$ 　　　　　　　　　　　　(b) $-4x^{-1}$ or $\dfrac{-4}{x}$

(c) 6^{-2} or $\dfrac{1}{6^2}$ 　　　　　　　　　　　　(d) $\dfrac{1}{3}x^3y^{-5}$ or $\dfrac{x^3}{3y^5}$

(e) $4x^4 - 3x + 2x^{-1}$ or $4x^4 - 3x + \dfrac{2}{x}$

Now turn to Exercises 7-7 for a set of problems on multiplication and division of algebraic expressions.

Exercises 7-7   Multiplying and Dividing Algebraic Expressions

A. Simplify by multiplying.

　　1. $5 \cdot 4x$ 　　　　　　　　　　　2. $(a^2)(a^3)$

　　3. $-3R(-2R)$ 　　　　　　　　　4. $3x \cdot 3x$

　　5. $(4x^2y)(-2xy^3)$ 　　　　　　　6. $2x \cdot 2x \cdot 2x$

　　7. $0.4a \cdot 1.5a$ 　　　　　　　　8. $x \cdot x \cdot A \cdot x \cdot 2x \cdot A^2$

　　9. $\frac{1}{2}Q \cdot \frac{1}{2}Q \cdot \frac{1}{2}Q \cdot \frac{1}{2}Q$ 　　　　10. $(pq^2)(\frac{1}{2}pq)(2.4p^2q)$

　11. $(-2M)(3M^2)(-4M^3)$ 　　　　12. $2x(3x - 1)$

　13. $-2(1 - 2y)$ 　　　　　　　　14. $ab(a^2 - b^2)$

　15. $p^2 \cdot p^4$ 　　　　　　　　　　16. $3 \cdot 2x^2$

　17. $5x^3 \cdot x^4$ 　　　　　　　　　18. $4x \cdot 2x^2 \cdot x^3$

19. $2y^5 \cdot 5y^2$

20. $2a^2b \cdot 2ab^2 \cdot a$

21. $xy \cdot x^2 \cdot xy^3$

22. $x(x + 2)$

23. $p^2(p + 2p)$

24. $xy(x + y)$

25. $5^6 \cdot 5^4$

26. $3^2 \cdot 3^7$

27. $10^2 \cdot 10^{-5}$

28. $2^{-6} \cdot 2^9$

29. $-3x^2(2x^2 - 5x^3)$

30. $6x^3(4x^4 + 3x^2)$

31. $2ab^2(3a^2 - 5ab + 7b^2)$

32. $-5xy^3(4x^2 - 6xy + 8y^2)$

B. Divide as indicated. Express all answers using positive exponents.

1. $\dfrac{4^5}{4^3}$

2. $\dfrac{10^9}{10^3}$

3. $\dfrac{x^4}{x}$

4. $\dfrac{y^6}{y^2}$

5. $\dfrac{10^4}{10^6}$

6. $\dfrac{m^5}{m^{10}}$

7. $\dfrac{8a^8}{4a^4}$

8. $\dfrac{-15m^{12}}{5m^6}$

9. $\dfrac{16y^2}{-24y^3}$

10. $\dfrac{-20t}{-5t^3}$

11. $\dfrac{-6a^2b^4}{-2ab}$

12. $\dfrac{36x^4y^5}{-9x^2y}$

13. $\dfrac{48m^2n^3}{-16m^6n^4}$

14. $\dfrac{15c^6d}{20c^2d^3}$

15. $\dfrac{-12a^2b^2}{20a^2b^5}$

16. $\dfrac{100xy}{-10x^2y^4}$

17. $\dfrac{6x^4 - 8x^2}{2x^2}$

18. $\dfrac{9y^3 + 6y^2}{3y}$

19. $\dfrac{12a^7 - 6a^5 + 18a^3}{-6a^2}$

20. $\dfrac{-15m^5 + 10m^2 + 5m}{5m}$

When you have completed these exercises, check your answers in the Appendix, then turn to Exercises 7-8 to learn about scientific notation.

7-8   SCIENTIFIC NOTATION

In technical work you will often deal with very small and very large numbers which may require a lot of space to write. For example, a certain computer function may take 5 nanoseconds (0.000000005 s), and in electricity, 1 kilowatt-hour (kWh) is equal to 3,600,000 joules (J) of work. Rather than take the time and space to write all the zeros, we use scientific notation to express such numbers.

Definition of Scientific Notation   In scientific notation 5 nanoseconds would be written as 5×10^{-9} s, and 1 kWh would be 3.6×10^6 J. As you can see, a number written in **scientific notation** is a product of a number between 1 and 10 and a power of 10. Stated formally, a number is expressed in scientific notation when it is in the form

$$P \times 10^k$$

where P is a number less than 10 and greater than or equal to 1, and k is an integer.

Numbers that are powers of ten are easy to write in scientific notation:

$$1000 = 1 \times 10^3$$
$$100 = 1 \times 10^2$$
$$10 = 1 \times 10^1 \quad \text{and so on}$$

In general, to write a positive decimal number in scientific notation, first write it as a number between 1 and 10 times a power of 10, then write the power using an exponent.

EXAMPLE

$$3,600,000 = 3.6 \times 1,000,000$$

A multiple of 10
A number between 1 and 10

$$= 3.6 \times 10^6$$

$$0.0004 = 4 \times 0.0001 = 4 \times \frac{1}{10,000} = 4 \times \frac{1}{10^4} = 4 \times 10^{-4}$$

Note ▶ If you need to review negative exponents, see page 403. ◀

Converting to Scientific Notation

For a shorthand way of converting to scientific notation, follow these steps.

Step 1 If the number is given without a decimal point, rewrite it with a decimal point.

Example: $57,400 = 57,400.$ 0.0038

Decimal point   Decimal point already shown

Step 2 Place a mark $_\wedge$ after the first nonzero digit.

Example: $5 _\wedge 7400.$ $0.003 _\wedge 8$

Step 3 Count the number of digits from the mark $_\wedge$ to the decimal point.

Example: $5 _\wedge 7400.$ $0.003 _\wedge 8$
4 digits   3 digits

Step 4 Place the decimal point in the marked position and use the resulting number as the multiplier P in the scientific notation form. Discard any right end zeros on this number. Use the number of digits from the mark $_\wedge$ to the original decimal point as the exponent. If the shift from the $_\wedge$ to the original decimal point is to the right, the exponent is positive; if the shift is to the left, the exponent is negative.

Example: $5 _\wedge 7400. = 5.74 \times 10^4$ $0.003 _\wedge 8 = 3.8 \times 10^{-3}$
Shift 4 digits right. Exponent is +4.   Shift 3 digits left. Exponent is −3.

(a) $150{,}000 = 1{\scriptstyle\wedge}50000. = 1.5 \times 10^5$ (b) $0.00000205 = 0.000002{\scriptstyle\wedge}05 = 2.05 \times 10^{-6}$

        Shift 5 digits   Shift 6 digits left.
        right. Exponent is $+5$.   Exponent is -6.

(c) $47 = 4{\scriptstyle\wedge}7. = 4.7 \times 10^1$

        Shift 1 digit
        right.

YOUR TURN

For practice, write the following numbers in scientific notation.

(a) 2900   (b) 0.006   (c) 1,960,100

(d) 0.0000028   (e) 600   (f) 0.0001005

SOLUTIONS

(a) $2{,}900 = 2{\scriptstyle\wedge}900. = 2.9 \times 10^3$

           Shift 3 digits right.

(b) $0.006 = 0.006{\scriptstyle\wedge} = 6 \times 10^{-3}$

           Shift 3 digits left.

(c) $1{,}960{,}100 = 1{\scriptstyle\wedge}960100. = 1.9601 \times 10^6$

           Shift 6 digits right.

(d) $0.0000028 = 0.000002{\scriptstyle\wedge}8 = 2.8 \times 10^{-6}$

           Shift 6 digits left.

(e) $600 = 6{\scriptstyle\wedge}00. = 6 \times 10^2$

           Shift 2 digits right.

(f) $0.0001005 = 0.0001{\scriptstyle\wedge}005 = 1.005 \times 10^{-4}$

           Shift 4 digits left.

Learning Help ▶ When a positive number greater than or equal to 10 is written in scientific notation, the exponent is positive. When a positive number less than 1 is written in scientific notation, the exponent is negative. When a number between 1 and 10 is written in scientific notation, the exponent is 0. ◀

Converting from Scientific Notation to Decimal Form   In order to convert a number from scientific notation to decimal form, shift the decimal point as indicated by the power of 10—to the right for a positive exponent, to the left for a negative exponent. Attach additional zeros as needed.

(a)

Attach additional zeros.

$$3.2 \times 10^4 = 3\,2000. \quad = \quad 32,000$$

Shift the
decimal point
4 places right.

For a positive exponent, shift the decimal point to the right.

(b)

Attach additional zeros.

$$2.7 \times 10^{-3} = 0.002\,7 = 0.0027$$

Shift the
decimal point
3 places left.

For a negative exponent, shift the decimal point to the left.

YOUR TURN

Write each of the following numbers in decimal form.

(a) 8.2×10^3

(b) 1.25×10^{-6}

(c) 2×10^{-4}

(d) 5.301×10^5

ANSWERS

(a) $8.2 \times 10^3 = 8\,200\,. = 8200$

(b) $1.25 \times 10^{-6} = 0.000001\,25 = 0.00000125$

(c) $2 \times 10^{-4} = 0.0002 = 0.0002$

(d) $5.301 \times 10^5 = 5\,30100. = 530,100$

Multiplying and Dividing in Scientific Notation

Scientific notation is especially useful when we must multiply or divide very large or very small numbers. Although most calculators have a means of converting decimal numbers to scientific notation and can perform arithmetic with scientific notation, it is important that you be able to do simple calculations of this kind quickly and accurately without a calculator.

To multiply or divide numbers given in exponential form, use the rules given in Section 7-7. (See page 401.)

$$a^m \times a^n = a^{m+n}$$

$$a^m \div a^n = \frac{a^m}{a^n} = a^{m-n}$$

To multiply numbers written in scientific notation, work with the decimal and exponential parts separately.

$$(A \times 10^B) \times (C \times 10^D) = (A \times C) \times 10^{B+D}$$

EXAMPLE

$26{,}000 \times 3{,}500{,}000 = ?$

Step 1 Rewrite each number in scientific notation.

$$= (2.6 \times 10^4) \times (3.5 \times 10^6)$$

Step 2 Regroup to work with the decimal and exponential parts separately.

$$= (2.6 \times 3.5) \times (10^4 \times 10^6)$$

Step 3 Multiply using the rule for multiplying exponential numbers.

$$= 9.1 \times 10^{4+6}$$
$$= 9.1 \times 10^{10}$$

When dividing numbers written in scientific notation, it may help to think of the division as a fraction.

$$(A \times 10^B) \div (C \times 10^D) = \frac{A \times 10^B}{C \times 10^D} = \frac{A}{C} \times \frac{10^B}{10^D}$$

Therefore,

$$(A \times 10^B) \div (C \times 10^D) = (A \div C) \times 10^{B-D}$$

EXAMPLE

$45{,}000 \div 0.0018 = ?$

Step 1 Rewrite. $= (4.5 \times 10^4) \div (1.8 \times 10^{-3})$

Step 2 Regroup. $= (4.5 \div 1.8) \times (10^4 \div 10^{-3})$

Step 3 Divide using the rules for exponential division.

$$= 2.5 \times 10^{4-(-3)}$$
$$= 2.5 \times 10^7$$

$\boxed{4 - (-3) = 4 + 3 = 7}$

If the result of the calculation is not in scientific notation, that is, if the decimal part is greater than 10 or less than 1, rewrite the decimal part so that it is in scientific notation.

EXAMPLE

$$0.0000072 \div 0.0009 = (7.2 \times 10^{-6}) \div (9 \times 10^{-4})$$
$$= (7.2 \div 9) \times (10^{-6} \div 10^{-4})$$
$$= 0.8 \times 10^{-6-(-4)}$$
$$= 0.8 \times 10^{-2}$$

$\boxed{-6 - (-4) = -6 + 4 = -2}$

But 0.8 is not a number between 1 and 10, so write it as

$0.8 = 8 \times 10^{-1}$. Then, $0.8 \times 10^{-2} = 8 \times 10^{-1} \times 10^{-2}$
$$= 8 \times 10^{-3} \quad \text{in scientific notation}$$

Perform the following calculations using scientific notation, and write the answer in scientific notation.

(a) $1600 \times 350{,}000$

(b) $64{,}000 \times 250{,}000$

(c) 2700×0.0000045

(d) $15{,}600 \div 0.0013$

(e) $0.000348 \div 0.087$

(f) $0.00378 \div 540{,}000{,}000$

SOLUTIONS

(a) $(1.6 \times 10^3) \times (3.5 \times 10^5) = (1.6 \times 3.5) \times (10^3 \times 10^5)$
$$= 5.6 \times 10^{3+5}$$
$$= 5.6 \times 10^8$$

(b) $(6.4 \times 10^4) \times (2.5 \times 10^5) = (6.4 \times 2.5) \times (10^4 \times 10^5)$
$$= 16 \times 10^{4+5}$$
$$= 16 \times 10^9$$
$$= 1.6 \times 10^1 \times 10^9 \qquad \boxed{16 = 1.6 \times 10^1}$$
$$= 1.6 \times 10^{10}$$

(c) $(2.7 \times 10^3) \times (4.5 \times 10^{-6}) = (2.7 \times 4.5) \times (10^3 \times 10^{-6})$
$$= 12.15 \times 10^{3 + (-6)} \qquad \boxed{3 + (-6) = -3}$$
$$= 12.15 \times 10^{-3}$$
$$= 1.215 \times 10^1 \times 10^{-3}$$
$$= 1.215 \times 10^{-2}$$

(d) $(1.56 \times 10^4) \div (1.3 \times 10^{-3}) = (1.56 \div 1.3) \times (10^4 \div 10^{-3})$
$$= 1.2 \times 10^{4-(-3)} \qquad \boxed{4 - (-3) = 4 + 3 = 7}$$
$$= 1.2 \times 10^7$$

(e) $(3.48 \times 10^{-4}) \div (8.7 \times 10^{-2}) = (3.48 \div 8.7) \times (10^{-4} \div 10^{-2})$
$$= 0.4 \times 10^{-4-(-2)} \qquad \boxed{-4 - (-2) = -4 + 2 = -2}$$
$$= 0.4 \times 10^{-2}$$
$$= 4 \times 10^{-1} \times 10^{-2} \qquad \boxed{0.4 = 4 \times 10^{-1}}$$
$$= 4 \times 10^{-3}$$

(f) $(3.78 \times 10^{-3}) \div (5.4 \times 10^8) = (3.78 \div 5.4) \times (10^{-3} \div 10^8)$
$$= 0.7 \times 10^{-3-8}$$
$$= 0.7 \times 10^{-11}$$
$$= 7 \times 10^{-1} \times 10^{-11}$$
$$= 7 \times 10^{-12}$$

CALCULATORS AND SCIENTIFIC NOTATION

If a very large or very small number contains too many digits to be shown on the display of a scientific calculator, it will be converted to scientific notation. For example, if you enter the product

6480000 $\boxed{\times}$ **75000** $\boxed{=}$

on a calculator, the answer will be displayed as

    $4.86\ 11$ or $4.86\ E\ 11$

(continued)

The space or the E between the 4.86 and the 11 indicates that the number is in scientific notation, and 11 is the power of 10. Interpret this as

$$6{,}480{,}000 \times 75{,}000 = 4.86 \times 10^{11}$$

Similarly, the division

.000006 $\boxed{\div}$ 48000000 $\boxed{=}$

gives the display $1.25\ {-}13$.

So $0.000006 \div 48{,}000{,}000 = 1.25 \times 10^{-13}$.

You may also enter numbers in scientific notation directly into your calculator using a key labeled $\boxed{\text{EXP}}$ or $\boxed{\text{EE}}$. To enter

$0.000006 \div 48{,}000{,}000$ or $(6 \times 10^{-6}) \div (4.8 \times 10^{7})$

enter

6 $\boxed{\text{EXP}}$ 6 $\boxed{+/-}$ $\boxed{\div}$ 4.8 $\boxed{\text{EXP}}$ 7 $\boxed{=}$

and your calculator will again display $1.25\ {-}13$.

Now turn to Exercises 7-8 for a set of problems on scientific notation.

Exercises 7-8

Scientific Notation

A. Rewrite each number in scientific notation.

1. 5000	2. 450	3. 90	4. 40,700
5. 0.003	6. 0.071	7. 0.0004	8. 0.0059
9. 6,770,000	10. 38,200	11. 0.0292	12. 0.009901
13. 1001	14. 0.0020	15. 0.000107	16. 810,000
17. 31.4	18. 0.6	19. 125	20. 0.74

21. Young's modulus for the elasticity of steel: 29,000,000 lb/in.^{2}

22. Thermal conductivity of wood: 0.00024 cal/cm·s

23. Power output: 95,500,000 watts

24. Speed of light: 658,800,000 mph

B. Rewrite each number in decimal form.

1. 2×10^5	2. 7×10^6	3. 9×10^{-5}	4. 3×10^{-4}
5. 1.7×10^{-3}	6. 3.7×10^{-2}	7. 5.1×10^4	8. 8.7×10^2
9. 4.05×10^4	10. 7.01×10^{-6}	11. 3.205×10^{-3}	12. 1.007×10^3
13. 2.45×10^6	14. 3.19×10^{-4}	15. 6.47×10^5	16. 8.26×10^{-7}

C. Rewrite each of the following in scientific notation and calculate the answer in scientific notation. If necessary, convert your answer to scientific notation. Round to one decimal place if necessary.

1. $2000 \times 40{,}000$	2. 0.0037×0.0000024
3. $460{,}000 \times 0.0017$	4. $0.0018 \times 550{,}000$
5. $0.0000089 \div 3200$	6. $0.000125 \div 5000$
7. $45{,}500 \div 0.0091$	8. $12{,}450 \div 0.0083$

9. $2,240,000 \div 16,000$

10. $25,500 \div 1,700,000$

11. $0.000045 \div 0.00071$

12. $0.000086 \div 0.000901$

13. $9,501,000 \times 2410$

14. 9800×0.000066

15. $0.0000064 \div 80,000$

16. 1070×0.0000055

17. $\dfrac{0.000056}{0.0020}$

18. $\dfrac{0.0507}{43,000}$

19. $\dfrac{0.00602 \times 0.000070}{72,000}$

20. $\dfrac{2,780,000 \times 512,000}{0.000721}$

21. $64,000 \times 2800 \times 370,000$

22. $0.00075 \times 0.000062 \times 0.014$

23. $\dfrac{0.0517}{0.0273 \times 0.00469}$

24. $\dfrac{893,000}{5620 \times 387,000}$

D. Solve.

1. A brick wall 15 m by 25 m is 0.48 m thick. Under particular temperature conditions, the rate of heat flow through the wall, in calories per second, is given by the expression

$$H = (1.7 \times 10^{-4}) \times \left(\frac{1500 \times 2500}{48} \right)$$

Calculate the value of this quantity to the nearest tenth.

2. If an atom has a diameter of 4×10^{-8} cm and its nucleus has a diameter of 8×10^{-13} cm, find the ratio of the diameter of the nucleus to the diameter of the atom.

3. **Electrical Technology** The capacitance of a certain capacitor (in microfarads) can be found using the following expression. Calculate this capacitance.

$$C = (5.75 \times 10^{-8}) \times \left(\frac{8.00}{3.00 \times 10^{-3}} \right)$$

4. The energy (in joules) of a photon of visible light with a wavelength of 5.00×10^{-7} m is given by the following expression. Calculate this energy.

$$E = \frac{(6.63 \times 10^{-34}) \times (3.00 \times 10^{8})}{5.00 \times 10^{-7}}$$

Check your answers in the Appendix, then turn to Problem Set 7 for some practice problems on the work of this chapter.

Basic Algebra

Answers are given in the Appendix.

A. Simplify.

1. $4x + 6x$

2. $6y + y + 5y$

3. $3xy + 9xy$

4. $6xy^3 + 9xy^3$

5. $3\frac{1}{3}x - 1\frac{1}{4}x - \frac{5}{8}x$

6. $8v - 8v$

7. $0.27G + 0.78G - 0.65G$

8. $7y - 8y^2 + 9y$

9. $3x \cdot 7x$

10. $(4m)(2m^2)$

11. $(2xy)(-5xyz)$

12. $3x \cdot 3x \cdot 3x$

13. $3(4x - 7)$

14. $2ab(3a^2 - 5b^2)$

15. $(4x + 3) + (5x + 8)$

16. $(7y - 4) - (2y + 3)$

17. $3(x + y) + 6(x - y)$

18. $(x^2 - 6) - 3(2x^2 - 5)$

19. $\dfrac{6^8}{6^4}$

20. $\dfrac{x^3}{x^8}$

21. $\dfrac{-12m^{10}}{4m^2}$

22. $\dfrac{6y}{-10y^3}$

23. $\dfrac{-16a^2b^2}{-4ab^3}$

24. $\dfrac{8x^4 - 4x^3 + 12x^2}{4x^2}$

B. Find the value of each of the following. Round to two decimal places when necessary.

1. $L = 2W - 3$ for $W = 8$

2. $M = 3x - 5y + 4z$ for $x = 3, y = 5, z = 6$

3. $I = PRt$ for $P = 800, R = 0.06, t = 3$

4. $I = \dfrac{V}{R}$ for $V = 220, R = 0.0012$

5. $V = LWH$ for $L = 3\frac{1}{2}, W = 2\frac{1}{4}, H = 5\frac{3}{8}$

Name

_____

Date

_____

Course/Section

_____

6. $N = (a + b)(a - b)$ for $a = 7, b = 12$

7. $L = \dfrac{s(P + p)}{2}$ for $s = 3.6, P = 38, p = 26$

8. $f = \dfrac{1}{8N}$ for $N = 6$

9. $t = \dfrac{D - d}{L}$ for $D = 12, d = 4, L = 2$

10. $V = \dfrac{gt^2}{2}$ for $g = 32.2, t = 4.1$

C. **Solve the following equations. Round your answer to two decimal places if necessary.**

1. $x + 5 = 17$ **2.** $m + 0.4 = 0.75$

3. $e - 12 = 32$ **4.** $a - 2.1 = -1.2$

5. $12 = 7 - x$ **6.** $4\frac{1}{2} - x = -6$

7. $5x = 20$ **8.** $-4m = 24$

9. $\frac{1}{2}y = 16$ **10.** $2x + 7 = 13$

11. $-4x + 11 = 35$ **12.** $0.75 - 5f = 6\frac{1}{2}$

13. $0.5x - 16 = -18$ **14.** $5y + 8 + 3y = 24$

15. $3g - 12 = g + 8$ **16.** $7m - 4 = 11 - 3m$

17. $3(x - 5) = 33$ **18.** $2 - (2x + 3) = 7$

19. $5x - 2(x - 1) = 11$ **20.** $2(x + 1) - x = 8$

21. $12 - 7y = y - 18$ **22.** $2m - 6 - 4m = 4m$

23. $9 + 5n + 11 = 10n - 25$ **24.** $6(x + 4) = 45$

25. $5y - (11 - 2y) = 3$ **26.** $4(3m - 5) = 8m + 20$

Solve the following formulas for the variable shown.

27. $A = bH$ for b **28.** $R = S + P$ for P

29. $P = 2L + 2W$ for L **30.** $P = \dfrac{w}{F}$ for F

31. $S = \frac{1}{2}gt - 4$ for g **32.** $V = \pi R^2 H - AB$ for A

D. **Rewrite in scientific notation.**

1. 7500   **2.** 12,800   **3.** 0.041   **4.** 0.000236

5. 0.00572   **6.** 0.00000482   **7.** 447,000   **8.** 2,127,000

9. 80,200,000   **10.** 46,710   **11.** 0.00000705   **12.** 0.001006

Rewrite as a decimal.

13. 9.3×10^5 **14.** 6.02×10^3 **15.** 2.9×10^{-4} **16.** 3.05×10^{-6}

17. 5.146×10^{-7} **18.** 6.203×10^{-3} **19.** 9.071×10^4 **20.** 4.006×10^6

Calculate using scientific notation, and write your answer in standard scientific notation. Round to one decimal place if necessary.

21. $45,000 \times 1,260,000$ **22.** $625,000 \times 12,000,000$

23. 0.0007×0.0043 **24.** 0.0000065×0.032

25. $56,000 \times 0.0000075$ **26.** 1020×0.00055

27. $0.0074 \div 0.00006$ **28.** $0.000063 \div 0.0078$

29. $96,000 \div 3,400,000$ **30.** $26,500,000 \div 12,000$

31. $0.00089 \div 37,000$ **32.** $123,500 \div 0.00077$

E. Practical Problems

1. **Aeronautical Mechanics** In weight and balance calculations, airplane mechanics are concerned with the center of gravity of an airplane. The center of gravity may be calculated from the formula

   $$CG = \frac{100(H - x)}{L}$$

   where H is the distance from the datum to the empty CG, x is the distance from the datum to the leading edge of the mean aerodynamic chord (MAC), and L is the length of the MAC. CG is expressed as a percent of the MAC.

   (All lengths are in inches.)
   (a) Find the center of gravity if $H = 180$, $x = 155$, and $L = 80$.
   (b) Solve the formula for L, and find L when CG = 30%, $H = 200$, and $x = 150$.
   (c) Solve the formula for H, and find H when CG = 25%, $x = 125$, and $L = 60$.
   (d) Solve the formula for x, and find x when CG = 28%, $H = 170$, and $L = 50$.

2. **Electrical Technology** Electricians use a formula which states that the level of light in a room, in foot-candles, is equal to the product of the fixture rating, in lumens, the coefficient of depreciation, and the coefficient of utilization, all divided by the area, in square feet.
   (a) State this as an algebraic equation.
   (b) Find the level of illumination for four fixtures rated at 2800 lumens each if the coefficient of depreciation is 0.75, the coefficient of utilization is 0.6, and the area of the room is 120 sq ft.
   (c) Solve for area.
   (d) Use your answer to (c) to determine the size of the room in which a level of 60 foot-candles can be achieved with 10,000 lumens, given that the coefficient of depreciation is 0.8 and the coefficient of utilization is 0.5.

3. **Sheet Metal Technology** The length of a piece of sheet metal is four times its width. The perimeter of the sheet is 80 cm. Set up an equation relating these measurements and solve the equation to find the dimensions of the sheet.

4. **Construction Technology** Structural engineers have found that a good estimate of the crushing load for a square wooden pillar is given by the formula

   $$L = \frac{25T^4}{H^2}$$ where L is the crushing load in tons, T is the thickness of the wood in inches, and H is the height of the post in feet

   Find the crushing load for a 6-in.-thick post 12 ft high.

5. **Machine Technology** Two shims must have a combined thickness of 0.048 in. One shim must be twice as thick as the other. Set up an equation and solve for the thickness of each shim.

6. **Office Services** Sal, Martin, and Gloria are partners in a welding business. Because of the differences in their contributions, it is decided that Gloria's share of the profits will be one-and-a-half times as much as Sal's share, and that Martin's share will be twice as much as Gloria's share. How should a profit of $48,200 be divided?

7. **Electrical Engineering** The calculation shown is used to find the velocity of an electron at an anode. Calculate this velocity (units are in meters per second).

$$V = \sqrt{\frac{2(1.6 \times 10^{-19})(2.7 \times 10^2)}{9.1 \times 10^{-31}}}$$

8. **Plumbing** The formula

$$N = \sqrt{\left(\frac{D}{d}\right)^5}$$

gives the approximate number N of smaller pipes of diameter d necessary to supply the same total flow as one larger pipe of diameter D. Unlike the formula in problem 22 on p. 347, this formula takes into account the extra friction caused by the smaller pipes. Use this formula to determine the number of $\frac{1}{2}$-in. pipes that will provide the same flow as one $1\frac{1}{2}$-in. pipe.

9. **Sheet Metal Technology** The length L of the stretchout for a square vent pipe with a grooved lock-seam is given by

$$L = 4s + 3w$$

where s is the side length of the square end and w is the width of the lock. Find the length of stretchout for a square pipe $12\frac{3}{4}$ in. on a side with a lock width of $\frac{3}{16}$ in.

10. **Refrigeration and Air Conditioning** In planning a solar energy heating system for a house, a contractor uses the formula

$$Q = 8.33GDT$$ to determine the energy necessary for heating water.

In this formula Q is the energy in Btu, G is the number of gallons heated per day, D is the number of days, and T is the temperature difference between tap water and the desired temperature of hot water. Find Q when G is 50 gallons, D is 30 days, and the water must be heated from 60° to 140°.

11. **Sheet Metal Technology** The bend allowance for sheet metal is given by the formula

$$BA = N(0.01743R + 0.0078T)$$ where N is the angle of the bend in degrees, R is the inside radius of the bend in inches, and T is the thickness of the metal in inches.

Find BA if N is 47°, R is 0.725 in., and T is 0.0625 in. Round to three significant digits.

12. **Machine Technology** To find the taper per inch of a piece of work, a machinist uses the formula

$$T = \frac{D - d}{L}$$ where D is the diameter of the large end, d is the diameter of the small end, and L is the length.

Find T if $D = 4.1625$ in., $d = 3.2513$ in., and $L = 8$ in.

13. **Construction Technology** The modulus of elasticity of a beam is 5,650,000 at a deflection limit of 0.360. If the modulus is directly proportional to the deflection limit, find the modulus at a deflection limit of 0.240. Round to three significant digits.

14. **Automotive Technology** The Static Stability Factor (SSF) is a measure of the relative rollover risk associated with an automobile.

$$\text{SSF} = \frac{1}{2} \times \frac{\text{Tire-track Width}}{\text{Height of Center of Gravity}}$$

An SSF of 1.04 or less indicates a 40% or higher rollover risk and an SSF of 1.45 or more indicates a 10% or less rollover risk. A model 2002 Avalunch SUV has a tire-track width of 56 in. and a center of gravity 26 in. above the road.
(a) Calculate the SSF for this SUV.
(b) By how much would auto designers need to lower the center of gravity to improve the SSF to 1.4?

15. **Agricultural Technology** Agricultural researchers developed the following formula from a study on the effects of the salinity of irrigation water on the yield potential of rice fields:

$$Y = 100 - 9.1\,(S - 1.9)$$

In this formula, Y is the percent yield of the rice fields, and S is the average salinity of the irrigation water in decisiemens per meter (dS/m).
(a) If $S = 5.2$ dS/m, find the percent yield of the rice fields. (Round to the nearest percent.)
(b) In order to achieve a yield of 80%, what must be the salinity of the irrigation water? (Round to the nearest 0.1 dS/m.)

Practical Plane Geometry

Objective	Sample Problems	For Help Go to Page

When you finish this unit you will be able to:

1. Measure angles with a protractor.

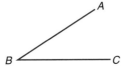

$\sphericalangle ABC =$ _____ 423

2. Use simple geometric relationships involving intersecting lines and triangles.

(a)

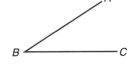

Assume that the two horizontal lines are parallel.

$\sphericalangle a =$ _____ 426

$\sphericalangle b =$ _____

$\sphericalangle c =$ _____

$\sphericalangle d =$ _____

$\sphericalangle e =$ _____

(b)

$\sphericalangle x =$ _____ 427

3. Identify polygons, including triangles (right, isosceles, equilateral), squares, rectangles, parallelograms, trapezoids, and hexagons.

(a)

_____ 448

(b)

_____ 435

(c)

_____

(d)

_____

(e)

_____ 458

_____
Name

_____
Date

_____
Course/Section

4. Use the Pythago-
   rean theorem.

$x = $ _____   450

Round to one decimal
place.

5. Find the area and
   perimeter of geo-
   metric figures.

(a)

Area = _____   454

(b)

Area = _____   441

(c)

Area = _____   459
(Round to two decimal places.)

Perimeter = _____

(d)

Area = _____   472

$r = 2.70"$

Circumference = _____   468

Use $\pi \approx 3.14$
Use $\pi \approx 3.14$
and round both
answers to the
nearest tenth.

(e)

Perimeter = _____   434

Area = _____   462

(f) Find the area of a
    ring with outside
    diameter 6.5 in.
    and inside diame-
    ter 4.2 in. Round
    to the nearest
    tenth.

_____   474

(Answers to these preview problems are given in the Appendix.)

If you are certain that you can work *all* these problems correctly, turn to page 481 for a set of practice problems. If you cannot work one or more of the preview problems, turn to the page indicated to the right of the problem. Those who wish to master this material with the greatest success should turn to Section 8-1 and begin work there.

8 Practical Plane Geometry

"NO! NO! I SAID BUILD AN ARK!"

© 1974 Reprinted by permission of *Saturda* and Orlando Busino and Omni International

Geometry is one of the oldest branches of mathematics. It involves study of the properties of points, lines, plane surfaces, and solid figures. Ancient Egyptian engineers used these properties when they built the Pyramids, and modern trades workers use them when cutting sheet metal, installing plumbing, building cabinets, and performing countless other tasks. When they study formal geometry, students concentrate on theory and actually prove the truth of various geometric theorems. In this chapter we concentrate on the applications of the ideas of plane geometry to do technical work and the trades.

8-1 ANGLE MEASUREMENT

In geometry, the word *plane* is used to refer to a mathematical concept describing an infinite set of points in space. For practical purposes, a plane surface can be thought of as a perfectly flat, smooth surface such as a tabletop or a window pane.

Labeling Angles   An *angle* is a measure of the size of the opening between two intersecting lines. The point of intersection is called the *vertex,* and the lines forming the opening are called the *sides*.

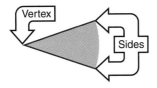

An angle may be identified in any one of the following ways:

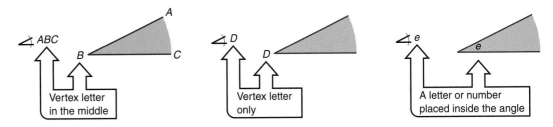

The angle symbol ∡ is simply a shorthand way to write the word "angle."

For the first angle, the middle letter is the vertex letter of the angle, while the other two letters represent points on each side. Notice that capital letters are used. The second angle is identified by the letter D near the vertex. The third angle is named by a small letter or number placed inside the angle.

The first and third methods of naming angles are most useful when several angles are drawn with the same vertex.

EXAMPLE

The interior of angle POR (or ROP) is shaded.

The interiors of angles a and 1 are shaded.

YOUR TURN

Name the angle shown in the margin using the three different methods.

ANSWERS

∡XYZ or ∡ZYX or ∡Y or ∡u

Measuring Angles

Naming or labeling an angle is the first step. Measuring it may be even more important. Angles are measured according to the size of their opening. The length of the sides of the angle is not related to the size of the angle.

and ◁ are the same size angle.

The basic unit of measurement of angle size is the *degree*. Because of traditions going back thousands of years, we define a degree as an angle that is $\frac{1}{360}$ of a full circle, or $\frac{1}{90}$ of a quarter circle.

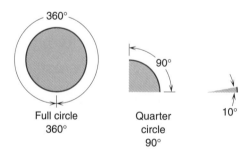

The degree can be further subdivided into finer angle units known as *minutes* and *seconds* (which are not related to the time units).

1 degree, $1° = 60$ minutes $= 60'$

1 minute, $1' = 60$ seconds $= 60''$

An angle of 62 degrees, 12 minutes, 37 seconds would be written $62°12'37''$. Most trade work requires precision only to the nearest degree.

As you saw in Chapter 5, page 282, a protractor is a device used to measure angles. To learn to measure angles, first look at the protractor below.

A protractor that you may cut out and use is available in the Appendix.

To measure an angle, place the protractor over it so that the zero-degree mark is lined up with one side of the angle, and the center mark is on the vertex. Read the measure in degrees where the other side of the angle intersects the scale of the protractor. When reading an angle clockwise, use the upper scale, and when reading counterclockwise, use the lower scale.

EXAMPLE

(a) In the drawing, $\angle EOB$ should be read counterclockwise from the $0°$ mark on the right. $\angle EOB = 135°$.

(b) $\angle AOD$ should be read clockwise from the $0°$ mark on the left. $\angle AOD = 120°$.

YOUR TURN

Find (a) $\angle EOD$ (b) $\angle AOB$

ANSWERS

(a) $\angle EOD = 60°$ Measure counterclockwise from the right $0°$ mark and read the lower scale.

(b) $\angle AOB = 45°$ Measure clockwise from the left $0°$ mark and read the upper scale.

Classifying Angles   An **acute angle** is an angle *less than* $90°$. Angles AOB and EOD are both less than $90°$, so both are acute angles.

Acute angles:

An **obtuse angle** is an angle *greater than* 90° and less than 180°. Angles *AOD* and *EOB* are both greater than 90°, so both are obtuse angles.

Obtuse angles:

C

X — E

Perpendicular
lines

A **right angle** is an angle exactly equal to 90°. Angle *EXC* in the drawing at the left is a right angle.

Two straight lines that meet in a right or 90° angle are said to be **perpendicular.** In this drawing, *CX* is perpendicular to *XE*.

Notice that a small square is placed at the vertex of a right angle to show that the sides are perpendicular.

A **straight angle** is an angle equal to 180°.

Angle *BOA* shown here is a straight angle.

$\overline{A \quad O \quad B}$

YOUR TURN

Use your protractor to measure each of the angles given and tell whether each is acute, obtuse, right, or straight. Estimate the size of the angle before you measure. (You will need to extend the sides of the angles to fit your protractor.)

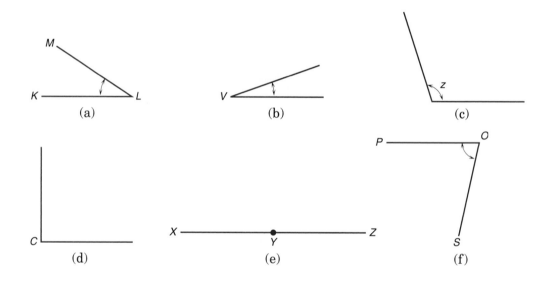

ANSWERS

(a) $\angle KLM = 33°$, an acute angle

(b) $\angle V = 20°$, an acute angle

(c) $\angle z = 108°$, an obtuse angle

(d) $\angle C = 90°$, a right angle

(e) $\angle XYZ = 180°$, a straight angle

(f) $\angle POS = 78°$, an acute angle

Two angles that add up to 90° are called **complementary angles.**

Since 40° + 50° = 90°, the 40° angle is called the *complement* of 50°, and the 50° angle is the complement of the 40° angle.

Two angles that add up to 180° are called **supplementary angles.**

Since 120° + 60° = 180°, the 120° angle is called the *supplement* of 60°, and the 60° angle is the supplement of the 120° angle.

Drawing Angles In addition to measuring angles that already exist, trades workers will also need to draw their own angles. To draw an angle of a given size follow these steps:

Step 1 Draw a line representing one side of the angle.

———————

Step 2 Place the protractor over the line so that the center mark is on the vertex, and the 0° mark coincides with the other end of the line. (**Note:** We could have chosen either end of the line for the vertex.)

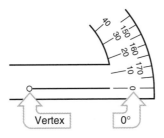

Step 3 Now place a small dot above the degree mark corresponding to the size of your angle. In this case the angle is 30°.

Step 4 Remove the protractor and connect the dot to the vertex.

⬛ YOUR TURN ▬▬▬▬▬▬▬▬▬▬▬▬▬▬▬

Now you try it. Use your protractor to draw the following angles.

$\angle A = 15°$ $\angle B = 112°$ $\angle C = 90°$

⬛ ANSWERS ▬▬▬▬▬▬▬▬▬▬▬▬▬▬▬

Here are our drawings:

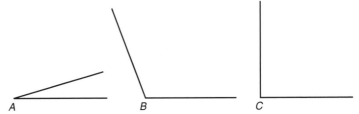

Angle Facts There are several important geometric relationships involving angles that you should know. For example, if we measure the four angles created by the intersecting lines shown,

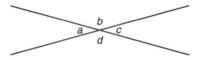

we find that

$\angle a = 30°$ $\qquad$ $\angle b = 150°$ $\qquad$ $\angle c = 30°$ $\quad$ and $\quad$ $\angle d = 150°$

Do you notice that opposite angles are equal?

For any pair of intersecting lines the opposite angles are called **vertical angles.** In the preceding drawing, angles a and c are a pair of vertical angles. Angles b and d are also a pair of vertical angles.

An important geometric rule is that

> When two straight lines intersect, the vertical angles are always equal.

$$\angle p = \angle r$$
and $\angle q = \angle s$

YOUR TURN

If you remember that a straight angle = 180°, you can complete the following statements for the preceding drawing.

$\angle p + \angle q =$ _____ $\qquad\qquad$ $\angle p + \angle s =$ _____

$\angle s + \angle r =$ _____ $\qquad\qquad$ $\angle q + \angle r =$ _____

ANSWERS

$\angle p + \angle q = 180°$ $\qquad$ $\angle p + \angle s = 180°$

$\angle s + \angle r = 180°$ $\qquad$ $\angle q + \angle r = 180°$

This leads to a second important geometry relation:

> When two lines intersect, each pair of adjacent angles sums to 180°.

YOUR TURN

Use these relationships to answer the following questions.

(a) $\angle a =$ _____   (b) $\angle b =$ _____   (c) $\angle c =$ _____

ANSWERS

(a) $\angle a = 40°$ since   vertical angles are equal
(b) $\angle b = 140°$ since $\angle a + \angle b = 180°$
(c) $\angle c = 140°$ since   vertical angles are equal, $\angle b = \angle c$

A triangle is formed from three distinct line segments. The line segments are called the *sides* of the triangle, and the points where they meet are called its *vertices*. Every triangle forms three angles, and these are related by an important geometric rule.

If we use a protractor to measure each angle in the following triangle, *ABC,* we find

$\angle A = 69°$ $\angle B = 36°$ $\angle C = 75°$

Notice that the three angles of the triangle sum to 180°.

$69° + 36° + 75° = 180°$

Measuring angles on many triangles of every possible size and shape leads to the third geometry relationship:

> The interior angles of a triangle always add to 180°.

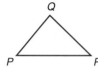

We can use this fact to solve the following problems.

(a) If $\angle P = 50°$ and $\angle Q = 85°$, find $\angle R$.

By the relationship given, $\angle P + \angle Q + \angle R = 180°$, so

$\angle R = 180° - \angle P - \angle Q$
$\angle R = 180° - 50° - 85°$
$\angle R = 45°$

(b) If $\angle P = 27° \, 42'$ and $\angle Q = 76° \, 25'$, find $\angle R$.

Step 1 $\angle R = 180° - 27° \, 42' - 76° \, 25'$

Step 2 Combine the two angles that are being subtracted.
$$27° \, 42' + 76° \, 25' = (27 + 76)° + (42 + 25)'$$
$$= 103° \, 67'$$

Step 3 To simplify $103° \, 67'$ use the fact that $60' = 1°$.
$$103° \, 67' = 103° + 60' + 7'$$
$$= 103° + 1° + 7' = 104° \, 7'$$

Step 4 Subtract this total from $180°$. First change $180°$ to $179° \, 60'$.
$$180° - 104° \, 7' = 179° \, 60' - 104° \, 7'$$
$$= (179 - 104)° + (60 - 7)'$$
$$= 75° \, 53'$$

Find the missing angle in each problem.

(a) If $\angle X = 34°$ and $\angle Y = 132°$, find $\angle Z$

(b) If $\angle X = 29° \, 51'$ and $\angle Z = 24° \, 34'$, find $\angle Y$

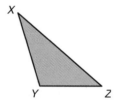

(a) $\angle Z = 180° - \angle X - \angle Y$
$ = 180° - 34° - 132°$
$ = 180° - 166° = 14°$

(b) $\angle Y = 180° - \angle X - \angle Z$
$ = 180° - 29° \, 51' - 24° \, 34'$
$ = 180° - 53° \, 85'$
$ = 180° - 54° \, 25'$
$ = 179° \, 60° - 54° \, 25'$
$ = 125° \, 35'$

$53°85' = 53° + 60' + 25'$
$ = 53° + 1° + 25'$
$ = 54°25'$

A fourth and final geometry fact you will find useful involves parallel lines. Two straight lines are said to be **parallel** if they always are the same distance apart and never meet. If a pair of parallel lines is cut by a third line called a **transversal,** several important pairs of angles are formed.

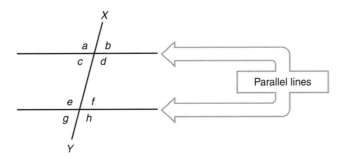

Angles *c* and *f* are called **alternate interior angles.** Angles *d* and *e* are also alternate interior angles.

The angles are *interior* (or inside) the original parallel lines and they are on *alternate* sides of the transversal *XY.* The alternate interior angles are equal when the lines cut by the transversal are parallel.

The angles *a* and *e* form one of four pairs of **corresponding angles.** The others are *c* and *g, b* and *f,* and *d* and *h.*

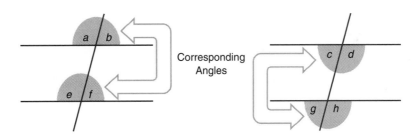

These angles are on the same side of the transversal, and they are on the same side of their respective parallel line. For example, *b* and *f* are both above their parallel lines while *c* and *g* are both below their lines. Corresponding angles are always equal when the lines cut by the transversal are parallel.

The geometric rule is:

> When parallel lines are cut by a third line, the corresponding angles are equal, and the alternate interior angles are equal.

YOUR TURN

Use this rule and the previous ones to answer the following questions. Assume that the two horizontal lines are parallel.

$\angle p = $ _____ $\angle q = $ _____

$\angle r = $ _____ $\angle s = $ _____

$\angle t = $ _____ $\angle u = $ _____

$\angle w = $ _____

ANSWERS

$\angle p = 110°$ since $70° + \angle p = 180°$. Adjacent angles sum to $180°$.

$\angle q = 110°$ since p and q are vertical angles.

$\angle r = 70°$ since it is a vertical angle to the $70°$ angle.

$\angle s = 70°$ since $\angle s = \angle r$, alternate interior angles.

$\angle t = 110°$ since $\angle t = \angle q$, alternate interior angles, and $\angle p = \angle t$, corresponding angles.

$\angle u = 110°$ since $\angle u = \angle q$, corresponding angles.

$\angle w = 70°$ since $\angle u + \angle w = 180°$ and $\angle w = \angle r$, corresponding angles.

Now, ready for some problems on angle measure? Turn to Exercises 8-1 for practice in measuring and working with angles.

Exercises 8-1

Angle Measurement

A. Solve the following. If you need a protractor, cut out and use the one in the Appendix.

1. Name each of the following angles using the three kinds of notation shown in the text.

(a)   (b)   (c)

(d)   (e)   (f)

2. For this figure,

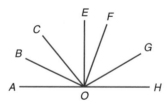

(a) Name an acute angle that has AO as a side.

(b) Name an acute angle that has HO as a side.

(c) Name an obtuse angle that has AO as a side.

(d) Name a right angle.

(e) Use your protractor to measure $\angle AOB$, $\angle GOF$, $\angle HOC$, $\angle FOH$, $\angle BOF$, $\angle COG$.

3. Use your protractor to draw the following angles.

(a) $\sphericalangle LMN = 29°$

(b) $\sphericalangle 3 = 100°$

(c) $\sphericalangle Y = 152°$

(d) $\sphericalangle PQR = 137°$

(e) $\sphericalangle t = 68°$

(f) $\sphericalangle F = 48°$

4. Measure the indicated angles on the following shapes. (You will need to extend the sides of the angles to improve the accuracy of your measurement.)

(a)

(b)

(c)

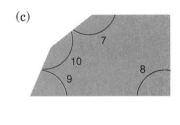

B. Use the geometry relationships to answer the following questions.

1. In each problem one angle measurement is given. Determine the others for each figure without using a protractor.

(a)

(b)

(c)

(d)

Where AB is parallel to CD

(e)

(f)

2. For each triangle shown, two angles are given. Find the third angle without using a protractor.

(a)

(b)

(c)

(d)

(e) (The angles labelled a are equal.)

(f)

(g) 90° 27°

(h) 109°43' 25°32'

(i) 46°12' 73°46'

3. In each figure, two parallel lines are cut by a third line. Find the angles marked.

(a) 100° w x y z

(b) b c f a 60° d e g

(c) n 80° p h q m k

(d) t x z s u w y 40°

Problem 1

Problem 2

Upper conduit

Lower conduit

Problem 5

Problem 6

C. **Applied Problems**

1. **Machine Technology** A machinist must punch holes A and B in the piece of steel shown by rotating the piece through a certain angle from vertex C. Use a protractor to find angle ACB.

2. **Plumbing** A plumber must connect the two pipes shown by first selecting the proper elbows. What angle elbows does he need?

3. **Carpentry** A carpenter needs to build a triangular hutch that will fit into the corner of a room as shown. Measure the three angles of the hutch.

Problem 3

Problem 4

4. **Sheet Metal Technology** A sheet metal worker must connect the two vent openings shown. At what angle must the sides of the connecting pieces flare out?

5. **Electrical Technology** An electrician wants to connect the two conduits shown with a 45° elbow. How far up must she extend the lower conduit before they will connect at that angle? Give the answer in inches. (*Hint:* Draw the 45° connection on the upper conduit first.) Each dot represents a 1-in. extension.

6. **Construction Technology** At what angle must the rafters be set to create the roof gable shown in the drawing?

7. **Machine Technology** A machinist receives a sketch of a part he must make as shown. If the indicated angle *BAC* is precisely half of angle *BAD*, find ⊾*BAD*.

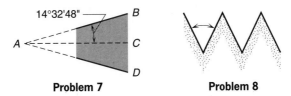

Problem 7   Problem 8

8. **Machine Technology** The drawing illustrates a cross section of a V-thread. Find the indicated angle.

9. **Carpentry** At what angles must drywall board be cut to create the wall shown in the figure?

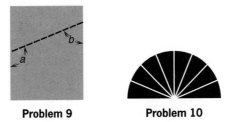

Problem 9   Problem 10

10. **Carpentry** A carpenter wishes to make a semicircular deck by cutting eight equal angular pieces of wood as shown. Measure the angle of one of the pieces. Was there another way to determine the angle measure without using a protractor?

11. **Carpentry** A carpenter's square is placed over a board as shown in the figure. If ⊾1 measures 74°, what is the size of ⊾2?

Problem 11   Problem 12

12. Board 1 must be joined to board 2 at a right angle. If ⊾*x* measures 42°, what must ⊾*y* measure?

When you have finished these problems check your answers in the Appendix, then continue in Section 8-2 with the study of plane figures.

8-2 AREA AND PERIMETER OF POLYGONS

Polygons A *polygon* is a closed plane figure containing three or more angles and bounded by three or more straight sides. The word "polygon" itself means "many sides." The following figures are all polygons.

A figure with a curved side is *not* a polygon.

 are not polygons

Every trades worker will find that an understanding of polygons is important. In this section you will learn how to identify the parts of a polygon, recognize some different kinds of polygons, and compute the perimeter and area of certain polygons.

In the general polygon shown, each *vertex* or corner is labeled with a letter: *A*, *B*, *C*, *D*, and *E*. The polygon is named simply "polygon *ABCDE*"—not a very fancy name, but it will do.

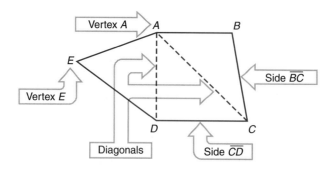

The *sides* of the polygon are named after the line segments that form each side: $\overline{AB}, \overline{BC}, \overline{CD}$, and so on. Notice that we place a bar over the letters to indicate that we are talking about a line segment. A *line segment* is a finite portion of a straight line; it has two endpoints.

The *diagonals* of the polygon are the line segments connecting nonconsecutive vertices such as $\overline{AC}$ and $\overline{AD}$. These are shown as dashed lines in the figure.

EXAMPLE

In the following polygon, identify all sides, vertices, and diagonals.

The sides are $\overline{UV}, \overline{VW}, \overline{WX}, \overline{XY}, \overline{YZ}$, and $\overline{ZU}$.

Note ▶ The letters naming a side can be reversed: side $\overline{VU}$ is the same as side $\overline{UV}$. ◀

The vertices are *U*, *V*, *W*, *X*, *Y*, and *Z*.

There are nine diagonals; we need to be certain to find them all.

The diagonals are $\overline{UW}, \overline{UX}, \overline{UY}, \overline{VX}, \overline{VY}, \overline{VZ}, \overline{WY}, \overline{WZ}$, and $\overline{XZ}$.

Perimeter For those who do practical work with polygons, the most important measurements are the lengths of the sides, the *perimeter,* and the *area*. The perimeter of any polygon is simply the sum of the lengths of its sides. It is the distance around the outside of the polygon.

EXAMPLE

In the polygon *KLMN* the perimeter is

3 in. + 5 in. + 6 in. + 4 in. = 18 in.

The perimeter is a length; therefore, it has length units, in this case inches.

434

Find the perimeter of each of the following polygons.

(a)

(b) 2"

(c) 24 yd

(d) 7.2 cm

(e)

(f) 2'8"

(g) 4 yd

(a) 25 ft   (b) 20 in.   (c) 80 yd   (d) 49.4 cm

(e) $42\frac{5}{8}$ in.   (f) 29 ft 10 in.   (g) 14 yd 2 ft or 44 ft

---

Careful ▶ As you may have discovered in problem (g) it is important that all sides be expressed in the same units before calculating the perimeter. ◀

Adding up the lengths of the sides will always give you the perimeter of a polygon, but handy formulas are needed to enable you to calculate the area of a polygon. Before you can use these formulas you must learn to identify the different types of polygons. Let's look at a few. In this section we examine the *quadrilaterals* or four-sided polygons.

Quadrilaterals Figure *ABCD* is a *parallelogram*. In a parallelogram opposite sides are parallel and equal in length.

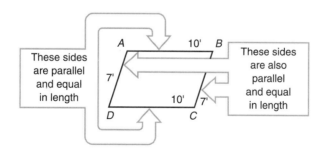

Figure *EFGH* is a *rectangle,* a parallelogram in which the four corner angles are right angles. The ∟ symbols at the vertices indicate right angles. Since a rectangle is a parallelogram, opposite sides are parallel and equal.

IJKL is a *square,* a rectangle in which *all* sides are the same length.

MNOP is called a *trapezoid.* A trapezoid contains two parallel sides and two non-parallel sides. $\overline{MN}$ and $\overline{OP}$ are parallel; $\overline{MP}$ and $\overline{NO}$ are not.

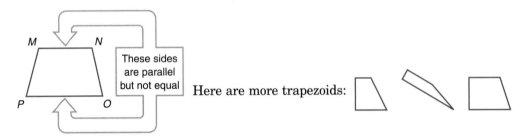

Here are more trapezoids:

If a four-sided polygon has none of these special features—no parallel sides—we call it a *quadrilateral. QRST* is a quadrilateral.

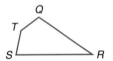

YOUR TURN

Name each of the following quadrilaterals. If more than one name fits, give the most specific name.

(a)

6'

6'   6'

6'

(b)   *AD* is parallel to *BC*

7 cm   *C*

D

15 cm   18.6 cm

A   6 cm   *B*

(c)

16"

13"   13"

16"

(d)

11 yd

16 yd

26 yd

22 yd

(e)

24"

18"   18"

24"

ANSWERS

(a) Square. (It is also a rectangle and a parallelogram, but "square" is a more specific name.)

(b) Trapezoid.

(c) Parallelogram.

(d) Quadrilateral.

(e) Rectangle. (It is also a parallelogram, but "rectangle" is a more specific name.)

Once you can identify a polygon, you can use a formula to find its area. Next let's examine and use some area formulas for these quadrilaterals.

Rectangles *The area of any plane figure is the number of square units of surface within the figure.*

In this rectangle

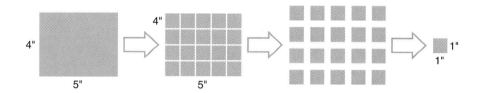

we can divide the surface into exactly 20 small squares each one inch on a side. By counting squares we can see that the area of this 4-in. by 5-in. rectangle contains 20 square inches. We abbreviate these area units as 20 square inches or 20 sq in.

Note ▶ As mentioned in Chapter 5, area units may be abbreviated using exponents. For example, the area unit square inches or sq in. results from multiplying inches × inches. Using the definition of exponents from Chapter 6, this can be expressed as inches^2 or in.^{2}. Similarly, square feet or sq ft can be written as ft^2, square centimeters as cm^2, and so on. Both types of units are used by trades workers, and we will use both in the remainder of this text. ◀

Of course there is no need to draw lines in the messy way shown above. We can find the area by multiplying the two dimensions of the rectangle.

AREA OF A RECTANGLE

A = length × width

$A = LW$

⬤ EXAMPLE

Find the area of rectangle *EFGH*.

L = 9 in.

W = 5 in.

A, area = LW = (9 in.)(5 in.)

A = 45 in.^{2}

Careful ▶ The formula is easy to use, but be careful. It is valid only for rectangles. If you use this formula with a different polygon, you will not get the correct answer. ◀

In technical work it is important to both state measurements and round calculations to the proper precision and accuracy. However, in the trades and in many occupations, people do not normally state quantities with the exact accuracy with which they have been measured. For example, the length and width of a box might very well be measured accurately to two significant digits, but a carpenter would write it as 9 in. rather than 9.0 in. To avoid confusion in calculations, either state your unrounded answer or follow the rounding instructions that accompany the problems.

Find the area of rectangle *IJKL*.

SOLUTION

First, write the formula: $A = LW$

Second, substitute *L* and *W:* $A = (42 \text{ cm})(18 \text{ cm})$

Third, calculate the answer $A = 756 \text{ cm}^2$
including the units.

 42 ⊗ **18** ⊜ →   756.

Of course the formula $A = LW$ may also be used to find *L* or *W* when the other quantities are known. As you learned in your study of basic algebra in Chapter 7, the following formulas are all equivalent.

$$A = LW \qquad L = \frac{A}{W} \qquad \text{or} \qquad W = \frac{A}{L}$$

EXAMPLE

A sheet metal worker must build a heating duct in which a rectangular vent 6 in. high must have the same area as a rectangular opening 8 in. by 9 in. Find the length of the vent.

The area of the opening is $A = LW$

$$A = (8 \text{ in.})(9 \text{ in.}) = 72 \text{ in.}^2$$

The length of the vent is $L = \dfrac{A}{W}$

$$L = \frac{72 \text{ in.}^2}{6 \text{ in.}} = \frac{72}{6} \frac{\text{in.} \times \text{in.}}{\text{in.}}$$

or $L = 12$ in.

Another formula that you may find useful allows you to calculate the perimeter of a rectangle from its length and width.

PERIMETER OF A RECTANGLE

$P = 2L + 2W$

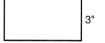

3"

YOUR TURN

Use this formula to find the perimeter of the rectangle shown here.

7"

SOLUTION

Perimeter $= 2L + 2W$
$= 2(7\text{ in.}) + 2(3\text{ in.})$
$= 14\text{ in.} + 6\text{ in.}$
$= 20\text{ in.}$

 2 ⊠ **7** ⊕ **2** ⊠ **3** ⊜ →   *20.*

PROBLEMS

For practice on using the area and perimeter formulas for rectangles, work the following problems.

(a) Find the area and perimeter of each of the following rectangles.

  (1)   8"   6"   (2)   12'   43'   (3) $9\frac{2}{3}$ yd $2\frac{2}{3}$ yd   (4)   1.80 m   1.80 m

(b) Find the area of a rectangle whose length is 4 ft 6 in. and whose width is 3 ft 3 in. (Give your answer in square feet in decimal form. Round to one decimal place.)

(c) **Sheet Metal Technology** A rectangular opening 12 in. wide must have the same total area as two smaller rectangular vents 6 in. by 4 in. and 8 in. by 5 in. What must be the height of the opening?

(d) **Carpentry** What is the cost of refinishing a wood floor in a room 35 ft by 20 ft at a cost of $2.25 per square foot?

(e) Find the area of a rectangular opening 20 in. wide and 3 ft long.

(f) **Landscaping** What is the cost of fencing a rectangular yard 45 ft by 58 ft at $11.45 per foot?

SOLUTIONS

(a) (1) $A = 48\text{ in.}^2, P = 28\text{ in.}$ (2) $A = 516\text{ ft}^2, P = 110\text{ ft}$

    (3) $A = 25\frac{7}{9}\text{ yd}^2, P = 24\frac{2}{3}\text{ yd}$ (4) $A = 3.24\text{ m}^2, P = 7.20\text{ m}$

(b) 4 ft 6 in. $= 4\frac{6}{12}\text{ ft} = 4.5\text{ ft}$
3 ft 3 in. $= 3\frac{3}{12}\text{ ft} = 3.25\text{ ft}$
4.5 ft $\times$ 3.25 ft $\approx 14.6\text{ ft}^2$

(c) Total area of smaller vents $= (6 \times 4) + (8 \times 5) = 64\text{ in.}^2$
Height of opening $= A/W = 64\text{ in.}^2 \div 12\text{ in.} = 5\frac{1}{3}\text{ in.}$

(d) Area $= 35\text{ ft} \times 20\text{ ft} = 700\text{ ft}^2$

Cost $= 700\text{ ft}^2 \times \$2.25\text{ per ft}^2 = 700\text{ ft}^2 \times \dfrac{\$2.25}{1\text{ ft}^2} = \$1575$

(e) 720 in.^{2} or 5 ft^2

(f) \$2358.70

 2 $\times$ **45** $+$ **2** $\times$ **58** $=$ $\times$ **11.45** $=$ $\rightarrow$ *2358.7*

Squares Did you notice that the rectangle in problem (a)(4) in the last set is actually a square? All sides are the same length. To save time when calculating the area or perimeter of a square, use the following formulas.

AREA OF A SQUARE

$A = s^2$

PERIMETER OF A SQUARE

$P = 4s$

YOUR TURN

 9' Use these formulas to find the area and perimeter of the square shown.

SOLUTION

Area, $A = s^2$
$A = (9\,\text{ft})^2$
$A = (9\,\text{ft})(9\,\text{ft}) = 81\,\text{ft}^2$

Remember s^2 means s times s and not $2s$.

Perimeter, $P = 4s$
$P = 4(9\,\text{ft})$
$P = 36\,\text{ft}$

A calculation that is very useful in carpentry, sheet metal work, and many other trades involves finding the length of one side of a square given its area. The following formula will help.

SIDE OF A SQUARE

$s = \sqrt{A}$

EXAMPLE

If a square opening is to have an area of 64 in.^{2} what must be its side length?

$s = \sqrt{A}$

$s = \sqrt{64\,\text{in.}^2}$

$s = 8\,\text{in.}$ $(\sqrt{\text{in.}^2} = \text{in.})$

PROBLEMS

Ready for a few problems involving squares? Try these.

(a) Find the area and perimeter of each of the following squares.

(b) Find the area of a square whose side length is $2\frac{1}{3}$ yd. (Round to three significant digits).

(c) Find the length of the side of a square whose area is (1) 144 m², (2) 927 in.² (Use a calculator and round to one decimal place.)

(d) **Landscaping** At 46 cents per square foot, how much will it cost to purchase sod for a square lawn 24 ft 6 in. on a side? (Round to the nearest cent.)

(e) **Sheet Metal Technology** A square hot-air duct must have the same area as two rectangular ones 5 in. × 8 in. and 4 in. × 6 in. How long are the sides of the square duct?

ANSWERS

(a) (1) $A = 36$ ft²; $P = 24$ ft

(2) $A = 306.25$ in.²; $P = 70$ in.

(3) $A = 6.25$ cm²; $P = 10$ cm

(4) $A \approx 11.11$ ft² (rounded) or $11\frac{1}{9}$ ft², using fractions; $P = 13$ ft 4 in.

(b) 5.44 yd²

(c) (1) 12 m   (2) 30.4 in.

(d) \$276.12

(e) $(5 \times 8) + (4 \times 6) = 40$ in.² $+ 24$ in.² $= 64$ in.²
$\sqrt{64 \text{ in.}^2} = 8$ in.

For problem (c)(2),
927 $\boxed{\sqrt{}}$ → 30.4466747

For problem (d),
24.5 $\boxed{x^2}$ $\boxed{\times}$ **.46** $\boxed{=}$ → 276.115 or 276.12, rounded

Parallelograms You should recall that a parallelogram is a four-sided figure whose opposite pairs of sides are equal and parallel. Here are a few parallelograms:

441

To find the area of a parallelogram use the following formula:

AREA OF A PARALLELOGRAM

$A = bh$

Notice that we do not use only the lengths of the sides to find the area of a parallelogram. The height h is the perpendicular distance between the base and the side opposite the base. The height h is perpendicular to the base b.

EXAMPLE

In parallelogram $ABCD$, the base is 7 in. and the height is 10 in. Find its area.

First, write the formula: $A = bh$.

Second, substitute the given values: $A = (7 \text{ in.})(10 \text{ in.})$

Third, calculate the area including the units: $A = 70$ sq in.

Notice that we ignore the length of the slant side 12 in.

YOUR TURN

Find the area of this parallelogram:

SOLUTION

Area $= bh$
$= (18 \text{ cm})(20 \text{ cm})$
$= 360 \text{ cm}^2$

Be careful to use the correct dimensions.

This problem could also be worked using the 24-cm side as the base.

In this case the new height is 15 cm and the area is

$A = (24 \text{ cm})(15 \text{ cm})$
$A = 360 \text{ cm}^2$

To find the perimeter of a parallelogram, simply add the lengths of the four sides.

EXAMPLE

Using the previous figure,

$P = 2a + 2b$
$P = 2(24 \text{ cm}) + 2(18 \text{ cm})$
$P = 48 \text{ cm} + 36 \text{ cm} = 84 \text{ cm}$

PROBLEMS

Find the perimeter and area for each parallelogram shown.

(a)

(b)

(c)

(d)

(e) A parallelogram-shaped opening has a base of 2 ft 9 in. and a height of 3 ft 3 in. Find its area. (Your answer should be in square feet rounded to three significant digits.)

(a) $P = 42$ in., $\qquad$ $A = 96$ in.^{2}

(b) $P = 50$ ft, $\qquad$ $A = 132$ ft^2

(c) $P = 82$ cm, $\qquad$ $A = 352$ cm^2

(d) $P = 58$ yd, $\qquad$ $A = 125\frac{2}{3}$ yd^2

(e) $A = 8.94$ ft^2

 2.75 $\boxed{\times}$ **3.25** $\boxed{=}$ → ⬚ 8.9375 ≈ 8.94

AREA OF A PARALLELOGRAM

You may be interested in where the formula for the area of a parallelogram came from. If so, follow this explanation.

Here is a typical parallelogram. As you can see, opposite sides $\overline{AB}$ and $\overline{DC}$ are parallel, and opposite sides $\overline{AD}$ and $\overline{BC}$ are also parallel.

Now let's cut off a small triangle from one side of the parallelogram by drawing the perpendicular line AE.

Next, reattach the triangular section to the other side of the figure, forming a rectangle. Notice that $EF = DC = b$.

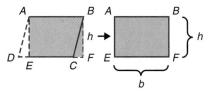

The area of the rectangle $ABFE$ is $A = bh$. This is exactly the same as the area of the original parallelogram $ABCD$.

Trapezoids　A trapezoid is a four-sided figure with only one pair of sides parallel. Here are a few trapezoids:

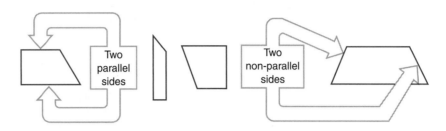

To find the area of a trapezoid use the following formula:

AREA OF A TRAPEZOID

$$A = \left(\frac{b_1 + b_2}{2}\right)h$$

or $\dfrac{h}{2}(b_1 + b_2)$

The factor $\left(\dfrac{b_1 + b_2}{2}\right)$ is the average length of the two parallel sides b_1 and b_2. The height h is the perpendicular distance between the two parallel sides.

EXAMPLE

Find the area of this trapezoid-shaped metal plate.

The parallel sides, b_1 and b_2, have lengths 7 in. and 19 in. The height h is 6 in.

$$A = \left(\frac{b_1 + b_2}{2}\right)h$$

$$A = \left(\frac{7 \text{ in.} + 19 \text{ in.}}{2}\right)(6 \text{ in.}) \quad \text{First, evaluate the quantity } \frac{7 + 19}{2}.$$

$$A = \left(\frac{26}{2} \text{ in.}\right)(6 \text{ in.})$$

$$A = (13 \text{ in.})(6 \text{ in.}) \qquad \text{Now multiply } 13 \times 6.$$

$$A = 78 \text{ in.}^2$$

 7 ⊞ **19** ⊟ ⊡ **2** ⊠ **6** ⊟ → ▨▨▨▨ 78. ▨▨▨▨

The **perimeter** of a trapezoid is simply the sum of the lengths of all four sides.

YOUR TURN

Find the area and perimeter of each of the following trapezoids.

(a)

(b)

(c)

(d)

(e)

(f)

SOLUTIONS

(a) $A = \left(\dfrac{8\text{ in.} + 26\text{ in.}}{2}\right)(12\text{ in.})$

$A = (17\text{ in.})(12\text{ in.})$

$A = 204\text{ in.}^2$

$P = 8\text{ in.} + 15\text{ in.} + 26\text{ in.} + 15\text{ in.}$

$P = 64\text{ in.}$

(b) $A = \left(\dfrac{7\text{ ft} + 14\text{ ft}}{2}\right)(11\text{ ft})$

$A = (10.5\text{ ft})(11\text{ ft})$

$A = 115.5\text{ ft}^2$

$P = 11\text{ ft} + 7\text{ ft} + 13\text{ ft} + 14\text{ ft}$

$P = 45\text{ ft}$

(c) $A = \left(\dfrac{4\text{ cm} + 19\text{ cm}}{2}\right)(20\text{ cm})$

$A = (11.5\text{ cm})(20\text{ cm})$

$A = 230\text{ cm}^2$

$P = 4\text{ cm} + 20\text{ cm} + 19\text{ cm} + 25\text{ cm}$

$P = 68\text{ cm}$

(d) $A = \left(\dfrac{9\frac{1}{2}\text{ in.} + 6\frac{1}{2}\text{ in.}}{2}\right)(2\frac{1}{4}\text{ in.})$

$A = (8\text{ in.})(2\frac{1}{4}\text{ in.})$

$A = (8\text{ in.})(2.25\text{ in.})$

$A = 18\text{ in.}^2$

$P = 3\frac{3}{4}\text{ in.} + 6\frac{1}{2}\text{ in.} + 2\frac{1}{4}\text{ in.} + 9\frac{1}{2}\text{ in.}$

$P = 22\text{ in.}$

(e) $A = \left(\dfrac{16\text{ yd} + 33\text{ yd}}{2}\right)(18\text{ yd})$

$A = (24.5\text{ yd})(18\text{ yd})$

$A = 441\text{ yd}^2$

$P = 16\text{ yd} + 19\text{ yd} + 33\text{ yd} + 21\text{ yd}$

$P = 89\text{ yd}$

(f) $A = \left(\dfrac{16\text{ in.} + 6\text{ in.}}{2}\right)(12\text{ in.})$

$A = (11\text{ in.})(12\text{ in.})$

$A = 132\text{ in.}^2$

$P = 13\text{ in.} + 16\text{ in.} + 13\text{ in.} + 6\text{ in.}$

$P = 48\text{ in.}$

Note ▶ In a real job situation you will usually need to measure the height of the figure. Remember, the bases of a trapezoid are the parallel sides and h is the distance between them. ◀

Turn to Exercises 8-2 for practice in finding the area and perimeter of polygons.

Exercises 8-2   Area and Perimeter of Polygons

A. Find the perimeter of each polygon.

1.

2.

3.

4.

5.

6.

7.

8.

9.

10.

11.

12.

B. Find the perimeter and area of each figure. Round to the nearest tenth if necessary. (Assume right angles and parallel sides except where obviously otherwise.)

1.

2.

3.

4.

5.

6.

7.

8.

9.

10.

11.

12.

C. Practical Problems

Problem 1

1. **Masonry** How many bricks will it take to build the wall shown if each brick measures $4\frac{1}{4}$ in. $\times$ $8\frac{3}{4}$ in. including mortar? (*Hint:* Change the wall dimensions to inches.)

1'9"   2'3"

3'6"

Problem 2

12'

19'

Problem 3

2. **Carpentry** How many square feet of wood are needed to build the cabinet in the figure? (Assume that wood is needed for all six surfaces.)

3. **Interior Decorating** The floor plan of a room is shown in the figure. If the entire room is to be carpeted, how many square *yards* of carpet are needed? At $39.95 per square yard, how much will it cost?

4. **Building Construction** The amount of horizontal wood siding (in sq) needed to cover a given wall area can be calculated from the formula

   Area of siding (sq) = K × total area to be covered (sq ft)

   where K is a constant that depends on the type and size of the siding used. How much 1 × 6 rustic shiplapped drop siding is needed to cover a rectangular wall 60 ft by 14 ft? Use $K = 2.19$ and round to three significant digits.

5. **Building Construction** How many bundles of strip flooring are needed to cover a rectangular floor area 16 ft by 12 ft 6 in. in size if a bundle of strip flooring contains 24 sq ft, and if 30% extra must be allowed for side and end matching?

6. **Masonry** A large rectangular window opening measures $72\frac{5}{8}$ in. by $60\frac{3}{8}$ in. Calculate the area of this opening. Round to the nearest square inch.

7. **Masonry** Find the area of one stretcher course of 20 concrete blocks if the blocks are $7\frac{5}{8}$ in. long and 4 in. high, and if the mortar joints are $\frac{3}{8}$ in. thick. Round to two significant digits. (*Hint:* Twenty blocks require 19 mortar joints.)

8. **Masonry** To determine the approximate number of concrete blocks needed to construct a wall, the following formula is often used.

   Number of blocks per course = N × area of one course (sq ft)

   where N is a number that depends on the size of the block. How many 8-in. by 8-in. by 16-in. concrete blocks are needed to construct a foundation wall 8 in. thick with a total length of 120 ft, if it is laid five courses high? Use $N = \frac{9}{8}$ for this block size. (*Hint:* The blocks are set lengthwise. Translate all dimensions to inches.)

9. **Auto Mechanics** A roll of gasket material is 18 in. wide. What length is needed to obtain 16 sq ft of the material? (*Careful:* The numbers are not expressed in compatible units.)

10. **Welding** A type of steel sheet weighs 3.125 lb per square foot. What is the weight of a piece measuring 24 in. by 108 in.? (*Hint:* Be careful of units.)

11. **Printing** A rule of thumb in printing says that the area of the typed page should be half the area of the paper page. If the paper page is 5 in. by 8 in., what should the length of the typed page be if the width is 3 in.?

12. **Printing** A ream of 17-in. by 22-in. paper weighs 16 lb. Find the weight of a ream of 19-in. by 24-in. sheets of the same density of paper.

13. **Printing** A sheet of 25-in. by 38-in. paper must be cut into 6-in. by 8-in. cards. The cuts can be made in either direction but they must be consistent. Which plan will result in the least amount of waste, cutting the 6-in. side of the card along the 25-in. side of the sheet or along the 38-in. side of the sheet? State the amount of waste created by each possibility. (*Hint:* Make a cutting diagram for each possibility.)

14. **Refrigeration and Air Conditioning** To determine the size of a heating system needed for a home, an installer needs to calculate the heat loss

through surfaces such as walls, windows, and ceilings exposed to the outside temperatures. The formula

$$L = kDA$$

gives the heat loss L (in BTU per hour) if D is the temperature difference between the inside and outside, A is the area of the surface in sq ft, and k is the insulation rating of the surface. Find the heat loss per hour for a 7-ft by 16-ft single-pane glass window ($k = 1.13$) if the outside temperature is 40° and the desired inside temperature is 68°. Round to two significant digits.

15. **Construction** A playground basketball court 94 ft long and 46 ft wide is to be resurfaced at a cost of $2.25 per sq ft. What will the resurfacing cost?

When you have finished these problems, check your answers in the Appendix, then continue in Section 8-3 with the study of other polygons.

8-3 TRIANGLES, HEXAGONS, AND IRREGULAR POLYGONS

A *triangle* is a polygon having three sides and therefore three angles.

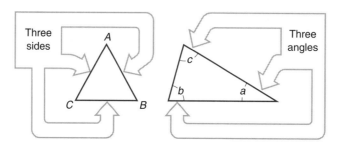

Just as there are several varieties of four-sided figures—squares, rectangles, parallelograms, and trapezoids—there are also several varieties of triangles. Fortunately, one area formula can be used with all triangles. First, you must learn to identify the many kinds of triangles that appear in practical work.

Triangles may be classified according to their sides.

An **equilateral triangle** is one in which all three sides have the same length. An equilateral triangle is also said to be *equiangular* since, if the three sides are equal, the three angles will also be equal. In fact, since the interior angles of any triangle always add up to 180°, each angle of an equilateral triangle must equal 60°.

An **isosceles triangle** is one in which at least two of the three sides are equal. It is always true that the two angles opposite the equal sides are also equal.

A **scalene triangle** is one in which no sides are equal.

Triangles may be classified also according to their angles. An **acute triangle** is one in which all three angles are acute (less than 90°). An **obtuse triangle** contains one obtuse angle (greater than 90°).

A **right triangle** contains a 90° angle. Two of the sides, called the *legs*, are perpendicular to each other. The longest side of a right triangle is always the side opposite to the right angle. This side is called the *hypotenuse* of the triangle. An isosceles or scalene triangle can be a right triangle, but an equilateral triangle can never be a right triangle.

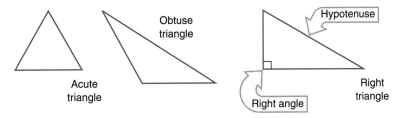

Acute triangle	Obtuse triangle	Right triangle

Hypotenuse

Right angle

PROBLEMS

Identify the following triangles as being equilateral, isosceles, or scalene. Also name the ones that are right triangles.

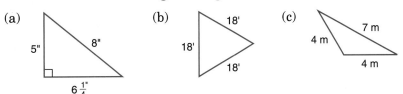

(a) 5" 8" $6\frac{1}{4}$"

(b) 18' 18' 18'

(c) 4 m 7 m 4 m

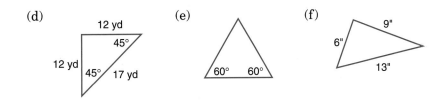

(d) 12 yd 45° 12 yd 45° 17 yd

(e) 60° 60°

(f) 6" 9" 13"

ANSWERS

(a) Scalene

(b) Equilateral

(c) Isosceles

(d) Isosceles

(e) Equilateral

(f) Scalene

(a) and (d) are also right triangles.

Did you have trouble with (d) or (e)? The answer to each depends on the fact that the interior angles of a triangle add up to 180°. In (d) the two given angles add up to 90°. Therefore, the missing angle must be 90° if the three angles sum to 180°. This makes (d) a right triangle.

In (e) the two angles shown sum to 120°, so the third angle must be 180° − 120° or 60°. Because the three angles are equal we know the triangle is an equiangular triangle, and this means it is also equilateral.

What does isosceles mean?

The prefix *iso* comes from a Greek word that means *same*, so an isosceles triangle is one that has two sides the same length.

Pythagorean Theorem

The Pythagorean theorem is a rule or formula that allows us to calculate the length of one side of a right triangle when we are given the lengths of the other two sides. Although the formula is named after the ancient Greek mathematician Pythagoras, it was known to Babylonian engineers and surveyors more than a thousand years before Pythagoras lived.

PYTHAGOREAN THEOREM

For any right triangle, the square of the hypotenuse is equal to the sum of the squares of the other two sides.

$$c^2 = a^2 + b^2$$

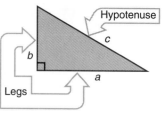

Geometrically, this means that if the squares are built on the sides of the triangle, the area of the larger square is equal to the sum of the areas of the two smaller squares. For the triangle shown

$$3^2 + 4^2 = 5^2$$

$$9 + 16 = 25$$

EXAMPLE

Let's use this formula to find the distance d between points A and B for this rectangular plot of land.

The Pythagorean theorem tells us that for the right triangle ABC,

$$d^2 = 36^2 + 20^2$$
$$d^2 = 1296 + 400$$
$$d^2 = 1696$$

Taking the square root of both sides of the equation, we have

$$d \approx 41 \text{ ft} \quad \text{rounded to the nearest foot}$$

 36 $\boxed{x^2}$ $\boxed{+}$ **20** $\boxed{x^2}$ $\boxed{=}$ $\boxed{\sqrt{}}$ → 41.18252056

This rule may be used to find any one of the three sides of a right triangle if the other two sides are given.

To solve for the hypotenuse c,

First, write the original formula

$$c^2 = a^2 + b^2$$

450

Then, take the square root of both sides of the formula.

$$\sqrt{c^2} = \sqrt{a^2 + b^2}$$

or

$$c = \sqrt{a^2 + b^2}$$

Note ▶ The square root of a sum is not equal to the sum of the square roots. For example $\sqrt{9 + 16} = \sqrt{25} = 5$, and this is not the same as $\sqrt{9} + \sqrt{16}$, which equals 7. The algebraic expression

$\sqrt{a^2 + b^2}$ cannot be simplified further. ◀

To solve for one of the legs,

First, write the original formula,

$$c^2 = a^2 + b^2$$

Next, subtract b^2 from both sides,

$$c^2 - b^2 = a^2$$

Finally, take the square root of both sides,

$$\sqrt{c^2 - b^2} = \sqrt{a^2}$$

or

$$a = \sqrt{c^2 - b^2}$$

Solving for the other leg, b, yields a similar result.

$$b = \sqrt{c^2 - a^2}$$

These results are summarized in the following table.

EQUIVALENT FORMS OF THE PYTHAGOREAN THEOREM

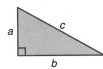

$$c = \sqrt{a^2 + b^2} \quad a = \sqrt{c^2 - b^2} \quad b = \sqrt{c^2 - a^2}$$

Careful ▶ These three formulas are valid only for a *right* triangle—a triangle with a right or 90° angle. ◀

YOUR TURN

Use these formulas to solve the problem.

A carpenter wants to use a 12-ft ladder to reach the top of a 10-ft wall. How far must the base of the ladder be from the base of the wall?

SOLUTION

Use the formula $a = \sqrt{c^2 - b^2}$ to get

$$x = \sqrt{12^2 - 10^2}$$

$$x = \sqrt{144 - 100}$$

$$x = \sqrt{44}$$

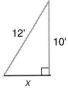

$x \approx 6.63$ ft or 6 ft 8 in., rounded to the nearest inch

If you need help with square roots, pause here and return to page 323 for a review.

For some right triangles, the three side lengths are all whole numbers rather than fractions or decimals. Such special triangles have been used in technical work since Egyptian surveyors used them 2000 years ago to lay out rectangular fields for farming.

Here are a few "Pythagorean triple" right triangles:

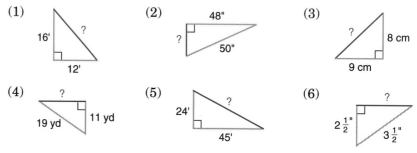

PROBLEMS

Solve the following problems, using the Pythagorean theorem. (If your answer does not come out exactly, round to the nearest tenth.)

(a) Find the missing side for each of the following right triangles.

(1) [16', 12', ?] (2) [48", 50", ?] (3) [8 cm, 9 cm, ?]

(4) [19 yd, 11 yd, ?] (5) [24', 45', ?] (6) [$2\frac{1}{2}$", $3\frac{1}{2}$", ?]

(7) Find the hypotenuse of a right triangle with legs measuring 4.6 m and 6.2 m.

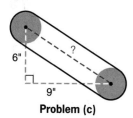

Problem (c)

(b) **Carpentry** A rectangular table measures 4 ft by 6 ft. Find the length of a diagonal brace placed beneath the table top.

(c) **Manufacturing** What is the distance between the centers of two pulleys if one is placed 9 in. to the left and 6 in. above the other?

(d) **Machine Technology** Find the length of the missing dimension x in the part shown in the figure.

Problem (d)

(e) **Carpentry** Find the diagonal length of the stairway shown in the drawing.

Problem (e)

(a) (1) 20 ft   (2) 14 in.   (3) 12.0 cm   (4) 15.5 yd

(5) 51 ft   (6) 2.4 in.   (7) 7.7 m

(b) 7.2 ft   or   7 ft $2\frac{1}{2}$ in.

(c) 10.8 in.

(d) 21.4 cm

(e) 12.9 ft or 12 ft 11 in., rounded to the nearest inch.

In feet: **8.5** $\boxed{x^2}$ $\boxed{+}$ **9.75** $\boxed{x^2}$ $\boxed{=}$ $\boxed{\sqrt{}}$ → 12.93493332 $\approx$ 12.9 ft

In inches: $\boxed{-}$ **12** $\boxed{=}$ $\boxed{\times}$ **12** $\boxed{=}$ → 11.21919984 $\approx$ 12 ft 11 in.

In problem (d), the first difficulty is to locate the correct triangle.

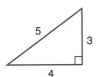

$x = \sqrt{22^2 - 5^2}$

$x = \sqrt{459}$

SOME SPECIAL RIGHT TRIANGLES

3-4-5 TRIANGLE

If you draw a triangle with legs 3, 4, and 5 in. long, or in the ratio 3:4:5, it will always be a right triangle.

30-60-90 TRIANGLE

In a triangle with angles 30°, 60°, and 90° the shortest side will be exactly half as long as the longest side—the hypotenuse.

Side b equals $a \times \sqrt{3}$, or about $1.7a$.

45-45-90 TRIANGLE

If the two legs of a right triangle are equal, the angles will be 45°, 45°, and 90°. The hypotenuse will be $\sqrt{2}$ times the length of the other sides.

This kind of triangle is called a right-isosceles triangle.

Hypotenuse = $a\sqrt{2}$ or about $1.4a$.

Area of a Triangle

The area of any triangle, no matter what its shape or size, can be found using the same simple formula.

AREA OF A TRIANGLE

Area, $A = \frac{1}{2}bh$ or $A = \frac{bh}{2}$

b = base
h = height

Any side of a triangle can be used as the base. For a given base, the height is the perpendicular distance from the opposite vertex to the base. The following three identical triangles show the three different base-height combinations.

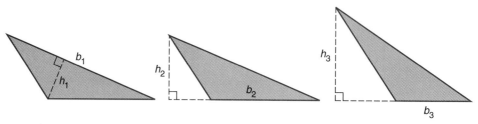

Notice that for the last two cases the height falls outside the triangle, and the base must be extended to meet the height at a right angle. For any such triangle, the three area calculations are equal:

$A = \frac{1}{2}b_1h_1 = \frac{1}{2}b_2h_2 = \frac{1}{2}b_3h_3$

EXAMPLE

In triangle PQR the base b is 13 in., and the height h is 8 in. Applying the formula, we have

Area, $A = \frac{1}{2}bh$
$\quad A = \frac{1}{2}(13 \text{ in.})(8 \text{ in.})$
$\quad A = 52 \text{ in.}^2$

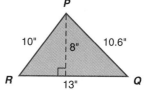

In triangle STU,

$b = 25$ cm

$h = 16$ cm

Area, $A = \frac{1}{2}(25 \text{ cm})(16 \text{ cm})$
$\qquad A = 200 \text{ cm}^2$

And in triangle VWX,

$b = 20$ cm

$h = 12$ cm

Area, $A = \frac{1}{2}(20 \text{ cm})(12 \text{ cm}) = 120 \text{ cm}^2$

YOUR TURN

Find the area of triangle ABC.

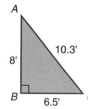

SOLUTION

Base, $b = 6.5$ ft

height, $h = 8$ ft

Area, $A = \frac{1}{2}bh$

$A = \frac{1}{2}(6.5 \text{ ft})(8 \text{ ft})$

$A = 26 \text{ ft}^2$

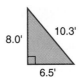

Notice that for a right triangle, the two sides meeting at the right angle can be used as the base and height.

AREA OF A TRIANGLE

The area of any triangle is equal to one-half of its base times its height, $A = \frac{1}{2}bh$.

To learn where this formula comes from, follow this explanation:

First, for any triangle ABC,

draw a second identical triangle DEF.

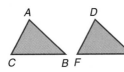

Second, flip the second triangle over and attach it to the first,

to get a parallelogram.

Finally, if the base and height of the original triangle are b and h, then these are also the base and height for the parallelogram.

The area of the parallelogram is

$A = bh$

so the area of the original triangle must be

$A = \frac{1}{2}bh$

YOUR TURN

Solve this slightly tricky problem. Find the area of the isosceles triangle shown.

(*Hint:* First use the Pythagorean theorem to find the height, then find the area. The perpendicular line from C to $\overline{AB}$ will divide side $\overline{AB}$ in half.)

Draw the height perpendicular to the 6-in. base. This will divide the base in half. Then in the triangle below

$h = \sqrt{9^2 - 3^2}$

$h = \sqrt{81 - 9}$

$h = \sqrt{72} \approx 8.485 \text{ in.}$

Now find the area:

$A = \frac{1}{2}bh$

$A \approx \frac{1}{2}(6 \text{ in.})(8.485 \text{ in.}) \approx 25.455 \text{ in.}^2$

$A \approx 25 \text{ in.}^2$ rounded

Using a calculator the entire calculation looks like this:

9 $\boxed{x^2}$ $\boxminus$ 3 $\boxed{x^2}$ $\boxed{=}$ $\boxed{\sqrt{}}$ $\boxtimes$ 6 $\boxed{\div}$ 2 $\boxed{=}$ → 25.45584412

| The height | | The area |

In general, to find the area of an isosceles triangle you may use the following formula.

AREA OF AN ISOSCELES TRIANGLE

$\text{Area}, A = \frac{1}{2}b \sqrt{a^2 - \left(\frac{b}{2}\right)^2}$

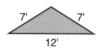

Construction   An attic wall has the shape shown. How many square feet of insulation are needed to cover the wall? (Round to the nearest square foot.)

The triangle is isosceles since two sides are equal. Substituting $a = 7$ ft and $b = 12$ ft into the formula, we have

$\text{Area}, A = \frac{1}{2}(12 \text{ ft}) \sqrt{(7 \text{ ft})^2 - \left(\frac{12 \text{ ft}}{2}\right)^2}$

$A = 6\sqrt{49 - 36}$

$A = 6\sqrt{13} \approx 21.63 \text{ ft}^2, \text{ or } 22 \text{ ft}^2$ rounded

7 $\boxed{x^2}$ $\boxminus$ 6 $\boxed{x^2}$ $\boxed{=}$ $\boxed{\sqrt{}}$ $\boxtimes$ 12 $\boxed{\div}$ 2 $\boxed{=}$ → 21.63330765

For an equilateral triangle an even simpler formula may be used.

AREA OF AN EQUILATERAL TRIANGLE

Area, $A = \dfrac{a^2\sqrt{3}}{4} \approx 0.433a^2,$ approximately

---

YOUR TURN

Machine Technology   What area of sheet steel is needed to make a triangular pattern where each side is 3 ft long?

Try it. Use the given formula and round to one decimal place.

SOLUTION

Area, $A \approx 0.433a^2,$ where $a = 3$ ft
$\qquad A \approx 0.433(3\text{ ft})^2$
$\qquad A \approx 0.433(9\text{ ft}^2)$
$\qquad A \approx 3.9\text{ ft}^2$ rounded

PROBLEMS

Work these problems for practice.

(a) In the following problems find the area of each triangle. (Round your answer to one decimal place if it does not come out exactly.)

(1)

(2)

(3)

(4)

(5)

(6)

(7)

(8)

Problem (b)

Problem (c)

(b) **Carpentry**   Find the area, to the nearest square foot, of the triangular patio deck shown.

(c) **Manufacturing**   At a cost of 97 cents per square foot, what would it cost to tile the triangular work space shown?

ANSWERS

(a) (1)   135 in.^{2} (2)   136.5 ft^2 (3)   359.1 cm^2

    (4)   43.6 yd^2 (5)   135 m^2 (6)   24.2 ft^2

    (7)   128 in.^{2} (8)   209.6 cm^2

For (a)(8): **.433** $\boxed{\times}$ **22** $\boxed{x^2}$ $\boxed{=}$ → 209.572

(b) 63 sq ft   (c) $62.08

In problem (a)(5) you should have used 18 m as the base and *not* 26 m. $b = 18$ m, $h = 15$ m.

In problem (c) you should have noticed that the height was given as 8 ft.

HERO'S FORMULA

It is possible to find the area of any triangle from the lengths of its sides. The formula that enables us to do this was first devised almost 2000 years ago by Hero or Heron, a Greek mathematician. It is a complicated formula, and you may feel like a hero yourself if you learn how to use it.

Area, $A = \sqrt{s(s - a)(s - b)(s - c)}$ For any triangle

where $s = \dfrac{a + b + c}{2}$

(*s* stands for "semiperimeter," or half of the perimeter.)

Example:

$a = 6$ in., $b = 5$ in., $c = 7$ in.

First, find s. $s = \dfrac{6 + 5 + 7}{2} = \dfrac{18}{2} = 9$

Next, substitute to find the area:

$A = \sqrt{9(9 - 6)(9 - 5)(9 - 7)}$

$A = \sqrt{9 \cdot 3 \cdot 4 \cdot 2}$

$A = \sqrt{216} \approx 14.7$ in.^{2} rounded

Example: Find the area of this triangle.

Step 1 $s = \dfrac{19 + 12 + 25}{2} = \dfrac{56}{2} = 28$

Step 2 $A = \sqrt{28(28 - 19)(28 - 12)(28 - 25)}$

$= \sqrt{28 \cdot 9 \cdot 16 \cdot 3}$

$= \sqrt{12096} = 109.98 \ldots \approx 110$ cm^2, rounded

Hexagons   A polygon is a plane geometric figure with three or more sides. A *regular polygon* is one in which all sides are the same length and all angles are equal. The equilateral triangle and the square are both regular polygons.

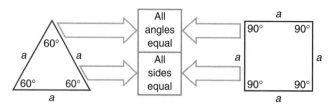

A *pentagon* is a five-sided polygon. A *regular pentagon* is one whose sides are all the same length and whose angles are equal.

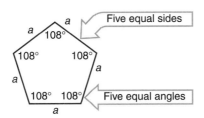

The regular *hexagon* is a six-sided polygon in which each interior angle is 120° and all sides are the same length.

Regular hexagon:
All sides equal
All angles 120°

If we draw the diagonals, six equilateral triangles are formed.

Since you already know that the area of one equilateral triangle is approximately $A \approx 0.433a^2$, the area of the complete hexagon will be 6 times $0.433a^2$ or $2.598a^2$.

AREA OF A REGULAR HEXAGON

Area, $A = \dfrac{3a^2\sqrt{3}}{2} \approx 2.598a^2$ approximately

---

EXAMPLE

If $a = 6.00$ cm, $A \approx 2.598(6 \text{ cm})^2$
First, square the number in parentheses. $A \approx 2.598(36 \text{ cm}^2)$
Then multiply. $A \approx 93.5 \text{ cm}^2$, rounded

  2.598 $\boxed{\times}$ **6** $\boxed{x^2}$ $\boxed{=}$ → 　　　*93.528*

YOUR TURN

Machine Technology   Use this formula to find the area of a hexagonal plate 4.0 in. on each side. Round to two significant digits.

Area, $A \approx 2.598(a^2)$ $a = 4$ in.
$\quad A \approx 2.598(4 \text{ in.})^2$
$\quad A \approx 2.598(16 \text{ in.}^2)$
$\quad A \approx 42 \text{ in.}^2$, rounded

Because the hexagon is used so often in practical and technical work, several other formulas may be helpful to you.

A hexagon is often measured by specifying the distance across the corners or the distance across the flats.

Distance across
the corners

Distance across
the flats

DIMENSIONS OF A REGULAR HEXAGON

Distance across the corners, $d = 2a$ or $a = 0.5d$

Distance across the flats, $f \approx 1.732a$ or $a \approx 0.577f$

EXAMPLE

For a piece of hexagonal bar stock where $a = \frac{1}{2}$ in., the distance across the corners would be

$d = 2a$ or $d = 1$ in.

The distance across the flats would be

$f \approx 1.732a$ or $f \approx 0.866$ in.

YOUR TURN

Machine Technology   If the cross section of a hexagonal nut measures $\frac{3}{4}$ in. across the flats, find

(a) The side length, a.

(b) The distance across the corners, d.

(c) The cross-sectional area.

Round to three significant digits.

SOLUTIONS

$f = \frac{3}{4}$ in.   or   0.75 in.

(a) $a \approx 0.577f$
$\quad a \approx 0.577(0.75 \text{ in.})$
$\quad a \approx 0.433$ in.

(b) $d = 2a$
      $d \approx 2(0.433 \text{ in.})$
      $d \approx 0.866 \text{ in.}$

(c) Area, $A \approx 2.598a^2$
      $A \approx 2.598(0.433 \text{ in.})^2$
      $A \approx 2.598(0.1875 \text{ in.}^2)$
      $A \approx 0.487 \text{ in.}^2$

PROBLEMS

Work these problems for practice.

(a) Fill in the blanks with the missing dimensions of each regular hexagon. (Round to two decimal places if necessary.)

	a	d	f	A
(1)	2 in.			
(2)		$\frac{3}{4}$ in.		
(3)			6 mm	

(b) A hex nut has a side length of $\frac{1}{4}$ in. From the following list, pick the smallest size wrench that will fit it.

   (1) $\frac{1}{4}$ in.   (2) $\frac{3}{8}$ in.   (3) $\frac{7}{16}$ in.   (4) $\frac{1}{2}$ in.   (5) $\frac{5}{8}$ in.

ANSWERS

	a	d	f	A
(a) (1)	2 in.	4 in.	3.46 in.	10.39 in.^{2}
(2)	$\frac{3}{8}$ in.	$\frac{3}{4}$ in.	0.65 in.	0.37 in.^{2}
(3)	3.46 mm	6.92 mm	6 mm	31.14 mm^2

For problem (a)(3),

6 $\times$ **.577** $=$ → 3.462 $\times$ **2** $=$ → 6.924

2.598 $\times$ **3.462** x^2 $=$ 31.13818351 Area

(b) $\frac{7}{16}$ in.

Let's take a closer look at problem (b).

By drawing a picture we can see that we must find f. Since $a = \frac{1}{4}$ in., we have

$f \approx 1.732a$
$f \approx 1.732(0.25)$
$f \approx 0.433 \text{ in.}$

By checking a list of wrench sizes in their decimal equivalents or converting the fractions given, we find that $\frac{7}{16}$ in. (0.4375 in.) is the smallest size that will fit the nut.

CLASSIFICATION OF POLYGONS

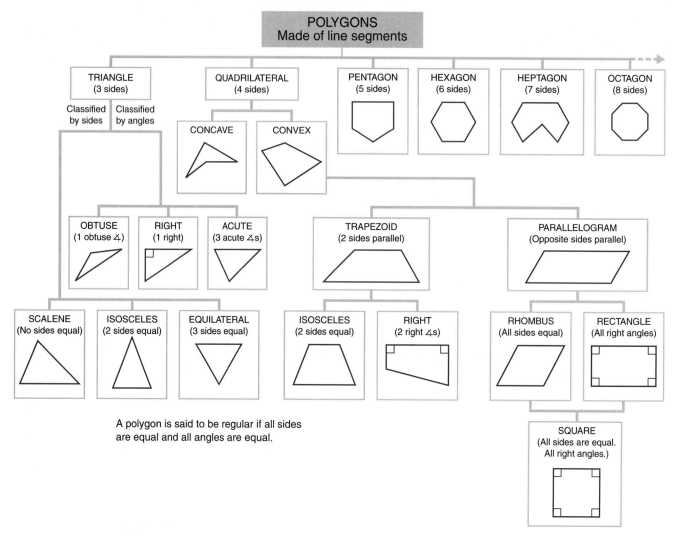

A polygon is said to be regular if all sides are equal and all angles are equal.

Irregular Polygons   Often, the shapes of polygons that appear in practical work are not the simple geometric figures that you have seen so far. The easiest way to work with irregular polygon shapes is to divide them into simpler, more familiar figures.

For example, look at this L-shaped figure:

There are two ways to find the area of this shape.

Method 1: Addition

By drawing the dashed line, we have divided this figure into two rectangles, I and II. I is 22 in. by 6 in. for an area of 132 in.^{2}. II is 13 in. by 12 in. for an area of 156 in.^{2} The total area is

$$132 + 156 = 288 \text{ in.}^2$$

462   **Chap. 8**   Practical Plane Geometry

Method 2: Subtraction

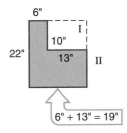

In this method we subtract the area of the small rectangle (I) from the area of the large rectangle (II). The area of I is

$$10 \times 13 = 130 \text{ in.}^2$$

The area of II is

$$22 \text{ in.} \times 19 \text{ in.} = 418 \text{ in.}^2$$
$$418 - 130 = 288 \text{ in.}^2$$

Notice that in the addition method we had to calculate the height of rectangle II as 22 in. − 10 in. or 12 in. In the subtraction method we had to calculate the width of rectangle II as 6 in. + 13 in. or 19 in.

Which method you use to find the area of an irregular polygon depends on your ingenuity and on which dimensions are given.

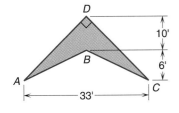

EXAMPLE

Find the area of the shape *ABCD*.

You may at first wish to divide this into two triangles and add their areas.

 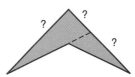

However, the dimensions provided do not allow you to do this.

The only option here is to subtract the area of triangle I from the area of triangle II.

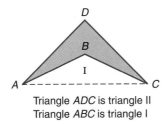

Triangle *ADC* is triangle II
Triangle *ABC* is triangle I

	Base	Height	Area
Triangle II	33 ft	16 ft	264 ft^2
Triangle I	33 ft	6 ft	− 99 ft^2
		Area =	165 ft^2

If both methods will work, pick the one that looks easier and requires the simpler arithmetic.

YOUR TURN

Find the area of this shape. (Round to the nearest square inch.)

You might have been tempted to split the shape up into the five regions indicated in the figure. This will work, but there is an easier way.

For the alternative shown in the next figure we need to compute only three areas, then we subtract II and III from I.

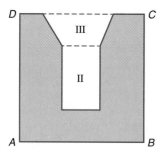

The outside rectangle *ABCD* is area I.

	Base	Height	Area	
Figure I	6 in. + 17 in. + 8 in. = 31 in.	8 in. + 15 in. + 8 in. = 31 in.	31 in. × 31 in.	$= 961 \text{ in.}^2$
Figure II	10 in.	15 in.	10 in. × 15 in.	$= 150 \text{ in.}^2$
Figure III (trapezoid)	17 in. and 10 in.	8 in.	$\left(\dfrac{17 \text{ in.} + 10 \text{ in.}}{2}\right)(8 \text{ in.})$	$= 108 \text{ in.}^2$

The total area is $961 \text{ in.}^2 - 150 \text{ in.}^2 - 108 \text{ in.}^2 = 703 \text{ in.}^2$

Learning Help ▶ When working with complex figures, organize your work carefully. Neatness will help eliminate careless mistakes. ◀

PROBLEMS

Now, for practice, find the areas of the following shapes. (Round to the nearest tenth.)

(a) 12' 14' 9' 8'

(b) 20" 32" 16"

(c) 14.5 cm 7.6 cm 3 cm 8.5 cm 7.6 cm 14.5 cm

(d) 6 yd 6 yd ←6 yd→ 6 yd 6 yd

(e) 3'9" 4'6" 9'6" 14'
(f) 4 mm 23 mm 14 mm 9 mm 21 mm
(g) 2' 2' 2' 2' 2' 6' 2'
(h) 40" 12" 16" 75" 8"

ANSWERS

(a) 192 ft^2

(b) 672 in.^{2}

(c) 200.3 cm^2

(d) 31.2 yd^2

(e) 70.7 ft^2

(f) 266 mm^2

(g) 22.4 ft^2

(h) 1488 in.^{2}

In problem (d) divide the figure into two equilateral triangles, then use the area formula $A \approx 0.433a^2$. Figure (g) is a hexagon plus a rectangle.

Now turn to Exercises 8-3 for a set of problems on the work of this section.

Exercises 8-3

Triangles, Hexagons, and Irregular Polygons

A. Find the missing dimensions of each figure shown. All hexagons are regular. (Round to the nearest tenth if necessary.)

1. 8" 7" ?

2.
18' ? 18' 24'

3.
10 cm ?

4.
41 yd 50 yd ?

5.
? 27 mm

6.
36" 15" ?

7.
32" ?

8.
5.1' ? 2.4'

9.
21' 35' ?

10.
8.6' 7.4' ?

11.
? 23" 9"

12.
?→ 56"

13.
? 14" Area = 42 in.^{2}

14.
? 10 cm 14 cm Area = 110 cm^2

15.
? 20 ft Area = 174 ft^2

16.
6" ? 11" Area = 45.9 in.^{2}

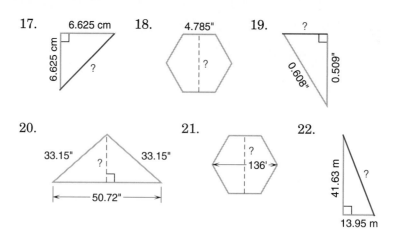

17. 6.625 cm 6.625 cm ?

18. 4.785" ?

19. ? 0.608" 0.509"

20. 33.15" 33.15" ? 50.72"

21. ? 136'

22. 41.63 m ? 13.95 m

B. Find the area of each figure. All hexagons are regular. (Round to the nearest tenth if necessary.)

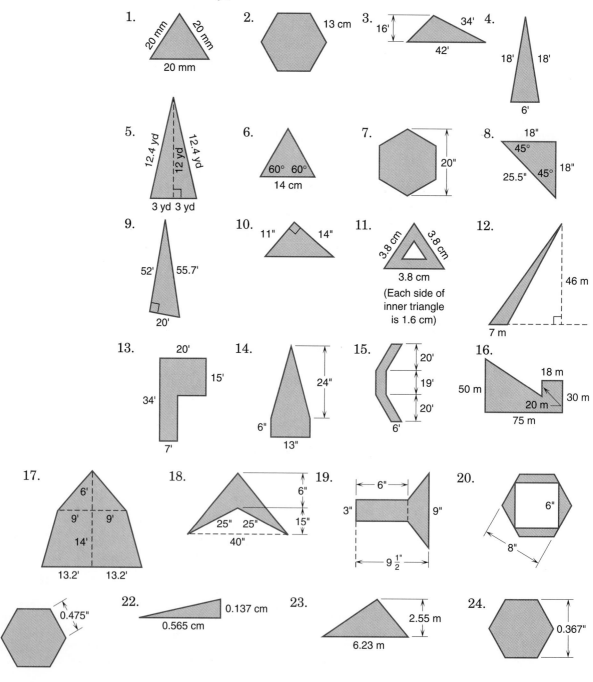

1. 20 mm 20 mm 20 mm

2. 13 cm

3. 16' 34' 42'

4. 18' 18' 6'

5. 12.4 yd 12 yd 12.4 yd 3 yd 3 yd

6. 60° 60° 14 cm

7. 20"

8. 18" 45° 45° 18" 25.5"

9. 52' 55.7' 20'

10. 11" 14"

11. 3.8 cm 3.8 cm 3.8 cm (Each side of inner triangle is 1.6 cm)

12. 46 m 7 m

13. 20' 15' 34' 7'

14. 24" 6" 13"

15. 20' 19' 20' 6'

16. 18 m 50 m 30 m 20 m 75 m

17. 6' 9' 9' 14' 13.2' 13.2'

18. 6" 25" 25" 15" 40"

19. 6" 3" 9" 9 1/2"

20. 6" 8"

21. 0.475"

22. 0.137 cm 0.565 cm

23. 2.55 m 6.23 m

24. 0.367"

C. Practical Problems

1. **Painting and Decorating** At 460 sq ft per gallon, how many gallons of paint are needed to cover the outside walls of a house as pictured below? (Do not count windows and doors and round up to the nearest gallon.)

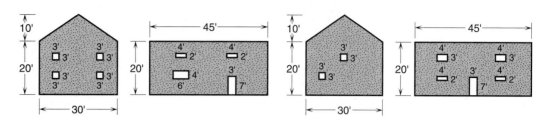

2. **Sheet Metal Technology** A four-sided vent connection must be made out of sheet metal. If each side of the vent is like the one pictured, how many total square inches of sheet metal will be used?

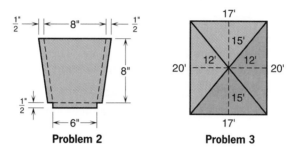

Problem 2   **Problem 3**

3. **Roofing** The aerial view of a roof is shown in the figure. How many squares (100 sq ft) of shingles are needed for the roof?

4. **Roofing** In problem 3 how many feet of gutters are needed?

5. **Carpentry** Allowing for a 3-ft overhang, how long a rafter is needed for the gable of the house in the figure? Round to the nearest tenth.

Problem 5

6. **Metalworking** A cut is to be made in a piece of metal as indicated by the dashed line. Find the length of the cut to the nearest tenth of an inch.

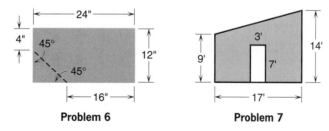

Problem 6   **Problem 7**

7. **Building Construction** How many square feet of drywall are needed for the wall shown in the figure?

8. **Building Construction** Find the area of the gable end of the house shown, not counting windows. Each window opening is 4 ft $3\frac{1}{4}$ in. by 2 ft $9\frac{1}{4}$ in. Round to the nearest square foot.

Problem 8

9. **Construction Technology** Find the missing dimensions x and y on the bridge truss shown.

Problem 9

When you have finished these problems, check your answers in the Appendix, then continue in Section 8-4 with the study of circles.

8-4 CIRCLES

A circle is probably the most familiar and the simplest plane figure. Certainly, it is the geometric figure that is most often used in practical and technical work. Mathematically, a circle is a closed curve representing the set of points some fixed distance from a given point called the *center*.

In this circle the center point is labeled O. The *radius* r is the distance from the center of the circle to the circle itself. The *diameter* d is the straight-line distance across the circle through the center point. It should be obvious from the drawing that the diameter is twice the radius.

Circumference

The *circumference* of a circle is the distance around it. It is a distance similar to the perimeter measure for a polygon. For all circles, the ratio of the circumference to the diameter, the distance around the circle to the distance across, is the same number.

$$\frac{\text{circumference}}{\text{diameter}} \approx 3.14159\ldots = \pi$$

The value of this ratio has been given the label π, the Greek letter "pi." The value of π is approximately 3.14, but the number has no simple decimal form—the digits continue without ending or repeating. Pressing the $\boxed{\pi}$ button on a scientific calculator will display the rounded ten-digit approximation of π:

$$3.141592654$$

The relationships between radius, diameter, and circumference can be summarized as follows:

CIRCLE FORMULAS

Diameter of a circle, $d = 2r$ or $r = \dfrac{d}{2}$

Circumference of a circle, $C = \pi d$ or $C = 2\pi r$

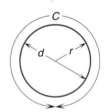

Note ▶ If you are performing mental math with π, it is important to remember that π is slightly larger than 3. If you are doing calculations by hand or with a simple calculator, it is important to know that $\pi \approx 3.14$ or 3.1416. To reinforce this knowledge, we will sometimes instruct you to use one of these decimal equivalents for π. Otherwise, you may use the $\boxed{\pi}$ button on your scientific calculator. ◀

> **EXAMPLE**

We can use these formulas to find the radius and circumference of a hole with a diameter of 4.00 in.

468

Radius, $r = \dfrac{d}{2}$

$$r = \frac{4 \text{ in.}}{2} = 2 \text{ in.}$$

Circumference, $C = \pi d$

$$\approx (\pi)(4 \text{ in.})$$

$$\approx 12.5664 \text{ in.} \approx 12.6 \text{ in.} \quad \text{rounded}$$

  π $\times$ **4** $=$ $\rightarrow$ *12.56637062*

YOUR TURN

Use these formulas to solve the following practical problems. (Round to two decimal places.)

(a) **Carpentry** If a hole is cut using a $\frac{3}{4}$-in.-diameter wood bit, what will be the radius and circumference of the hole? (Use 3.14 for π.)

(b) **Sheet Metal Technology** An 8-ft strip of sheet metal is bent into a circular tube. What will be the diameter of this tube?

(c) **Carpentry** A circular redwood hot tub has a diameter of 5 ft. What length of steel band is needed to encircle the tub to hold the individual boards in position? Allow 4 in. for fastening. (Use 3.1416 for π.)

(d) **Machine Technology** If eight bolts are to be equally spaced around a circular steel plate at a radius of 9 in., what must be the spacing between the centers of the bolts along the curve?

Problem (d)

Problem (e)

(e) **Construction** At $18.50 per foot, what will be the cost of the molding around the semicircular window shown? (Compute for the curved portion only.)

SOLUTIONS

(a) $r = \dfrac{d}{2}$

$$r = \frac{\frac{3}{4} \text{ in.}}{2} = \frac{3}{8} \text{ in.} \qquad \frac{\frac{3}{4}}{2} = \frac{3}{4} \cdot \frac{1}{2} = \frac{3}{8}$$

$C = \pi d$
$C \approx (3.14)(\frac{3}{4} \text{ in.})$
$C \approx (3.14)(0.75 \text{ in.})$ Change the fraction to a decimal.
$C \approx 2.36 \text{ in.}$

(b) If $C = \pi d$ then $d = \dfrac{C}{\pi}$ Divide both sides by π.

and
$$d \approx \frac{8 \text{ ft}}{\pi}$$
$$d \approx 2.55 \text{ ft} \qquad \text{rounded}$$

(c) $C = \pi d$
$C \approx (3.1416)(5 \text{ ft})$
$C \approx 15.71 \text{ ft} \approx 15 \text{ ft } 8\frac{1}{2} \text{ in.}$

Adding 4 in., the length needed is 16 ft $\frac{1}{2}$ in.

Notice that 0.7 ft is $(0.7)(12 \text{ in.}) = 8.4$ in. or about $8\frac{1}{2}$ in.

 3.1416 $\times$ **5** $=$ $\rightarrow$ *15.708* $-$ **15** $=$ $\times$ **12** $=$ $\rightarrow$ *8.496*

| Decimal feet | 0.708′ changed to inches |

(d) $C = 2\pi r$
$C \approx 2(\pi)(9 \text{ in.})$
$C \approx 56.55 \text{ in.}$

Since eight bolts must be spaced evenly around the 9-in. circle, divide by 8.
Spacing $= 56.55 \div 8 \approx 7.07$ in.

(e) $C = \pi d$ for a complete circle
$C \approx (\pi)(6 \text{ ft})$
$C \approx 18.8496 \text{ ft}$

The length of molding is half of this or 9.4248 ft. The cost is

$(\$18.50)(9.4248 \text{ ft}) \approx \$174.36$, rounded to the nearest cent.

Parts of Circles   Carpenters, plumbers, sheet metal workers, and other trades people often work with parts of circles. A common problem is to determine the length of a piece of material that contains both curved and straight segments.

Suppose that we need to determine the total length of steel rod needed to form the curved piece shown. The two straight segments are no problem: We will need a total of 8 in. + 6 in. or 14 in. of rod for these. When a rod is bent, the material on the outside of the curve is stretched, and the material on the inside is compressed. We must calculate the circumference for the curved section from a *neutral line* midway between the inside and outside radius.

The inside radius is given as 2 in. The stock is 1 in. thick, so the midline or neutral line is at a radius of $2\frac{1}{2}$ in. The curved section is one-quarter of a full circle, so the length of bar needed for the curved arc is

$C = \dfrac{2\pi r}{4}$ Divide the circumference by 4 to find the length of the quarter-circle arc.

$C = \dfrac{\pi r}{2}$

$C \approx \dfrac{(\pi)(2.5 \text{ in.})}{2}$

$C \approx 3.93$ in.   rounded to two decimal places.

The total length of bar needed for the piece is

$L \approx 3.93$ in. $+ 14$ in.
$L \approx 17.93$ in.

YOUR TURN

Find the total length of ornamental iron $\frac{1}{2}$ in. thick needed to bend into this shape. Round to two decimal places.

SOLUTION

Measuring to the midline we find that the upper curve (A) has a radius of $4\frac{1}{4}$ in. and the lower curve (C) has a radius of $3\frac{3}{4}$ in. Calculate the lengths of the two curved pieces this way:

$4\frac{1}{4}$ in. to midline

A

Upper Curve (A)

(Half circle)

$C = \frac{1}{2} \cdot 2\pi r = \pi r \leftarrow$ Multiply by $\frac{1}{2}$ for half circle.

$C \approx (\pi)(4\frac{1}{4}$ in.$)$

$C \approx 13.35$ in.

Lower Curve (C)

(Three-fourths circle)

$C = \frac{3}{4} \cdot 2\pi r = \frac{3\pi r}{2} \leftarrow$ Multiply by $\frac{3}{4}$ for three fourths of a circle.

$C \approx \frac{(3)(\pi)(3\frac{3}{4}\text{ in.})}{2}$

$C \approx 17.67$ in.

6 in.

B

Adding the length of the straight piece (B), the total length of straight stock needed to create this shape is

$3\frac{3}{4}$ in. to midline 13.35 in. $+ 17.67$ in. $+ 6$ in. ≈ 37.02 in.

Once you have your work organized, you can perform the entire calculation on a calculator as follows:

C

$\boxed{\pi}$ $\boxed{\times}$ **4.25** $\boxed{+}$ **3** $\boxed{\times}$ $\boxed{\pi}$ $\boxed{\times}$ **3.75** $\boxed{\div}$ **2** $\boxed{=}$ $\boxed{+}$ **6** $\boxed{=}$ $\rightarrow$ **37.02322746**

PROBLEMS

Find the length of stock needed to create each of the following shapes. Round to one decimal place.

(a)

1"

9"

(b)

$\frac{1}{2}$"

$6\frac{1}{2}$

(c)

100 mm

12.5 mm

70 mm

120 mm

(d)

16"   16"

18"

$1\frac{1}{2}$"

(e)

38.8 cm

12.7 cm

34.3 cm

2.8 cm

(a) 31.4 in.   (b) 31.8 in.   (c) 339.8 mm

(d) 62.6 in.   (e) 118.9 cm

On problem (e) a bit of careful reasoning will convince you that the dimensions are as follows:

Each of the four corner arcs has radius 9.4 cm to the midline, and each is a quarter circle.

Radius = 10.8 cm − 1.4 cm

         = 9.4 cm

Area of a Circle   To find the *area* of a circle use one of the following formulas.

AREA OF A CIRCLE

$$A = \pi r^2$$

or $$A = \frac{\pi d^2}{4} \approx 0.7854d^2$$

If you are curious about how these formulas are obtained, don't miss the box on page 474.

If you are curious about how these formulas are obtained, don't miss the box on page 474.

EXAMPLE

Let's use the formula to find the area of a circle with a diameter of 8 in.

$$A = \frac{\pi d^2}{4}$$

$$A \approx \frac{(\pi)(8 \text{ in.})^2}{4}$$

$A \approx 50.2655 \text{ in.}^2$ or 50.3 in.^2 rounded

 $\boxed{\pi}$ $\boxed{\times}$ **8** $\boxed{x^2}$ $\boxed{\div}$ **4** $\boxed{=}$ → 50.26548246

Of course you would find exactly the same area if you used the radius $r = 4$ in. in the first formula. Most people who need the area formula in their work memorize the formula $A = \pi r^2$ and calculate the radius r if they are given the diameter d.

Now for practice try these problems.

(a) Find the areas of the following circles. (Round to three significant digits.)

(1)

(2)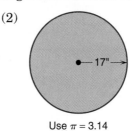

Use $\pi = 3.14$

(3) Find the area of the head of a piston with a radius of 6.5 cm.

(4)

19 yd

Use $\pi = 3.14$

(5)

2.2 mm

(b) **Painting** Find the area of wood that must be stained to finish the top and bottom of a circular table 6 ft in diameter. Use $\pi \approx 3.14$ and round to three significant digits.

(c) **Construction** Find the area of stained glass used in a semicircular window with a radius of 9 in. Use $\pi \approx 3.14$ and round to three significant digits.

(d) **Sheet Metal Technology** What is the area of the largest circle that can be cut out of a square piece of sheet metal 4 ft 6 in. on a side? Round to the nearest hundredth.

(e) **Hydraulic Technology** A pressure of 860 lb per square foot is exerted on the bottom of a cylindrical water tank. If the bottom has a diameter of 15 ft, what is the total force on the bottom? Round the final answer to three significant digits.

SOLUTIONS

(a) (1) $A = \pi r^2 = (\pi)(5 \text{ ft})^2 = (\pi)(25 \text{ ft}^2)$

$A \approx 78.5 \text{ ft}^2$ rounded

(2) $A = \pi r^2 \approx (3.14)(17 \text{ in.})^2$

$A \approx 907 \text{ in.}^2$

(3) $A = \pi r^2 = (\pi)(6.5 \text{ cm})^2$

$A \approx 133 \text{ cm}^2$ rounded

(4) $A = \frac{1}{2}(\pi r^2) \approx \frac{1}{2}(3.14)(9.5 \text{ yd})^2$

$A \approx 142 \text{ yd}^2$ rounded

(5) $A = \frac{1}{4}(\pi r^2) = \frac{1}{4}(\pi)(2.2 \text{ mm})^2$

$A \approx 3.80 \text{ mm}^2$ rounded

(b) $A = 2\pi r^2 \approx 2(3.14)(3 \text{ ft})^2$

$A \approx 56.5 \text{ ft}^2$

(c) $\quad A = \frac{1}{2}(\pi r^2) \approx \frac{1}{2}(3.14)(9 \text{ in.})^2$

$\quad\quad A \approx 127 \text{ in.}^2$

(d) $\quad A = \pi r^2 = (\pi)(2 \text{ ft } 3 \text{ in.})^2$

$\quad\quad A = (\pi)(2.25 \text{ ft})^2 \approx 15.90 \text{ ft}^2$

(e) $\quad A = \pi r^2 = (\pi)(7.5 \text{ ft})^2$

$\quad\quad A \approx 176.715 \text{ ft}^2$

$\quad\quad \text{Force} = 860\frac{\text{lb}}{\text{ft}^2} \times 176.715 \text{ ft}^2$

$\quad\quad \text{Force} = 151{,}974.5 \text{ or } 152{,}000 \text{ lb} \quad \text{rounded}$

Problem (d)

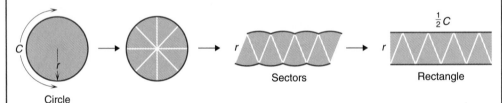

AREA OF A CIRCLE

To find the formula for the area of a circle, first divide the circle into many pie-shaped sectors, then rearrange them like this:

When the sectors are rearranged, they fit into a rectangle whose height is r, the radius of the original circle, and whose base length is roughly half the circumference of the circle. The more sectors the circle is divided into, the better the fit. The area of the sectors is approximately equal to the area of the rectangle.

Area of circle, $A = r \cdot \frac{1}{2}C$

$\quad\quad\quad\quad A = r \cdot \frac{1}{2}(2\pi r)$

$\quad\quad\quad\quad A = \pi r^2$

Rings A circular *ring*, or annulus, is defined as the area between two concentric circles—two circles having the same center point. A washer is a ring; so is the cross section of a pipe, a collar, or a cylinder.

To find the area of a ring, subtract the area of the inner circle from the area of the outer circle. Use the following formula.

AREA OF A RING (ANNULUS)

Shaded area = area of outer circle − area of inner circle

$$A = \pi(R^2 - r^2) \quad \text{or} \quad A = \frac{1}{4}\pi(D^2 - d^2)$$

$$A \approx 0.7854(D^2 - d^2)$$

EXAMPLE

To find the cross-sectional area of the wall of a ceramic pipe whose I.D. (inside diameter) is 8 in. and whose O.D. (outside diameter) is 10 in., substitute $d = 8$ in. and $D = 10$ in. into the second formula.

$A = \frac{1}{4}(\pi)[(10 \text{ in.})^2 - (8 \text{ in.})^2]$

$A \approx (0.7854)(100 \text{ in.}^2 - 64 \text{ in.}^2)$

$A \approx (0.7854)(36 \text{ in.}^2)$

$A \approx 28.3 \text{ in.}^2$ rounded

Note ▶ Notice that $\frac{\pi}{4}$ is equal to 0.7854 rounded to four decimal places. Also notice that the diameters are *squared first* and *then* the results are *subtracted*. ◀

YOUR TURN

Try it. Find the area of a washer with an outside radius of $\frac{7}{16}$ in. and a thickness of $\frac{1}{8}$ in.

Round to three significant digits.

SOLUTION

$R = \frac{7}{16}$ in. $r = \frac{7}{16}$ in. $- \frac{1}{8}$ in.   or $r = \frac{5}{16}$ in.

then

$A = \pi(R^2 - r^2)$

$A = (\pi)\left[\left(\frac{7}{16} \text{ in.}\right)^2 - \left(\frac{5}{16} \text{ in.}\right)^2\right]$

$A = (\pi)[(0.4375 \text{ in.})^2 - (0.3125 \text{ in.})^2]$

$A \approx (\pi)[0.1914 \text{ in.}^2 - 0.0977 \text{ in.}^2]$

$A \approx (\pi)[0.0937 \text{ in.}^2]$

$A \approx 0.295 \text{ in.}^2$ rounded

 (7 ÷ 16) x^2 − (5 ÷ 16) x^2 = × π = → 0.294524311

Most people find it easiest to convert the fractions to decimals before squaring, especially if they are working with a calculator.

PROBLEMS

Solve the following problems involving rings. (Round to two decimal places.)

Problem (a)

(a) **Plumbing**   Find the cross-sectional area of the wall of this pipe shown in the figure.

(b) Find the area of a washer with an inside diameter (I.D.) of $\frac{1}{2}$ in. and an outside diameter (O.D.) of $\frac{7}{8}$ in.

(c) **Machine Technology**   What is the cross-sectional area of a steel collar 0.6 cm thick with an inside diameter of 9.4 cm?

(d) **Landscaping**   At $1.95 per square foot find the cost of sodding a circular lawn 6 ft wide surrounding a flower garden with a radius of 20 ft.

(e) **Plumbing**   For a cross section of pipe with an inside diameter of 8 in., how much more material is needed to make the pipe with a $\frac{1}{2}$-in.-thick wall than to make it $\frac{3}{8}$ in. thick?

(a) 351.86 in.^{2} (b) 0.40 in.^{2} (c) 18.85 cm^2

(d) $1690.81   (e) 3.49 in.^{2}

In problem (c),

Radius of inner hole $= \dfrac{9.4 \text{ cm}}{2} = 4.7$ cm

Radius of outer edge $= 4.7 \text{ cm} + 0.6 \text{ cm} = 5.3$ cm

Then,

Area $= \pi \times (5.3^2 - 4.7^2) \approx 18.85$ cm^2

To solve problem (e) using a calculator, first calculate the cross-sectional area of the pipe with the smaller thickness:

 3 $\div$ 8 $\times$ 2 $+$ 8 $=$ x^2 $-$ 8 x^2 $=$ $\times$ .7854 $=$ STO $\rightarrow$ 9.8665875

Then calculate the cross-sectional area of the thicker pipe and subtract the value in memory:

 9 x^2 $-$ 8 x^2 $=$ $\times$ .7854 $-$ RCL $=$ $\rightarrow$ 3.4852125

Now turn to Exercises 8-4 for a set of review problems on circles.

COMBINING CIRCULAR AREAS

In many practical situations it is necessary to determine what size circle contains the same area as two or more smaller circles.

Area of larger circle = combined areas of the smaller circles

$$\pi R^2 = \pi r_1^2 + \pi r_2^2 + \cdots$$

Dividing all terms by π, we have

$$R^2 = r_1^2 + r_2^2 + \cdots$$

Taking the square root of both sides of the equation yields

$$R = \sqrt{r_1^2 + r_2^2 + \cdots}$$

As an example of how to apply this formula, suppose the flow from a 4-in.-radius pipe and a 6-in.-radius pipe must be combined into one large pipe of area equal to the total of the smaller two. To find the radius of the larger pipe, substitute into the formula

$$R = \sqrt{r_1^2 + r_2^2}$$
$$R = \sqrt{(4 \text{ in.})^2 + (6 \text{ in.})^2}$$
$$R = \sqrt{16 \text{ in.}^2 + 36 \text{ in.}^2}$$
$$R = \sqrt{52 \text{ in.}^2}$$

$R \approx 7.2$ in.   rounded

A similar relationship holds true for the diameters of circles.

$$D = \sqrt{d_1^2 + d_2^2 + \cdots}$$

Try these problems for practice.

1. **Construction**   Find the radius of a steel rod whose area must equal that of four rods each of radius 2 in.

2. What diameter gas line would be needed to replace two 6-in.-diameter lines?

3. **Hydraulic Technology**   Find the diameter of a water main with the same total area as two mains of diameter 3 in. and 4 in.

4. **Sheet Metal Technology**   What radius vent will have the same cross-sectional area as three 6-in.-radius vents?

5. **Plumbing**   Find the diameter of a water pipe that has the same area as four pipes, two with 2-in. diameters and two with 3-in. diameters.

All answers are given in the Appendix.

Exercises 8-4   Circles

A. Find the circumference and area of each circle. Round to three significant digits.

1.

2.

3.

4.

5. Circle of
    radius
    12 ft.

6. Circle of
    diameter
    0.50 m.

7. Circle
    of diameter
    1.2 cm.

8. Circle of
    radius
    0.4 in.

B. Find the area of each figure. Round to the nearest tenth.

1.

2.

3.

4.

5.

6.

7.

8.

9.

10.

11.

12.

Find the length of stock needed to create each shape. Round to the nearest tenth.

13.   14. 15.

C. Practical Problems

Problem 1

1. **Metalworking** What length of 1-in. stock is needed to bend a piece of steel into this shape? Use $\pi \approx 3.14$ and round to the nearest tenth.

2. **Landscaping** (a) Calculate, to the nearest square foot, the cultivable area of a circular garden 38 ft in diameter if the outer 2 ft are to be used for a path. (b) Find the area of the path.

Problem 3

3. **Machine Technology** What diameter must a circular piece of stock be to mill a hexagonal shape with a side length of 0.3 in.?

4. **Machine Technology** In problem 3 how many square inches of stock will be wasted? Round to the nearest hundredth.

5. **Plumbing** What is the cross-sectional area of a cement pipe 2 in. thick with an I.D. of 2 ft? Use $\pi \approx 3.14$ and round to the nearest whole number.

Problem 6

6. **Machine Technology** In the piece of steel shown, $4\frac{1}{2}$-in.-diameter holes are drilled in a 22-in.-diameter circular plate. Find the area remaining. Use $\pi \approx 3.14$ and round to three significant digits.

7. **Plumbing** How much additional cross-sectional area is there in a 30-in.-I.D. pipe 2 in. thick than one 1 in. thick? Round to one decimal place.

8. **Machine Technology** What diameter round stock is needed to mill a hexagonal nut measuring $1\frac{1}{2}$ in. across the flats? (Round up to the nearest quarter of an inch.)

9. **Landscaping** How many plants spaced every 6 in. are needed to surround a circular walkway with a 25-ft radius? Use $\pi \approx 3.14$.

10. **Machine Technology** What diameter circular stock is needed to mill a square end 3 cm on a side? Round to the nearest hundredth.

Problem 10   Problem 11

11. **Carpentry** At $11.25 per foot for the curved portion and $1.75 per foot for the straight portion, how much will it cost to put molding around the window pictured? Use $\pi \approx 3.14$.

Problem 12

12. Given the radius of the semicircular ends of the track is 32 m, how long must each straightaway be to make a 400-m track? Use $\pi \approx 3.14$.

Problem 13

13. **Manufacturing** What is the total length of belting needed for the pulley shown? Round to three significant digits.

Problem 14

14. **Electrical Technology** An electrician bends a $\frac{1}{2}$-in. conduit as shown in the figure to follow the bend in a wall. Find the total length of conduit needed between A and B. Use $\pi \approx 3.14$.

15. **Metalworking** How many inches apart will eight bolts be along the circumference of the metal plate shown in the figure? Round to the nearest tenth of an inch.

Problem 15   **Problem 16**

16. **Landscaping** How many square *yards* of concrete surface are there in a walkway 8 ft wide surrounding a tree if the inside diameter is 16 ft? Round to one decimal place. See the figure.

17. **Carpentry** The circular saw shown has a diameter of 18 cm and 22 teeth. What is the spacing between teeth? Round to two significant digits.

Problem 17

18. **Auto Mechanics** A hose must fit over a cylindrical opening 57.2 mm in diameter. If the hose wall is 3.0 mm thick, what length of clamp strap is needed to go around the hose connection? Round to two significant digits.

19. **Auto Mechanics** What length on the circumference would a 20° angle span on a 16-in. flywheel? Round to the nearest hundredth.

20. **Auto Mechanics** In making a certain turn on a road, the outside wheels of a car must travel 8.75 ft farther than the inside wheels. If the tires are 23 in. in diameter, how many additional turns do the outside wheels make during this turn? (*Hint:* In one "turn," a tire travels a distance equal to its circumference. Convert 8.75 ft to inches.)

21. **Wastewater Technology** To find the velocity of flow (in feet per second, or fps) through a pipe, divide the rate of flow (in cubic feet per second, or cfs) by the cross-sectional area of the pipe (in square feet). If the rate of flow through an 8.0-in.-diameter pipe is 0.75 cfs, find the velocity of flow. Round to the nearest tenth.

22. **Plumbing** The pressure exerted by a column of water amounts to 0.434 psi of surface for every foot of height or *head*. What is the total force (pressure times area) exerted on the bottom surface of a cylindrical container 16 in. in diameter if the head is 4.5 ft? Round to three significant digits.

23. **Plumbing** In order to construct a bracket for a drain pipe, Jerry the plumber needs to know its circumference. Calculate the circumference of a pipe 42.3 cm in diameter. Round to the nearest centimeter.

24. **Culinary Arts** A medium pizza at Papa's Pizza Place is 9 in. in diameter and a large pizza is 12 in. in diameter. If a medium pizza will serve four customers, how many servings will a large pizza provide?

Check your answers in the Appendix, then turn to Problem Set 8 for practice over the work of Chapter 8.

A summary of all the formulas you have learned for plane figures appears in the box. Be certain you know these before you continue.

SUMMARY OF FORMULAS FOR PLANE FIGURES

Figure	**Area**	**Perimeter**
Rectangle W L	$A = LW$	$P = 2L + 2W$
Square s	$A = s^2$	$P = 4s$
Parallelogram 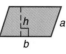 h a b	$A = bh$	$P = 2a + 2b$
Triangle a h c b	$A = \dfrac{1}{2}bh$ or $\dfrac{bh}{2}$ $A = \sqrt{s(s-a)(s-b)(s-c)}$ where $s = \dfrac{a+b+c}{2}$	$P = a + b + c$
Equilateral triangle a	$A = \dfrac{a^2\sqrt{3}}{4} \approx 0.433a^2$	$P = 3a$
Isosceles triangle a a b	$A = \dfrac{b}{2}\sqrt{a^2 - \left(\dfrac{b}{2}\right)^2}$	$P = 2a + b$
Right triangle 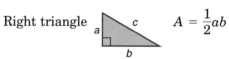 a c b	$A = \dfrac{1}{2}ab$	$P = a + b + c$
Trapezoid b_1 a h c b_2	$A = \left(\dfrac{b_1 + b_2}{2}\right)h$ or $\dfrac{h}{2}(b_1 + b_2)$	$P = a + c + b_1 + b_2$
Regular hexagon a	$A = \dfrac{3a^2\sqrt{3}}{2} \approx 2.598a^2$	$P = 6a$
Circle C r d	$A = \pi r^2$ $A = \dfrac{\pi d^2}{4} \approx 0.7854d^2$	$C = \pi d$ $C = 2\pi r$
Ring C r R d D	$A = \pi(R^2 - r^2)$ $A = \dfrac{\pi}{4}(D^2 - d^2) \approx 0.7854(D^2 - d^2)$	$C = 2\pi R$, the outside circumference

---

Practical
Plane Geometry

Answers are given in the Appendix.

A. Solve the following problems involving angles.

Name each angle and tell whether it is acute, obtuse, or right.

1.

2.

3.

4.

5.

6.

Measure the indicated angles using a protractor.

7.

8.

9.

10.

11.

12.

13. Use a protractor to draw angles of the following measures.
(a) 65°   (b) 138°   (c) 12°   (d) 90°

14. Find the size of each indicated angle without using a protractor.

(a)

(b)

(c)

The horizontal lines are parallel.

Name _____

Date _____

Course/Section _____

(d)

(e)

(f)

(g)

(h)

(i)

B. Solve the following problems involving plane figures.

Find the perimeter (or circumference) and area of each figure. (Round to the nearest tenth.)

1.

2.

3.

4.

5.

6.

7.

8.

9.

10.

11.

12.

13.

14.

15.

16.

17.

18.
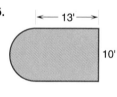

Find the missing dimensions of the following figures. (Round to the nearest hundredth if necessary.)

19.

20.

21.

22.

23.

24.

C. Practical Problems

1. **Building Construction** What will it cost to pave a rectangular parking lot 220 ft by 85 ft at $1.55 per square foot?
2. **Metalworking** A steel brace is used to strengthen the table leg as shown. What is the length of the brace? (Round to the nearest hundredth of an inch.)

Problem 2

Problem 3

Problem 4

3. **Machine Technology** Holes are punched in a regular octagonal steel plate as indicated. Through what angle must the plate be rotated to locate the second hole?
4. **Masonry** How many square feet of brick are needed to lay a patio in the shape shown?
5. **Electrical Technology** A coil of wire has an average diameter of 8.0 in. How many feet of wire are there if it contains 120 turns? (Round to two significant digits.)
6. **Sheet Metal Technology** A sheet metal worker needs to make a vent connection in the shape of the trapezoidal prism shown in the drawing. Use a protractor to find the indicated angle.

Problem 6

Problem 7

7. **Landscaping** At $21.35 per foot, how much does it cost to fence the yard shown in the figure?
8. **Metalworking** Find the length of straight stock needed to bend $\frac{1}{2}$-in. steel into the shape indicated. Use $\pi \approx 3.14$.

Problem 8

Problem 9

9. **Machine Technology** A hexagonal piece of steel 9 in. on a side must be milled by a machinist. What diameter round stock does he need? (Round to the nearest tenth of an inch.)

10. How much guy wire is needed to anchor a 30-ft pole to a spot 20 ft from its base? Allow 4 in. for fastening. (Round to the nearest inch.)

Problem 10

11. **Carpentry**   How many 4 ft by 8 ft sheets of exterior plywood must be ordered for the 4 walls of a building 20 ft long, 32 ft wide, and 12 ft high? Assume that there are 120 sq ft of window and door space. (*Note:* You can cut the sheets to fit, but you may not order a fraction of a sheet.)

12. **Metalworking**   Find the perimeter of the steel plate shown in the figure. (Round to the nearest tenth of an inch.)

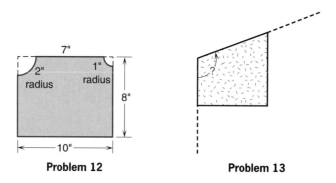

| Problem 12 | Problem 13 |

13. **Building Construction**   At what angle must sheet rock be cut to conform to the shape of the wall shown? Measure with a protractor.

14. **Metalworking**   A triangular shape with a base of 6 in. and a height of 9 in. is cut from a steel plate weighing 5.1 lb/sq ft. Find the weight of the shape (1 sq ft = 144 sq in.) (Round to the nearest ounce.)

15. **Mining Technology**   An air shaft is drilled from the indicated spot on the hill to the mine tunnel below. How long is the shaft to the nearest foot?

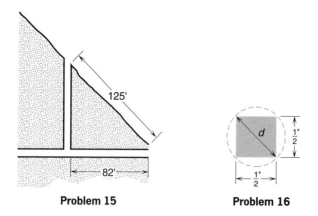

| Problem 15 | Problem 16 |

16. **Machine Technology**   What diameter circular stock is needed to mill a square bolt $\frac{1}{2}$ in. on a side? (Round to three significant digits.)

17. **Interior Design**   At 43 cents a foot how much will it cost to purchase weatherstripping for the following: 6 windows measuring 4 ft by 6 ft; 8 windows measuring 3 ft by 2 ft; and all sides except the bottom of two doors 3 ft wide and 7 ft high?

18. **Building Construction** How many square feet must be plastered to surface the wall shown? (Round to three significant digits.)

Problem 18

Problem 19

19. **Manufacturing** What length of belt is needed for the pulley shown? (Round to the nearest tenth.)

20. **Electrical Technology** Electrical conduit must conform to the shape and dimensions shown. Find the total length of conduit in inches. (Use $\pi \approx 3.14$ and round to the nearest tenth.)

Problem 20

Problem 21

21. **Building Construction** At $3.75 per square foot, how much will it cost to pave the circular pathway shown? (Use $\pi \approx 3.14$.)

22. **Carpentry** Find the size of $\angle 2$ when the carpenter's square is placed over the board as shown.

Problem 22

23. **Welding** The top to a container is made of steel weighing 4.375 lb/sq ft. If the top is a 30.0 in. by 30.0 in. square, find its weight to three significant digits.

24. **Machine Technology** A square bolt must be machined from round stock. If the bolt must be $2\frac{1}{8}$ in. on each side, what diameter stock is needed? (See the figure.) Round to the nearest thousandth.

Problem 24

25. **Auto Mechanics** A fan pulley has a diameter of $3\frac{7}{8}$ in. What is its circumference? Round to four significant digits.

26. **Fire Protection** Firefighters report that a wildfire with a front about 10 miles long has consumed approximately 25,000 acres. What is the average width of the destruction in miles? (1 sq mi = 640 acres. Round to the nearest mile.)

Solid Figures

Objective	Sample Problems		For Help Go to Page

Objective

When you finish this unit you will be able to:

Sample Problems

For Help Go to Page

1. Identify solid figures, including prisms, cubes, cones, cylinders, pyramids, spheres, and frustums.

(a) _____ 489

(b) _____

(c) _____

2. Find the surface area and volume of solid objects. (Round to one decimal place.)

(a) 2' 4'

Volume = _____ 508

Total surface area = _____

(b) Sphere
$r = 2.1$ cm

Volume = _____ 511

Surface area = _____

(c) 4" 0.6"

Volume = _____ 491

(d) 8" 6"

Lateral surface area = _____ 516

Volume = _____

(e) 6 cm 6 cm
12 cm s 12 cm
$h = 4$ cm
$s = 5$ cm

Lateral surface area = _____ 504

Volume = _____

Name _____

Date _____

Course/Section _____

3. Solve practical problems involving solid figures.

(a) Calculate the amount of water, in gallons, needed to fill the swimming pool shown. Round to the nearest 100 gallons. (1 cu ft ≈ 7.48 gal)

494

(b) Find the weight, in ounces, of the steel pin shown if its density is 0.2963 lb/cu in. Round to three significant digits.

495

(c) The roof of a building is in the shape of a pyramid having a square base 84 ft on each side. If the slant height of the roof is 60 ft, what would be the cost of the roofing material needed, at $0.45 per sq ft?

500

(Answers to these preview problems are given in the Appendix.)

If you are certain that you can work *all* these problems correctly, turn to page 525 for a set of practice problems. If you cannot work one or more of the preview problems, turn to the page indicated to the right of the problem. Those who wish to master this material with the greatest success should turn to Section 9-1 and begin work there.

9 Solid Figures

By permission Jonny Hart and Creators Syndicate, Inc.

Plane figures have only two dimensions: length and width or base and height. A plane figure can be drawn exactly on a flat plane surface. Solid figures are three-dimensional: they have length, width, and height. Making an exact model of a solid figure requires shaping it in clay, wood, or paper. A square is a two-dimensional or plane figure; a cube is a three-dimensional or solid figure.

Trades workers encounter solid figures in the form of tanks, pipes, ducts, boxes, birthday cakes, and buildings. They must be able to identify the solid and its component parts and compute its surface area and volume.

9-1 PRISMS

A *prism* is a solid figure having at least one pair of parallel surfaces that create a uniform cross section. The figure shows a hexagonal prism.

Hexagonal prism

Cutting the prism anywhere parallel to the hexagonal surfaces produces the same hexagonal cross section.

All the polygons that form the prism are called *faces*. The faces that create the uniform cross section are called the *bases*. They give the prism its name. The others are called lateral faces.

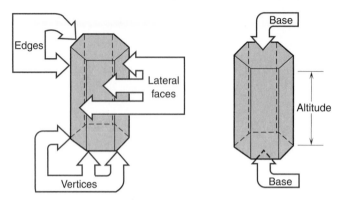

The sides of the polygons are the *edges* of the prism, and the corners are still referred to as *vertices*. The perpendicular distance between the bases is called the *altitude* of the prism.

In a *right prism* the lateral edges are perpendicular to the bases, and the lateral faces are rectangles. In this section you will learn about right prisms only.

Right prisms   Not right prisms

YOUR TURN

The components of a prism are its faces, vertices, and edges. Count the number of faces, vertices, and edges in this prism:

Number of faces = _____

Number of vertices = _____

Number of edges = _____

ANSWERS

There are 7 faces (2 bases and 5 lateral faces), 15 edges, and 10 vertices on this prism.

Here are some examples of simple prisms.

Rectangular Prism   The most common solid in practical work is the *rectangular prism*. All opposite faces are parallel, so any pair can be chosen as bases. Every face is a rectangle. The angles at all vertices are right angles.

Cube   A *cube* is a rectangular prism in which all edges are the same length.

Triangular Prism In a *triangular prism* the bases are triangular.

 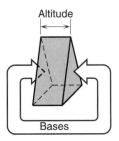

Trapezoidal Prism In a *trapezoidal prism* the bases are identical trapezoids.

We know that the trapezoids and not the other faces are the bases because cutting the prism anywhere parallel to the trapezoid faces produces another trapezoidal prism.

Three important quantities may be calculated for any prism: the lateral surface area, the total surface area, and the volume. The following formulas apply to all right prisms.

RIGHT PRISMS

Lateral surface area $L = ph$ h = altitude

Total surface area $S = L + 2A$ p = perimeter of the base

Volume $V = Ah$ A = area of the base

The lateral surface area L is the area of all surfaces *excluding* the two bases.

The total surface area S is the lateral surface area *plus* the area of the two bases.

The *volume* or capacity of a prism is the total amount of space inside it. Volume is measured in cubic units. (You may want to review these units in Chapter 5, page 244.)

Note ▶ As with units of area, units of volume can be written using exponents. For example, cubic inches (cu in.) result from multiplying in. × in. × in., which in exponent form is in.^{3}. Similarly, cubic yards (cu yd) can be expressed as yd^3, cubic meters (cu m) as m^3, and so on. Workers in the trades use both types of notation, and we shall use both in the remainder of this text. ◄

EXAMPLE

Let's look at an example. Find the (a) lateral surface area, (b) total surface area, and (c) volume of the triangular cross-section duct shown.

(a) The perimeter of the base is

$p = 6$ in. $+ 8$ in. $+ 10$ in. $= 24$ in. and $h = 15$ in.

Therefore, the lateral surface area is

$L = ph = (24$ in.$)(15$ in.$)$

$L = 360$ in.^{2}

(b) In this prism the bases are right triangles, so that the area of each base is

$A = \frac{1}{2}bh$

$A = \frac{1}{2}(6$ in.$)(8$ in.$)$

$A = 24$ in.^{2}

Therefore, the total surface area is

$S = L + 2A$

$S = 360$ in.$^2 + 2(24$ in.$^2)$

$S = 408$ in.^{2}

(c) The volume of the prism is

$V = Ah$

$V = (24$ in.$^2)(15$ in.$)$

$V = 360$ in.^{3}

YOUR TURN

Find L, S, and V for this trapezoidal right prism.

SOLUTION

$h = 10$ ft   and $p = 4$ ft $+ 15$ ft $+ 12$ ft $+ 17$ ft $= 48$ ft

The area of each trapezoidal base is

$A = \left(\dfrac{b_1 + b_2}{2}\right)h$

$A = \left(\dfrac{4 \text{ ft} + 12 \text{ ft}}{2}\right)(15 \text{ ft})$

$A = 120 \text{ ft}^2$

Therefore,

$L = ph$	$S = L + 2A$	$V = Ah$
$L = (48 \text{ ft})(10 \text{ ft})$	$S = 480 \text{ ft}^2 + 2(120 \text{ ft}^2)$	$V = (120 \text{ ft}^2)(10 \text{ ft})$
$L = 480 \text{ ft}^2$	$S = 720 \text{ ft}^2$	$V = 1200 \text{ ft}^3$

PROBLEMS

Find the lateral surface area, total surface area, and volume for each of the following prisms. Round to the nearest whole number.

(a)

(b)

(c)

(d)

SOLUTIONS

(a) $h = 4\frac{1}{2}$ in.

$p = 3 \text{ in.} + 2 \text{ in.} + 3 \text{ in.} + 2 \text{ in.} = 10 \text{ in.}$ $A = (3 \text{ in.})(2 \text{ in.}) = 6 \text{ in.}^2$

$L = ph = (10 \text{ in.})(4.5 \text{ in.})$ $S = L + 2A$

$L = 45 \text{ in.}^2$ $S = 45 \text{ in.}^2 + 2(6 \text{ in.}^2)$

$S = 57 \text{ in.}^2$

$V = Ah$

$V = (6 \text{ in.}^2)(4.5 \text{ in.})$

$V = 27 \text{ in.}^3$

(b) $L = 1080 \text{ ft}^2$ $S = 1464 \text{ ft}^2$ $V = 3456 \text{ ft}^3$

(c) $L = 416.25 \text{ in.}^2$ $S = 476.25 \text{ in.}^2$ $V = 450 \text{ in.}^3$

$\approx 416 \text{ in.}^2$, rounded $\approx 476 \text{ in.}^2$, rounded

(d) For a cube, $L = 4s^2 = 196 \text{ cm}^2$

$S = 6s^2 = 294 \text{ cm}^2$

$V = s^3 = 343 \text{ cm}^3$

In practical problems it is often necessary to convert volume units from cubic inches or cubic feet to gallons or similar units. Use the following conversion factors and set up unity fractions as shown in Chapter 5.

1 cu ft $(\text{ft}^3) = 1728$ cu in. (in.^3)	1 cu yd $(\text{yd}^3) = 27$ cu ft (ft^3)
≈ 7.48 gallons	
≈ 28.32 liters	1 cu in. $(\text{in}^3) \approx 16.39$ cu cm (cm^3)
≈ 0.0283 cu m (m^3)	
≈ 0.806 bushel	1 gal $= 231$ cu in. (in.^3)
	1 liter $= 1000$ ml $= 1000$ cu cm (cm^3)

Calculate the volume of water needed to fill this lap pool in gallons. Round to three significant digits.

The pool is a prism with trapezoidal bases. The height of the trapezoid is 25 yd or 75 ft and the altitude of the prism is 12 ft.

$V = Ah$

$V = \left(\dfrac{3\,\text{ft} + 11\,\text{ft}}{2}\right)(75\,\text{ft})(12\,\text{ft})$

$V = (7\,\text{ft})(75\,\text{ft})(12\,\text{ft})$

$V = 6300\,\text{cu ft}$

In gallons,

$V \approx 6300\,\text{ft}^3 \times \dfrac{7.48\,\text{gal}}{1\,\text{ft}^3}$

$V \approx 47{,}124\,\text{gal}$ or $47{,}100\,\text{gal}$, rounded

 3 $\boxed{+}$ **11** $\boxed{=}$ $\boxed{\div}$ **2** $\boxed{\times}$ **75** $\boxed{\times}$ **12** $\boxed{\times}$ **7.48** $\boxed{=}$ → 47124.

YOUR TURN

What volume of liquid, in liters, is needed to fill the container shown? Assume that the base is a regular hexagon and round your answer to two significant digits.

SOLUTION

As shown in Chapter 8, the area of a hexagon can be calculated as

$A \approx 2.598a^2$

and the volume of the prism is

$V = Ah \approx 2.598a^2h$

$\approx 2.598\,(6.5\,\text{in.})^2\,(16\,\text{in.})$

$\approx 1756.248\,\text{in.}^3$

To convert this to liters, first convert to either cu ft or cu cm, then use a second unity fraction to convert to liters.

$V \approx 1756.248\,\text{in.}^3 \times \dfrac{1\,\text{ft}^3}{1728\,\text{in}^3} \times \dfrac{28.32\,\text{liters}}{1\,\text{ft}^3}$

$\approx 28.78\ldots\text{liters} \approx 29\,\text{liters}$, rounded.

If we convert in.^{3} to cm^3 first,

$V \approx 1756.248\,\text{in.}^3 \times \dfrac{16.39\,\text{cm}^3}{1\,\text{in.}^3} \times \dfrac{1\,\text{liter}}{1000\,\text{cm}^3}$

$\approx 28.78\ldots\text{liters} \approx 29\,\text{liters}$, rounded

In many practical problems the prism encountered has an irregular polygon for its base.

EXAMPLE

Steel weighs approximately 0.283 lb/in.³ Find the weight of the steel bar shown. Round to one decimal place.

First, we must calculate the volume of this prism. To find this, we first calculate the cross-sectional area by subtracting.

$$A = (6'' \times 1.5'') - (3'' \times \frac{3}{4}'')$$

$$A = 9\,\text{in.}^2 - 2.25\,\text{in.}^2$$

$$A = 6.75\,\text{in.}^2$$

and the volume of the bar is

$$V = Ah$$
$$V = (6.75\,\text{in.}^2)(18\,\text{in.})$$
$$V = 121.5\,\text{in.}^3$$

The weight of the bar is

$$W = 121.5\,\cancel{\text{in.}^3} \times 0.283\,\frac{\text{lb}}{\cancel{\text{in.}^3}}$$

$$W = 34.3845\,\text{lb} \qquad \text{or} \qquad 34.4\,\text{lb} \quad \text{rounded}$$

Learning Help ▶ The most common source of errors in most calculations of this kind is carelessness. Organize your work neatly. Work slowly and carefully. ◀

PROBLEMS

Now try these practice problems.

(a) How many cubic *yards* of concrete does it take to pour a square pillar 2 ft on a side if it is 12 ft high? (Round to two significant digits.)

(b) What is the capacity to the nearest gallon of a septic tank in the shape of a rectangular prism 12 ft by 16 ft by 6 ft?

(c) What is the weight of the steel V-block in the figure at 0.0173 lb/cm³? (Round to the nearest pound.)

Problem (c)

Problem (d)   **Problem (e)**

(d) Find the weight of the piece of brass pictured at 0.2963 lb/in.³ (Round to the nearest tenth.)

(e) Find the weight (in ounces) of the steel pin shown if its density is 0.0173 lb/cm³. (Round to two significant digits.)

(a) $V = Ah = (2 \text{ ft})(2 \text{ ft})(12 \text{ ft})$

$= 48 \text{ ft}^3$ But $1 \text{ yd}^3 = 27 \text{ ft}^3$

$= 48 \text{ ft}^3 \times \dfrac{1 \text{ yd}^3}{27 \text{ ft}^3}$

$= \dfrac{48}{27} \text{ yd}^3 \approx 1.8 \text{ yd}^3$

(b) $V = Ah = (12 \text{ ft})(16 \text{ ft})(6 \text{ ft})$

$= 1152 \text{ ft}^3$ But $1 \text{ ft}^3 \approx 7.48 \text{ gallons}$

$\approx 1152 \text{ ft}^3 \times \dfrac{7.48 \text{ gal}}{1 \text{ ft}^3}$

$\approx 8616.96 \text{ gal} \approx 8617 \text{ gal}$

(c) **First,** find the area of the base. The cross-sectional view shown indicates that the base of the block is a rectangle with a triangle cut out. The area of the base is

$$A = \underbrace{(10 \text{ cm})(19 \text{ cm})}_{\text{Rectangle}} - \underbrace{\tfrac{1}{2}(10 \text{ cm})(5 \text{ cm})}_{\text{Triangle}}$$

$A = 190 \text{ cm}^2 - 25 \text{ cm}^2 = 165 \text{ cm}^2$

Next, find the volume.

$V = Ah = (165 \text{ cm}^2)(38 \text{ cm})$

$= 6270 \text{ cm}^3$

Finally, calculate the weight, W, by multiplying the volume by the density.

$W \approx (6270 \text{ cm}^3)(0.0173 \text{ lb/cm}^3)$

$\approx 108.471 \text{ lb} \approx 108 \text{ lb}$

(d) The brass object is in the form of a prism with base as shown. The area of the base is the area of the narrow rectangle plus the area of the trapezoid.

$$A = \underbrace{(10 \text{ in.})(1 \text{ in.})}_{\text{Rectangle}} + \underbrace{\tfrac{1}{2}(1 \text{ in.} + 8 \text{ in.})(6 \text{ in.})}_{\text{Trapezoid}}$$

$= 10 \text{ in.}^2 + 27 \text{ in.}^2 = 37 \text{ in.}^2$

The altitude of the prism is the thickness of the brass or 2 in. The volume of the object is

$V = Ah = (37 \text{ in.}^2)(2 \text{ in.}) = 74 \text{ in.}^3$

The weight of the object is

$W \approx 74 \text{ in.}^3 \times 0.2963 \text{ lb/in.}^3 \approx 21.9262 \text{ lb} \approx 21.9 \text{ lb}$

(e) The steel pin is in the form of a hexagonal prism. The volume is

$V = Ah \approx (2.598)(0.9 \text{ cm})^2(6.5 \text{ cm}) \approx 13.67847 \text{ cm}^3$

The weight in pounds is

$W \approx 13.67847 \text{ cm}^3 \times 0.0173 \text{ lb/cm}^3 \approx 0.2366 \ldots \text{ lb}$

496

The weight in ounces is

$$W \approx 0.2366 \text{ lb} \times \frac{16 \text{ oz}}{1 \text{ lb}} \approx 3.8 \text{ oz, rounded to two significant digits}$$

 2.598 ✕ **0.9** x^2 ✕ **6.5** ✕ **.0173** ✕ **16** = → ▦ *3.786200496*

Now turn to Exercises 9-1 for more practice on calculating the surface area and volume of prisms.

Exercises 9-1   Prisms

A. Find the volume of each of the following shapes. (Round to the nearest tenth unless indicated otherwise.)

1.   2.   3. 4.

Round to nearest hundred

Round to nearest cu in.

5.   6. 7. 8.

Round to the nearest hundredth

B. Find the lateral surface area and the volume of each of the following solids. (Round to the nearest tenth.)

1. 2. 3. 4.

5.   6.

C. Find the total outside surface area and the volume of each of the following solids. (Round to the nearest tenth.)

1.
5 yd 5 yd
5 yd

2.
9.3 m
7.4 m
5.2 m

3.
16' 16'
22'

4.
$\frac{1}{2}$"
$3\frac{1}{2}$"
The base is a regular hexagon.

5.
13 cm
35 cm
25 cm
37 cm
36 cm

6.
6 yd
12 yd
8 yd
2 yd
15 yd

D. Practical Problems (Round to the nearest tenth.)

1. **Manufacturing** How many cubic *feet* of warehouse space are needed for 450 boxes 16 in. by 8 in. by 10 in.?

2. **Metalworking** Find the weight of the piece of steel pictured. (Steel weighs 0.2833 lb/cu in.)

3"
6"
3"
2"
10"
28'
3'
75'
9'
28'

Problem 2   **Problem 3**

3. **Building Construction** How many gallons of water are needed to fill a swimming pool that approximates the shape in the figure? (Round to the nearest gallon.)

4. **Painting and Decorating** How many gallons of paint are needed for the outside walls of a building 26 ft high by 42 ft by 28 ft if there are 480 sq ft of windows? One gallon covers 400 sq ft. (Assume that you cannot buy a fraction of a gallon.)

5. **Building Construction** Dirt must be excavated for the foundation of a building 30 yards by 15 yards to a depth of 3 yards. How many trips will it take to haul the dirt away if a truck with a capacity of 3 cu yd is used?

6. **Masonry** If brick has a density of 103 lb/cu ft, what will be the weight of 500 bricks, each $3\frac{3}{4}$ in. by $2\frac{1}{4}$ in. by 8 in.? Round to the nearest pound.

7. **Landscaping** Dirt cut 2 ft deep from a section of land 38 ft by 50 ft is used to fill a section 25 ft by 25 ft to a depth of 3 ft. How many cubic *yards* of dirt are left after the fill?

8. Find the weight of the steel V-block in the figure if its density is 0.0173 lb/cm³.

15 cm
15 cm
8 cm
48 cm
24 cm

9. **Metalworking** Find the weight of the cast iron shape in the figure. (Cast iron weighs 0.2607 lb/cu in. The base is an equilateral triangle.)

10. **Metalworking** Find the weight of the piece of brass shown. Brass weighs 0.296 lb/cu in.

Problem 10

11. **Manufacturing** How high must a 400-gallon rectangular tank be if the base is a square 3 ft 9 in. on a side? (1 cu ft $\approx$ 7.48 gallons.)

12. **Landscaping** How many pounds of rock will you need to fill an area 25 ft by 35 ft to a depth of 2 in. if the rock weighs 1050 lb/cu yd? Round to the nearest ten lb.

13. **Wastewater Technology** A rectangular sludge bed measuring 80 ft by 120 ft is filled to a depth of 2 ft. If a truck with a capacity of 3 cu yd is used to haul away the sludge, how many trips must the truck make to empty the sludge bed?

14. **Wastewater Technology** A reservoir with a surface area of 2.5 acres loses 0.3 in. per day to evaporation. How many gallons does it lose in a week? (1 acre = 43,560 sq ft, 1 sq ft = 144 in., 1 gal = 231 in.^{3}) (Round to the nearest thousand.)

15. **Carpentry** A carpenter needs to pour a rectangular slab 28 ft long, 16 ft wide, and 6 in. thick. How many cubic yards of concrete does the carpenter need? (Round to the nearest tenth.)

16. **Construction Technology** Concrete weighs 150 lb/cu ft. What is the weight of a concrete wall 18 ft long, 4 ft 6 in. high, and 8 in. thick?

17. **Masonry** The concrete footings for piers for a raised foundation are cubes measuring 18 in. by 18 in. by 18 in. How many cubic yards of concrete are needed for 28 footings?

18. **Plumbing** A flush tank 21 in. by 6.5 in. contains water to a depth of 12 in. How many gallons of water will be saved per flush if a conservation device reduces the capacity to two-thirds of this amount? (Round to the nearest tenth of a gallon.)

19. **Roofing** A flat roof 28 ft long and 16 ft wide has a 3-in. depth of water sitting on it. What is the weight of the water on the roof? (Water weighs 62.5 lb/cu ft.)

20. **Building Construction** In a 4-hour percolation test, an 8-ft depth of water seeped out of a square pit 4 ft on a side. The plumbing code says that the soil must be able to absorb 5000 gallons in 24 hours. At this rate, will the soil be up to code?

21. **Plumbing** The BiltRite Construction Company installs a septic tank 10 ft long by 5 ft wide and 6 ft deep. Calculate the capacity of the tank in gallons.

22. **Plumbing** A flush tank 8 in. wide by 19 in. long is filled to a depth of 14 in. Calculate its capacity in gallons. Round to the nearest 0.1 gal.

23. **Plumbing** A pipe trench is roughly rectangular with dimensions 75 ft long by 2 ft wide and with an average depth of $3\frac{1}{2}$ ft. What volume of dirt was removed?

24. **Building Construction** BiltWell Construction needs to create a concrete foundation 5 ft deep measuring 60 ft by 32 ft, outside dimensions, with walls 7 in. thick. How many cubic yards of concrete will they need?

When you have completed these exercises, check your answers in the Appendix, then turn to Section 9-2 to study pyramids.

9-2 PYRAMIDS AND FRUSTUMS OF PYRAMIDS

Pyramids A *pyramid* is a solid object with one base and three or more *lateral faces* that taper to a single point opposite the base. This single point is called the *apex*.

A *right pyramid* is one whose base is a regular polygon and whose apex is centered over the base. We will examine only right pyramids here.

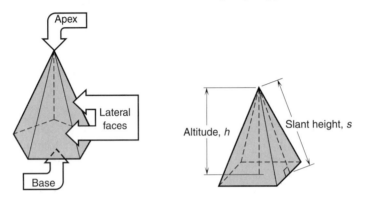

The *altitude* of a pyramid is the perpendicular distance from the apex to the base. The *slant height* of the pyramid is the height of one of the lateral faces.

Several kinds of right pyramids can be constructed:

Square Pyramid   **Triangular Pyramid** or **Tetrahedron**

There are several very useful formulas relating to pyramids.

PYRAMIDS

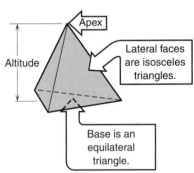

Lateral surface area $L = \frac{1}{2}ps$

Volume $V = \frac{1}{3}Ah$

---

where p = perimeter of the base
s = slant height of a lateral face
A = area of the base
h = altitude

EXAMPLE

Find the lateral surface area and volume of this square pyramid:

$s \approx 10.77$ ft, $h = 10$ ft   Round to the nearest tenth.

$p = 8$ ft $+ 8$ ft $+ 8$ ft $+ 8$ ft $= 32$ ft $A = (8$ ft$)(8$ ft$) = 64$ ft^2

The lateral surface area is

$L = \frac{1}{2}ps$

$L \approx \frac{1}{2}(32$ ft$)(10.77$ ft$)$

$L \approx 172.3$ ft^2 This is the area of the four triangular lateral faces.

The volume is

$V = \frac{1}{3}Ah$

$V = \frac{1}{3}(64$ ft$^2)(10$ ft$)$

$V \approx 213.3$ ft^3 rounded

YOUR TURN

36.3 cm

16 cm
$h = 36$ cm

Find the lateral surface area and volume of this triangular pyramid. (*Remember: The base is an equilateral triangle.*) Round to the nearest whole unit.

SOLUTION

$p = 16$ cm $+ 16$ cm $+ 16$ cm
$p = 48$ cm   perimeter of the base
$s \approx 36.3$ cm

Therefore, $L = \frac{1}{2}ps$

$L \approx \frac{1}{2}(48$ cm$)(36.3$ cm$)$

$L \approx 871.2$ cm^2 or 871 cm^2 rounded

$A \approx 0.433s^2$ for an equilateral triangle
$A \approx 0.433(16$ cm$)^2$
$A \approx 110.8$ cm^2

The volume is

$V = \frac{1}{3}Ah$

$V \approx \frac{1}{3}(110.8$ cm$^2)(36$ cm$)$

$V \approx 1330$ cm^3 rounded

The entire calculation for volume can be done as follows using a calculator:

.433 $\times$ **16** x^2 $\times$ **36** $\div$ **3** $=$ → *1330.176*

A

Multiplying by $\frac{1}{3}$ is the same as dividing by 3

To calculate the **total** surface area, S, of a pyramid, add the area of the base to the lateral surface area.

3.50 yd

12.0 yd

slant ht≈12.38 yd

EXAMPLE

Find the total surface area of the hexagonal pyramid shown. Round to three significant digits.

First, find the lateral surface area, L.

$p = 6(3.5 \text{ yd}) = 21 \text{ yd}$

$s \approx 12.38 \text{ yd}$

$L = \frac{1}{2}ps \approx \frac{1}{2}(21 \text{ yd})(12.38 \text{ yd})$

$\approx 129.99 \text{ yd}^2$

Then find the area of the hexagonal base.

$A \approx 2.598\,(3.5 \text{ yd})^2 \approx 31.8255 \text{ yd}^2$

Finally, add these areas and round.

$S = L + A \approx 129.99 \text{ yd}^2 + 31.8255 \text{ yd}^2$

$\approx 161.8155 \text{ yd}^2 \approx 162 \text{ yd}^2$, rounded.

PROBLEMS

More practice.

(a) Find the lateral surface area and volume of each pyramid. (Round to three significant digits.)

(1)

6"

2.5" 2.5"

Slant height
≈ 6.13"
Altitude = 6"

(2)

4'

3'

Slant height ≈ 4.58'
Altitude = 3'

(b) Find the total surface area and volume of this pyramid. Round to three significant digits.

13 cm

Altitude = 16 cm
Slant height ≈ 16.4 cm

(c) Find the volume of an octagonal pyramid if the area of the base is 245 in.^{2} and the altitude is 16 in. (Round to the nearest cubic inch.)

SOLUTIONS

(a) (1) $p = 4(2.5 \text{ in.}) = 10 \text{ in.}$ $s \approx 6.13 \text{ in.}$

$L = \frac{1}{2}ps \approx \frac{1}{2}(10)(6.13)$

$\approx 30.65 \text{ in.}^2 \approx 30.7 \text{ in.}^2$, rounded

$$V = \tfrac{1}{3}Ah = \tfrac{1}{3}(2.5 \text{ in.})^2(6 \text{ in.})$$
$$= 12.5 \text{ in.}^3$$

(2) $\quad p = 6(4 \text{ ft}) = 24 \text{ ft} \qquad s \approx 4.58 \text{ ft}$

$$L = \tfrac{1}{2}ps \approx \tfrac{1}{2}(24 \text{ ft})(4.58 \text{ ft})$$
$$\approx 54.96 \text{ ft}^2 \approx 55.0 \text{ ft}^2, \text{ rounded}$$

$$V = \tfrac{1}{3}Ah \approx \tfrac{1}{3}\overbrace{(2.598)(4 \text{ ft})^2}^{\text{Area of hexagonal base}}(3 \text{ ft})$$
$$\approx 41.568 \text{ ft}^3 \approx 41.6 \text{ ft}^3, \text{ rounded}$$

(b) **First,** find the lateral surface area, L.

$$p = 3(13 \text{ cm}) = 39 \text{ cm} \qquad s \approx 16.4 \text{ cm}$$
$$L = \tfrac{1}{2}ps \approx \tfrac{1}{2}(39 \text{ cm})(16.4 \text{ cm})$$
$$\approx 319.8 \text{ cm}^2$$

Next, find A, the area of the base, an equilateral triangle.

$$A \approx 0.433a^2$$
$$A \approx 0.433(13 \text{ cm})^2 \approx 73.177 \text{ cm}^2$$

Finally, add $L + A$ to obtain the total surface area, S.

$$S \approx 319.8 \text{ cm}^2 + 73.177 \text{ cm}^2$$
$$\approx 392.977 \text{ cm}^2 \approx 393 \text{ cm}^2, \text{ rounded}$$

Having calculated the area of the base we can now find the volume.

$$V = \tfrac{1}{3}Ah \approx \tfrac{1}{3}(73.177 \text{ cm}^2)(16 \text{ cm})$$
$$\approx 390.277\ldots \text{ cm}^3 \approx 390 \text{ cm}^3, \text{ rounded}$$

(c) $\quad V = \tfrac{1}{3}Ah = \tfrac{1}{3}(245 \text{ in.}^2)(16 \text{ in.})$
$$= 1306.666\ldots \text{ in.}^3 \approx 1307 \text{ in.}^3$$

Frustum of a Pyramid

A *frustum* of a pyramid is the solid figure remaining after the top of the pyramid is cut off parallel to the base. Frustum shapes appear as containers, as building foundations, and as transition sections in ducts.

Most frustums used in practical work, such as sheet metal construction, are frustums of square pyramids or cones.

Every frustum has two bases, upper and lower, that are parallel and different in size. The altitude is the perpendicular distance between the bases.

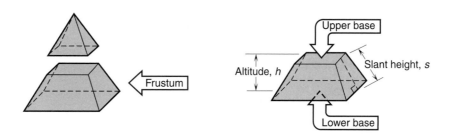

The following formulas enable you to find the lateral area and volume of any pyramid frustum.

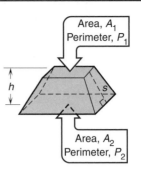

FRUSTUM OF A PYRAMID

Lateral surface area

$L = \frac{1}{2}(P_1 + P_2)s$

Volume

$V = \frac{1}{3}h(A_1 + A_2 + \sqrt{A_1 A_2})$

where P_1 and P_2 = upper and lower perimeters for a pyramid frustum
   A_1 and A_2 = upper and lower base areas for any frustum

EXAMPLE

Find the lateral surface area and volume of the pyramid frustum shown.

Upper perimeter, P_1 = 9 in. + 9 in. + 9 in. + 9 in. = 36 in.

Lower perimeter, P_2 = 15 in. + 15 in. + 15 in. + 15 in. = 60 in.

$L = \frac{1}{2}(36 \text{ in.} + 60 \text{ in.})(5 \text{ in.})$

$L = 240 \text{ in.}^2$

Upper area, A_1 = (9 in.)(9 in.) = 81 in.^{2}

Lower area, A_2 = (15 in.)(15 in.) = 225 in.^{2}

$V = \frac{1}{3}(4 \text{ in.})(81 \text{ in.}^2 + 225 \text{ in.}^2 + \sqrt{81 \cdot 225})$

$V = \frac{1}{3}(4 \text{ in.})(81 + 225 + 135) \text{ in.}^2$

$V = 588 \text{ in.}^3$

Use the following calculator sequence to find the volume:

 81 ⊠ **225** ⊟ ☑ ⊞ **225** ⊞ **81** ⊟ ⊠ **4** ÷ **3** ⊟ →   588.

To find the total surface area of the frustum, add the areas A_1 and A_2 to the lateral surface area.

YOUR TURN

Find the total surface area and volume of the frustum shown.

SOLUTION

P_1 = 4(8 ft) = 32 ft P_2 = 4(14 ft) = 56 ft

$L = \frac{1}{2}(32 \text{ ft} + 56 \text{ ft})(5 \text{ ft}) = 220 \text{ ft}^2$

A_1 = (8 ft)^2 = 64 ft^2 A_2 = (14 ft)^2 = 196 ft^2

Total surface area, $S = L + A_1 + A_2$

$S = 220 \text{ ft}^2 + 64 \text{ ft}^2 + 196 \text{ ft}^2$

  $= 480 \text{ ft}^2$

$V = \frac{1}{3}(4 \text{ ft})(64 \text{ ft}^2 + 196 \text{ ft}^2 + \sqrt{64 \times 196})$

  $= 496 \text{ ft}^3$

PROBLEMS

Practice with these problems.
Round to three significant digits.

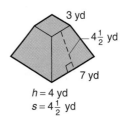

(a) Find the lateral surface area
and volume of this frustum.

3 yd

$4\frac{1}{2}$ yd

7 yd

$h = 4$ yd
$s = 4\frac{1}{2}$ yd

(b) Find the total surface area
and volume of this frustum.

6.5 cm

7.6 cm

3 cm

$h = 7$ cm
$s = 7.6$ cm

(c) A connection must be made between two square vent openings in an air conditioning system. Find the total amount of sheet metal needed using the dimensions shown in the figure.

6"

6"

9"

9"

$s = 28"$

Problem (c)

22'

22'

6"

35'

35'

$h = 6"$

Problem (d)

(d) How many cubic *yards* of concrete are needed to pour the foundation shown in the figure? (Watch your units!) Round to the nearest tenth.

SOLUTIONS

(a) $P_1 = 4(3\ \text{yd}) = 12\ \text{yd}$ $P_2 = 4(7\ \text{yd}) = 28\ \text{yd}$

$L = \frac{1}{2}(12\ \text{yd} + 28\ \text{yd})(4\frac{1}{2}\ \text{yd}) = 90\ \text{yd}^2$

$A_1 = (3\ \text{yd})^2 = 9\ \text{yd}^2$ $A_2 = (7\ \text{yd})^2 = 49\ \text{yd}^2$

$V = \frac{1}{3}(4\ \text{yd})(9\ \text{yd}^2 + 49\ \text{yd}^2 + \sqrt{9 \times 49})$

$= 105.33 \ldots \text{yd}^3 \approx 105\ \text{yd}^3$, rounded

(b) $P_1 = 6(6.5\ \text{cm}) = 39\ \text{cm}$ $P_2 = 6(3\ \text{cm}) = 18\ \text{cm}$

$L = \frac{1}{2}(39\ \text{cm} + 18\ \text{cm})(7.6\ \text{cm}) = 216.6\ \text{cm}^2$

$A_1 \approx 2.598\,(6.5\ \text{cm})^2 \approx 109.7655\ \text{cm}^2$

$A_2 \approx 2.598\,(3\ \text{cm})^2 \approx 23.382\ \text{cm}^2$

$S = L + A_1 + A_2$

$S \approx 216.6\ \text{cm}^2 + 109.7655\ \text{cm}^2 + 23.382\ \text{cm}^2$

$\approx 349.7475\ \text{cm}^2 \approx 350\ \text{cm}^2$

$V \approx \frac{1}{3}(7\ \text{cm})(109.7655 + 23.382 + \sqrt{109.7655 \times 23.382})$

$\approx 428.8865\ \text{cm}^3 \approx 429\ \text{cm}^3$, rounded

(c) $P_1 = 24$ in. $P_2 = 36$ in.

$L = \frac{1}{2}(24\ \text{in.} + 36\ \text{in.})(28\ \text{in.}) = 840\ \text{in.}^2$

(d) $A_1 = (22 \text{ ft})^2 = 484 \text{ ft}^2$ $A_2 = (35 \text{ ft})^2 = 1225 \text{ ft}^2$

$V = \frac{1}{3}(0.5 \text{ ft})(484 \text{ ft}^2 + 1225 \text{ ft}^2 + \sqrt{484 \times 1225})$

$= 413.166\ldots \text{ ft}^3$

$\approx 413.166 \text{ ft}^3 \times \dfrac{1 \text{ yd}^3}{27 \text{ ft}^3} \approx 15.3 \text{ yd}^3$

Notice that the height, 6 in., was converted to 0.5 ft to agree with the other units. The answer was then converted from cu ft to cu yd.

Learning Help ▶   Always be certain that your units agree before entering them into a calculation. ◀

Now turn to Exercises 9-2 for more practice on finding the surface area and volume of pyramids and frustums of pyramids.

Exercises 9-2

Pyramids and Frustums of Pyramids

A. Find the lateral surface area and volume of the solid objects shown. Round to three significant digits, if necessary.

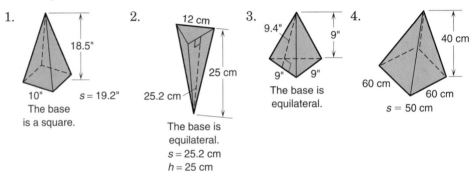

1. 18.5" 10" s = 19.2" The base is a square.

2. 12 cm 25 cm 25.2 cm The base is equilateral. s = 25.2 cm h = 25 cm

3. 9.4" 9" 9" 9" The base is equilateral.

4. 40 cm 60 cm 60 cm s = 50 cm

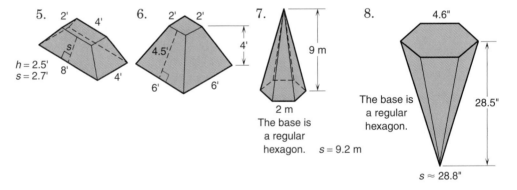

5. 2' 4' s 8' 4' h = 2.5' s = 2.7'

6. 2' 2' 4' 4.5' 6' 6'

7. 9 m 2 m The base is a regular hexagon. s = 9.2 m

8. 4.6" The base is a regular hexagon. 28.5" s ≈ 28.8"

B. Find the total outside surface area and volume of the following solid objects. Round to three significant digits, if necessary. (Assume that all bases are regular.)

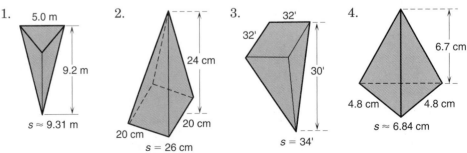

1. 5.0 m 9.2 m s ≈ 9.31 m

2. 24 cm 20 cm 20 cm s = 26 cm

3. 32' 32' 30' s = 34'

4. 6.7 cm 4.8 cm 4.8 cm s ≈ 6.84 cm

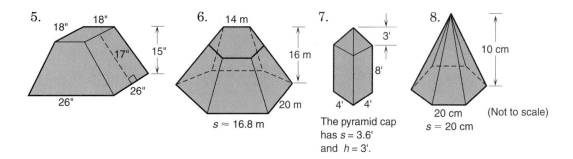

5.   6.   7.   8.

    C.   Practical Problems

        1. **Roofing** The roof of a building is in the shape of a square pyramid 25 m on each side. If the slant height of the pyramid is 18 m, how much will roofing material cost at \$4.25 per sq m?

        2. **Machine Technology** A pyramid-shaped piece of steel has a triangular base measuring 6.8″ on each side. If the height of the piece is 12.4″, find the weight of two dozen of these pyramids. Assume that the steel has a density of 0.283 lb/cu in. Round to the nearest pound.

        3. **Air Conditioning** The Last Chance gambling casino is designed in the shape of a square pyramid with a side length of 75 ft. The height of the building is 32 ft. An air filtering system can circulate air at 20,000 cu ft per hr. How long will it take the system to filter all the air in the room?

        4. **Manufacturing** How many bushels will the bin in the figure hold? (1 cu ft = 1.24 bushels.) Round to the nearest bushel.

        5. **Sheet Metal Technology** How many square inches of sheet metal are used to make the vent transition shown? (The ends are open.)

Problem 4

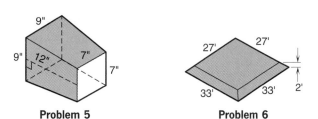

Problem 5   Problem 6

        6. **Building Construction** How many cubic yards of concrete are needed to pour the building foundation shown in the figure?

    When you have completed these exercises, check your answers in the Appendix, then turn to Section 9-3 to study cylinders and spheres.

9-3   CYLINDERS AND SPHERES

Cylinders A *cylinder* is a solid object with two identical circular bases. The *altitude* of a cylinder is the perpendicular distance between the bases.

A *right cylinder* is one whose curved side walls are perpendicular to its circular base. Whenever we mention the radius, diameter, or circumference of a cylinder, we are referring to those dimensions of its circular base.

Two important formulas enable us to find the lateral surface area and volume of a cylinder.

CYLINDERS

Lateral surface area

$L = Ch$ or $L = 2\pi rh = \pi dh$

Volume

$V = \pi r^2 h$ or $V = \frac{1}{4}\pi d^2 h \approx 0.7854 d^2 h$

where C is the circumference of the base, r is the radius of the base, d is the diameter of the base, and h is the altitude of the cylinder. To find the total surface area of a cylinder, simply add the areas of the two circular bases to the lateral area. The formula is

Total surface area $= 2\pi r^2 + 2\pi rh = 2\pi r(r + h)$

Learning Help ▶ You can visualize the lateral surface area by imagining the cylinder wall "unrolled" into a rectangle as shown.

Lateral surface area $=$ area of rectangle
$\qquad\qquad\qquad = L \times w$
$\qquad\qquad\qquad = \pi d \times h$
$\qquad\qquad\qquad = \pi dh$ ◀

EXAMPLE

Use these formulas to find the lateral surface area and volume of a cylindrical container whose altitude is 24 in. and whose diameter is 22 in.

$d = 22$ in. $r = 11$ in. $h = 24$ in.

Then the lateral surface area is

$L = \pi dh$
$L \approx (\pi)(22\text{ in.})(24\text{ in.})$
$L \approx 1659$ in.^{2} rounded

and the volume is

$V = \pi r^2 h$
$V \approx (\pi)(11\text{ in.})^2(24\text{ in.})$
$V \approx (\pi)(121\text{ in.}^2)(24\text{ in.})$
$V \approx 9123$ in.^{3} rounded

YOUR TURN

Find the lateral surface area and the volume in gallons of the cylindrical storage tank shown. Round answers to the nearest unit.

SOLUTION

$L = \pi dh$
$L \approx (\pi)(14\text{ ft})(30\text{ ft})$
$L \approx 1319$ ft^2 rounded

$V \approx 0.7854d^2h$ approximately
$V \approx (0.7854)(14\text{ ft})^2(30\text{ ft})$
$V \approx (0.7854)(196\text{ ft}^2)(30\text{ ft})$
$V \approx 4618.152\text{ ft}^3$ or 4618 ft^3 rounded

In gallons,

$$V \approx 4618.152\ \cancel{\text{ft}^3} \times \frac{7.48\text{ gal}}{1\ \cancel{\text{ft}^3}}$$

$V \approx 34{,}544\text{ gal}$ rounded

After organizing your work, you can key it into your calculator this way:

$\boxed{\pi}$ $\boxed{\div}$ **4** $\boxed{\times}$ **14** $\boxed{x^2}$ $\boxed{\times}$ **30** $\boxed{\times}$ **7.48** $\boxed{=}$ → **34543.69618**

Careful ▶ When using an intermediate value in a later calculation, do not round it prematurely. For example, if you had rounded V to 4618 and used this value to calculate V in gallons, the result would be 34,543, which is not correct to the nearest gallon. ◀

PROBLEMS

Try these problems for practice in working with cylinders.

(a) Find the lateral surface area and volume of each cylinder. (Use 3.1416 for π and round to the nearest whole number.)

(1) 20", 9"   (2) 14", 6"   (3) 26.5 cm, 14 cm

(b) Find the total surface area of a cylinder 9 yd high and 16 yd in diameter. (Round to the nearest whole number.)

(c) How many cubic *yards* of concrete will it take to pour a cylindrical column 14 ft high with a diameter of 2 ft? (Round to the nearest tenth.)

(d) What is the capacity in gallons of a cylindrical tank with a radius of 8 ft and an altitude of 20 ft? (Round to the nearest 100 gallons.)

(e) A pipe 3 in. in diameter and 40 ft high is filled to the top with water. If 1 cu ft of water weighs 62.4 lb, what is the weight at the base of the pipe? (Be careful of your units.) (Round to the nearest pound.)

(f) The *bottom* and the *inside walls* of the cylindrical tank shown in the figure must be lined with sheet copper. How many square feet of sheet copper are needed? (Round to the nearest tenth.)

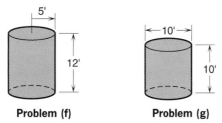

Problem (f)   Problem (g)

(g) How many quarts of paint are needed to cover the outside wall (not including the top and bottom) of a cylindrical water tank 10 ft in diameter and 10 ft in height if one quart covers 100 sq ft? Use 3.14 for π. (Assume that you cannot buy a fraction of a quart.)

(a) (1) 565 in.^{2}, 1272 in.^{3} (2) 528 in.^{2}, 3695 in.^{3}

    (3) 1166 cm^2, 4079 cm^3

(b) $d = 16$ yd $r = 8$ yd $h = 9$ yd

    $S = 2\pi r(r + h)$

    $S = 2\pi(8 \text{ yd})(8 \text{ yd} + 9 \text{ yd})$

    $S \approx 854.513\ldots \text{ yd}^2 \approx 855 \text{ yd}^2$, rounded

(c) **First,** find the volume in ft^3.

    $V \approx 0.7854 d^2 h$

       $\approx 0.7854(2 \text{ ft})^2(14 \text{ ft})$

       $\approx 43.9824 \text{ ft}^3$

    Then, convert this result to yd^3.

    $V \approx 43.9824 \ \cancel{\text{ft}^3} \times \dfrac{1 \text{ yd}^3}{27 \ \cancel{\text{ft}^3}}$

    $V \approx 1.6289\ldots \text{ yd}^3 \approx 1.6 \text{ yd}^3$, rounded.

With a calculator, the entire operation may be keyed in as follows:

 $\boxed{\pi}$ $\boxed{\div}$ **4** $\boxed{\times}$ **2** $\boxed{x^2}$ $\boxed{\times}$ **14** $\boxed{\div}$ **27** $\boxed{=}$ $\rightarrow$ ▨ *1.628973969*

(d) **First,** find the volume in ft^3.

    $V = \pi r^2 h = \pi(8 \text{ ft})^2(20 \text{ ft})$

       $\approx 4021.24 \text{ ft}^3$

    Then, convert this quantity to gallons.

    $V \approx 4021.24 \ \cancel{\text{ft}^3} \times \dfrac{7.48 \text{ gal}}{1 \ \cancel{\text{ft}^3}}$

       $\approx 30{,}078.8752 \text{ gal} \approx 30{,}100 \text{ gal}$, rounded

(e) **First,** convert the diameter, 3 in., to 0.25 ft to agree with the other units in the problem. **Next,** find the volume in ft^3.

    $V \approx 0.7854 \, d^2 h \approx 0.7854(0.25 \text{ ft})^2(40 \text{ ft})$

       $\approx 1.9635 \text{ ft}^3$

    Finally, convert the volume to lb units.

    $V \approx 1.9635 \ \cancel{\text{ft}^3} \times \dfrac{62.4 \text{ lb}}{1 \ \cancel{\text{ft}^3}}$

       $\approx 123 \text{ lb}$, rounded

 $\boxed{\pi}$ $\boxed{\div}$ **4** $\boxed{\times}$ **.25** $\boxed{x^2}$ $\boxed{\times}$ **40** $\boxed{\times}$ **62.4** $\boxed{=}$ $\rightarrow$ ▨ *122.5221135*

(f) The inside wall is the lateral surface area.

    $L = 2\pi r h = 2\pi(5 \text{ ft})(12 \text{ ft})$

       $= 120\pi \text{ ft}^2$

    The bottom wall is a circle.

    $A = \pi r^2 = \pi(5 \text{ ft})^2$

       $= 25\pi \text{ ft}^2$

    The total area of sheet copper is

    $L + A = 120\pi + 25\pi = 145\pi \text{ ft}^2$

          $\approx 455.53\ldots \text{ ft}^2 \approx 455.5 \text{ ft}^2$, rounded

Learning Help ▶ When you are adding two or more circular areas, it is usually simpler to express each in terms of π and substitute for π after finding their sum. ◀

(g) **First,** find the lateral surface area.

$$L = \pi dh \approx 3.14(10 \text{ ft})(10 \text{ ft})$$
$$\approx 314 \text{ ft}^2$$

Then, use a unity fraction to determine the amount of paint used.

$$L \approx 314 \text{ ft}^2 \times \frac{1 \text{ qt}}{100 \text{ ft}^2} \approx 3.14 \text{ qt}$$

Because you cannot buy a fraction of a quart, you will need 4 qt of paint.

Spheres The *sphere* is the simplest of all solid geometric figures. Geometrically, it is defined as the surface whose points are all equidistant from a given point called the *center*. The *radius* is the distance from the center to the surface. The *diameter* is the straight-line distance across the sphere on a line through its center.

The following formulas enable you to find the surface area and volume of any sphere.

SPHERE

Surface area

$$S = 4\pi r^2 \quad \text{or} \quad S = \pi d^2$$

Volume

$$V = \frac{4\pi r^3}{3} \quad \text{or} \quad V = \frac{\pi d^3}{6}$$

The volume formulas can be written approximately as

$$V \approx 4.1888 r^3 \qquad \text{and} \qquad V \approx 0.5236 d^3$$

EXAMPLE

Find the surface area and volume of a spherical basketball of diameter $9\frac{1}{2}$ in. Round to the nearest tenth.

The surface area is

$$S = \pi d^2$$
$$S \approx \pi(9.5 \text{ in.})^2$$
$$S \approx \pi(90.25 \text{ in.}^2)$$
$$S \approx 283.5 \text{ in.}^2$$

The volume is

$$V = \frac{\pi d^3}{6}$$

$$V \approx \frac{\pi(9.5 \text{ in.})^3}{6}$$

$$V \approx 448.9 \text{ in.}^3 \quad \text{rounded}$$

Using a calculator to find the volume, we have

$\boxed{\pi}$ $\boxed{\times}$ **9.5** $\boxed{y^x}$ **3** $\boxed{\div}$ **6** $\boxed{=}$ → `448.9205002`

YOUR TURN

Find the surface area and the volume in gallons of a spherical tank 8 ft in radius. Round to the nearest whole unit.

SOLUTION

The surface area is

$S = 4\pi r^2$

$S = 4\pi(8 \text{ ft})^2$

$S = 4\pi(64 \text{ ft}^2) = 804.2477\ldots \text{ ft}^2$

$S \approx 804 \text{ ft}^2$, rounded

The volume is

$V = \dfrac{4\pi r^3}{3}$

$V \approx \dfrac{4\pi(8 \text{ ft})^3}{3} \approx 2144.66\ldots \text{ ft}^3$

$V \approx 2145 \text{ ft}^3$, rounded

$V \approx 2144.7 \text{ ft}^3 \times \dfrac{7.48 \text{ gallons}}{1 \text{ ft}^3}$ Convert to gallons using a unity fraction.

$V \approx 16{,}042 \text{ gallons}$ rounded to the nearest gallon

PROBLEMS

For practice, try the following problems.

(a) Find the surface area and volume of each of the following spheres. (Use $\pi = 3.14$ and round to the nearest whole unit.)

    (1) radius = 15 in.   (2) diameter = 22 ft

    (3) radius = 6.5 cm   (4) diameter = 8 ft 6 in.

(b) How many gallons of water can be stored in a spherical tank 50 in. in diameter? (Use 1 gallon = 231 in.^{3} Round to the nearest gallon.)

(c) What will be the weight of a spherical steel tank of radius 9 ft when it is full of water? The steel used weighs approximately 127 lb/ft^2 and water weighs 62.4 lb/ft^3. (Round to the nearest 100 lb.)

ANSWERS

(a) (1) 2827 in.^{2}, 14,137 in.^{3}
      (2) 1521 ft^2, 5575 ft^3
      (3) 531 cm^2, 1150 cm^3
      (4) 227 ft^2, 322 ft^3

(b) 283 gallons

(c) 319,800 lb

In problem (b), the volume is approximately 65,450 in.^{3} Multiply by a unity fraction to convert to gallons.

$$65{,}450 \text{ in.}^3 \times \frac{1 \text{ gal}}{231 \text{ in.}^3} = \frac{65{,}450}{231} \text{ gal or about 283 gal}$$

In problem (c) the area is 1017.9 ft^2. To find its weight, multiply by a unity fraction:

$$1017.9 \text{ ft}^2 \times \frac{127 \text{ lb}}{1 \text{ ft}^2} \approx 129{,}300 \text{ lb} \quad \text{rounded}$$

The volume is 3053.6 ft^3 or the weight of water is

$$3053.6 \text{ ft}^3 \times \frac{62.4 \text{ lb}}{1 \text{ ft}^3} \approx 190{,}500 \text{ lb} \quad \text{rounded}$$

We can perform the entire calculation on a calculator as follows.

First, calculate the weight of the tank and store it in memory.

4 $\times$ π $\times$ 9 x^2 $\times$ 127 $=$ STO → `129270.2545`

Then, calculate the weight of the water and add it to the tank weight in memory.

4 $\times$ π $\times$ 9 y^x 3 $\div$ 3 $\times$ 62.4 $+$ RCL $=$ → `319816.6454`

Now turn to Exercises 9-3 for a set of problems on cylinders and spheres.

Exercises 9-3

Cylinders and Spheres

A. Find the lateral surface area and volume of each of the following cylinders. (Round to three significant digits and use $\pi \approx 3.14$.)

1. 46.2 cm / 85.4 cm
2. 28 mm / 17 mm
3. 3" / 13"
4. ←12 yd→ / 8 yd

B. Find the total surface area and volume of each of the following solid objects. (Round to three significant digits.)

1. Diameter = 1.5"
2. $r = 18'$
3. 8.25' / 16.5'
4. 3¼" / 1⅜"
5. 44 m / 17 m
6. 18 yd / 52 yd

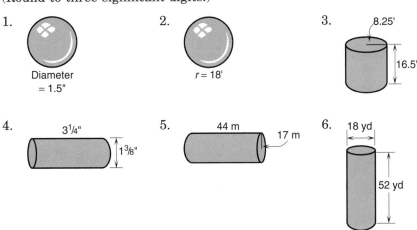

C. Practical Problems (Round to the nearest tenth unless otherwise directed.)

1. **Painting**   A cylindrical tank has a radius of 6.5 ft and an altitude of 14 ft. If a gallon of paint will cover 120 ft² of surface, how much paint is needed to put two coats of paint on the entire surface of the tank? (Round to the nearest gallon.)

2. **Plumbing**   A 4-ft-high cylindrical pipe has a diameter of 3 in. If water weighs 62.4 lb/ft³, what is the weight of water in the pipe when filled to the top? (*Hint:* Change the diameter to feet.)

3. **Landscaping**   A layer of crushed rock must be spread over a circular area 23 ft in diameter. How deep a layer will be obtained using 100 ft³ of rock? (Round to the nearest hundredth of a foot.)

4. How many liters of water can be stored in a cylindrical tank 8 m high with a radius of 0.5 m? (1 m³ contains 1000 liters.)

5. **Metalworking**   What is the weight of the piece of aluminum shown at 0.0975 lb/in.³? (*Hint:* Subtract the volume of the cylindrical hole.)

Problem 5

6. How high should a 50-gallon cylindrical tank be if it must fit into an area allowing a 16-in. diameter? (1 gallon = 231 in.³)

7. **Plumbing**   A marble-top bathroom lavatory has the shape of a hemispherical (a half sphere) porcelain bowl with an inside diameter of 15 in. How much water, in gallons, will the bowl hold? (Round to the nearest gallon.)

8. **Sheet Metal Technology**   How many square inches of sheet copper are needed to cover the inside wall of a cylindrical tank 6 ft high with a 9-in. radius? (*Hint:* Change the altitude to inches.)

9. **Machine Technology**   What is the weight of the bushing shown if it is made of steel weighing 0.2833 lb/in.³? (The inner cylinder is hollow.)

10. **Manufacturing**   Find the capacity in gallons of the oil can in the figure. (1 gallon = 231 in.³)

11. **Plumbing**   A septic tank has the shape shown. How many gallons does it hold? (1 ft³ ≈ 7.48 gallons.) (Round to the nearest gallon.)

12. **Building Construction**   How many cubic yards of concrete are needed to pour eight cylindrical pillars 12 ft 6 in. high, each with a diameter of 1 ft 9 in.?

Problem 9

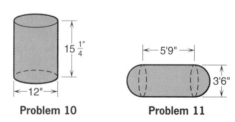

Problem 10   Problem 11

514

13. **Machine Technology** At a density of 0.0925 lb/in.^{3}, calculate the weight of the aluminum piece shown. (Round to the nearest hundredth.)

Diameter of four outside holes: $\frac{1}{4}$ in.

Diameter of inside hole: $\frac{1}{3}$ in.

Thickness of piece: $\frac{1}{2}$ in.

14. **Industrial Technology** A spherical tank has a diameter of 16.5 ft. If a gallon of paint will cover 80 ft^2, how many gallons are needed to cover the tank? Round to the nearest gallon.

15. **Plumbing** A cylindrical tank 28 in. in diameter must have a capacity of 75 gallons. How high must it be?

16. **Plumbing** The water tower shown consists of a 16-ft-high cylinder and a 10-ft-radius hemisphere. How many gallons of water does it contain when it is filled to the top of the cylinder? Round to the nearest hundred.

Problem 16

17. **Plumbing** Calculate the volume of a cylindrical pump housing 10.4 inches in diameter and 10.2 inches high. (Round to three significant digits.)

18. **Plumbing** A water trough is constructed with semicircular ends. Calculate its volume in gallons if the diameter of the end is 15 inches and the length of the trough is 8 feet. (Round to the nearest 0.1 gal.) *Hint:* Be careful of units.

19. **Construction** The metal silo shown has a hemispherical top. It is to be painted with exterior weatherproofing paint. If a gallon of this paint covers 280 ft^2, how much paint is needed for two coats? (Do not include the bottom of the silo. Round your answer to the nearest gallon.)

When you have completed these exercises, check your answers in the Appendix, then turn to Section 9-4 to learn about cones.

9-4 CONES AND FRUSTUMS OF CONES

Cones

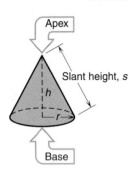

A *cone* is a pyramid-like solid figure with a circular base. The radius and diameter of a cone refer to its circular base. The *altitude* is the perpendicular distance from the apex to the base. The *slant height* is the apex-to-base distance along the surface of the cone.

The following formulas enable us to find the lateral surface area and volume of a cone.

CONES

Lateral surface area

$L = \pi r s$ or $L = \frac{1}{2}\pi d s$

Volume

$V = \frac{1}{3}\pi r^2 h$ or $V = \frac{1}{12}\pi d^2 h \approx 0.2618 d^2 h$

$d = 2r$

---

where r = radius of the base
   d = diameter of the base
   h = altitude
   s = slant height

To find the total surface area of a cone, simply add the area of the circular base to the lateral area. The formula would be

---

Total surface area $= \pi r^2 + \pi r s = \pi r(r + s)$

EXAMPLE

Find the lateral surface area and volume of the cone shown here. (Round to the nearest whole unit.)

Note that $h = 15$ in., $s = 18$ in., $r = 10$ in.

Then the lateral surface area is

$L = \pi r s$
$L = \pi(10 \text{ in.})(18 \text{ in.})$
$L = 565.486\ldots \text{ in.}^2$ or 565 in.^2 rounded

and the volume is

$V = \frac{1}{3}\pi r^2 h$

$V = \frac{1}{3}\pi(10 \text{ in.})^2(15 \text{ in.})$

$V = 1570.79\ldots \text{ in.}^3$ or 1571 in.^3 rounded

Using a calculator to find the volume, we obtain

π ✕ **10** x^2 ✕ **15** ÷ **3** = → *1570.796327*

Find the lateral surface area, total surface area, and volume of this cone. (Round to the nearest whole unit.)

SOLUTION

$h = 12$ cm $\qquad r = 5$ cm $\qquad s = 13$ cm

Lateral surface area $= L = \pi r s$
$$L \approx \pi(5 \text{ cm})(13 \text{ cm})$$
$$L \approx 204.2035\ldots \text{ cm}^2 \quad \text{or} \quad 204 \text{ cm}^2 \quad \text{rounded}$$

Total surface area $= \pi r^2 + \pi r s$
$$\approx \pi(5 \text{ cm})^2 + 204.204 \text{ cm}^2$$
$$\approx 78.54 \text{ cm}^2 + 204.204 \text{ cm}^2$$
$$\approx 282.744 \quad \text{or} \quad 283 \text{ cm}^2 \quad \text{rounded}$$

Volume $= V = \frac{1}{3}\pi r^2 h$
$$V \approx \frac{1}{3}\pi(5 \text{ cm})^2(12 \text{ cm})$$
$$V \approx 314.159\ldots \text{ cm}^3 \quad \text{or} \quad 314 \text{ cm}^3 \quad \text{rounded}$$

You should realize that the altitude, radius, and slant height for any cone are always related by the Pythagorean theorem.

If you are given any two of these quantities, you can use this formula to find the third.

Notice that the volume of a cone is exactly one-third the volume of the cylinder that just encloses it.

PROBLEMS

Solve the following practice problems.

(a) Find the lateral surface area and volume of each of the following cones. (Use $\pi = 3.14$ and round to the nearest whole number.)

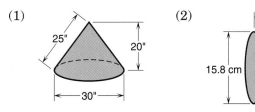

(b) Find the total surface area of a cone with radius $\frac{3}{4}$ in., altitude 1 in., and slant height $1\frac{1}{4}$ in. (Round to two decimal places.)

6"

13"

←10"→

Problem (e)

(c) A pile of sand dumped by a hopper is cone-shaped. If the diameter of the base is 18 ft 3 in. and the altitude is 8 ft 6 in., how many cubic feet of sand are in the pile? (Round to the nearest tenth.)

(d) What is the capacity in gallons of a conical drum 4 ft high with a radius of 3 ft? (1 cubic foot $\approx$ 7.48 gallons.) (Round to the nearest gallon.)

(e) Find the weight of the cast iron shape shown at 0.26 lb/in.^{3} (Round to the nearest pound.)

(f) Find the capacity in gallons (to the nearest tenth) of a conical oil container 15 in. high with a diameter of 16 in. (1 gallon = 231 in.^{3}).

SOLUTIONS

(a) (1) 1178 in.^{2}, 4712 in.^{2} (2) 432 cm^2, 1013 cm^3

    (3) **First,** we must calculate the slant height using the Pythagorean Theorem.

$$s^2 = h^2 + r^2$$
$$= 15^2 + 6^2 = 261$$
$$s = \sqrt{261} \approx 16.155 \text{ ft}$$

The lateral surface area, L, is

$$L = \pi r s \approx 3.14(6 \text{ ft})(16.155 \text{ ft})$$
$$L \approx 304.3602 \text{ ft}^2 \approx 304 \text{ ft}^2$$

The volume, V, is

$$V = \tfrac{1}{3}\pi r^2 h \approx \tfrac{1}{3}(3.14)(6 \text{ ft})^2(15 \text{ ft})$$
$$\approx 565.2 \text{ ft}^3 \approx 565 \text{ ft}^3$$

(b) $S = \pi r(r + s) = \pi(\tfrac{3}{4} \text{ in.})(\tfrac{3}{4} \text{ in.} + 1\tfrac{1}{4} \text{ in.})$

$$= \pi(0.75 \text{ in.})(2 \text{ in.})$$
$$\approx 4.71 \text{ in.}^2$$

(c) $d = 18 \text{ ft } 3 \text{ in.} = 18.25 \text{ ft}$ $h = 8 \text{ ft } 6 \text{ in.} = 8.5 \text{ ft}$

$$V = \tfrac{1}{12}\pi d^2 h = \tfrac{1}{12}\pi(18.25 \text{ ft})^2(8.5 \text{ ft})$$
$$= 741.16 \ldots \text{ ft}^3 \approx 741.2 \text{ ft}^3, \text{ rounded}$$

(d) **First,** find the volume in cubic feet.

$$V = \tfrac{1}{3}\pi r^2 h = \tfrac{1}{3}(\pi)(3 \text{ ft})^2(4 \text{ ft})$$
$$= 37.699 \ldots \text{ ft}^3$$

Then convert this to gallons.

$$V \approx 37.699 \text{ ft}^3 \times \frac{7.48 \text{ gal}}{1 \text{ ft}^3} \approx 282 \text{ gal}$$

Of course, with a calculator you would not need to round the intermediate result.

 $\boxed{\pi}$ $\boxed{\div}$ **3** $\boxed{\times}$ **3** $\boxed{x^2}$ $\boxed{\times}$ **4** $\boxed{\times}$ **7.48** $\boxed{=}$ → $\boxed{281.9893566}$

(e) **First,** find the volume in cubic inches.

$$d = 10'' \text{ so } r = 5''$$

For the cylinder: **For the cone:**

$V = \pi r^2 h = \pi(5 \text{ in.})^2(13 \text{ in.})$ $V = \tfrac{1}{3}\pi r^2 h = \tfrac{1}{3}\pi(5 \text{ in.})^2(6 \text{ in.})$

$\approx 1021.02 \text{ in.}^3$ $\approx 157.08 \text{ in.}^3$

Total volume, $V \approx 1178.1$ in.^{3}

Then, convert to pounds.

$$\text{Weight} \approx 1178.1 \text{ in.}^3 \times \frac{0.26 \text{ lb}}{1 \text{ in.}^3} \approx 306 \text{ lb}$$

(f) The capacity, in cubic inches, is

$$V \approx 0.2618 d^2 h \approx 0.2618(16 \text{ in.})^2(15 \text{ in.}) \quad (\tfrac{\pi}{12} \approx 0.2618)$$

$$\approx 1005.312 \text{ in.}^3$$

The capacity in gallons is

$$V \approx 1005.312 \text{ in.}^3 \times \frac{1 \text{ gal}}{231 \text{ in.}^3}$$

$$\approx 4.352 \text{ gal} \approx 4.4 \text{ gal, rounded to the nearest tenth}$$

 $\boxed{\pi}$ $\boxed{\div}$ **12** $\boxed{\times}$ **16** $\boxed{x^2}$ $\boxed{\times}$ **15** $\boxed{\div}$ **231** $\boxed{=}$ → `4.351989898`

Learning Help ▶ The three formulas for the volume of a cone were provided to give you a choice. You might want to memorize only the formula $V = \frac{1}{3}\pi r^2 h$, from which the others were derived. ◀

Frustum of a Cone The frustum of a cone, like the frustum of a pyramid, is the figure remaining when the top of the cone is cut off parallel to its base.

Frustum
of a cone

Frustum shapes appear as containers, as funnels, and as transition sections in ducts.

Every frustum has two bases, upper and lower, that are parallel and different in size. The altitude is the perpendicular distance between bases.

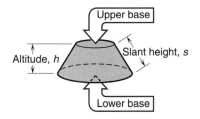

The following formulas enable you to find the lateral area and volume of any frustum of a cone.

FRUSTUM OF A CONE

Lateral surface area

$$L = \pi s(r_1 + r_2) \quad \text{or} \quad \tfrac{1}{2}\pi s(d_1 + d_2)$$

Volume

$$V = \tfrac{1}{3}\pi h(r_1^2 + r_2^2 + r_1 r_2) \quad \text{or} \quad \tfrac{1}{12}\pi h(d_1^2 + d_2^2 + d_1 d_2)$$

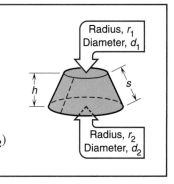

These formulas can be obtained from the equation for the lateral surface area and volume of the frustum of a pyramid, using πr^2 or $\frac{1}{4}\pi d^2$ for the area of a circle.

To find the *total surface area* of the frustum, add the areas of the circular ends to the lateral surface area.

EXAMPLE

To find the lateral surface area, total surface area, and volume of the frustum shown, use $s = 9.0$ ft, $r_1 = 2$ ft, $r_2 = 5$ ft, $h = 8.5$ ft and $\pi = 3.1416$.

Lateral surface area, $L \approx 3.1416(9)(2 + 5)$
$$\approx 197.9208 \text{ ft}^2 \quad \text{or} \quad 198 \text{ ft}^2 \quad \text{rounded}$$

Total surface area, $A \approx L + \pi r_1^2 + \pi r_2^2$
$$A \approx 197.9208 + 3.1416(2^2) + 3.1416(5^2)$$
$$A \approx 289.0272 \text{ ft}^2 \quad \text{or} \quad 289 \text{ ft}^2 \quad \text{rounded}$$

Volume, $V \approx \frac{1}{3}(3.1416)(8.5)(2^2 + 5^2 + 2 \cdot 5)$
$$V \approx 347.1468 \text{ ft}^3 \quad \text{or} \quad 347 \text{ ft}^3 \quad \text{rounded}$$

Using a calculator to find the volume, we have

 3.1416 ÷ **3** × **8.5** × (**2** x^2 + **5** x^2 + **2** × **5**) = → 347.1468

PROBLEMS

Practice with these problems.

(a) Find the lateral surface area and volume of each of the following frustums. (Round to the nearest whole unit and use $\pi \approx 3.14$.)

(1)

(2)

$s = 10.8"$
$h = 10"$

(3)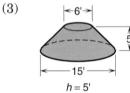

$h = 5'$

Hint: Calculate the slant height s using the Pythagorean theorem.

(b) What is the capacity in gallons of the oil can shown in the figure? (1 gallon $\approx$ 231 in.^{3}) Round to the nearest tenth.

$h = 12"$

SOLUTIONS

(a) (1) 449 in.^{2}, 1218 in.^{3} (2) 271 in.^{2}, 544 in.^{3}

(3) **First,** find the slant height this way:

  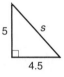 $s = \sqrt{5^2 + 4.5^2} = \sqrt{45.25} \approx 6.73$ ft

Then, find the lateral surface area.

$$L = \tfrac{1}{2}\pi s(d_1 + d_2) \approx \tfrac{1}{2}(3.14)(6.73 \text{ ft})(6 \text{ ft} + 15 \text{ ft})$$
$$\approx 221.8881 \text{ ft}^2 \approx 222 \text{ ft}^2$$

The volume is

$$V = \tfrac{1}{12}\pi h(d_1^2 + d_2^2 + d_1 \cdot d_2)$$
$$\approx \tfrac{1}{12}(3.14)(5)(6^2 + 15^2 + 6 \times 15)$$
$$\approx 459.225 \text{ ft}^3 \approx 459 \text{ ft}^3$$

(b) **First,** find the volume in cubic inches.

$$V = \tfrac{1}{12}\pi h(d_1^2 + d_2^2 + d_1 \cdot d_2)$$
$$= \tfrac{1}{12}(\pi)(12)(16^2 + 8^2 + 16 \times 8)$$
$$\approx 1407.43 \text{ in.}^3$$

Then, translate this to gallons.

$$V \approx 1407.43 \text{ in.}^3 \times \frac{1 \text{ gal}}{231 \text{ in.}^3}$$
$$\approx 6.1 \text{ gal, rounded}$$

Using a calculator on problem (b),

 π ÷ **12** × **12** × (**16** x^2 + **8** x^2 + **16** × **8**) ÷ **231** = → **6.092785752**

Now turn to Exercises 9-4 for more practice with cones and frustums of cones.

Exercises 9-4

Cones and Frustums of Cones

A. Find the lateral surface area and volume of each solid object. Round to three significant digits.

1.

16.5 cm

|←11.6 cm→|

2.

15'

33.5'   30'

s = 33.5'
h = 30'

3.

4.20 m

9.79 m   9.70 m

|←6.80 m→|

4.

|← 26.0" →|

25.1"   24.0"

11.0"

B. Find the total surface area and volume of each solid object. Round to three significant digits.

1.

2.

3.

4.

s = 5.0'
h = 4.0'

C. Find the volume of each figure. (Use π ≈ 3.14 and round to two significant digits.)

1.

2.

s = 16.3"
h = 16"

3.

4.

d = 2'

D. Practical Problems (Round to the nearest tenth if necessary.)

1. **Landscaping** How many cubic meters of dirt are there in a pile, conical in shape, 10 m in diameter and 4 m high?

2. **Sheet Metal Technology** How many square centimeters of sheet metal are needed for the sides and bottom of the pail shown in the figure?

Problem 2   Problem 3

Problem 5

3. **Manufacturing** How many bushels will the bin in the figure hold? (1 cu ft ≈ 1.24 bushels.)

4. **Landscaping** Decorative Moon Rocks cost $92 per cu yd. What will be the cost of a conical pile of this rock 3⅓ yd in diameter and 2¼ yd high? (Round to the nearest dollar.)

5. **Metalworking** Calculate the weight of the piece of steel shown in the drawing. (This steel weighs 0.283 lb/cu in.)

Problem 6

$s = 6.3$
4.2
15.7
14.6

6. **Sheet Metal Technology** How many square centimeters of sheet metal will it take to make the open-top container shown in the figure? All measurements are in centimeters. (*Hint:* No metal is needed for the small opening at the top.)

7. **Manufacturing** A conical oil cup with a radius of 3.8 cm must be designed to hold 64 cu cm of oil. What should be the altitude of the cup?

8. **Landscaping** At an excavation site a conical pile of earth 9 ft high and 14 ft in diameter at the base must be hauled away by a small truck with a capacity of 2.6 cu yd. How many truckloads will it take to transport the soil? (Be careful with units!)

9. **Metalworking** Find the weight of the steel rivet shown in the figure. (Steel weighs 0.0173 lb/cu cm.)

6.4 cm   2.0 cm
           3.2 cm
      10.0 cm

Problem 9

When you have completed these exercises, go to Problem Set 9 for a set of problems covering the work of this chapter.

The following table provides a handy summary of the formulas for solid figures presented in this chapter.

SUMMARY OF FORMULAS FOR SOLID FIGURES

Figure	Lateral Surface Area	Volume
Prism	$L = ph$	$V = Ah$
Pyramid	$L = \dfrac{1}{2}ps$	$V = \dfrac{1}{3}Ah$
Cylinder	$L = \pi dh$	$V = \pi r^2 h$
	or $\ \ L = 2\pi rh$	or $\ \ V \approx 0.7854 d^2 h$
Cone	$L = \dfrac{1}{2}\pi ds$	$V = \dfrac{1}{3}\pi r^2 h$
	or $\ \ L = \pi rs$	or $\ \ V \approx 0.2618 d^2 h$
Frustum of pyramid	$L = \dfrac{1}{2}(P_1 + P_2)s$	$V = \dfrac{1}{3}h(A_1 + A_2 + \sqrt{A_1 A_2})$
Frustum of cone	$L = \dfrac{1}{2}\pi s(d_1 + d_2)$	$V = \dfrac{1}{12}\pi h(d_1^2 + d_2^2 + d_1 d_2)$
	or $\ \ L = \pi s(r_1 + r_2)$	or $\ \ V = \dfrac{1}{3}\pi h(r_1^2 + r_2^2 + r_1 r_2)$
Sphere	$S = 4\pi r^2$	$V = \dfrac{4}{3}\pi r^3$
	or $\ \ S = \pi d^2$	or $\ \ V \approx 4.1888 r^3$
		$V = \dfrac{1}{6}\pi d^3$
		or $\ \ V \approx 0.5236 d^3$

Solid Figures

Answers are given in the Appendix.

A. Solve the following problems involving solid figures.

Find the lateral surface area and volume of each of the following. (Round to the nearest tenth.)

1.
4"
9"

2.
6"
14"
6"

3.
5.4'
5'
4' 4'

4.
21' 12'
3' 10'
13.9'

5.
11 mm
14 mm 15 mm

6.
7"
15"

7.
6"
10" 8"
18"

8.
12' 12'
4'
5'
18' 18'
$h = 4'$
$s = 5'$

Find the total surface area and volume of each of the following. (Round to the nearest tenth.)

9.
8'
8' 8'

10.
8"
$8\frac{1}{4}$"
7" 7"

11.
30 cm
15 cm

12.
12"
5"
Slant height
= $12\frac{3}{4}$"

13.
7'
12' 5'
3'
15' 12'

14.
18" 15"
10"

15. A sphere with a radius of 10 cm.

16.
12" 14"
6"

17. A sphere with a diameter of 13 in.

Name

Date

Course/Section

18.

19.

$h = 30"$
$s = 30.1"$

20.

21.

B. Practical Problems

1. **Manufacturing**   How many boxes 16 in. by 12 in. by 10 in. will fit in 1250 ft^3 of warehouse space? (1 ft$^3 = 1728$ in.^{3})

2. **Construction Technology**   How many cubic *yards* of concrete are needed to pour the highway support shown in the figure? (1 yd$^3 = 27$ ft^3.) (Round to nearest cubic yard.)

Problem 2

Problem 3

3. **Metalworking**   What is the weight of the 10-ft-long steel I-beam if the density of steel is 0.2833 lb/in.^{3}? (Round to the nearest pound.)

4. **Plumbing**   What is the capacity in gallons of a cylindrical water tank 3 ft in diameter and 8 ft high? (1 ft$^3 \approx 7.48$ gal.) (Round to one decimal place.)

5. **Painting and Decorating**   How many quarts of paint will it take to cover a spherical water tank 25 ft in diameter if one quart covers 50 ft^2?

6. **Construction Technology**   A hole must be excavated for a swimming pool in the shape shown in the figure. How many trips will be needed to haul the dirt away if the truck has a capacity of 9 m^3? (*Remember:* A fraction of a load constitutes a trip.)

Problem 6

Problem 7

7. **Manufacturing**   How many cubic feet of propane will the tank in the figure contain? (1 ft$^3 = 1728$ in.^{3}) (Round to the nearest tenth of a cubic foot.)

8. **Manufacturing**   What must be the height of a cylindrical 750-gallon tank if it is 4 ft in diameter? (1 ft$^3 \approx 7.48$ gallons.) (Round to the nearest inch.)

9. At a density of 42 lb/ft^3, what is the weight of fuel in the rectangular gas tank pictured?

Problem 9   **Problem 10**

10. **Manufacturing** How many square inches of paper are needed to produce 2000 conical cups like the one in the figure? (Round to the nearest 100 in.^{2})

11. **Building Construction** How many cubic yards of concrete are needed to pour the monument foundation shown in the figure? (Round to the nearest cubic yard.)

Problem 11   **Problem 12**

12. **Sheet Metal Technology** How many square inches of sheet metal are needed to make the bottom and sides of the pail in the figure? (Round to the nearest square inch and use $\pi \approx 3.14$.)

13. Find the capacity in gallons of the oil can in the figure. It is a cylinder with a cone on top. (1 gallon = 231 in.^{3}) (Round to the nearest tenth of a gallon.)

Problem 13   **Problem 14**

14. **Machine Technology** At a density of 7.7112 g/cm^3, find the weight of 1200 hex nuts, each 1.00 cm on a side and 0.50 cm thick with a hole 1.00 cm in diameter. (Round to two significant digits.)

15. A cylindrical tank can be no more than 6 ft high. What diameter must it have in order to hold 1400 gallons?

PREVIEW

10

Triangle Trigonometry

Objective	Sample Problems		For Help Go to Page

When you finish this unit you will be able to:

1. Determine whether or not a figure is a right triangle, given certain information.

(a) _____ 539

(b) _____ 544

(c) _____ 543

(d) _____ 541

(e) _____ 540

2. Find the values of trig ratios.

(a) $\sin 26°$ _____ 550

(b) $\cos 84°$ _____ 552

(c) $\tan 43°20'$ _____ 553

(Round to three decimal places.)

3. Find the angle when given the value of a trig ratio.

Find $\angle x$ to the nearest minute.

(a) $\sin x = 0.242$ $\angle x =$ _____ 556

(b) $\cos x = 0.549$ $\angle x =$ _____

Find $\angle x$ to the nearest tenth of a degree.

(c) $\tan x = 3.821$ $\angle x =$ _____

(d) $\sin x = 0.750$ $\angle x =$ _____

_____
Name

_____
Date

_____
Course/Section

4. Work with angles in radian measure.

Express each angle in radians.

(a) 46°   _____ 536

(b) 8.4°   _____

(c) 80°15′   _____

(d) Calculate the arc length and area of the circular sector shown. (Round to one decimal place.)   _____ 537

(e) What is the angular velocity in radians per second of a large flywheel that rotates through an angle of 200° in 2 seconds? (Round to the nearest tenth.)   _____ 538

5. Solve problems involving triangles. (Round sides to the nearest tenth and angles to the nearest minute.)

(a)

$X =$ _____ 558

$Y =$ _____

(b)

$\sphericalangle m =$ _____

$X =$ _____

(c) In a pipe-fitting job, what is the run if the offset is $22\frac{1}{2}°$ and the length of set is $16\frac{1}{2}$ in.?   run = _____ 561

(d) Find the angle of taper on the figure.   angle = _____

6. Solve oblique triangles. (Round sides to the nearest tenth and angles to the nearest degree.)

Solve these triangles.

(a) $a = 6.5$ in., $b = 6.0$ in., $c = 3.5$ in.   _____   566

(b) $A = 68°$, $B = 79°$, $b = 8.0$ ft   _____

(Answers to these preview problems are given in the Appendix.)

If you are certain that you can work all of these problems correctly, turn to page 579 for a set of practice problems. If you cannot work one or more of the preview problems, turn to the page indicated to the right of the problem. Those who wish to master this material with the greatest success should turn to Section 10-1 and begin work there.

10 Triangle Trigonometry

"Not the Bermuda triangle?"

Drawing by Richter © 1977. The New Yorker Magazine, Inc

10-1 ANGLES AND TRIANGLES

The word *trigonometry* means simply "triangle measurement." We know that the ancient Egyptian engineers and architects of 4000 years ago used practical trigonometry in building the pyramids. By 140 B.C. the Greek mathematician Hipparchus had made trigonometry a part of formal mathematics and taught it as astronomy. In this chapter we look at only the simple practical trigonometry used in electronics, drafting, machine tool technology, aviation mechanics, and similar technical work.

In Chapter 8 you learned some of the vocabulary of angles and triangles. To be certain you remember this information, try these problems.

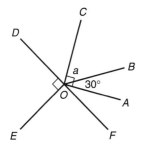

(a) Find each of the angles shown.

(1) $\angle AOB =$ _____   (2) $\angle DOF =$ _____

(3) $\angle a$ = _____   (4) $\angle DOE =$ _____

(b) Label each of these angles with the correct name: acute angle, obtuse angle, straight angle, right angle.

(1)

(2)

(3)

(4)

(5)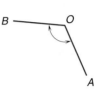

(6)

ANSWERS

(a) (1) $\angle AOB = 30°$

(2) $\angle DOF = 180°$

(The symbol ∟ means "right angle" or 90°.)

(3) $\angle a = 60°$

(4) $\angle DOE = 90°$

(b) (1) right angle

(2) acute angle

(3) obtuse angle

(4) straight angle

(5) acute angle

(6) obtuse angle

If you missed any of these, you should return to Section 8-1 on page 421 for a quick review; otherwise, continue here.

Angle Measurement In most practical work angles are measured in degrees and fractions of a degree. Smaller units have been defined as follows:

60 minutes = 1 degree abbreviated, $60' = 1°$
60 seconds = 1 minute abbreviated, $60'' = 1'$

For most technical purposes angles can be rounded to the nearest minute. In the trades angles are usually rounded to the nearest degree.

Sometimes in your work you will need to convert an angle measured in decimal degrees to its equivalent in degrees and minutes.

EXAMPLE

$17\frac{1}{2}° = 17° + \frac{1}{2}°$

$= 17° + \left(\frac{1°}{2} \times \frac{60'}{1°} \right)$ Use a unity fraction to convert.

$= 17°30'$

YOUR TURN

Write this angle to the nearest minute:

$36.25° = \underline{\hspace{1cm}}$

$36.25° = 36° + 0.25°$

$$= 36° + \left(0.25° \times \frac{60'}{1°}\right) \quad \text{Multiply the decimal part by a unity fraction.}$$

$$= 36°15'$$

The reverse procedure is also useful.

EXAMPLE

To convert the angle $72°6'$ to decimal form,

$72°6' = 72° + 6'$

$$= 72° + \left(6' \times \frac{1°}{60'}\right) \quad \text{Use a unity fraction to convert.}$$

$$= 72° + 0.1°$$

$$= 72.1°$$

CONVERTING ANGLE UNITS

Scientific calculators have special keys designed to perform angle conversions.

- To convert $36.25°$ to degrees and minutes, use a key labeled ▸DMS or ° ' ''

    36.25 ▸DMS → ▨▨▨ *36°15'00''* ▨▨▨ $36.25° = 36°15'$

    ↖ DMS stands for "degrees, minutes, seconds"

- To convert $72°6'$ to decimal degrees, either enter the angle as 72.6 or use a special ° ' '' key to enter the angle as $72°6'$. Then use the decimal degrees key ▸DD to convert.

    72.6 ▸DD → ▨▨▨ *72.1* ▨▨▨

    or **72** ° ' '' **6** ° ' '' ▸DD → ▨▨▨ *72.1* ▨▨▨ $72°6' = 72.1°$

- To convert to decimal degrees when the angle is given to the nearest second, enter one of the following sequences. For example, to convert $36°14'46''$ to decimal form,

    32.1446 ▸DD → ▨▨▨ *32.24611111* ▨▨▨

    or **32** ° ' '' **14** ° ' '' **46** ° ' '' ▸DD → ▨▨▨ *32.24611111* ▨▨▨

                       $32°14'46'' \approx 32.246°$

PROBLEMS

For practice in working with angles, rewrite each of the following angles as shown.

(a) Write in degrees and minutes.

     (1) $24\frac{1}{3}° = $ _____   (2) $64\frac{3}{4}° = $ _____

     (3) $15.3° = $ _____   (4) $38.6° = $ _____

     (5) $46\frac{3}{8}° = $ _____   (6) $124.8° = $ _____

(b) Write in decimal form.

(1) $10°48' = $ _____

(2) $96°9' = $ _____

(3) $57°36' = $ _____

(4) $168°54' = $ _____

(5) $33' = $ _____

(6) $1°12' = $ _____

ANSWERS

(a) (1) 24°20'   (2) 64°45'   (3) 15°18'
     (4) 38°36'   (5) $46°22\frac{1}{2}'$ (6) 124°48'

(b) (1) 10.8°   (2) 96.15°   (3) 57.6°
     (4) 168.9°   (5) 0.55°   (6) 1.2°

Radian Measure In some scientific and technical work angles are measured or described in an angle unit called the *radian*. By definition,

$$1 \text{ radian} = \frac{180°}{\pi} \text{ or about } 57.296°$$

One radian is the angle at the center of a circle that corresponds to an arc exactly one radius in length.

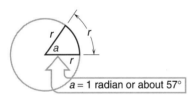

$a = 1$ radian or about 57°

To understand where the definition comes from, set up a proportion:

$$\frac{\text{angle } a}{\text{total angle in the circle}} = \frac{\text{arc length}}{\text{circumference of the circle}}$$

$$\frac{a}{360°} = \frac{r}{2\pi r}$$

$$a = \frac{360° \, \cancel{r}}{2\pi \cancel{r}}$$

$$a = \frac{180°}{\pi} \approx 57.295779°$$

Degrees to Radians To convert angle measurements from degrees to radians (rad), simply multiply by $\pi/180$ and round as needed. Use the π button on your calculator unless you are told otherwise.

EXAMPLE

(a) $43° = 43\left(\dfrac{\pi}{180}\right) = \dfrac{43 \cdot \pi}{180}$

$\approx 0.75 \text{ rad}$ rounded

536

(b) $60° = 60\left(\dfrac{\pi}{180}\right) = \dfrac{60 \cdot \pi}{180}$

$= \dfrac{\pi}{3}\,\text{rad} \approx 1.0472\,\text{rad}$

Radians to Degrees   To convert angles from radians to degrees, multiply by $180/\pi$ and round.

EXAMPLE

(a) $1.3\,\text{rad} = 1.3\left(\dfrac{180}{\pi}\right) = \dfrac{1.3 \cdot 180}{\pi}$

$\approx 74.5°$

(b) $0.5\,\text{rad} = 0.5\left(\dfrac{180}{\pi}\right) = \dfrac{0.5 \cdot 180}{\pi}$

$= \dfrac{90}{\pi}\ \text{degrees or about } 28.6°$

PROBLEMS

Try these problems for practice in using radian measure.

(a) Convert to radians. (Round to two decimal places.)

    (1) 10°   (2) 35°   (3) 90°   (4) 120°

(b) Convert to degrees. (Round to the nearest tenth.)

    (1) 0.3 rad   (2) 1.5 rad   (3) 0.8 rad   (4) 0.05 rad

ANSWERS

(a) (1) $\dfrac{10 \cdot \pi}{180} \approx 0.17\,\text{rad}$ (2) 0.61 rad   (3) 1.57 rad   (4) 2.09 rad

(b) (1) $\dfrac{0.3 \cdot 180}{\pi} \approx 17.2°$ (2) 85.9°   (3) 45.8°   (4) 2.9°

Sectors   The wedge-shaped portion of a circle shown is called a **sector.** Both its area and the length of arc S can be expressed most directly using radian measure.

SECTORS		
Arc length	$S = ra$	
Area	$A = \frac{1}{2}r^2a$	where a is the central angle given in radians.

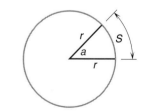

EXAMPLE

A sheet metal worker wants to know the arc length and area of a sector with central angle 150° cut from a circular sheet of metal with radius 16.0 in.

First, calculate angle a in radians: $a = \dfrac{150 \times \pi}{180} \approx 2.618 \text{ rad}$

Second, calculate the arc length S: $S \approx 16 \text{ in.} \times 2.618$
$$\approx 41.9 \text{ in.}$$

Finally, calculate the area A: $A \approx \frac{1}{2}(16^2)(2.618)$
$$\approx 335 \text{ in.}^2$$

Using a calculator, we have

S: **16** $\times$ **150** $\times$ $\boxed{\pi}$ $\div$ **180** $\boxed{=}$ → `41.88790205`

A: **.5** $\times$ **16** $\boxed{x^2}$ $\times$ **150** $\times$ $\boxed{\pi}$ $\div$ **180** $\boxed{=}$ → `335.1032164`

YOUR TURN

Calculate the arc length and area of a sector with central angle 80° and radius 25.0 ft. Round to three significant digits.

SOLUTION

The central angle is $a = \dfrac{80 \times \pi}{180} \approx 1.396 \text{ rad}$.

The arc length is $S \approx 25 \times 1.396 \approx 34.9 \text{ ft}$.

The area is $A \approx \frac{1}{2} \times 25^2 \times 1.396 \approx 436 \text{ ft}^2$.

Linear and Angular Speed

Radian units are also used in science and technology to describe the rotation of an object. When an object moves along a straight line for a distance d in time t, its **average linear speed** v is defined as

$$v = \frac{d}{t}$$

EXAMPLE

If a car travels 1800 ft in 24 seconds, its average linear speed for the trip is

$$v = \frac{1800 \text{ ft}}{24 \text{ sec}} = 75 \text{ ft/sec}$$

When an object moves along a circular arc for a distance S in time t, it goes through an angle a in radians, and its **average angular speed** w is defined as

$$w = \frac{a}{t}$$

EXAMPLE

If a 24-in.-diameter flywheel rotates through an angle of 140° in 1.1 seconds, what is the average angular speed of the flywheel?

First, calculate the angle in radians. The angle traveled is $a = \dfrac{140 \times \pi}{180} \approx 2.44 \text{ rad}$.

538

Then, calculate the average angular speed, $w \approx \dfrac{2.44}{1.1} \approx 2.2$ rad/sec.

 $140 \; \boxed{\times} \; \boxed{\pi} \; \boxed{\div} \; 180 \; \boxed{\div} \; 1.1 \; \boxed{=} \; \rightarrow \; \boxed{2.221328139}$

PROBLEMS

For practice in working with linear and angular speeds, try these problems.

(a) If a bicycle travels 24 miles in 1.5 hr, find its average linear speed.

(b) The blade on a shop fan rotates at 8.0 revolutions per second. Calculate its average angular speed. Round to two significant digits. (*Hint:* Each revolution is 360°.)

(c) A belt-driven drum of radius 28.4 cm makes one revolution every 0.250 sec. What is the linear speed of the belt driving the drum? Round to three significant digits. (*Hint:* If the belt doesn't slip, the distance traveled by the belt in one revolution is equal to the circumference of the drum.)

(d) A Ferris wheel in an amusement park takes an average of 40 seconds for one non-stop revolution. Calculate its average angular speed. Round to two significant digits.

SOLUTIONS

(a) $v = \dfrac{d}{t} = \dfrac{24 \text{ mi}}{1.5 \text{ hr}} = 16 \text{ mi/hr}$

(b) Each revolution is 360°, therefore the fan blade travels through an angle of $8 \times 360° = 2880°$ in one second. Converting to radians,

$$\dfrac{2880° \times \pi}{180} \approx 50.265 \text{ rad}$$

Using the formula for angular speed,

$$w = \dfrac{a}{t} \approx \dfrac{50.265 \text{ rad}}{1 \text{ sec}} \approx 50 \text{ rad/sec}$$

(c) The circumference of the drum is

$$C = 2\pi r = 2\pi(28.4 \text{ cm}) \approx 178.4 \text{ cm}$$

This is the distance traveled by any point on the belt in 0.250 sec. Substitute these values into the linear speed formula.

$$v = \dfrac{d}{t} \approx \dfrac{178.4 \text{ cm}}{0.25 \text{ sec}} \approx 714 \text{ cm/sec}$$

(d) One revolution is 360°. In radians this is

$$\dfrac{360 \times \pi}{180} \approx 6.28 \text{ rad}$$

Using the formula for angular speed,

$$w = \dfrac{a}{t} \approx \dfrac{6.28 \text{ rad}}{40 \text{ sec}} \approx 0.16 \text{ rad/sec}$$

Right Triangles   Trigonometry is basically triangle measurement, and, to keep it as simple as possible, we can begin by studying only *right* triangles. You should remember from

Chapter 8 that a right triangle is one that contains a right angle, a 90° angle. All of the following are right triangles. In each triangle the right angle is marked.

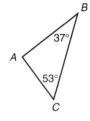

YOUR TURN

Is triangle *ABC* a right triangle?

SOLUTION

You should remember that the sum of the three angles of any triangle is 180°. For triangle *ABC* two of the angles total 37° + 53° or 90°; therefore, angle *A* must equal 180° − 90° or 90°. The triangle *is* a right triangle.

To make your study of triangle trigonometry even easier, you may want to place a right triangle in *standard position*. For this collection of right triangles

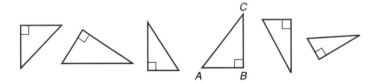

only triangle *ABC* is in standard position.

Place the triangle so that the right angle is on the right side, one leg (*AB*) is horizontal, and the other leg (*BC*) is vertical. The hypotenuse (*AC*) will slope up from left to right.

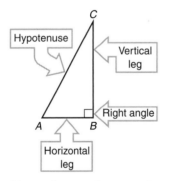

If you do not place a triangle in standard position, remember that the legs are the two sides forming the right angle, and the hypotenuse is the side opposite the right angle. The hypotenuse is always the longest side in a right triangle.

Pythagorean Theorem You should also recall from Chapter 8 that the *Pythagorean theorem* is an equation relating the lengths of the sides of any right triangle.

PYTHAGOREAN THEOREM

$$H^2 = A^2 + B^2$$

540

This equation is true for every right triangle.

EXAMPLE

The triangle

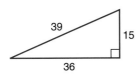

gives $39^2 = 36^2 + 15^2$
or $1521 = 1296 + 225$ which is correct

YOUR TURN

For review, use this rule to find the missing side for each of the following triangles.

(a)   (b)   (c)

(Round to 1 decimal place)

SOLUTIONS

(a) $H^2 = 21^2 + 20^2$ (b) $18.1^2 = B^2 + 6.5^2$ (c) $6.29^2 = 6.21^2 + A^2$
  $H^2 = 441 + 400$ $327.61 = B^2 + 42.25$ $39.5641 = 38.5641 + A^2$
  $H^2 = 841$ $B^2 = 327.61 - 42.25$ $A^2 = 39.5641 - 38.5641$
  $H = 29 \text{ in.}$ $B^2 = 285.36$ $A^2 = 1$
                 $B \approx 16.9$ $A = 1 \text{ cm}$

Here's how to work problem (b) on a calculator:

 18.1 $\boxed{x^2}$ $\boxed{-}$ **6.5** $\boxed{x^2}$ $\boxed{=}$ $\boxed{\sqrt{}}$ → **16.89260193**

Special Right Triangles

Three very special right triangles are used very often in practical work. Here they are:

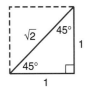

45°–45°–90°△   The first triangle is a *45° right triangle*. Its angles are 45°, 45°, and 90°, and it is formed by the diagonal and two sides of a square. Notice that it is an isosceles triangle—the two shorter sides or legs are the same length.

Our triangle has legs one unit long, but we can draw a 45° right triangle with any length sides. If we increase the length of one side, the others increase in proportion while the angles stay the same size.

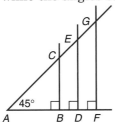

Triangle *ABC* is a 45° right triangle.
Triangle *ADE* is also a 45° right triangle.
Triangle *AFG* is also a 45° right triangle, . . . and so on.

EXAMPLE

(a) If $AD = 2$ in., then

$$DE = 2 \text{ in.}$$

The legs are always equal in length. If AD is 2 in. long, DE is also 2 in. long.

and

$$AE = 2\sqrt{2} \text{ in. or about } 2.8 \text{ in.}$$

The hypotenuse of a $45°$ right triangle is $\sqrt{2}$ times the length of each leg.

(b) If $AC = 5$ in., then

$$AC = AB\sqrt{2}, \text{ so } AB = \frac{AC}{\sqrt{2}} = \frac{5}{\sqrt{2}}$$

Using a calculator gives

5 ÷ **2** √ = → $\boxed{3.535533906}$ ≈ 3.5 in.

No matter what the size of the triangle, if it is a right triangle and if one angle is $45°$, the third angle is also $45°$. The two legs are equal, and the hypotenuse is $\sqrt{2} \approx 1.4$ times the length of a leg.

(c) **Plumbing** Parallel pipes are connected by $45°$ fittings as shown. Calculate the length of the connecting pipe from A to B. (Round to the nearest tenth.)

The connecting pipe AB is the hypotenuse of a $45°–45°–90°$ triangle. The length L is, therefore, $8.5\sqrt{2}$ in. ≈ 12.0 in.

YOUR TURN

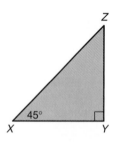

(a) If $XY = 4.6$ cm, find YZ and XZ.

(b) If $XZ = 15.8$ ft, find XY and YZ.

Round to the nearest tenth.

SOLUTIONS

(a) $YZ = 4.6$ cm

$$XZ = 4.6\sqrt{2} \text{ cm} \approx 6.5 \text{ cm}$$

(b) $XY = YZ = \dfrac{XZ}{\sqrt{2}} = \dfrac{15.8 \text{ ft}}{\sqrt{2}} \approx 11.2 \text{ ft}$

30°–60°–90° △ The second of our special triangles is the 30°–60° right triangle. In a 30°–60° right triangle the length of the side opposite the 30° angle, the smallest side, is exactly one-half the length of the hypotenuse. The length of the third side is $\sqrt{3}$ times the length of the smaller side.

To see where this relationship comes from, start with an equilateral triangle with sides 2 units long.

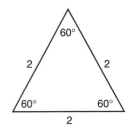

Draw a perpendicular from one vertex to the opposite side. This segment also bisects the opposite side.

Then in the 30°–60° right triangle the Pythagorean theorem gives

$$a^2 + 1^2 = 2^2$$
$$a^2 + 1 = 4$$
$$a^2 = 3$$
$$a = \sqrt{3}$$

EXAMPLE

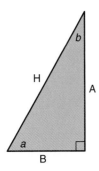

(a) In this triangle, if $a = 60°$ and $B = 8$ ft, then

$$H = 2B = 2 \cdot 8 \, \text{ft} = 16 \, \text{ft}$$

and $A = B\sqrt{3} = 8\sqrt{3} \approx 13.9 \, \text{ft}$

(b) If $a = 60°$ and $H = 24$ in., then

$$B = \tfrac{1}{2}H = \tfrac{1}{2}(24) = 12 \, \text{in.}$$

and $A = B\sqrt{3} = 12\sqrt{3} \approx 20.8 \, \text{in.}$

In a 30°–60° right triangle,

$A = B\sqrt{3}$, therefore, dividing both sides by $\sqrt{3}$, $B = \dfrac{A}{\sqrt{3}}$

If $a = 60°$ and $A = 32$ yd, then

$$B = \frac{32}{\sqrt{3}} \approx 18.5 \, \text{yd}$$

and $H = 2B \approx 37.0 \, \text{yd}.$

YOUR TURN

(a) In this triangle, if $a = 30°$ and $A = 2$ in., calculate
   H, B, and b

(b) If $a = 30°$, $H = 3$ in., then $B = ?$, $A = ?$

(c) If $a = 30°$, $B = 18$ cm, then $A = ?$, $H = ?$

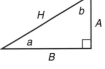

(a) $H = 4$ in., $B = 2\sqrt{3}$ in. or about 3.5 in., $b = 60°$

(b) In a 30°–60° right triangle, $A = \frac{1}{2}H = \frac{1}{2}(3) = 1.5$ in.

and $B = A\sqrt{3} = 1.5 \times \sqrt{3} \approx 2.6$ in.

(c) $A = \dfrac{18}{\sqrt{3}} \approx 10.4$ cm

$H = 2A \approx 20.8$ cm

Again, we may form any number of 30°–60° right triangles, and in each triangle the hypotenuse is twice the length of the smallest side. The third side is $\sqrt{3}$ or about 1.7 times the length of the smallest side. When you find a right triangle with one angle equal to 30° or 60°, you automatically know a lot about it.

3-4-5 $\triangle$ The third of our special triangles is the 3–4–5 triangle that was discussed in Chapter 8. If a triangle is found to have sides of length 3, 4, and 5 units (any units—in., ft, cm), it must be a right triangle. The 3–4–5 triangle is the smallest right triangle with sides whose lengths are whole numbers.

Construction workers or surveyors can use this triangle to set up right angles.

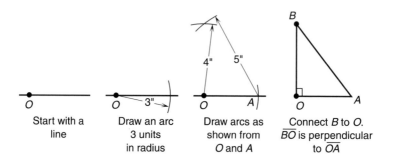

| Start with a line | Draw an arc 3 units in radius | Draw arcs as shown from O and A | Connect B to O. $\overline{BO}$ is perpendicular to $\overline{OA}$ |

Notice that in any 3–4–5 triangle the acute angles are approximately 37° and 53°. The smallest angle is always opposite the smallest side.

You will find these three triangles appearing often in triangle trigonometry. Now, for some practice in working with angles and triangles, turn to Exercises 10-1.

Exercises 10-1

Angles and Triangles

A. Label the shaded angles as acute, obtuse, right, or straight.

1.

2.

3.

4. 90°

5.

6.

7.

8.

9. 45° 45°

10.

B. Write the following angles in degrees and minutes. (Round to the nearest minute.)

1. $36\frac{1}{4}°$ 2. $73\frac{3}{5}°$ 3. $65.45°$ 4. $84.24°$

5. $17\frac{1}{2}°$ 6. $47\frac{3}{8}°$ 7. $16.11°$ 8. $165.37°$

Write the following angles in decimal degrees. (Round to the nearest hundredth if necessary.)

9. $27°30'$ 10. $80°15'$ 11. $154°39'$ 12. $131°6'$

13. $57°3'$ 14. $44°20'$ 15. $16°53'$ 16. $16'$

Write the following angles in radians and round to the nearest hundredth.

17. $24.75°$ 18. $185.8°$ 19. $1.054°$ 20. $14\frac{3}{4}°$

21. $67\frac{3}{8}°$ 22. $9.65°$ 23. $80.60°$ 24. $216.425°$

Write the following angles in degrees and round to the nearest hundredth.

25. 0.10 rad 26. 0.84 rad 27. 2.1 rad 28. 3.5 rad

C. Solve the following problems involving triangles.

Find the lengths of the missing sides on the following triangles using the Pythagorean theorem. (Round to one decimal place.)

1.   2.   3.   4.   5.

6.   7.   8.   9.   10.

Each of the following right triangles is an example of one of the three "special triangles" described in this chapter. Find the quantities indicated without using the Pythagorean theorem. (Round the sides to the nearest tenth if necessary.)

11.   12.

(a) $a =$ _____   (a) $B =$ _____

(b) $A =$ _____   (b) $C =$ _____

13.

(a) $X = \underline{\hspace{1.5cm}}$

(b) $a = \underline{\hspace{1.5cm}}$

(c) $b = \underline{\hspace{1.5cm}}$

14.

(a) $E = \underline{\hspace{1.5cm}}$

(b) $F = \underline{\hspace{1.5cm}}$

(c) $t = \underline{\hspace{1.5cm}}$

15.

(a) $K = \underline{\hspace{1.5cm}}$

(b) $a = \underline{\hspace{1.5cm}}$

16.

(a) $R = \underline{\hspace{1.5cm}}$

(b) $q = \underline{\hspace{1.5cm}}$

17.

(a) $B = \underline{\hspace{1.5cm}}$

(b) $A = \underline{\hspace{1.5cm}}$

(c) $d = \underline{\hspace{1.5cm}}$

18.

(a) $M = \underline{\hspace{1.5cm}}$

(b) $N = \underline{\hspace{1.5cm}}$

(c) $t = \underline{\hspace{1.5cm}}$

19.

(a) $Q = \underline{\hspace{1.5cm}}$

(b) $u = \underline{\hspace{1.5cm}}$

20.

(a) $X = \underline{\hspace{1.5cm}}$

(b) $a = \underline{\hspace{1.5cm}}$

(c) $b = \underline{\hspace{1.5cm}}$

Calculate the arc length S and area A for each of the following sectors. (Round to the nearest whole number if necessary.)

21. $a = 40°$	**22.** $a = 120°$	**23.** $a = 0.8$ rad	**24.** $a = 2$ rad
$r = 30$ cm	$r = 24$ ft	$r = 110$ ft	$r = 18$ m
25. $a = 72°$	**26.** $a = 3$ rad	**27.** $a = 0.5$ rad	**28.** $a = 180°$
$r = 5$ in.	$r = 80$ cm	$r = 10$ m	$r = 30$ in.

D. Practical Problems

1. **Sheet Metal Technology** A transition duct is constructed from a circular sector of radius 18 in. and central angle 115°. (a) What is the area of the

metal used? (b) Calculate the arc length of the sector. (Round to the nearest tenth.)

2. **Landscaping** Calculate the area (in square feet) of a flower garden shaped like a circular sector with radius 60 ft and central angle 40°.

3. **Landscaping** In problem 2, if shrubs are planted every 2 ft along the outer border of the garden, how many shrubs are needed?

4. **Construction** The outfield fencing for a Little League field forms a circular sector with home plate as the center. The fence is placed at a uniform distance of 215 ft from home plate. The boundaries of the fence, which extends partway into foul territory, create an angle of 105° with home plate. At $22 per foot, how much will the fence cost? (Round to the nearest $10.)

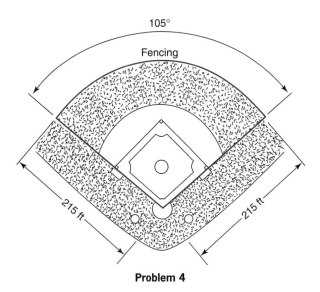

Problem 4

5. **Construction** In problem 4, topsoil must be bought to cover the sector to a depth of 4 in. How many cubic yards of topsoil are needed? (Be careful with the units.)

6. If a car travels 725 miles in 12.5 hr, find its average linear speed.

7. **Manufacturing** The blades of a fan have a radius of 16.5 in. and they turn one revolution every 0.125 sec. Calculate the linear speed of the tip of a blade. (Round to three significant digits.)

8. **Manufacturing** A pulley with a radius of 36.2 cm rotates at 6.5 revolutions per second. Find the linear speed of the belt driving the pulley. (Round to three significant digits.)

9. If the indicator on a flow rate meter moves 75° in 16.4 seconds, what is its average angular speed during this time? (Round to two decimal places.)

10. **Machine Technology** If a lathe makes 25 revolutions per second, calculate its average angular speed. (Round to the nearest whole number.)

11. A Ferris wheel in an amusement park takes an average of 16 seconds for each revolution. What is its average angular speed? (Round to two decimal places.)

12. **Plumbing** For the connection shown on page 548, calculate the rise, the distance *H*, and the length *AB*. The elbows are 45° fittings. (Round to the nearest tenth.)

Problem 12

Check your answers in the Appendix, then continue in Section 10-2.

10-2 TRIGONOMETRIC RATIOS

In Section 10-1 we showed that in *all* 45° right triangles, the adjacent side is equal in length to the opposite side, and the hypotenuse is roughly 1.4 times this length. Similarly, in *all* 30°–60° right triangles, the hypotenuse is twice the length of the side opposite the 30° angle, and the adjacent side is about 1.7 times this length.

The first skill needed for trigonometry is the ability to label the three sides of a right triangle according to convention: the hypotenuse, the side *adjacent* to a given acute angle, and the side *opposite* that angle.

In triangle ABC, side $\overline{AB}$ is the hypotenuse. It is the longest side and it is opposite the right angle.

For acute angle A, side $\overline{AC}$ is the adjacent side and side $\overline{BC}$ is the opposite side.

For acute angle B, side $\overline{BC}$ is the adjacent side and side $\overline{AC}$ is the opposite side.

Note ▶ The hypotenuse is never considered to be either the adjacent side or the opposite side. It is not a leg of the triangle. ◀

YOUR TURN

Check your understanding of this vocabulary by putting triangle XYZ in standard position and by completing the following statements.

(a) The hypotenuse is _____.

(b) The opposite side for angle Y is _____.

(c) The adjacent side for angle Y is _____.

(d) The opposite side for angle Z is _____.

(e) The adjacent side for angle Z is _____.

ANSWERS

(a) YZ (b) XZ (c) XY

(d) XY (e) XZ

Of course, the phrases "adjacent side," "opposite side," and "hypotenuse" are used only with right triangles.

PROBLEMS

For each of the following right triangles, mark the hypotenuse with the letter H, mark the side adjacent to the shaded angle with the letter A, and mark the side opposite the shaded angle with the letter B.

(a)   (b)   (c)   (d)   (e)

ANSWERS

(a)   (b)   (c)   (d)   (e)

Learning Help ▶   If you find it difficult to identify the adjacent side and the opposite side of a given angle in a right triangle, try placing the triangle in standard position. Position the triangle so that the given angle is at the lower left and the right angle is at the lower right. When the triangle is in this position, the adjacent side is always the horizontal leg, A, and the opposite side will always be the vertical leg, B. ◀

The key to understanding trigonometry is to realize that in *every* right triangle there is a fixed relationship connecting the length of the adjacent side, opposite side, hypotenuse, and the angle that characterizes the triangle.

For example, in all right triangles that contain the angle $20°$,

the ratio $\dfrac{\text{opposite side}}{\text{hypotenuse}}$ will always have the same value. Therefore,

$$\frac{B}{H} = \frac{B_1}{H_1} = \frac{B_2}{H_2} = \frac{B_3}{H_3} = \ldots \text{and so on}$$

This ratio will be approximately 0.342.

Sine Ratio   This kind of ratio of the side lengths of a right triangle is called a *trigonometric ratio*. It is possible to write down six of these ratios, but we will use only three here. Each ratio is given a special name.

For the triangle

the **sine ratio,** abbreviated sin *x*, is defined as

$$\sin x = \frac{\text{opposite side}}{\text{hypotenuse}} = \frac{B}{H}$$

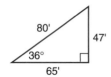

(Pronounce "sin" as "sign" to rhyme with "dine" not with "sin." *Sin* is an abbreviation for *sine*. Sin *x* is read "sine of angle *x*.")

EXAMPLE

In the triangle

$$\sin 36° = \frac{47 \text{ ft}}{80 \text{ ft}} \approx 0.59$$

YOUR TURN

Use this triangle to find sin 60°. (Round to two decimal places.)

$$\sin 60° = \underline{\hspace{1.5cm}}$$

SOLUTION

The hypotenuse has length 2.0″ and the side opposite the 60° angle has length 1.73″.

$$\sin 60° = \frac{\text{opposite side}}{\text{hypotenuse}}$$

$$\sin 60° = \frac{1.73 \text{ in.}}{2.0 \text{ in.}}$$

$$\sin 60° \approx 0.87 \quad \text{approximately}$$

Every possible angle *x* will have some number sin *x* associated with it, and we can calculate the value of sin *x* from any right triangle that contains the angle *x*.

EXAMPLE

Sin 40° can be calculated from the following triangle.

95.0 / 61.1 / 40°

$$\sin 40° = \frac{61.1}{95.0}$$ ⟵ Opposite side / ⟵ Hypotenuse

$$\approx 0.643 \quad \text{rounded}$$

YOUR TURN

What is the value of sin 0°?

(*Hint:* Construct a right triangle with a very small angle and guess.)

SOLUTION

sin 0° = 0   In this triangle

side B is much smaller than H; therefore, the number $\frac{B}{H}$ is very small.

As angle a decreases to zero, side B gets smaller and smaller, and sin a gets closer and closer to zero.

In a similar way you can show that sin 90° = 1. (Try it.)

Now, remember those three special right triangles from Section 10-1.

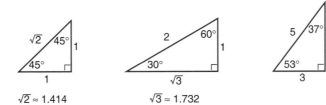

$\sqrt{2} \approx 1.414$ $\sqrt{3} \approx 1.732$

YOUR TURN

Use these drawings to calculate the following values of sin x. (Round to the nearest hundredth.)

x	30°	37°	45°	53°	60°
sin x					

x	30°	37°	45°	53°	60°
sin x	0.50	0.60	0.71	0.80	0.87

Note ▶ Sin x is a trigonometric *ratio* or trigonometric *function* associated with the angle x. The value of sin x can never be greater than 1, since the legs of any right triangle can never be longer than the hypotenuse. ◀

Cosine Ratio A second useful trigonometric ratio is the **cosine** of angle x, abbreviated cos x. This ratio is defined as

$$\cos x = \frac{\text{adjacent side}}{\text{hypotenuse}} = \frac{A}{H}$$

EXAMPLE

In the triangle

$$\cos 36° = \frac{65 \text{ ft}}{80 \text{ ft}}$$

← Adjacent side

← Hypotenuse

$$\cos 36° \approx 0.81 \quad \text{approximately}$$

YOUR TURN

Use the following triangle to find cos 60°.

SOLUTION

The hypotenuse is 2.0″ and the adjacent side to the 60° angle is 1.0″.

$$\cos 60° = \frac{\text{adjacent side}}{\text{hypotenuse}}$$

$$\cos 60° = \frac{1.0 \text{ in.}}{2.0 \text{ in.}} = 0.50$$

PROBLEMS

Now use the drawing of the three special right triangles on page 551 to help you to complete this table. (Round to two decimal places.)

x	0°	30°	37°	45°	53°	60°	90°
cos x	1						0

We have entered the values for 0° and 90°.

x	0°	30°	37°	45°	53°	60°	90°
cos x	1	0.87	0.80	0.71	0.60	0.50	0

Tangent Ratio   A third trigonometric ratio may also be defined. The ratio **tangent x,** abbreviated tan x, is defined as

$$\tan x = \frac{\text{opposite side}}{\text{adjacent side}} = \frac{B}{A}$$

EXAMPLE

In the triangle

$$\tan 33° = \frac{13\text{ ft}}{20\text{ ft}}$$

Opposite side ⟵
Adjacent side ⟵

$$\tan 33° \approx 0.65 \quad \text{approximately}$$

YOUR TURN

Use the following triangle to find tan 30°.

SOLUTION

$$\tan 30° = \frac{\text{opposite side}}{\text{adjacent side}}$$

$$\tan 30° = \frac{1.0\text{ in.}}{1.73\text{ in.}} \approx 0.58$$

PROBLEMS

Complete this table for tan x using the three special right triangles. (Round to two decimal places.)

x	0°	30°	37°	45°	53°	60°	90°
tan x							

(tan x for $x = 90°$ is very tricky.)

x	0°	30°	37°	45°	53°	60°	90°
$\tan x$	0	0.58	0.75	1.00	1.33	1.73	not defined

To find $\tan 90°$ draw a right triangle with x very large.

As angle x gets larger and larger, the opposite side B gets longer and longer.

The ratio $\tan x = \dfrac{B}{A}$ gets larger and larger. As x approaches 90°, $\tan x$ becomes infinitely large, and we have no number large enough to name it.

Finding Values of Trigonometric Functions

You have been asked to calculate these "trig" ratios to help you get a feel for how the ratios are defined and to encourage you to memorize a few values. In actual practical applications of trigonometry, the values of the ratios are found by using a calculator with built-in trigonometric function keys.

To find the value of a trigonometric ratio using most scientific calculators, enter the value of the angle, then press the key for the given function.

EXAMPLE

To find $\sin 38°$, enter this sequence:

38 (sin) → | *0.615661475* | $\sin 38° \approx 0.616$

For some scientific and all graphing calculators, we need to enter the trig function first, then the angle.

Careful ▶ Angles can be measured in units of radians and gradients as well as degrees, and most calculators are capable of accepting all three units of measure. Before entering an angle in degrees, make certain the calculator is in "degree mode" by either using the (DRG) key or the (MODE) key. Look for the "DEG" indication on the display. On some calculators absence of any mode indication means that it is in degree mode. ◀

YOUR TURN

Now you try it. Find the value of each of the following to two decimal places.

(a) $\cos 26°$

(b) $\tan 67.8°$

(c) $\sin 8°$

(d) $\tan 14.25°$

(e) $\cos 47\frac{1}{2}°$

(f) $\sin 39.46°$

(g) $\sin 138°$

(h) $\cos 151.6°$

(i) $\tan 141.12°$

(a) **26** $\boxed{\cos}$ → `0.898794046` $\cos 26° \approx 0.90$

(b) **67.8** $\boxed{\tan}$ → `2.450425198` $\tan 67.8° \approx 2.45$

(c) **8** $\boxed{\sin}$ → `0.139173101` $\sin 8° \approx 0.14$

(d) **14.25** $\boxed{\tan}$ → `0.253967646` $\tan 14.25° \approx 0.25$

(e) **47.5** $\boxed{\cos}$ → `0.675590207` $\cos 47\frac{1}{2}° \approx 0.68$

(f) **39.46** $\boxed{\sin}$ → `0.635539369` $\sin 39.46° \approx 0.64$

(g) **138** $\boxed{\sin}$ → `0.669130606` $\sin 138° \approx 0.67$

(h) **151.6** $\boxed{\cos}$ → `-0.879648572` $\cos 151.6° \approx -0.88$

(i) **141.12°** $\boxed{\tan}$ → `-0.806322104` $\tan 141.12° \approx -0.81$

- Notice in problems (g), (h), and (i) that the trigonometric functions exist for angles greater than 90°. The values of the cosine and tangent functions are negative for angles between 90° and 180°.

- Notice in problem (g) that

$$\sin 138° = 0.669130606\ldots$$
$$\text{and } \sin 42° = 0.669130606\ldots \quad 138° + 42° = 180°$$

This suggests that the following formula is true:

$$\sin A = \sin(180° - A)$$

This fact will be useful in Section 10-4.

Angles in Degrees and Minutes   When finding the sine, cosine, or tangent of an angle given in degrees and minutes, either enter the angle as given if your calculator has a degrees/minutes/seconds key, or first convert the angle to decimal form. You may want to review the material on page 535.

EXAMPLE

Find $\cos 59°37'$.

Here are two possible calculator key sequences for this calculation.

59 $\boxed{°'''}$ **37** $\boxed{°'''}$ $\boxed{\cos}$

59.37 $\boxed{\triangleright DD}$ $\boxed{\cos}$ $\Big\} \to$ `0.505782847`

Therefore, $\cos 59°37' \approx 0.506$.

Be careful to consult your calculator instruction manual to learn how your calculator works.

YOUR TURN

Try these problems for practice. Find each value to three decimal places.

(a) $\sin 71°26'$ (b) $\tan 18°51'$ (c) $\cos 42°16'$ (d) $\cos 117°22'$

ANSWERS

(a) 0.948   (b) 0.341   (c) 0.740   (d) −0.460

Finding the Angle

In some applications of trigonometry it may happen that we know the value of the trigonometric ratio number and we need to find the angle associated with it. This is called finding the **inverse** of the trigonometric function. To find the angle associated with any given value of a trigonometric function, enter the function value followed by the [INV] or [2ndF] key, and then press the key of the trigonometric function given. (For some calculators this sequence may be reversed.)

These inverse trigonometric functions will be labeled $\sin^{-1}, \cos^{-1},$ and $\tan^{-1}$ on your calculator.

EXAMPLE

To find the angle whose sine is 0.728, enter the following:

.728 [INV] [sin] → 46.718988

The angle is approximately 46.7°; therefore, $\sin 46.7° \approx 0.728.$* We can also express this result as $\sin^{-1} 0.728 \approx 46.7°.$

Careful ▶ Be certain your calculator is in the degree mode before you enter this kind of calculation. ◀

YOUR TURN

Use your calculator to find the angle x for the following trigonometric function values. (Round to the nearest tenth of a degree.)

(a) $\cos x = 0.589$ (b) $\sin x = 0.248$ (c) $\tan x = 1.75$

ANSWERS

(a) $x = \cos^{-1} 0.589 \approx 53.9°$ (b) $x = \sin^{-1} 0.248 \approx 14.4°$

(c) $x = \tan^{-1} 1.75 \approx 60.3°$

---

*We have found the angle between 0 and 90° that corresponds to the given trigonometric function value, but there are many angles greater than 90° whose sine ratio has this same value. Since such values of the inverse are generally not very important in the trades, we will not discuss them in detail here.

To find the angle in degrees and minutes, follow the procedure described to find the angle in decimal degrees, and then use the [▸DMS] key to convert.

EXAMPLE

To find x in degrees and minutes for $\cos x = 0.296$, follow this sequence:

Since 56.9″ is larger than half of a minute, we round up.

Therefore, $x \approx 72°47'$.

YOUR TURN

Now try these problems. Find x to the nearest minute.

(a) $\sin x = 0.818$ (b) $\tan x = 0.654$ (c) $\cos x = 0.513$

ANSWERS

(a) $x \approx 54°53'$ (b) $x \approx 33°11'$ (c) $x \approx 59°8'$

Now turn to Exercises 10-2 for a set of practice problems on trigonometric ratios.

Exercises 10-2   Trigonometric Ratios

A. For each of the following triangles *calculate* the indicated trigonometric ratios of the given angle. Do not use the trigonometric function keys on your calculator. (Round to nearest hundredth.)

1.

sin 33°

2.

cos 42°

3.

tan 23°

4.

tan 28°

5.

cos 64°

6.

sin 44°

B. Find each of the following trig values. (Round to three decimal places.)

1. $\sin 27°$ 2. $\cos 38°$ 3. $\tan 12°$ 4. $\sin 86°$

5. $\cos 79°$ 6. $\tan 6°$ 7. $\cos 87°$ 8. $\cos 6°30'$

9. $\tan 50°20'$ 10. $\sin 75°40'$ 11. $\cos 41.25°$ 12. $\cos 81°25'$

13. sin 50.4°   14. tan 74.15°   15. tan 81.06°   16. sin 12.6°

17. cos 98°   18. sin 106°   19. sin 144.2°   20. cos 134.5°

C. Find the acute angle *x*. (Round to the nearest minute.)

1. sin *x* = 0.974   2. cos *x* = 0.719

3. tan *x* = 2.05   4. sin *x* = 0.077

5. cos *x* = 0.262   6. tan *x* = 0.404

7. sin *x* = 0.168   8. cos *x* = 0.346

Find the acute angle *x*. (Round to the nearest tenth of a degree.)

9. tan *x* = 1.165   10. cos *x* = 0.662

11. cos *x* = 0.437   12. tan *x* = 0.225

13. tan *x* = 0.872   14. sin *x* = 0.472

15. sin *x* = 0.605   16. cos *x* = 0.154

Check your answers in the Appendix, then turn to Section 10-3 to learn how to use trigonometric ratios to solve problems.

10-3 SOLVING RIGHT TRIANGLES

In the first two sections of this chapter you learned about angles, triangles, and the trigonometric ratios that relate the angles to the side lengths in right triangles. Now it's time to look at a few of the possible applications of these trig ratios.

If you know the length of one of the three sides of a right triangle and the size of either of the two acute angles, it is always possible to determine the other side lengths and the third angle.

Suppose you need to measure the distance between two points *A* and *B* on opposite sides of a river. You have a tape measure and a protractor but no way to stretch the tape measure across the river. How can you measure the distance *AB*?

Try it this way:

1. Choose a point *C* in line with points *A* and *B*. If you stand at *C*, *A* and *B* will appear to "line up."

2. Construct a perpendicular to line *BC* at *A* and pick any point *D* on this perpendicular.

3. Find a point *E* where *B* and *D* appear to "line up." Draw *DE* and extend *DE* to *F* to form angle *a*.

4. Triangle ABD is a right triangle. You can measure AD and angle a, and, using the trigonometric ratios, you can calculate AB.

Suppose $AD = 100$ ft and $\angle a = 49°$. To find AB, first draw the following right triangle.

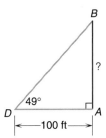

AD is the adjacent side to angle a.

AB is the side opposite angle a.

The trig ratio that relates the opposite to the adjacent side is the tangent.

$$\tan a = \frac{\text{opposite side}}{\text{adjacent side}}$$

$$\tan a = \frac{AB}{AD}$$

$$\tan 49° = \frac{AB}{100 \text{ ft}}$$

Multiply both sides of this equation by 100 to get

$$100 \tan 49° = AB$$

Use a calculator to find that $\tan 49° \approx 1.150$ approximately.

Therefore,

$$AB \approx 100 \times 1.150$$
$$AB \approx 115 \text{ ft} \quad \text{rounded to the nearest foot}$$

Or, using the calculator,

100 ⊠ **49** (tan) (=) → ▢ *115.0368407*

Using trig ratios, you have managed to calculate the distance from A to B without ever coming near point B.

EXAMPLE

Find the length of side A in the triangle shown.

Step 1   Decide which trig ratio is the appropriate one to use. In this case, we are given the hypotenuse (21 cm), and we need to find the adjacent side (A). The trig ratio that relates the adjacent side and the hypotenuse is the cosine.

$$\cos x = \frac{\text{adjacent side}}{\text{hypotenuse}}$$

Step 2   Write the cosine of the given angle in terms of the sides of the given right triangle.

$$\cos 50° = \frac{A}{21 \text{ cm}}$$

Step 3 Solve for the unknown quantity. In this case multiply both sides of the equation by 21.

$$A = 21 \cdot \cos 50°$$
$$A \approx 21 \cdot 0.64278 \ldots$$
$$A \approx 13.4985 \text{ cm} \quad \text{or} \quad 13.5 \text{ cm} \quad \text{rounded}$$

 21 ✕ 50 cos = → ▓ *13.4985398*

YOUR TURN

Find the length of the hypotenuse H in the triangle at the left. Round to the nearest tenth.

Work it out using the three steps shown above.

SOLUTION

Step 1 We are given the side opposite to the angle 71°, and we need to find the hypotenuse. The trig ratio relating the opposite side and the hypotenuse is the sine.

$$\sin x = \frac{\text{opposite side}}{\text{hypotenuse}}$$

Step 2 $\quad \sin 71° = \dfrac{10 \text{ in.}}{H}$

Step 3 Solve for H. $\qquad H \cdot \sin 71° = 10 \text{ in.}$

$$\text{or} \qquad H = \frac{10 \text{ in.}}{\sin 71°} \approx \frac{10}{0.9455 \ldots}$$

$$H \approx 10.6 \text{ in.}$$

 10 ÷ 71 sin = → ▓ *10.57620681*

PROBLEMS

Here are a few problems for practice in solving right triangles. Round to the nearest tenth.

(a) H 12' 42° Find H

(b) 16 m 34° X Find X

(c) 80° A 35" Find A

(d) Q P 20° a 15 cm Find P, Q, and a

SOLUTIONS

(a) $\quad \sin 42° = \dfrac{12 \text{ ft}}{H}$

$$H = \frac{12}{\sin 42°} \approx 17.9 \text{ ft}$$

(b) $\cos 34° = \dfrac{X}{16 \text{ m}}$

$X = 16 \cdot \cos 34° \approx 13.3 \text{ m}$

(c) $\tan 80° = \dfrac{35 \text{ in.}}{A}$

$A = \dfrac{35 \text{ in.}}{\tan 80°} \approx 6.2 \text{ in.}$

(d) $\sin 20° = \dfrac{P}{15 \text{ cm}}$

$P = 15 \cdot \sin 20° \approx 5.1 \text{ cm}$

$\cos 20° = \dfrac{Q}{15 \text{ cm}}$

$Q = 15 \cdot \cos 20° \approx 14.1 \text{ cm}$

$a = 90° - 20° = 70°$

On problems (b), (c), and (d) the triangles were redrawn in standard position to help you determine the adjacent and opposite sides.

Practical problems are usually a bit more difficult to set up, but they are solved with the same three-step approach. Try the following practical problems.

PROBLEMS

(a) Find the angle a in the casting shown. Round to the nearest minute.

Problem (a)   **Problem (b)**

(b) In a pipe-fitting job, what is the length of set if the offset angle is $22\frac{1}{2}°$ and the travel is $32\frac{1}{8}$ in.? Round to the nearest tenth.

(c) Find the angle of taper, x. Round to the nearest minute.

Problem (c)   **Problem (d)**

(d) Find the width, w, of the V-slot shown. (Round to 0.001 cm.)

(e) Find the dimensions d and h on the metal plate shown. (Round to the nearest tenth.)

Problem (e)

(a) In this problem the angle a is unknown. The information given allows us to calculate $\tan a$:

$$\tan a = \frac{\text{opposite side}}{\text{adjacent side}} = \frac{3\frac{1}{8}\text{ in.}}{4\frac{1}{4}\text{ in.}}$$

$$\tan a \approx 0.7353$$

$$a \approx \tan^{-1} 0.7353 \approx 36°20'$$

  3.125 ⌹ **4.25** ⌸ (INV) (tan) → `36.32682595` (▸DMS) → `36°19'36.5"`

(b) In this triangle the opposite side is unknown and the hypotenuse is given. Use the formula for sine:

$$\text{sine } 22\tfrac{1}{2}° = \frac{\text{opposite side}}{\text{hypotenuse}}$$

Then substitute, $\sin 22.5° = \dfrac{S}{32\frac{1}{8}\text{ in.}}$

And solve, $S = (32\frac{1}{8}\text{ in.})(\sin 22.5°) \approx 12.3$ in.

(c) Notice that angle x, the *taper angle*, is twice as large as angle m. We can find angle m from triangle ABC.

$$\tan m = \frac{\text{opposite side}}{\text{adjacent side}} = \frac{AC}{BC}$$

$$\tan m = \frac{1\text{ in.}}{11\text{ in.}} \approx 0.090909\ldots$$

$$m \approx \tan^{-1} 0.090909\ldots \approx 5°11.7'$$

$$x = 2m \approx 10°23'$$

(d) The V-slot creates an isosceles triangle whose height is 5.500 cm. The height divides the triangle into two identical right triangles as shown in the figure.

To find y,

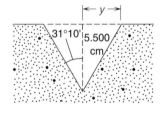

$$\tan 31°10' = \frac{y}{5.500}$$

or

$$y = (5.500)(\tan 31°10')$$

$$y \approx 3.3265\text{ cm}$$

$$w = 2y \approx 6.653\text{ cm}$$

  5.5 (×) **31** (°'") **10** (°'") (tan) (×) **2** ⌸ or **5.5** (×) **31.1** (▸DD) (tan) (×) **2** ⌸

→ `6.653093129`

(e) $\sin 72° = \dfrac{22}{h}$

$h = \dfrac{22}{\sin 72°} \approx 23.1$ ft

$\tan 72° = \dfrac{22}{d}$

$d = \dfrac{22}{\tan 72°} \approx 7.1$ ft

Now turn to Exercises 10-3 for a set of problems on solving right triangles using trigonometric ratios.

Exercises 10-3   Solving Right Triangles

A. Find the missing quantities as indicated. (Round all angles to the nearest minute and sides to the nearest tenth.)

1.

2.

3.

4.

$X = \underline{\hspace{1.5cm}}$

$X = \underline{\hspace{1.5cm}}$

$X = \underline{\hspace{1.5cm}}$

$\angle a = \underline{\hspace{1.5cm}}$

$Y = \underline{\hspace{1.5cm}}$

$Y = \underline{\hspace{1.5cm}}$

$Y = \underline{\hspace{1.5cm}}$

$X = \underline{\hspace{1.5cm}}$

5.

6.

7.

8.

$\angle a = \underline{\hspace{1.5cm}}$

$X = \underline{\hspace{1.5cm}}$

$X = \underline{\hspace{1.5cm}}$

$\angle a = \underline{\hspace{1.5cm}}$

$X = \underline{\hspace{1.5cm}}$

$Y = \underline{\hspace{1.5cm}}$

$Y = \underline{\hspace{1.5cm}}$

$X = \underline{\hspace{1.5cm}}$

Find the missing quantities as indicated. (Round all angles to the nearest tenth of a degree and sides to the nearest hundredth.)

9.

10.

11.

12.

$\angle a = \underline{\hspace{1.5cm}}$

$X = \underline{\hspace{1.5cm}}$

$X = \underline{\hspace{1.5cm}}$

$\angle a = \underline{\hspace{1.5cm}}$

$X = \underline{\hspace{1.5cm}}$

$Y = \underline{\hspace{1.5cm}}$

$Y = \underline{\hspace{1.5cm}}$

$X = \underline{\hspace{1.5cm}}$

13.

14.

15.

16.

$\angle m = \underline{\hspace{1.5cm}}$

$X = \underline{\hspace{1.5cm}}$

$X = \underline{\hspace{1.5cm}}$

$\angle m = \underline{\hspace{1.5cm}}$

$X = \underline{\hspace{1.5cm}}$

$Y = \underline{\hspace{1.5cm}}$

$Y = \underline{\hspace{1.5cm}}$

$X = \underline{\hspace{1.5cm}}$

Problem 1

B. Practical Problems

1. **Manufacturing** The most efficient operating angle for a certain conveyor belt is 31°. If the parts must be moved a vertical distance of 16 ft, what length of conveyor is needed? (Round to the nearest tenth.)

2. **Metalworking** Find the angle m in the casting shown. (Round to the nearest tenth of a degree.)

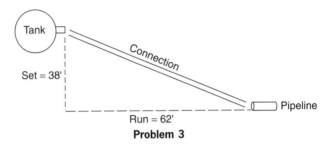

Problem 2

3. **Plumbing** A pipe fitter must connect a pipeline to a tank as shown. The run from the pipeline to the tank is 62 ft, while the set (rise) is 38 ft.

   (a) How long is the connection? (Round to the nearest inch.)

   (b) Will the pipe fitter be able to use standard pipe fittings (i.e., $22\frac{1}{2}°$, 30°, 45°, 60°, or 90°)?

Problem 3

4. A helicopter, flying directly over a fishing boat at an altitude of 1200 ft, spots an ocean liner at a 15° angle of depression. How far from the boat is the liner? (Round to the nearest foot.)

Problem 4

5. **Metalworking** Find the angle of taper on the steel bar shown if it is equal to twice m. (Round to the nearest minute.)

Problem 5   **Problem 6**

6. **Construction** A road has a rise of 6 ft in 80 ft. What is the gradient angle of the road? (Round to the nearest tenth of a degree.)

7. **Carpentry** Find the length of the rafter shown. (Round to the nearest inch.)

Problem 7

8. **Metalworking**   Three holes are drilled into a steel plate. Find the distances A and B as shown in the figure. (Round to the nearest hundredth of an inch.)

Problem 8

Problem 9

9. **Machine Technology**   Find the depth of cut x needed for the V-slot shown. (Assume that the V-slot is symmetric. Round to the nearest hundredth.)

Problem 10

10. **Masonry**   A 20-ft ladder leans against a building at a 65° angle with the ground. Will it reach a window 17 ft above the ground?

11. **Metalworking**   Find the distance B in the special countersink shown. (Round to the nearest tenth.)

Problem 11   **Problem 12**

12. Find angle a in the figure. (Round to the nearest hundredth.)

13. **Construction**   A road has a slope of 2°25′. Find the rise in 6500 ft of horizontal run. (Round to the nearest foot.)

14. Find the missing dimension d in the $\frac{5}{16}$-in. flathead screw shown. (Round to 0.0001 in.)

Problem 14

15. **Plumbing**   What total length of conduit is needed for sections A and B of the pipe connection shown? (Round to one decimal place.)

16. **Machine Technology**   Find the included angle m of the taper shown. (Round to the nearest minute.)

Problem 15

Problem 16

17. **Machine Technology** Ten holes are spaced equally around a flange on a 5.0-in.-diameter circle. Find the center-to-center distance x in the figure. Round to two decimal places.

Problem 17

18. **Police Science** Detectives investigating a crime find a bullet hole in a wall at a height of 7 ft 6 in. from the floor. The bullet passed through the wall at an angle of 34°. If they assume that the gun was fired from a height of 4 ft above the floor, how far away from the wall was the gun when it was fired? (Round to the nearest foot.)

Problem 18

19. **Construction** The Santa Teresa municipal building code specifies that public buildings have a handicapped access ramp with an incline of no more than 10°. What length of ramp is needed if the ramp climbs 3 ft 8 in.? (Express the answer in feet and inches to the nearest inch. *Hint:* Work the problem in inches only, then convert the answer to feet and inches.)

20. **Building Construction** A gable roof is to be constructed with a slope of 28° and a run of 27 ft 6 in. Calculate the rise of the roof. (Express the answer in feet and inches to the nearest inch.)

Turn to the Appendix to check your answers to these problems. Then go to Section 10-4 to study oblique triangles.

10-4 OBLIQUE TRIANGLES (OPTIONAL)

We have already seen how the trigonometric functions can be used to solve right triangles. To solve for the missing parts of a triangle that does not contain a right angle, we need two new formulas, the *law of sines* and the *law of cosines*. In this section we will derive these formulas and show how they may be used to solve triangles.

Oblique Triangles Any triangle that is not a right triangle is called an **oblique triangle.** There are two types of oblique triangles: acute and obtuse. In an **acute triangle** all three angles are acute—each is less than 90°. In an **obtuse triangle** one angle is obtuse—that is, greater than 90°. Notice in these figures that the angles have been labeled A, B, and C, and the sides opposite these angles have been labeled a, b, and c, respectively.

Acute triangle   Obtuse triangle

Law of Sines   The law of sines states that the lengths of the sides of a triangle are proportional to the sines of the corresponding angles. To show this for an acute triangle, first construct a perpendicular from C to side AB. Then in triangle ACD,

$$\sin A = \frac{h}{b} \quad \text{or} \quad h = b \sin A$$

and in triangle BCD,

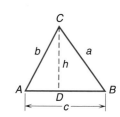

$$\sin B = \frac{h}{a} \quad \text{or} \quad h = a \sin B$$

Then $a \sin B = b \sin A$ or $\dfrac{a}{\sin A} = \dfrac{b}{\sin B}$

By repeating this process with a perpendicular from B to side AC, we can obtain a similar equation involving angles A and C and sides a and c. Combining these results gives the **law of sines.**

LAW OF SINES

$$\frac{a}{\sin A} = \frac{b}{\sin B} = \frac{c}{\sin C}$$

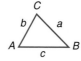

This law is sometimes written as

$$\frac{\sin A}{a} = \frac{\sin B}{b} = \frac{\sin C}{c}$$

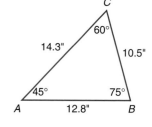

YOUR TURN

For the triangle shown, calculate the ratios

$$\frac{a}{\sin A}, \frac{b}{\sin B}, \text{and} \frac{c}{\sin C}$$

Round to one decimal place.

SOLUTION

All three ratios are equal to 14.8, rounded to one decimal place. Using a calculator, we find the first ratio like this:

    10.5 ÷ **45** [sin] [=] → ⬛ *14.8492424* ≈ 14.8 rounded

To *solve* a triangle means to calculate the values of all the unknown sides and angles from the information given. The law of sines enables us to solve any oblique triangle if the following information is given.

Case 1. Two angles and a side are known.

Case 2. Two sides and the angle opposite one of them are known.

Case 1

EXAMPLE

For the triangle shown, $A = 47°$, $B = 38°$, and $a = 8.0$ in. Find all its unknown parts, in this case angle C and sides b and c.

First, sketch the triangle as shown. Then since the angles of a triangle sum to 180°, find angle C by subtracting.

$$C = 180° - A - C = 180° - 47° - 38° = 95°$$

Next, apply the law of sines. Substitute A, B, and a to find b.

$$\frac{b}{\sin B} = \frac{a}{\sin A} \quad \text{gives} \quad b = \frac{a}{\sin A} \cdot \sin B = \frac{a \sin B}{\sin A}$$

$$= \frac{8.0 \sin 38°}{\sin 47°} \approx 6.7 \text{ in.}$$

Using a calculator, we have **8** $\times$ **38** [sin] $\div$ **47** [sin] [=] → `6.734486736`

Finally, apply the law of sines, again using A, C, and a to find c.

$$\frac{c}{\sin C} = \frac{a}{\sin A} \quad \text{gives} \quad c = \frac{a \sin C}{\sin A}$$

$$= \frac{8.0 \sin 95°}{\sin 47°} \approx 10.9 \text{ in.}$$

In any triangle the largest angle is opposite the longest side, and the smallest angle is opposite the shortest side. Always check to make certain that this is true for your solution.

Careful ▶ Never use the Pythagorean theorem with an oblique triangle. It is valid only for the sides of a right triangle. ◀

YOUR TURN

Solve the triangle for which $c = 6.0$ ft, $A = 52°$, and $C = 98°$. Round sides to the nearest tenth. $B = ?$, $a = ?$, $b = ?$

SOLUTION

$$B = 180° - 52° - 98° = 30°$$

$$\frac{a}{\sin A} = \frac{c}{\sin C} \quad \text{or} \quad a = \frac{c}{\sin C} \cdot \sin A = \frac{c \sin A}{\sin C}$$

$$= \frac{6 \sin 52°}{\sin 98°} \approx 4.8 \text{ ft}$$

$$\frac{b}{\sin B} = \frac{c}{\sin C} \quad \text{or} \quad b = \frac{c \sin B}{\sin C}$$

$$= \frac{6 \sin 30°}{\sin 98°} \approx 3.0 \text{ ft}$$

Case 2

In case 2, when two sides and the angle opposite one of these sides are known, the numbers may not always lead to a solution. For example, if $A = 50°$, $a = 6$, and $b = 10$, we may make a sketch of the triangle like this:

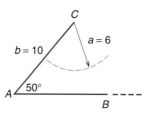

No matter how we draw side a it will not intersect line AB. There is no triangle with these measurements.

By looking at the given information before trying to solve the problem you can get an idea of what the solution triangle looks like. Let's examine all the possibilities.

- If $A = 90°$, then the triangle is a right triangle. Solve it using the methods of Section 10-3.
- If $A > 90°$ and $a > b$, then the triangle looks like this $\longrightarrow$
- If $A > 90°$ and $a \le b$, then there will be no solution, like this

- If $A < 90°$ and $a = b \sin A$, then the triangle is a right triangle. You can solve it using the methods of Section 10-3. $\longrightarrow$

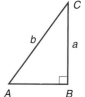

- If $A < 90°$ and $a \ge b$, then the triangle looks like this $\longrightarrow$

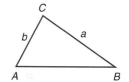

- If $A < 90°$ and $a < b$ and $a > b \sin A$, then the triangle looks either like this

 or like this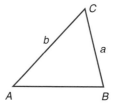

Because two different triangles may satisfy the given information, this is called the **ambiguous** case.

- If $A < 90°$ and $a < b \sin A$, there will be no solution.

Now, let's look at some examples.

EXAMPLE

Given $A = 30.0°$ and $b = 10.0$ ft, solve the triangle where (a) $a = 12.0$ ft, (b) $a = 5.00$ ft, (c) $a = 4.50$ ft, (d) $a = 8.00$ ft.

(a) **First,** note that $a > b$ so the triangle will look like this:

Make a quick sketch of the triangle.

Second, solve for angle B using the law of sines.

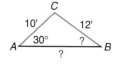

$$\frac{b}{\sin B} = \frac{a}{\sin A} \text{ or } a\sin B = b\sin A, \text{ then}$$

$$\sin B = \frac{b\sin A}{a} = \frac{10\,(\sin 30°)}{12} \qquad \text{Calculate } \sin B.$$

$$\sin B = 0.4166\ldots \qquad\qquad \text{Find angle } B \text{ using inverse sine.}$$

$$B \approx 24.62° \approx 24.6°$$

Note ▶ When using an intermediate answer in the next step of a problem, always carry at least one additional decimal place to avoid rounding errors. In this case the rounded answer is 24.6°, but we use 24.62° in performing the next calculation. ◀

Using a calculator, we obtain **10** $\times$ **30** (sin) $\div$ **12** (=) (INV) (sin) $\rightarrow$ **24.62431835**

Next, calculate angle C. $\qquad C = 180° - A - B \approx 180° - 30.00° - 24.62°$

$$C \approx 125.38° \approx 125.4°$$

Finally, use the law of sines to find side c.

$$\frac{c}{\sin C} = \frac{a}{\sin A} \quad \text{gives} \quad c = \frac{a\sin C}{\sin A} \approx \frac{12\,(\sin 125.38°)}{\sin 30°}$$

$$c \approx 19.6 \text{ ft}$$

(b) **First,** note that $a < b$. Then test to see if a is less than $b\sin A$. $b\sin A = 10\sin 30° = 5$. Since $a = b\sin A$, the triangle is a right triangle and there is only one possible solution. We can show that the triangle is a right triangle by using the law of sines:

$$\frac{b}{\sin B} = \frac{a}{\sin A} \quad \text{gives} \quad \sin B = \frac{b\sin A}{a} = \frac{10\sin 30°}{5}$$

$$\sin B = 1$$

$$B = 90°$$

Next, calculate angle C: $\qquad C = 180° - 30° - 90° = 60°$

Finally, calculate side c: $\qquad \cos A = \dfrac{c}{b}$

$$c = b\cos A = 10\cos 30°$$

$$c \approx 8.66 \text{ ft}$$

(c) Note that $a < b$, then test to see if a is less than $b\sin A$. $b\sin A = 5$, and $4.5 < 5$; therefore, there is *no* triangle that satisfies the given conditions.

We can check this by using the law of sines:

$$\sin B = \frac{b\sin A}{a} = \frac{10\sin 30°}{4.5} \approx 1.11$$

But the value of $\sin B$ can never be greater than 1; therefore, no such triangle exists. Using a calculator, we get

10 $\times$ **30** (sin) $\div$ **4.5** (=) (INV) (sin) $\rightarrow$ _____ **E**

The error message tells us that there is no solution.

(d) **First,** note that $a < b$. Because $8 > 5$, $a > b\sin A$, and we have the ambiguous case. There will be two possible triangles that satisfy these given conditions. Sketch them.

and

As given earlier, p. 555, $\sin A = \sin(180° - A)$. Therefore we know that $B_2 = 180° - B_1$.

Second, apply the law of sines to find angle B.

$$\frac{a}{\sin A} = \frac{b}{\sin B} \qquad \text{or} \qquad a\sin B = b\sin A$$

$$\sin B = \frac{b\sin A}{a} = \frac{10\sin 30°}{8}$$

$$\sin B = 0.625$$

$$B \approx 38.68° \approx 38.7°$$

The two possible angles are

$$B_1 \approx 38.68° \qquad \text{and} \qquad B_2 \approx 180° - 38.68° \approx 141.32° \approx 141.3°$$

- For the acute value of angle B,

$$C_1 \approx 180° - 30.00° - 38.68° \approx 111.32° \approx 111.3°$$

and from the law of sines,

$$\frac{c_1}{\sin 111.32°} = \frac{8.00}{\sin 30°} \qquad \text{gives} \qquad c_1 \approx 14.9 \text{ ft}$$

- For the obtuse value of angle B,

$$C_2 \approx 180° - 30.00° - 141.32° \approx 8.68° \approx 8.7°$$

and from the law of sines,

$$\frac{c_2}{\sin 8.68°} = \frac{8.00}{\sin 30°} \qquad \text{gives} \qquad c_2 \approx 2.41 \text{ ft}$$

Summary of Procedures for Case 2

Step 1 Check the given angle. Is it $=, <,$ or $> 90°$?

Step 2 If the angle is **equal to 90°,** the triangle is a right triangle.

Step 3 If the angle is **greater than 90°,** compare a and b. (a is the side opposite the given angle A.)

- If $a > b$,

 Use the law of sines to find this solution.

- If $a = b$, or $a < b$, there is no solution. Your calculator will display an error message when you try to calculate angle B.

Step 4 If the angle is **less than 90°,** compare a and b.

- If $a = b$ or $a > b$,

 Use the law of sines to find this solution.

- If $a < b$, compare a and $b\sin A$.
- If $a = b\sin A$, the triangle is a right triangle. Solve using either the law of sines or right triangle trigonometry.
- If $a > b\sin A$, there are two solutions.

After finding B_1, subtract from $180°$ to find B_2, then complete both solutions.

- If $a < b\sin A$, there is no solution.

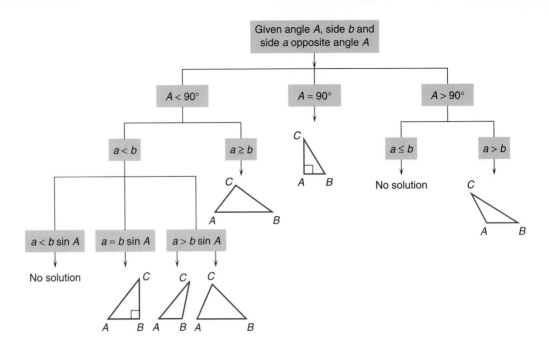

Given angle A, side b and side a opposite angle A

A < 90°   A = 90°   A > 90°

a < b   a ≥ b   a ≤ b   a > b

a < b sin A   a = b sin A   a > b sin A   No solution

No solution

EXAMPLE

If $A = 105.0°$ and $b = 125$ ft, solve the triangle where
      (a) $a = 95.0$ ft   (b) $a = 156$ ft.

(a) The given angle is greater than 90°, and since $a < b$, from **step 3,** we know there will be no solution. Check this using the law of sines:

$$\frac{a}{\sin A} = \frac{b}{\sin B} \qquad \text{gives} \qquad \sin B = \frac{b \sin A}{a} = \frac{125 \sin 105°}{95}$$

$$\sin B \approx 1.27$$

But this is not an acceptable value; $\sin B$ cannot be greater than 1.

(b) In this case $a > b$, and we expect from **step 3** to find a unique solution. Using the law of sines, we have

$$\sin B = \frac{b \sin A}{a} = \frac{125 \sin 105°}{156}$$

$$= 0.7739 \ldots$$

$$B \approx 50.71° \approx 50.7°$$

Then $C \approx 180° - 105.00° - 50.71° \approx 24.29° \approx 24.3°$

Use the law of sines to calculate side c:

$$\frac{c}{\sin C} = \frac{a}{\sin A} \qquad \text{gives} \qquad c = \frac{a \sin C}{\sin A} \approx \frac{156 \sin 24.29°}{\sin 105°}$$

$$c \approx 66.4 \text{ ft}$$

YOUR TURN

Use these steps and the law of sines to solve the triangle where $B = 36.0°$, $c = 16.4$ cm, and   (a) $b = 20.5$ cm,   (b) $b = 15.1$ cm. (Round angles to the nearest 0.1° and sides to three significant digits.)

(a) The given angle is less than 90°, so by **step 4** we need to compare the two sides. The side opposite the given angle is b, and b is greater than the other given side, c, so by **step 4** we know there is only one solution. Sketch it.

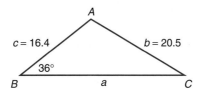

Now use the law of sines to solve for angle C.

$$\frac{c}{\sin C} = \frac{b}{\sin B} \qquad \text{gives} \qquad \sin C = \frac{c \sin B}{b} = \frac{16.4 \sin 36°}{20.5}$$

$$\sin C = 0.4702 \ldots$$

$$C \approx 28.049° \approx 28.0°$$

$$A \approx 180° - 28.05° - 36.00° \approx 115.95°$$

$$\frac{a}{\sin A} = \frac{b}{\sin B} \qquad \text{gives} \qquad a = \frac{b \sin A}{\sin B} \approx \frac{20.5 \sin 115.95°}{\sin 36°}$$

$$a \approx 31.4 \text{ cm}$$

(b) For this case $b < c$, so we need to compare b and $c \sin B$.

$$c \sin B = 16.4 \, (\sin 36°)$$

$$\approx 9.64$$

so $b > c \sin B$ and we know by **step 4** that there are two solutions.

$$\sin C = \frac{c \sin B}{b} = \frac{16.4 \sin 36°}{15.1}$$

$$\sin C = 0.6383 \ldots$$

$$C \approx 39.67° \approx 39.7°$$

The two possible solutions are

$$C_1 \approx 39.67° \qquad \text{and} \qquad C_2 \approx 180° - 39.67° \approx 140.33° \qquad \text{Sketch the triangles.}$$

$$\approx 140.3°$$

For the first value of angle C,

$$A_1 \approx 180° - 39.67° - 36.00° \approx 104.33° \approx 104.3°$$

and $\dfrac{a}{\sin A_1} = \dfrac{b}{\sin B}$ or $a \approx \dfrac{15.1 \sin 104.33°}{\sin 36°} \approx 24.9 \text{ cm}$

For the second value of $C, A_2 \approx 180° - 140.33° - 36.00° \approx 3.67°$ and

$$a \approx \frac{15.1 \sin 3.67°}{\sin 36°} \approx 1.64 \text{ cm}$$

Law of Cosines   In addition to the two cases already mentioned, it is also possible to solve any oblique triangle if the following information is given.

Case 3.   Two sides and the angle included by them.

Case 4.   All three sides.

But the law of sines is not sufficient for solving triangles given this information. We need an additional formula called the **law of cosines.**

To obtain this law, in the triangle shown construct the perpendicular from C to side AB. Using the Pythagorean theorem, in triangle BCD,

$$a^2 = (c - x)^2 + h^2$$
$$= c^2 - 2cx + x^2 + h^2$$

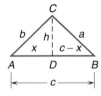

and in triangle ACD,

$$b^2 = x^2 + h^2$$

Therefore,

$$a^2 = c^2 - 2cx + b^2$$
$$= b^2 + c^2 - 2cx$$

In triangle ACD, $\cos A = \dfrac{x}{b}$ or $x = b \cos A$; therefore,

$$a^2 = b^2 + c^2 - 2bc \cos A$$

By rotating the triangle ABC, we can derive similar equations for b^2 and c^2.

LAW OF COSINES

$$a^2 = b^2 + c^2 - 2bc \cos A$$
$$b^2 = c^2 + a^2 - 2ca \cos B$$
$$c^2 = a^2 + b^2 - 2ab \cos C$$

Learning Hint ▶   Notice the similarity in these formulas. Rather than memorize all three formulas, you may find it easier to remember the following statement from which they may be written. ◀

LAW OF COSINES

The square of the length of any side of a triangle equals the sum of the squares of the lengths of the other two sides minus twice the product of the lengths of these two sides and the cosine of the angle between them.

The law of cosines may be used to solve any oblique triangle for which the information in case 3 and case 4 is given.

EXAMPLE

Solve the triangle with sides $a = 106$ m, $b = 135$ m, and $c = 165$ m.

First, use the law of cosines to find the largest angle, the angle opposite the longest side. Side c is the longest side, so find C.

$$c^2 = a^2 + b^2 - 2ab \cos C$$

Solve for $\cos C$.

$$c^2 - a^2 - b^2 = -2ab \cos C$$

or

$$2ab \cos C = a^2 + b^2 - c^2$$

$$\cos C = \frac{a^2 + b^2 - c^2}{2ab} = \frac{106^2 + 135^2 - 165^2}{2(106)(135)}$$

Using a calculator, we get

 106 $\boxed{x^2}$ $\boxed{+}$ **135** $\boxed{x^2}$ $\boxed{-}$ **165** $\boxed{x^2}$ $\boxed{=}$ $\boxed{\div}$ $\boxed{(}$ **2** $\boxed{\times}$ **106** $\boxed{\times}$ **135** $\boxed{)}$ $\boxed{=}$ $\boxed{\text{INV}}$ $\boxed{\cos}$

$$\rightarrow \quad \boxed{85.51907571}$$

$C \approx 85.5°$

Next, use the law of sines to find one of the remaining angles. To find angle A,

$$\frac{a}{\sin A} = \frac{c}{\sin C} \qquad \text{or} \qquad \sin A = \frac{a \sin C}{c}$$

These are both known

$$\approx \frac{106 \sin 85.52°}{165} \approx 0.6404\ldots$$

$$A \approx 39.8°$$

Finally, since the angles of a triangle must sum to 180°, we can calculate the remaining angle by subtracting.

$$B \approx 180° - 85.5° - 39.8° \approx 54.7°$$

Careful ▶ To avoid difficulties that can arise in the second step of the last example, you must find the largest angle first. Remember, the largest angle is always opposite the longest side. ◀

YOUR TURN

Now you try it. Solve the triangle with sides $a = 9.5$ in., $b = 4.2$ in., and $c = 6.4$ in.

SOLUTION

First, find the largest angle A using the law of cosines.

$$a^2 = b^2 + c^2 - 2bc \cos A$$

$$\cos A = \frac{b^2 + c^2 - a^2}{2bc} = \frac{4.2^2 + 6.4^2 - 9.5^2}{2(4.2)(6.4)}$$

$$A \approx 126.07° \approx 126.1°$$

Next, use the law of sines to find angle B.

$$\frac{b}{\sin B} = \frac{a}{\sin A} \qquad \text{gives} \qquad \sin B = \frac{b \sin A}{a} \approx \frac{4.2 \sin 126.07°}{9.5}$$

$$B \approx 20.9°$$

Finally, subtract to find angle C.

$$C \approx 180° - 126.1° - 20.9° \approx 33.0°$$

Next, let's solve a triangle given two sides and the included angle.

Solve the triangle with $a = 22.8$ cm, $b = 12.3$ cm, and $C = 42.0°$.

First, make a sketch and notice that the given angle C is the angle formed by the given sides a and b. Use the form of the law of cosines that involves the given angle to find the third side c.

$$c^2 = a^2 + b^2 - 2ab \cos C$$
$$= 22.8^2 + 12.3^2 - 2(22.8)(12.3) \cos 42°$$

Using a calculator, we get

22.8 $\boxed{x^2}$ $\boxed{+}$ **12.3** $\boxed{x^2}$ $\boxed{-}$ **2** $\boxed{\times}$ **22.8** $\boxed{\times}$ **12.3** $\boxed{\times}$ **42** $\boxed{\cos}$ $\boxed{=}$ $\boxed{\checkmark}$

$$\rightarrow \quad 15.94725463$$

$c \approx 15.9$ cm

Next, since we now know all three sides and one angle, we can use the law of cosines to find one of the remaining angles.

$$a^2 = b^2 + c^2 - 2bc \cos A$$

or $\quad \cos A = \dfrac{b^2 + c^2 - a^2}{2bc} \approx \dfrac{12.3^2 + 15.95^2 - 22.8^2}{2(12.3)(15.95)}$

Using a calculator gives us

12.3 $\boxed{x^2}$ $\boxed{+}$ **15.95** $\boxed{x^2}$ $\boxed{-}$ **22.8** $\boxed{x^2}$ $\boxed{=}$ $\boxed{\div}$ $\boxed{(}$ **2** $\boxed{\times}$ **12.3** $\boxed{\times}$ **15.95** $\boxed{)}$ $\boxed{=}$ $\boxed{\text{INV}}$ $\boxed{\cos}$

$$\rightarrow \quad 106.912924$$

$A \approx 106.9°$

Finally, find the remaining angle by subtracting.

$$B \approx 180° - 106.9° - 42.0° \approx 31.1°$$

Triangle figure: vertex C with angle $42°$; side 12.3 cm from C to A; side 22.8 cm from C to B; angle at A is $107.2°$, angle at B is $30.8°$; side $AB = 15.9$ cm.

Practice using the law of cosines by solving the triangle with $B = 34.4°$, $a = 145$ ft, and $c = 112$ ft.

Use the law of cosines to find the third side b.

$$b^2 = a^2 + c^2 - 2ac \cos B$$
$$= 145^2 + 112^2 - 2(145)(112) \cos 34.4°$$

$b \approx 82.28$ ft ≈ 82.3 ft

Now find angle A. $\quad \cos A = \dfrac{b^2 + c^2 - a^2}{2bc} \approx \dfrac{82.28^2 + 112^2 - 145^2}{2(82.28)(112)}$

$$A \approx 95.3°$$
$$C \approx 180° - 34.4° - 95.3° \approx 50.3°$$

The following table provides a summary showing when to use each law in solving oblique triangles. In each situation, remember that when two of the angles are known, the third may be found by subtracting from 180°.

When Given . . .	Use . . .	To Find . . .
Any two angles and a side	Law of sines	A second side opposite one of the given angles. You may need to find the third angle first.
	Law of sines	The third side, opposite the third angle.
Two sides and the angle opposite one of them	Law of sines	A second angle, opposite one of the given sides. This is the ambiguous case.
	Law of sines	The third side, after finding the third angle by subtraction.
Two sides and the angle included by them	Law of cosines	The third side.
	Law of cosines	Any one of the remaining angles.
All three sides	Law of cosines	The largest angle (opposite the longest side).
	Law of sines	One of the remaining angles.

Now turn to Exercises 10-4 for a set of problems on solving oblique triangles.

Exercises 10-4   Oblique Triangles

A. Solve each triangle.

1. $a = 6.5$ ft, $A = 43°$, $B = 62°$

2. $b = 17.2$ in., $C = 44.0°$, $B = 71.0°$

3. $b = 165$ m, $B = 31.0°$, $C = 110.0°$

4. $c = 2300$ yd, $C = 120°$, $B = 35°$

5. $a = 96$ in., $b = 58$ in., $B = 30°$

6. $b = 265$ ft, $c = 172$ ft, $C = 27.0°$

7. $a = 8.5$ m, $b = 6.2$ m, $C = 41°$

8. $b = 19.3$ m, $c = 28.7$ m, $A = 57.0°$

9. $a = 625$ ft, $c = 189$ ft, $B = 102.0°$

10. $b = 1150$ yd, $c = 3110$ yd, $A = 125.0°$

11. $a = 27.2$ in., $b = 33.4$ in., $c = 44.6$ in.

12. $a = 4.8$ cm, $b = 1.6$ cm, $c = 4.2$ cm

13. $a = 7.42$ m, $c = 5.96$ m, $B = 99.7°$

14. $b = 0.385$ in., $c = 0.612$ in., $A = 118.5°$

15. $a = 1.25$ cm, $b = 5.08$ cm, $c = 3.96$ cm

16. $a = 6.95$ ft, $b = 9.33$ ft, $c = 7.24$ ft

B. Practical Problems

1. **Landscape Design** To measure the height of a tree, Steve measures the angle of elevation of the treetop as 46°. He then moves 15 ft closer to the tree and from this new point measures the angle of elevation to be 59°. How tall is the tree?

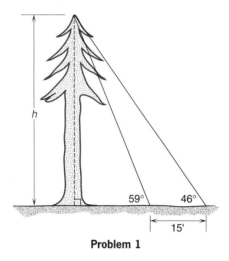

Problem 1

2. **Construction Technology** In the channel shown, angle $B = 122.0°$, angle $A = 27.0°$, and side $BC = 32.0$ ft. Find the length of the slope AB.

Problem 2

3. **Construction Technology** A triangular traffic island has sides 21.5, 46.2, and 37.1 ft. What are the angles at the corners?

4. **Surveying** The lot shown is split along a diagonal as indicated. What length of fencing is needed for the boundary line?

Problem 4

5. **Carpentry** Two sides of a sloped ceiling meet at an angle of 105.5°. If the distances along the sides to the opposite walls are 11.0 and 13.0 ft, what length of beam is needed to join the walls?

Problem 5

6. **Machine Technology** For the crankshaft shown, $AB = 4.2$ in., and the connecting rod $AC = 12.5$ in. Calculate the size of angle A when the angle at C is 12°.

Problem 6   **Problem 7**

7. **Carpentry** If the chord of a circle is 23.5 in. long and subtends a central angle of 55.0°, what is the radius of the circle?

Turn to the Appendix to check your answers to these problems. Then go to Problem Set 10 for a set of problems covering the work of this chapter.

Triangle Trigonometry

Answers are given in the Appendix.

A. Which of the following are right triangles?

1.

2.

3.

4.

5.

6.

7.

8.

9.

10.

B. Find each of the following trig values using a calculator. (Round to three decimal places.)

1. cos 67°
2. tan 81°
3. sin 4°
4. cos 63°10′

5. tan 35.75°
6. sin 29.2°
7. sin 107°
8. cos 123°

Use a calculator to find the acute value of x to the nearest minute.

9. sin x = 0.242
10. tan x = 1.54
11. sin x = 0.927

12. cos x = 0.309
13. tan x = 0.194
14. cos x = 0.549

15. tan x = 0.823
16. sin x = 0.672
17. cos x = 0.118

Find the acute value of x to the nearest tenth of a degree.

18. tan x = 0.506
19. cos x = 0.723
20. sin x = 0.488

21. sin x = 0.154
22. cos x = 0.273
23. tan x = 2.041

24. tan x = 0.338
25. cos x = 0.608
26. sin x = 0.772

Express in radians. (Round to three decimal places.)

27. 35°
28. 21.4°
29. 74°30′
30. 112.2°

Name _____

Date _____

Course/Section _____

Express in degrees. (Round to the nearest tenth.)

31. 0.45 rad   **32.** 1.7 rad   **33.** 2.3 rad   **34.** 0.84 rad

C. Find the missing dimensions in the right triangles as indicated. (Round all distances to one decimal place, and all angles to the nearest degree.)

1.

2.

3.

4.

$X =$ _____ $X =$ _____   ∡$a =$ _____ $X =$ _____

$Y =$ _____ $Y =$ _____ $X =$ _____ $Y =$ _____

Round all distances to the nearest tenth and all angles to one decimal place.

5.

6.

7.

8.

∡$a =$ _____ $X =$ _____   ∡$t =$ _____ $X =$ _____

$X =$ _____ $Y =$ _____ $X =$ _____ $Y =$ _____

D. Solve the following oblique triangles.

1. $a = 17.9$ in., $A = 65.0°$, $B = 39.0°$

2. $a = 721$ ft, $b = 444$ ft, $c = 293$ ft

3. $a = 260$ yd, $c = 340$ yd, $A = 37°$

4. $b = 87.5$ in., $c = 23.4$ in., $A = 118.5°$

5. $a = 51.4$ m, $b = 43.1$ m, $A = 64.3°$

6. $a = 166$ ft, $c = 259$ ft, $B = 47.0°$

7. $a = 1160$ m, $c = 2470$ m, $C = 116.2°$

8. $a = 7.6$ in., $b = 4.8$ in., $B = 30°$

E. Practical Problems

Round all angles to the nearest minute, and all distances to one decimal place, unless told otherwise.

1. Machine Technology   What height of gauge blocks is required to set an angle of 7°15′ for 10-in. plots?

Problem 1

Problem 2

2. The destination of an airplane is due north, and the plane flies at 220 mph. There is a crosswind of 20.0 mph from the west. What heading angle a should the plane take? What is its relative ground speed v? (Round a to the nearest minute and v to three significant digits.)

3. **Woodworking**   Six holes are spaced evenly around a 4-in.-diameter circle. Find the distance, d, between the holes.

Problem 3

(*Hint:* The dashed line creates two identical right triangles.)

Problem 4

4. **Machine Technology**   Find the angle of taper of the shaft in the figure.

5. A TV technician installs a 50-ft antenna on a flat roof. Safety regulations require a minimum angle of 30° between mast and guy wires.
   (a)   Find the minimum value of X.
   (b)   Find the length of the guy wires.

Problem 5

Problem 6

$\left(Hint: \dfrac{15}{100} = \dfrac{12}{x} \right)$

6. **Construction**   A bridge approach is 12 ft high. The maximum slope allowed is 15%—that is, $\frac{15}{100}$.
   (a)   What is the length of the approach?
   (b)   What is the angle a of the approach?

7. **Machine Technology**   The *helix angle* of a screw is the angle at which the thread is cut. The *lead* is the distance advanced by one turn of the screw. The circumference refers to the circumference of the screw as given by $C = \pi d$. The following formula applies:

Problem 7

$$\text{Tangent of helix angle} = \frac{\text{lead}}{\text{circumference}}$$

   (a)   Find the helix angle for a 2-in.-diameter screw if the lead is $\frac{1}{8}$ in.
   (b)   Find the lead of a 3-in.-diameter screw if the helix angle is 2°.

8. **Construction**   Find the lengths x and y of the beams shown in the bridge truss.

Problem 8

9. **Machine Technology**   Find the width X of the V-thread shown. (Round to the nearest thousandth.)

Problem 9

Problem 10

10. **Machine Technology**   Find the head angle a of the screw shown.

11. **Machine Technology** A machinist makes a cut 13.8 cm long in a piece of metal. Then, another cut is made that is 18.6 cm long and at an angle of 62°0′ with the first cut. How long a cut must be made to join the two endpoints?

12. **Drafting** Determine the center-to-center measurement x in the adjustment bracket shown in the figure. Round to the nearest hundredth.

Problem 12

Objective	Sample Problems		For Help Go to Page
When you finish this unit you will be able to:			
1. Solve a system of two linear equations in two variables.	(a) $5x + 2y = 20$	$x =$ _____	586
	$3x - 2y = 12$	$y =$ _____	
	(b) $x = 3y$	$x =$ _____	
	$2y - 4x = 30$	$y =$ _____	
	(c) $3x - 5y = 7$	$x =$ _____	
	$6x - 10y = 14$	$y =$ _____	
	(d) $2y = 5x - 7$	$x =$ _____	
	$10x - 4y = 9$	$y =$ _____	
2. Solve word problems involving systems of equations in two variables.	(a) The sum of two numbers is 35. Twice the smaller decreased by the larger is 10. Find them.	_____	601
	(b) The perimeter of a rectangular vent is 28 in. The length of the vent is 2 in. greater than its width. Find the dimensions.	_____	
	(c) A mixture of 800 drywall screws costs $84. If the mixture consists of one type costing 8 cents each and another type costing 12 cents each, how many of each kind are there?	_____	

_____
Name

_____
Date

_____
Course/Section

3. Solve quadratic equations.

   (a) $x^2 = 16$ $x =$ _____   606

   (b) $x^2 - 7x = 0$ $x =$ _____

   (c) $x^2 - 5x = 14$ $x =$ _____

   (d) $3x^2 + 2x - 16 = 0$ $x =$ _____

   (e) $2x^2 + 3x + 11 = 0$ $x =$ _____

4. Solve word problems involving quadratic equations.

   (a) Find the side length of a square opening whose area is 36 cm^2.   _____   615

   (b) Find the radius of a circular pipe whose cross-sectional area is 220 sq in.   _____

   (c) In the formula $P = RI^2$, find I (in amperes) if $P = 2500$ watts and $R = 15$ ohms.   _____

   (d) The length of a rectangular vent is 3 in. longer than its width. Find the dimensions if the cross-sectional area is 61.75 sq in.   _____

(Answers to these preview problems are given in the Appendix.)

If you are certain that you can work *all* these problems correctly, turn to page 621 for a set of practice problems. If you cannot work one or more of the preview problems, turn to the page indicated to the right of the problem. Those who wish to master this material with the greatest success should turn to Section 11-1 and begin work there.

Advanced Algebra

Reprinted by permission: Tribune Media Services

11-1 SYSTEMS OF EQUATIONS

This chapter is designed for students who are interested in highly technical occupations. We explain how to solve systems of linear equations and how to solve quadratic equations. Before beginning, you may want to return to Chapter 7 to review the basic algebraic operations explained there. When you have had the review you need, return here and continue.

Before you can solve a system of two equations in two unknowns, you must be able to solve a single linear equation in one unknown. Let's review what you learned in Chapter 7 about solving linear equations.

A *solution* to an equation such as $3x - 4 = 11$ is a number that we may use to replace x in order to make the equation a true statement. To find such a number, we *solve* the equation by changing it to an *equivalent* equation with only x on the left. The process goes like this:

> **EXAMPLE**

Solve: $3x - 4 = 11$

Step 1 Add 4 to both sides of the equation

$$3x \underbrace{- 4 + \boxed{4}}_{0} = \underbrace{11 + \boxed{4}}_{15}$$

$$3x = 15$$

Step 2 Divide both sides of the equation by 3.

$$\frac{3x}{3} = \frac{15}{3}$$

$$x = 5$$ This is the solution and you can check it by substituting 5 for x in the original equation.

☑ $3(5) - 4 = 11$
$15 - 4 = 11$ which is true

YOUR TURN

Solve each of the following equations and check your answer.

(a) $\dfrac{3x}{2} = 9$ (b) $17 - x = 12$

(c) $2x + 7 = 3$ (d) $3(2x + 5) = 4x + 17$

ANSWERS

(a) 6   (b) 5   (c) -2 (d) 1

A *system of equations* is a set of equations in two or more variables that may have a common solution.

EXAMPLE

The pair of equations

$$2x + y = 11$$
$$4y - x = 8$$

has the common solution $x = 4$, $y = 3$. This pair of numbers will make each equation a true statement. If we substitute 4 for x and 3 for y, the left side of the first equation becomes

$2(4) + 3 = 8 + 3$ or 11, which equals the right side,

and the left side of the second equation becomes

$4(3) - 4 = 12 - 4$ or 8, which equals the right side.

The pair of values $x = 4$ and $y = 3$ satisfies *both* equations. This solution is often written $(4, 3)$, where the x-value is listed first and the y-value is listed second.

Note ▶ Each of these equations by itself has an infinite number of solutions. For example, the pairs $(0, 11)$, $(1, 9)$, $(2, 7)$, and so on, all satisfy $2x + y = 11$, and the pairs $(0, 2)$, $\left(1, \frac{9}{4}\right)$, $\left(2, \frac{5}{2}\right)$, and so on, all satisfy $4y - x = 8$. However, $(4, 3)$ is the only pair that satisfies both equations. ◀

YOUR TURN

By substituting, show that the numbers $x = 2$, $y = -5$ give the solution to the pair of equations

$$5x - y = 15$$
$$x + 2y = -8$$

       Chap. 11 Advanced Algebra

The first equation is

$$5(2) - (-5) = 15$$
$$10 + 5 = 15, \text{ which is correct.}$$

The second equation is

$$(2) + 2(-5) = -8$$
$$2 - 10 = -8, \text{ which is also correct.}$$

Solution by Substitution

In this chapter we will show you two different methods of solving a system of two linear equations in two unknowns. The first method is called the method of *substitution*.

EXAMPLE

To solve the pair of equations

$$y = 3 - x$$
$$3x + y = 11$$

follow these steps.

Step 1 *Solve* the first equation for x or y.

The first equation is already solved for y:

$$y = 3 - x$$

Step 2 *Substitute* this expression for y in the second equation.

$$3x + y = 11 \quad \text{becomes}$$

$$3x + \boxed{(3 - x)} = 11$$

Substitute this for y

Step 3 *Solve* the resulting equation.

$$3x + (3 - x) = 11 \quad \text{becomes}$$
$$2x + 3 = 11 \quad \text{Subtract 3 from each side.}$$
$$2x = 8 \quad \text{Divide each side by 2.}$$
$$x = 4$$

Step 4 *Substitute* this value of x into the first equation and find a value for y.

$$y = 3 - x$$
$$y = 3 - (4) \qquad \boxed{x \text{ value}}$$
$$y = -1$$

The solution is $x = 4, y = -1$ or $(4, -1)$.

$\boxed{y \text{ value}}$

✓ **Step 5** *Check* your solution by substituting it back into the second equation.

$$3x + y = 11 \quad \text{becomes}$$
$$3(4) + (-1) = 11$$
$$12 - 1 = 11 \quad \text{which is correct}$$

Try it. Use this substitution procedure to solve the system of equations

$x - 2y = 3$
$2x - 3y = 7$

SOLUTION

Step 1 *Solve* the first equation for x by adding $2y$ to both sides of the equation.

$x - 2y + \boxed{2y} = 3 + \boxed{2y}$

$x = 3 + 2y$

Step 2 *Substitute* this expression for x in the second equation.

$2x - 3y = 7$ becomes

$2\boxed{(3 + 2y)} - 3y = 7$

Step 3 *Solve:*

$6 + 4y - 3y = 7$ Simplify by combining the y-terms.
$6 + y = 7$ Subtract 6 from each side.
$y = 1$

Step 4 *Substitute* this value of y in the first equation to find x.

$x - 2y = 3$ becomes
$x - 2(1) = 3$
$x - 2 = 3$ Add 2 to each side.
$x = 5$

The solution is $x = 5$, $y = 1$ or $(5, 1)$.

☑ **Step 5** *Check* the solution in the second equation.

$2x - 3y = 7$ becomes
$2(5) - 3(1) = 7$
$10 - 3 = 7$ which is correct.

Of course, it does not matter which variable, x or y, we solve for in **Step 1,** or which equation we use in **Step 4.**

EXAMPLE

In the pair of equations

$2x + 3y = 22$

$x - y = 1$

the simplest procedure is to solve the *second* equation for x to get

$x = 1 + y$ and substitute this expression for x into the first equation.

The entire solution looks like this:

Step 1 In the second equation, add y to both sides to get $x = 1 + y$.

Step 2 When we substitute into the first equation,

$2x + 3y = 22$ becomes

$2\boxed{(1 + y)} + 3y = 22$

Step 3 *Solve:*

$$2 + 2y + 3y = 22 \qquad \text{Combine terms.}$$
$$2 + 5y = 22 \qquad \text{Subtract 2 from each side.}$$
$$5y = 20 \qquad \text{Divide each side by 5.}$$
$$y = 4$$

Step 4 *Substitute* 4 for y in the second equation.

$$x - y = 1 \qquad \text{becomes}$$
$$x - (4) = 1$$

or

$$x = 5 \qquad \text{The solution is } x = 5, y = 4.$$

☑ **Step 5** *Check* the solution by substituting these values into the first equation.

$$2x + 3y = 22 \qquad \text{becomes}$$
$$2(5) + 3(4) = 22$$
$$10 + 12 = 22 \qquad \text{which is correct.}$$

PROBLEMS

Solve the following systems of equations by using the substitution method.

(a) $x = 1 + y$
$2y + x = 7$

(b) $3x + y = 1$
$y + 5x = 9$

(c) $x - 3y = 4$
$3y + 2x = -1$

(d) $y + 2x = 1$
$3y + 5x = 1$

(e) $x - y = 2$
$y + x = 1$

(f) $y = 4x$
$2y - 6x = 0$

SOLUTIONS

(a) **Step 1** The first equation is already solved for x.

Step 2 Substitute this expression for x into the second equation.

$$2y + \boxed{1 + y} = 7$$

Step 3 Solve for y.

$$3y + 1 = 7$$
$$3y = 6$$
$$y = 2$$

Step 4 Substitute 2 for y in the first equation.

$$x = 1 + (2)$$
or $x = 3$

Step 5 The solution is $x = 3$, $y = 2$.
Be sure to check it.

(b) **Step 1** You can easily solve either equation for y. Using the first equation, subtract $3x$ from both sides to obtain

$$y = 1 - 3x$$

Step 2 Substitute this expression for y into the second equation.

$$\boxed{1 - 3x} + 5x = 9$$

Step 3 Solve for x.

$$1 + 2x = 9$$
$$2x = 8$$
$$x = 4$$

Step 4 Substitute 4 for x in the equation from **step 1**.

$$y = 1 - 3(4)$$
$$y = 1 - 12 = -11$$

Step 5 The solution is $x = 4$, $y = -11$. Check your answer.

(c) **Step 1** Use the first equation to solve for x.

$$x = 3y + 4$$

Step 2 Substitute this expression for x into the second equation.

$$3y + 2 \,\boxed{(3y + 4)}\, = -1$$

Step 3 Solve for y.

$$3y + 6y + 8 = -1$$
$$9y + 8 = -1$$
$$9y = -9$$
$$y = -1$$

Step 4 Substitute -1 for y in the equation from **step 1**.

$$x = 3(-1) + 4$$
$$x = 1$$

Step 5 The solution is $x = 1$, $y = -1$.

(d) $x = 2$, $y = -3$ (e) $x = 1\frac{1}{2}$, $y = -\frac{1}{2}$ (f) $x = 0$, $y = 0$

Dependent and Inconsistent Systems

So far we have looked only at systems of equations with a single solution—one pair of numbers. Such a system of equations is called a *consistent* and *independent* system. However, it is possible for a system of equations to have no solution at all or to have very many solutions.

> **EXAMPLE**

The system of equations

$$y + 3x = 5$$
$$2y + 6x = 10$$

has *no unique* solution. If we solve for y in the first equation

$$y = 5 - 3x$$

and substitute this expression into the second equation

$$2y + 6x = 10$$

or $\quad 2\,\boxed{(5 - 3x)}\, + 6x = 10$

This resulting equation simplifies to

$$10 = 10$$

There is no way of solving to get a unique value of x or y.

A system of equations that does not have a single unique number-pair solution but has an unlimited number of solutions is said to be *dependent*. The two equations are essentially the same. For the system shown in this example, the second equation is simply twice the first equation. There are infinitely many pairs of numbers that will satisfy this pair of equations. For example,

$x = 0, y = 5$
$x = 1, y = 2$
$x = 2, y = -1$
$x = 3, y = -4$

and so on.

If a system of equations is such that our efforts to solve it produce a false statement, the equations are said to be *inconsistent*. The system of equations has no solution.

EXAMPLE

The pair of equations

$y - 1 = 2x$
$2y - 4x = 7$ is inconsistent.

If we solve the first equation for y

$y = 2x + 1$

and substitute this expression into the second equation,

$$2y - 4x = 7 \quad \text{becomes}$$

$$2 \boxed{(2x + 1)} - 4x = 7$$

$$4x + 2 - 4x = 7$$

or $2 = 7$ which is false.

All the variables have dropped out of the equation, and we are left with an incorrect statement. The original pair of equations is said to be inconsistent, and it has no solution.

YOUR TURN

Try solving the following systems of equations.

(a) $2x - y = 5$
$\quad\;\; 2y - 4x = 3$

(b) $3x - y = 5$
$\quad\;\; 6x - 10 = 2y$

SOLUTIONS

(a)   Solve the first equation for y.

$y = 2x - 5$

Substitute this into the second equation.

$$2y - 4x = 3 \quad \text{becomes}$$

$$2 \boxed{(2x - 5)} - 4x = 3$$

$$4x - 10 - 4x = 3$$

$$-10 = 3 \quad \text{This is impossible. There is no solution for this system of equations. The equations are inconsistent.}$$

(b) Solve the first equation for y.

$$y = 3x - 5$$

Substitute this into the second equation.

$6x - 10 = 2y$ becomes

$6x - 10 = 2\ \boxed{(3x - 5)}$

$6x - 10 = 6x - 10$

$0 = 0$ This is true, but all of the variables have dropped out and we cannot get a single unique solution. The equations are dependent.

Solution by Elimination

The second method for solving a system of equations is called the *elimination method* or the *addition method*. When it is difficult or "messy" to solve one of the two equations for either x or y, the method of elimination may be the simplest way to solve the system of equations.

EXAMPLE

In the system of equations

$2x + 3y = 7$
$4x - 3y = 5$

neither equation can be solved for x or y without introducing fractions that are difficult to work with. But we can simply add these equations, and the y-terms will be eliminated.

$$\begin{array}{l} 2x + 3y = 7 \\ \underline{4x - 3y = 5} \quad \text{Add like terms.} \\ 6x + 0 \ = 12 \\ 6x = 12 \\ x = 2 \end{array}$$

Now substitute this value of x back into either one of the original equations to obtain a value for y.

The first equation

$2x + 3y = 7$ becomes
$2(2) + 3y = 7$

$4 + 3y = 7$ Subtract 4 from each side.
$3y = 3$ Divide each side by 3.
$y = 1$

The solution is $x = 2$, $y = 1$, or $(2, 1)$.

☑ Check the solution by substituting it back into the second equation.

$4x - 3y = 5$
$4(2) - 3(1) = 5$
$8 - 3 = 5$ which is correct.

YOUR TURN

Try it. Solve the following system of equations by adding.

$2x - y = 3$
$y + x = 9$

$2x - y = 3$
$y + x = 9$

becomes

$2x - y = 3$
$x + y = 9$ by rearranging the terms in the second equation.

Add to get

$3x + 0 = 12$
$\quad 3x = 12$
$\quad\ \ x = 4$

Substitute 4 for x in the first equation.

$\quad\ 2x - y = 3$
$2(4) - y = 3$ Simplify.
$\quad 8 - y = 3$ Subtract 8 from each side.
$\quad\ \ -y = 3 - 8$ Combine terms.
$\quad\ \ -y = -5$ Multiply both sides by -1.
$\quad\ \ \ \ y = 5$

The solution is $x = 4$, $y = 5$, or (4, 5). Check it.

Careful ▶ It is important to rearrange the terms in the equations so that the x and y terms appear in the same order in both equations.

PROBLEMS

Solve the following systems of equations by this process of elimination.

(a) $x + 5y = 17$ (b) $x + y = 16$ (c) $3x - y = -5$ (d) $\frac{1}{2}x + 2y = 10$
      $-x + 3y = 7$ $x - y = 4$ $y - 5x = 9$ $y + 1 = \frac{1}{2}x$

SOLUTIONS

(a) **Solve:**

$\quad\ x + 5y = 17$

$\underline{-x + 3y = 7}$

$\quad 0 + 8y = 24$ adding like terms
$\qquad\quad 8y = 24$
$\qquad\quad\ \ y = 3$

Substitute 3 for y in the first equation.

$$x + 5(3) = 17$$
$$x + 15 = 17$$
$$x = 2 \qquad \text{This solution is } x = 2, y = 3, \text{ or } (2, 3).$$

☑
$$-x + 3y = 7$$
$$-(2) + 3(3) = 7$$
$$-2 + 9 = 7 \qquad \text{which is correct.}$$

(b) **Solve:**

$$x + y = 16$$
$$\underline{x - y = 4}$$
$$2x + 0 = 20 \qquad \text{adding like terms}$$
$$2x = 20$$
$$x = 10$$

Substitute 10 for x in the first equation.

$$(10) + y = 16$$
$$y = 6 \qquad \text{The solution is } x = 10, y = 6, \text{ or } (10, 6).$$

☑
$$x - y = 4$$
$$(10) - (6) = 4$$
$$10 - 6 = 4 \qquad \text{which is correct.}$$

(c) **Solve:**

$$3x - y = -5$$
$$y - 5x = 9$$

Rearrange the order of the terms in the second equation.

$$3x - y = -5$$
$$\underline{-5x + y = 9}$$
$$-2x + 0 = 4 \qquad \text{adding like terms}$$
$$-2x = 4$$
$$x = -2$$

Substitute -2 for x in the first equation.

$$3(-2) - y = -5$$
$$-6 - y = -5$$
$$-y = -5 + 6 = 1$$
$$y = -1 \qquad \text{The solution is } x = -2, y = -1, \text{ or } (-2, -1).$$

☑
$$y - 5x = 9$$
$$(-1) - 5(-2) = 9$$
$$-1 + 10 = 9 \qquad \text{which is correct.}$$

(d) **Solve:**

$$\tfrac{1}{2}x + 2y = 10$$
$$y + 1 = \tfrac{1}{2}x$$

Rearrange terms in the second equation so that they are in the same order as in the first equation. Subtract 1 from both sides, then subtract $\tfrac{1}{2}x$ from both sides.

$$\tfrac{1}{2}x + 2y = 10$$
$$\underline{-\tfrac{1}{2}x + y = -1}$$
$$0 + 3y = 9 \qquad \text{adding like terms}$$
$$3y = 9$$
$$y = 3$$

Substitute 3 for y in the first equation.

$$\tfrac{1}{2}x + 2(3) = 10$$
$$\tfrac{1}{2}x + 6 = 10$$
$$\tfrac{1}{2}x = 4$$
$$x = 8 \qquad \text{The solution is } x = 8, y = 3, \text{ or } (8, 3).$$

$$y + 1 = \tfrac{1}{2}x$$
$$(3) + 1 = \tfrac{1}{2}(8)$$
$$3 + 1 = 4 \qquad \text{which is correct.}$$

Multiplication with the Elimination Method

With some systems of equations neither x nor y can be eliminated by simply adding like terms.

EXAMPLE

In the system

$$3x + y = 17$$
$$x + y = 7$$

adding like terms will not eliminate either variable. To solve this system of equations, multiply all terms of the second equation by -1 so that

$$\boxed{x + y = 7} \quad \boxed{\text{becomes}} \Longrightarrow \quad \boxed{-x - y = -7}$$

and the system of equations becomes

$$3x + y = 17$$
$$-x - y = -7$$

The system of equations may now be solved by adding the like terms as before.

$$\begin{array}{l} 3x + y = 17 \\ \underline{-x - y = -7} \\ 2x + 0 = 10 \qquad \text{adding like terms} \\ 2x = 10 \\ x = 5 \end{array}$$

Substitute 5 for x in the first equation.

$$3(5) + y = 17$$
$$15 + y = 17$$
$$y = 17 - 15 = 2 \qquad \text{The solution is } x = 5, y = 2, \text{ or } (5, 2).$$

$$x + y = 7$$
$$(5) + (2) = 7$$
$$5 + 2 = 7 \qquad \text{which is correct.}$$

Careful ▶ When you multiply an equation by some number, be careful to multiply *all* terms on *both* sides by that number. It is very easy to forget to multiply on the right side. ◀

Use this "multiply and add" procedure to solve the following system of equations.

$2x + 7y = 29$
$2x + y = 11$

$2x + 7y = 29$
$2x + y = 11$

Multiply all terms in the second equation by -1.

$$\begin{array}{r} 2x + 7y = 29 \\ -2x - y = -11 \\ \hline 0 + 6y = 18 \\ 6y = 18 \\ y = 3 \end{array} \quad \text{adding like terms}$$

Substitute 3 for y in the first equation.

$2x + 7(3) = 29$
$2x + 21 = 29$
$2x = 8$
$x = 4 \qquad$ The solution is $x = 4$, $y = 3$, or $(4, 3)$.

Check the solution by substituting in the original equations.

$$\begin{array}{ll} 2x + y = 11 & 2x + 7y = 29 \\ 2(4) + (3) = 11 & 2(4) + 7(3) = 29 \\ 8 + 3 = 11 & 8 + 21 = 29 \end{array}$$

Solving by the multiply and add procedure may involve multiplying by constants other than -1, of course.

EXAMPLE

Let's use this method to solve the following system of equations.

$7x + 4y = 25$
$3x - 2y = 7$

First, look at these equations carefully. Notice that the y-terms can be eliminated easily if we multiply all terms in the second equation by 2.

The y column $\boxed{\begin{array}{c} +4y \\ -2y \end{array}}$ becomes $\Rightarrow$ $\boxed{\begin{array}{c} +4y \\ -4y \end{array}}$ when we multiply by 2.

$$\text{sum} = 0$$

The second equation becomes $2\,(3x) - 2\,(2y) = 2\,(7)$

or $ 6x - 4y = 14$

and the system of equations is converted to the equivalent system

$$\begin{array}{r} 7x + 4y = 25 \\ 6x - 4y = 14 \\ \hline 13x + 0 = 39 \\ 13x = 39 \\ x = 3 \end{array} \quad \text{adding like terms}$$

Substitute 3 for x in the first equation.

$$7(3) + 4y = 25$$
$$21 + 4y = 25$$
$$4y = 4$$
$$y = 1 \qquad \text{The solution is } x = 3, y = 1, \text{ or } (3, 1).$$

Check the solution by substituting it back into the original equations.

PROBLEMS

Try these problems to make certain you understand this procedure.

(a) $3x - 2y = 14$
    $5x - 2y = 22$

(b) $6x + 5y = 14$
    $-2x + 3y = -14$

(c) $5y - x = 1$
    $2x + 3y = 11$

(d) $-x - 2y = 1$
    $19 - 2x = -3y$

SOLUTIONS

(a) **Solve:**

$$3x - 2y = 14$$
$$5x - 2y = 22$$

Multiply each term in the first equation by -1.

$$\boxed{(-1)} (3x) - \boxed{(-1)} (2y) = \boxed{(-1)} (14)$$
$$-3x + 2y = -14$$

The system of equations is therefore

$$\begin{aligned} -3x + 2y &= -14 \\ \underline{5x - 2y} &= \underline{22} \\ 2x + 0 &= 8 \qquad \text{adding like terms} \\ x &= 4 \end{aligned}$$

Substitute 4 for x in the first equation.

$$3(4) - 2y = 14$$
$$12 - 2y = 14$$
$$-2y = 2$$
$$-y = 1$$
$$y = -1 \qquad \text{The solution is } x = 4, y = -1, \text{ or } (4, -1).$$

☑
$$\begin{aligned} 3x - 2y &= 14 & 5x - 2y &= 22 \\ 3(4) - 2(-1) &= 14 & 5(4) - 2(-1) &= 22 \\ 12 + 2 &= 14 & 20 + 2 &= 22 \\ 14 &= 14 & 22 &= 22 \end{aligned}$$

(b) **Solve:**

$$6x + 5y = 14$$
$$-2x + 3y = -14$$

Multiply each term in the second equation by 3.

$$\boxed{(3)} (-2x) + \boxed{(3)} (3y) = \boxed{(3)} (-14)$$
$$-6x + 9y = -42$$

The system of equations is now

$$6x + 5y = 14$$
$$\underline{-6x + 9y = -42}$$
$$0 + 14y = -28 \quad \text{adding like terms}$$
$$y = -2$$

Substitute -2 for y in the first equation.

$$6x + 5(-2) = 14$$
$$6x - 10 = 14$$
$$6x = 24$$
$$x = 4 \quad \text{The solution is } x = 4, y = -2, \text{ or } (4, -2).$$

Be certain to check your solution.

(c) **Solve:**

$$5y - x = 1$$
$$2x + 3y = 11$$

Rearrange to put the terms in the first equation in the same order as they are in the second equation.

$$-x + 5y = 1$$
$$2x + 3y = 11$$

Multiply each term in the first equation by 2.

$$-2x + 10y = 2$$
$$\underline{2x + 3y = 11}$$
$$0 + 13y = 13 \quad \text{adding like terms}$$
$$13y = 13$$
$$y = 1$$

Substitute 1 for y in the first equation.

$$5(1) - x = 1$$
$$5 - x = 1$$
$$-x = 1 - 5 = -4$$
$$x = 4 \quad \text{The solution is } x = 4, y = 1, \text{ or } (4, 1).$$

Check it.

(d) **Solve:**

$$-x - 2y = 1$$
$$19 - 2x = -3y$$

Rearrange the terms in the second equation in the same order as they are in the first equation.

$$-x - 2y = 1$$
$$-2x + 3y = -19$$

Multiply each term in the first equation by -2.

$$\boxed{(-2)}\,(-x) - \boxed{(-2)}\,(2y) = \boxed{(-2)}\,(1)$$
$$2x + 4y = -2$$

The system of equations is now

$$2x + 4y = -2$$
$$\underline{-2x + 3y = -19}$$
$$0 + 7y = -21 \quad \text{adding like terms}$$
$$7y = -21$$
$$y = -3$$

Substitute -3 for y in the first equation of the original problem.

$$-x - 2(-3) = 1$$
$$-x + 6 = 1$$
$$-x = 1 - 6 = -5$$
$$x = 5 \qquad \text{The solution is } x = 5, y = -3, \text{ or } (5, -3).$$

Check your solution.

Multiplying Both Equations

If you examine the system of equations

$$3x + 2y = 7$$
$$4x - 3y = -2$$

you will find that there is no single integer we can use as a multiplier that will allow us to eliminate one of the variables when the equations are added. Instead we must convert each equation separately to an equivalent equation, so that when the new equations are added one of the variables is eliminated.

EXAMPLE

With the system of equations given, if we wish to eliminate the y variable, we must multiply the first equation by 3 and the second equation by 2.

First equation: $\boxed{3x + 2y = 7}$ $\boxed{\text{Multiply by 3}}$ $\Rightarrow$ $\boxed{9x + 6y = 21}$

Second equation: $\boxed{4x - 3y = -2}$ $\boxed{\text{Multiply by 2}}$ $\Rightarrow$ $\boxed{8x - 6y = -4}$

The new system of equations is

$$\begin{array}{l} 9x + 6y = 21 \\ \underline{8x - 6y = -4} \\ 17x + 0 = 17 \qquad \text{adding like terms} \\ x = 1 \end{array}$$

Substitute 1 for x in the original first equation.

$$3x + 2y = 7$$
$$3(1) + 2y = 7$$
$$3 + 2y = 7$$
$$2y = 4$$
$$y = 2 \qquad \text{The solution is } x = 1, y = 2, \text{ or } (1, 2).$$

Check this solution by substituting it back into the original pair of equations.

$\checkmark$

$3x + 2y = 7$	$4x - 3y = -2$
$3(1) + 2(2) = 7$	$4(1) - 3(2) = -2$
$3 + 4 = 7$	$4 - 6 = -2$
$7 = 7$	$-2 = -2$

YOUR TURN

Use this same procedure to solve the following system of equations:

$$2x - 5y = 9$$
$$3x + 4y = 2$$

We can eliminate x from the two equations as follows:

First equation: $\boxed{2x - 5y = 9}$ $\boxed{\text{Multiply by } -3}$ ⟹ $\boxed{-6x + 15y = -27}$

Second equation: $\boxed{3x + 4y = 2}$ $\boxed{\text{Multiply by } 2}$ ⟹ $\boxed{6x + 8y = 4}$

The new system of equations is:

$$\begin{array}{l} -6x + 15y = -27 \\ \underline{6x + 8y = 4} \\ 0 + 23y = -23 \qquad \text{adding like terms} \\ y = -1 \end{array}$$

Substitute -1 for y in the first original equation.

$$\begin{array}{rcl} 2x - 5y &=& 9 \\ 2x - 5(-1) &=& 9 \\ 2x + 5 &=& 9 \\ 2x &=& 4 \\ x &=& 2 \end{array}$$

The solution is $x = 2$, $y = -1$, or $(2, -1)$.

Note ▶ Of course, we could have chosen to eliminate the y-variable and we would have arrived at the same solution. Try it. ◀

PROBLEMS

When you are ready to continue, practice your new skills by solving the following systems of equations.

(a) $\begin{aligned} 2x + 2y &= 4 \\ 5x + 7y &= 18 \end{aligned}$ 　　　(b) $\begin{aligned} 5x + 2y &= 11 \\ 6x - 3y &= 24 \end{aligned}$

(c) $\begin{aligned} 3x + 2y &= 10 \\ 2x &= 5y - 25 \end{aligned}$ 　　　(d) $\begin{aligned} -7x - 13 &= 2y \\ 3y + 4x &= 0 \end{aligned}$

ANSWERS

(a) $x = -2$, $y = 4$ 　　　(b) $x = 3$, $y = -2$

(c) $x = 0$, $y = 5$ 　　　(d) $x = -3$, $y = 4$

In (a) multiply the first equation by -5 and the second equation by 2.

In (b) multiply the first equation by 3 and the second equation by 2.

In (c) rearrange the terms of the second equation, then multiply the first equation by -2 and the second equation by 3.

In (d) rearrange the terms of the first equation to agree with the second equation, then multiply the first equation by 3 and the second equation by 2.

Word Problems In many practical situations not only must you be able to solve a system of equations, you must also be able to write the equations in the first place. You must be able to set up and solve word problems. In Chapter 7 we listed some "signal words" and showed how to translate English sentences and phrases to mathematical equations and expressions. If you need to review the material on word problems in Chapter 7, turn to page 389 now; otherwise, continue here.

EXAMPLE

Translate the following sentence into *two* equations.

The sum of two numbers is 26 and their difference is 2. (Let x and y represent the two numbers.)

The $\underbrace{\text{sum of two numbers}}_{\downarrow}$ $\underset{\downarrow\,\downarrow}{\text{is 26}}$

$$x + y \qquad = 26$$

$\ldots$ $\underbrace{\text{their difference}}_{\downarrow}$ $\underset{\downarrow\,\downarrow}{\text{is 2.}}$

$$x - y \qquad = 2$$

The two equations are

$x + y = 26$
$x - y = 2$

We can solve this pair of equations using the elimination method.

The solution is $x = 14$, $y = 12$. Check the solution by seeing if it fits the original problem. The sum of these numbers is 26 ($14 + 12 = 26$) and their difference is 2 ($14 - 12 = 2$).

YOUR TURN

Ready for another word problem? Translate and solve this one:

The difference of two numbers is 14, and the larger number is three more than twice the smaller number.

SOLUTION

The first phrase in the sentence should be translated as

"The $\underbrace{\text{difference of two numbers}}_{\downarrow}$ $\underset{\downarrow\,\downarrow}{\text{is 14 }\ldots}$"   The larger number is L; the
$$L - S \qquad = 14$$ smaller number is S; the
difference must be $L - S$.

and the second phrase should be translated as

"$\ldots$ the larger number is three more than twice the smaller $\ldots$"
$$\quad\;\; L \quad = \;\; 3 \quad + \qquad 2S$$

The system of equations is

$L - S = 14$
$L = 3 + 2S$

To solve this system of equations, substitute the value of L from the second equation into the first equation. Then the first equation becomes

$(3 + 2S) - S = 14$

or $3 + S = 14$
$ S = 11$

Now substitute this value of S into the first equation to find L.

$L - (11) = 14$
$ L = 25$ The solution is $L = 25$, $S = 11$.

Check the solution by substituting it back into both of the original equations. Never neglect to check your answer.

PROBLEMS

Translating word problems into systems of equations is a very valuable and very practical algebra skill. Translate each of the following problems into a system of equations, then solve.

(a) **Machine Technology** The total value of an order of nuts and bolts is $1.40. The nuts cost 5 cents each and the bolts cost 10 cents each. If the number of bolts is four more than twice the number of nuts, how many of each are there? (*Hint:* Keep all money values in cents to avoid decimals.)

(b) **Landscaping** The perimeter of a rectangular lot is 350 ft. The length of the lot is 10 ft more than twice the width. Find the dimensions of the lot.

(c) **Construction** A materials yard wishes to make a 900-cu ft mixture of two different types of rock. One type of rock costs $6.80 per cubic foot and the other costs $8.50 per cubic foot. If the cost of the mixture is to run $7200, how many cubic feet of each should go into the mixture? (Round to the nearest tenth.)

(d) **Medical Technology** A lab technician wishes to mix a 5% salt solution and a 15% salt solution to obtain 4 liters of a 12% salt solution. How many liters of each solution must be added?

SOLUTIONS

(a) In problems of this kind it is often very helpful to first set up a table:

Item	Number of Items	Cost per Item	Total Cost
Nuts	N	5	$5N$
Bolts	B	10	$10B$

We can write the first equation as

"The total value of an order of nuts and bolts is 140 cents."

Chap. 11 Advanced Algebra

The second equation would be

The system of equations to be solved is

$5N + 10B = 140$
$B = 4 + 2N$

Use substitution. Since B is equal to $4 + 2N$, replace B in the first equation with $4 + 2N$.

$5N + 10 \boxed{(4 + 2N)} = 140$
$5N + 40 + 20N = 140$
$25N + 40 = 140$ collecting like terms
$25N = 100$
$N = 4$

When we replace N with 4 in the second equation,

$B = 4 + 2(4)$
$B = 4 + 8$
$B = 12$

There are 4 nuts and 12 bolts.

☑
$$5N + 10B = 140 \qquad B = 4 + 2N$$
$$5(4) + 10(12) = 140 \qquad 12 = 4 + 2(4)$$
$$20 + 120 = 140 \qquad 12 = 4 + 8$$
$$140 = 140 \qquad 12 = 12$$

(b) Recalling the formula for the perimeter of a rectangle, we have

The second sentence gives us

"The length of the lot is 10 ft more than twice the width."

$$L \qquad = 10 \qquad + \qquad 2W$$

The system of equations to be solved is

$2L + 2W = 350$
$L = 10 + 2W$

Using substitution, we replace L with $10 + 2W$ in the first equation.

$2 \boxed{(10 + 2W)} + 2W = 350$
$20 + 4W + 2W = 350$
$20 + 6W = 350$
$6W = 330$
$W = 55 \text{ ft}$

Now substitute 55 for W in the second equation:

$L = 10 + 2(55)$
$L = 10 + 110$
$L = 120 \text{ ft}$

The lot is 120 ft long and 55 ft wide.

☑

$2L + 2W = 350$	$L = 10 + 2W$
$2(120) + 2(55) = 350$	$120 = 10 + 2(55)$
$240 + 110 = 350$	$120 = 10 + 110$
$350 = 350$	$120 = 120$

(c) First set up the following table:

Item	Amount (cu ft)	Cost per cu ft	Total Cost
Cheaper rock	x	$6.80	$6.80x$
More expensive rock	y	$8.50	$8.50y$

The first equation comes from the statement:

"... a 900 cubic foot mixture of two different types of rock."

$$900 \qquad = \qquad x + y$$

Consulting the table, we write the second equation as follows:

"... the cost of the mixture is to run $7200."

$$6.80x + 8.50y \qquad = \qquad 7200$$

Multiply this last equation by 10 to get the system of equations

$$x + \ y = 900$$
$$68x + 85y = 72{,}000$$

Multiply the first equation by -68 and add it to the second equation:

$$-68x + (-68y) = -61{,}200$$
$$\underline{\ \ 68x + \ \ \ 85y \ = \ \ 72{,}000}$$
$$17y \ = \ \ 10{,}800$$
$$y \approx 635.3 \text{ cu ft}$$

Replacing y with 635.3 in the first equation, we have:

$$x + (635.3) = 900$$
$$x \approx 264.7 \text{ cu ft}$$

There should be approximately 635.3 cu ft of the $8.50 per cu ft mixture, and 264.7 cu ft of the $6.80 per cu ft mixture.

☑

$x + y = 900$	$6.80x + 8.50y = 7200$
$264.7 + 635.3 = 900$	$6.80(264.7) + 8.50(635.3) = 7200$
$900 = 900$	$1799.96 + 5400.05 = 7200$
	$7200 = 7200$

(d)

Solution	Amount (liters)	Salt Fraction	Total Salt
5%	A	0.05	$0.05A$
15%	B	0.15	$0.15B$
12%	4	0.12	$0.12(4)$

Since the final solution is to contain 4 liters, we have

$$A + B = 4$$

The second equation represents the total amount of salt:

$$0.05A + 0.15B = (0.12)4$$

Multiplying this last equation by 100 to eliminate the decimals, we have the system

$$A + B = 4$$
$$5A + 15B = 48$$

To solve this system, multiply each term in the first equation by -5 and add to get

$$
\begin{array}{rl}
-5A + (-5B) = & -20 \\
5A + 15B = & 48 \\
\hline
10B = & 28 \\
B = & 2.8 \text{ liters}
\end{array}
$$

Substituting back into the first equation, we have

$$A + 2.8 = 4$$
$$A = 1.2 \text{ liters}$$

The technician must mix 1.2 liters of the 5% solution with 2.8 liters of the 15% solution.

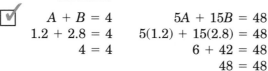

$$
\begin{array}{ll}
A + B = 4 & 5A + 15B = 48 \\
1.2 + 2.8 = 4 & 5(1.2) + 15(2.8) = 48 \\
4 = 4 & 6 + 42 = 48 \\
& 48 = 48
\end{array}
$$

Now turn to Exercises 11-1 for a set of problems on systems of equations.

Exercises 11-1   Systems of Equations

A. Solve each of the following systems of equations using the method of substitution. If the system is inconsistent or dependent, say so.

1. $y = 10 - x$
   $2x - y = -4$

2. $3x - y = 5$
   $2x + y = 15$

3. $2x - y = 3$
   $x - 2y = -6$

4. $2y - 4x = -3$
   $y = 2x + 4$

5. $3x + 5y = 26$
   $x + 2y = 10$

6. $x = 10y + 1$
   $y = 10x + 1$

7. $3x + 4y = 5$
   $x - 2y = -5$

8. $2x - y = 4$
   $4x - 3y = 11$

B. Solve each of the following systems of equations. If the system is inconsistent or dependent, say so.

1. $x + y = 5$
   $x - y = 13$

2. $2x + 2y = 10$
   $3x - 2y = 10$

3. $2y = 3x + 5$
   $2y = 3x - 7$

4. $2y = 2x + 2$
   $4y = 5 + 4x$

5. $x = 3y + 7$
   $x + y = -5$

6. $3x - 2y = -11$
   $x + y = -2$

7. $5x - 4y = 1$
   $3x - 6y = 6$

8. $y = 3x - 5$
   $6x - 3y = 3$

9. $y - 2x = -8$
   $x - \frac{1}{2}y = 4$

10. $x + y = a$
    $x - y = b$

11. $3x + 2y = 13$
    $5x + 3y = 20$

12. $3x + 4y = 11$
    $2x + 2y = 7$

C. Practical Problems

Translate each problem statement into a system of equations and solve.

1. The sum of two numbers is 39 and their difference is 7. What are the numbers?

2. The sum of two numbers is 14. The larger is two more than three times the smaller. What are the numbers?

3. Separate a collection of 20 objects into two parts so that twice the larger amount equals three times the smaller amount.

4. The average of two numbers is 25 and their difference is 8. What are the numbers?

5. Four bleebs and three freems cost $11. Three bleebs and four freems cost $10. What does a bleeb cost?

6. **Carpentry** The perimeter of a rectangular window is 14 ft, and its length is 2 ft less than twice the width. What are the dimensions of the window?

7. Harold exchanged a $1 bill for change and received his change in nickels and dimes, with seven more dimes than nickels. How many of each coin did he receive?

8. If four times the larger of two numbers is added to three times the smaller, the result is 26. If three times the larger number is decreased by twice the smaller, the result is 11. Find the numbers.

9. Mr. Brown bought five cans of peas and four cans of corn, but he forgot what each cost. He knows that the total cost was $8.41 and he recalls that a can of peas cost 10 cents less than a can of corn. How much did each can cost?

10. **Sheet Metal Technology** The length of a piece of sheet metal is twice the width. The difference in length and width is 20 in. What are the dimensions?

11. **Electrical Technology** A 30-in. piece of wire is to be cut into two parts, one part being four times the length of the other. Find the length of each.

12. **Painting and Decorating** A painter wishes to mix paint worth $15 per gallon with paint worth $18 per gallon to make a 12-gallon mixture worth $16.75 per gallon. How many gallons of each should he mix?

13. A chemist wishes to mix a 10% salt solution with a 2% salt solution to obtain 6 liters of a 4% salt solution. How many liters of each should be added?

14. The perimeter of a rectangular field is 520 ft. The length of the field is 20 ft more than three times the width. Find the dimensions.

15. **Landscaping** A 24-ton mixture of crushed rock is needed in a construction job; it will cost $800. If the mixture is composed of rock costing $30 per ton and $40 per ton, how many tons of each should be added?

When you have completed these problems check your answers in the Appendix. Then turn to Section 11-2 to learn about quadratic equations.

11-2 QUADRATIC EQUATIONS

Thus far in your study of algebra you have worked only with linear equations. In a linear equation the variable appears only to the first power. For example, $3x + 5 = 2$ is a linear equation. The variable appears as x or x^1. No powers of x such as x^2, x^3, or x^4 appear in the equations.

An equation in which the variable appears to the second power, but to no higher power, is called a *quadratic equation*.

YOUR TURN

Which of the following are quadratic equations?

(a) $x^2 = 49$

(b) $2x - 1 = 4$

(c) $3x - 2y = 19$

(d) $5x^2 - 8x + 3 = 0$

(e) $x^3 + 3x^2 + 3x + 1 = 0$

ANSWERS

Equations (a) and (d) are quadratic equations. Equation (b) is a linear equation in one variable. Equation (c) is a linear equation in two variables, x and y. Equation (e) is a cubic or third-order equation. Because (e) contains an x^3 term, it is not a quadratic.

Standard Form Every quadratic equation can be put into a *standard quadratic form*.

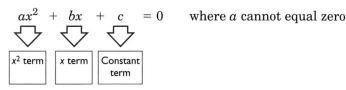

$$ax^2 + bx + c = 0 \qquad \text{where } a \text{ cannot equal zero}$$

Every quadratic equation must have an x^2 term, but the x term and the constant term may be missing.

EXAMPLE

$2x^2 + x - 5 = 0$ is a quadratic equation in standard form: the x^2 term is first, the x term second, and the constant term last on the left.

$x^2 + 4 = 0$ is also a quadratic equation in standard form. The x term is missing, but the other terms are in the proper order. We could rewrite this equation as

$$x^2 + 0 \cdot x + 4 = 0$$

YOUR TURN

Which of the following quadratic equations are written in standard form?

(a) $7x^2 - 3x + 6 = 0$ (b) $8x - 3x^2 - 2 = 0$

(c) $2x^2 - 5x = 0$ (d) $x^2 = 25$

(e) $x^2 - 5 = 0$ (f) $4x^2 - 5x = 6$

ANSWERS

Equations (a), (c), and (e) are in standard form.

To solve a quadratic equation, the first step is usually to rewrite it in standard form.

The equation

$x^2 = 25x$	becomes	$x^2 - 25x = 0$	in standard form
$8x - 3x^2 - 2 = 0$	becomes	$-3x^2 + 8x - 2 = 0$	in standard form
	or	$3x^2 - 8x + 2 = 0$	if we multiply all terms by -1
$4x^2 - 5x = 6$	becomes	$4x^2 - 5x - 6 = 0$	in standard form

In each case we add or subtract a term on both sides of the equation until all terms are on the left, then rearrange terms until the x^2 is first on the left, the x term next, and the constant term third.

YOUR TURN

Try it. Rearrange the following quadratic equations in standard form.

(a) $5x - 19 + 3x^2 = 0$ (b) $7x^2 = 12 - 6x$

(c) $9 = 3x - x^2$ (d) $2x - x^2 = 0$

(e) $5x = 7x^2 - 12$ (f) $x^2 - 6x + 9 = 49$

(g) $3x + 1 = x^2 - 5$ (h) $1 - x^2 + x = 3x + 4$

ANSWERS

(a) $3x^2 + 5x - 19 = 0$ (b) $7x^2 + 6x - 12 = 0$

(c) $x^2 - 3x + 9 = 0$ (d) $-x^2 + 2x = 0$ or $x^2 - 2x = 0$

(e) $7x^2 - 5x - 12 = 0$ (f) $x^2 - 6x - 40 = 0$

(g) $x^2 - 3x - 6 = 0$ (h) $x^2 + 2x + 3 = 0$

Solutions to Quadratic Equations   The solution to a linear equation is a single number. The solution to a quadratic equation is usually a *pair* of numbers, each of which satisfies the equation.

EXAMPLE

The quadratic equation

$$x^2 - 5x + 6 = 0$$

has the solutions

$$x = 2 \quad \text{or} \quad x = 3$$

To see that either 2 or 3 is a solution, substitute each into the equation.

For $x = 2$	For $x = 3$
$(2)^2 - 5(2) + 6 = 0$	$(3)^2 - 5(3) + 6 = 0$
$4 - 10 + 6 = 0$	$9 - 15 + 6 = 0$
$10 - 10 = 0$	$15 - 15 = 0$

YOUR TURN

Show that $x = 5$ or $x = 3$ gives a solution of the quadratic equation

$x^2 - 8x + 15 = 0$

SOLUTION

$x^2 - 8x + 15 = 0$

For $x = 5$ For $x = 3$

$(5)^2 - 8(5) + 15 = 0$ $(3)^2 - 8(3) + 15 = 0$

$25 - 40 + 15 = 0$ $9 - 24 + 15 = 0$

$40 - 40 = 0$ $24 - 24 = 0$

Solving $x^2 - a = 0$ The easiest kind of quadratic equation to solve is one in which the x term is missing.

EXAMPLE

To solve the quadratic equation

$x^2 - 25 = 0$

simply rewrite it as

$x^2 = 25$

and take the square root of both sides of the equation.

$\sqrt{x^2} = \sqrt{25}$

or $x = \pm \sqrt{25}$

or $x = 5$ or $x = -5$

Both 5 and -5 satisfy the original equation.

For $x = 5$ For $x = -5$

$x^2 - 25 = 0$ and $x^2 - 25 = 0$

$(5)^2 - 25 = 0$ $(-5)^2 - 25 = 0$

$25 - 25 = 0$ $25 - 25 = 0$

Every positive number has two square roots, one positive and the other negative. Both of them may be important in solving a quadratic equation.

YOUR TURN

Solve each of the following quadratic equations and check *both* solutions. (Round to two decimal places if necessary.)

(a) $x^2 - 36 = 0$ (b) $x^2 = 8$ (c) $x^2 - 64 = 0$

(d) $4x^2 = 81$ (e) $3x^2 = 27$ (f) $x^2 - 1 = 2$

SOLUTIONS

(a) $x^2 = 36$ (b) $x^2 = 8$

    $x = \pm \sqrt{36}$ $x = \pm \sqrt{8}$

    $x = 6$ or $x = -6$ $x \approx 2.83$ or $x \approx -2.83$ rounded

(c) $x^2 = 64$
$x = \pm \sqrt{64}$
$x = 8$ or $x = -8$

(d) $4x^2 = 81$

$x^2 = \dfrac{81}{4}$

$x = \pm \sqrt{\dfrac{81}{4}}$

$x = \dfrac{9}{2}$ or $x = -\dfrac{9}{2}$

(e) $3x^2 = 27$
$x^2 = 9$
$x = \pm \sqrt{9}$
$x = 3$ or $x = -3$

(f) $x^2 = 3$
$x = \pm \sqrt{3}$
$x \approx 1.73$ or $x \approx -1.73$ rounded

Notice in each case that first we rewrite the equation so that x^2 appears alone on the left and a number appears alone on the right. Second, take the square root of both sides. The equation will have two solutions.

You should also notice that the equation

$x^2 = -4$

has no solution. There is no number x whose square is a negative number.

PROBLEMS

Solve the following quadratic equations. (Round to two decimal places if necessary.)

(a) $x^2 - 3.5 = 0$

(b) $x^2 = 18$

(c) $6 - x^2 = 0$

(d) $9x^2 = 49$

(e) $7x^2 = 80$

(f) $\dfrac{3x^2}{5} = 33.3$

SOLUTIONS

(a) $x^2 = 3.5$

$x = \pm \sqrt{3.5}$

$x \approx 1.87$ or $x \approx -1.87$

(b) $x^2 = 18$

$x = \pm \sqrt{18}$

$x \approx 4.24$ or $x \approx -4.24$

(c) $6 - x^2 = 0$
$x^2 = 6$
$x = \pm \sqrt{6}$
$x \approx 2.45$ or ≈ -2.45

(d) $9x^2 = 49$

$x^2 = \dfrac{49}{9}$

$x = \pm \sqrt{\dfrac{49}{9}}$

$x = \dfrac{7}{3}$ or $x = -\dfrac{7}{3}$

(e) $7x^2 = 80$

$x^2 = \dfrac{80}{7}$

$x = \pm \sqrt{\dfrac{80}{7}}$

$x \approx \pm \sqrt{11.42857}$

$x \approx 3.38$ or $x \approx -3.38$

(f) $\dfrac{3x^2}{5} = 33.3$

$x^2 = \dfrac{(33.3)(5)}{3} = 55.5$

$x = \pm\sqrt{55.5}$

$x \approx 7.45$ or $x \approx -7.45$

Do problem (f) this way on a calculator:

33.3 $\boxed{\times}$ **5** $\boxed{\div}$ **3** $\boxed{=}$ $\boxed{\surd}$ → | 7.449832213 |

Quadratic Formula

In general, a quadratic equation will contain all three terms, an x^2 term, an x term, and a constant term. The solution of any quadratic equation

$$ax^2 + bx + c = 0$$

is

$$x = \frac{-b + \sqrt{b^2 - 4ac}}{2a} \quad \text{or} \quad x = \frac{-b - \sqrt{b^2 - 4ac}}{2a}$$

or

THE QUADRATIC FORMULA

$$x = \frac{-b \pm \sqrt{b^2 - 4ac}}{2a}$$

Careful ▶ Note that the entire expression in the numerator is divided by $2a$. ◀

EXAMPLE

To solve the quadratic equation

$$2x^2 + 5x - 3 = 0$$

follow these steps.

Step 1 *Identify* the coefficients a, b, and c for the quadratic equation.

$$2x^2 \quad + \quad 5x \quad - \quad 3 \quad = 0$$

$$\boxed{a = 2} \qquad \boxed{b = 5} \qquad \boxed{c = -3}$$

Step 2 *Substitute* these values of a, b, and c into the quadratic formula.

$$x = \frac{-(5) \pm \sqrt{(5)^2 - 4(2)(-3)}}{2(2)}$$

The $\pm$ sign means that there are two solutions, one to be calculated using the $+$ sign and the other solution calculated using the $-$ sign.

Step 3 *Simplify* these equations for x.

$$x = \frac{-5 \pm \sqrt{25 + 24}}{4}$$

$$x = \frac{-5 \pm \sqrt{49}}{4}$$

$$x = \frac{-5 \pm 7}{4}$$

$$x = \frac{-5 + 7}{4} \quad \text{or} \quad x = \frac{-5 - 7}{4}$$

$$x = \frac{2}{4} = \frac{1}{2} \quad \text{or} \quad x = \frac{-12}{4} = -3$$

The solution is $x = \dfrac{1}{2}$ or $x = -3$.

Step 4 *Check* the solution numbers by substituting them into the original equation.

☑ $2x^2 + 5x - 3 = 0$

For $x = \frac{1}{2}$ For $x = -3$

$$2\left(\frac{1}{2}\right)^2 + 5\left(\frac{1}{2}\right) - 3 = 0 \qquad 2(-3)^2 + 5(-3) - 3 = 0$$

$$\qquad\qquad\qquad\qquad\qquad\qquad 2(9) + 5(-3) - 3 = 0$$

$$2\left(\frac{1}{4}\right) + 5\left(\frac{1}{2}\right) - 3 = 0 \qquad 18 - 15 - 3 = 0$$

$$\qquad\qquad\qquad\qquad\qquad\qquad 18 - 18 = 0$$

$$\frac{1}{2} + 2\frac{1}{2} - 3 = 0$$

$$3 - 3 = 0$$

YOUR TURN

Use the quadratic formula to solve $x^2 + 4x - 5 = 0$.

SOLUTION

Step 1 x^2 + $4x$ − 5 = 0.

⬇   ⬇   ⬇

$a = 1$ $b = 4$ $c = -5$

Step 2 $x = \dfrac{-(4) \pm \sqrt{(4)^2 - 4(1)(-5)}}{2(1)}$

Step 3 Simplify

$$x = \frac{-4 \pm \sqrt{36}}{2}$$

$$x = \frac{-4 \pm 6}{2}$$

$$x = \frac{-4 + 6}{2} \qquad \text{or} \qquad x = \frac{-4 - 6}{2}$$

$$x = 1 \qquad\qquad \text{or} \qquad x = -5 \qquad \text{The solution is } x = 1 \text{ or } x = -5.$$

Step 4 ☑ $x^2 + 4x - 5 = 0$

For $x = 1$ For $x = -5$

$$(1)^2 + 4(1) - 5 = 0 \qquad (-5)^2 + 4(-5) - 5 = 0$$

$$1 + 4 - 5 = 0 \qquad\qquad 25 - 20 - 5 = 0$$

$$5 - 5 = 0 \qquad\qquad\qquad 25 - 25 = 0$$

EXAMPLE

Here is one that is a bit tougher. Solve $3x^2 - 7x = 5$.

Step 1 Rewrite the equation in standard form.

$$3x^2 \quad - \quad 7x \quad - \quad 5 \quad = 0$$

$$\boxed{a = 3} \quad \boxed{b = -7} \quad \boxed{c = -5}$$

$$-(-7) = +7$$

Step 2 $x = \dfrac{-(-7) \pm \sqrt{(-7)^2 - 4(3)(-5)}}{2(3)}$

$$(-7)^2 = (-7) \cdot (-7) = 49$$

Step 3 Simplify

$$x = \frac{7 \pm \sqrt{109}}{6}$$

$$x = \frac{7 + \sqrt{109}}{6} \qquad \text{or} \qquad x = \frac{7 - \sqrt{109}}{6}$$

$$x \approx \frac{7 + 10.44}{6} \qquad \text{or} \qquad x \approx \frac{7 - 10.44}{6} \qquad \text{rounded to two decimal places}$$

$$x \approx \frac{17.44}{6} \qquad \text{or} \qquad x \approx \frac{-3.44}{6}$$

$$x \approx 2.91 \qquad \text{or} \qquad x \approx -0.57 \qquad \begin{array}{l}\text{The solution is}\\ x \approx 2.91 \quad \text{or} \quad x \approx -0.57.\end{array}$$

Step 4 Check both answers by substituting them back into the original quadratic equation.

Use the following calculator sequence to find this solution.

7 ⊕⁄⊖ ⌊x²⌋ ⊖ 4 ⊗ 3 ⊗ 5 ⊕⁄⊖ ⊜ √ ⌊STO⌋ ⊕ 7 ⊜ ÷ 6 ⊜ → **2.906717751** ≈ 2.91

7 ⊖ ⌊RCL⌋ ⊜ ÷ 6 ⊜ → **−0.573384418** ≈ −0.57

PROBLEMS

Use the quadratic formula to solve each of the following equations. (Round to two decimal places if necessary.)

(a) $6x^2 - 13x + 2 = 0$ (b) $3x^2 - 13x = 0$

(c) $2x^2 - 5x + 17 = 0$ (d) $8x^2 = 19 - 5x$

SOLUTIONS

(a) $6x^2 \quad - \quad 13x \quad + \quad 2 \quad = 0$

$$\boxed{a = 6} \quad \boxed{b = -13} \quad \boxed{c = 2}$$

$$x = \frac{-(-13) \pm \sqrt{(-13)^2 - 4(6)(2)}}{2(6)}$$

$$x = \frac{13 \pm \sqrt{121}}{12} = \frac{13 \pm 11}{12}$$

The solution is $x = \dfrac{13 + 11}{12}$ or $x = \dfrac{13 - 11}{12}$

$$x = 2 \qquad \text{or} \qquad x = \frac{1}{6}$$

The solution is $x = 2$ or $x = \frac{1}{6}$.

Check it.

(b) $3x^2 - 13x = 0$

$$3x^2 \quad - \quad 13x \quad + \quad 0 \quad = 0$$

$\boxed{a = 3}$ $\boxed{b = -13}$ $\boxed{c = 0}$

$$x = \frac{-(-13) \pm \sqrt{(-13)^2 - 4(3)(0)}}{2(3)}$$

$$x = \frac{13 \pm \sqrt{169}}{6}$$

The solution is $x = \dfrac{13 + 13}{6}$ or $x = \dfrac{13 - 13}{6}$

$$x = \frac{13}{3} \qquad \text{or} \qquad x = 0$$

The solution is $x = 0$ or $x = \frac{13}{3}$.

(c) $\quad 2x^2 \quad - \quad 5x \quad + \quad 17 \quad = 0$

$\boxed{a = 2}$ $\boxed{b = -5}$ $\boxed{c = 17}$

$$x = \frac{-(-5) \pm \sqrt{(-5)^2 - 4(2)(17)}}{2(2)}$$

$$x = \frac{5 \pm \sqrt{-111}}{4}$$

But the square root of a negative number is not acceptable if our answer must be a real number. This quadratic equation has no real number solution.

(d) $8x^2 = 19 - 5x$ or

$$8x^2 \quad + \quad 5x \quad - \quad 19 \quad = 0$$

$\boxed{a = 8}$ $\boxed{b = 5}$ $\boxed{c = -19}$

$$x = \frac{-(5) \pm \sqrt{(5)^2 - 4(8)(-19)}}{2(8)}$$

$$x = \frac{-5 \pm \sqrt{633}}{16}$$

$$x \approx \frac{-5 \pm 25.159}{16} \qquad$$ Find the square root to three decimal places to ensure an accurate answer.

The solution is $x \approx -1.88$ or $x \approx 1.26$ rounded

When you check a solution that includes a rounded value, the check may not give an exact fit. The differences should be very small if you have the correct solution.

Word Problems and Quadratic Equations

EXAMPLE

In a dc circuit the power P dissipated in the circuit is given by the equation

$$P = RI^2 \qquad \text{where } R \text{ is the circuit resistance in ohms and } I \text{ is the current in amperes.}$$

What current will produce 1440 watts of power in a 10-ohm resistor?

Substituting into the equation yields

$$1440 = 10I^2$$

or $\qquad I^2 = 144$

$$I = \pm\sqrt{144}$$

The solution is $I = 12$ amperes or $I = -12$ amperes.

EXAMPLE

One side of a rectangular opening for a heating pipe is 3 in. longer than the other side. If the total cross-sectional area is 70 sq in., find the dimensions of the cross section.

Let L = length, W = width.

Then $L = 3 + W$

and the area is

$$\text{area} = LW$$

or $\quad 70 = LW$

Substituting $L = \boxed{3 + W}$ into the area equation, we have

$$70 = \boxed{(3 + W)}\ W$$

or $\qquad 70 = 3W + W^2$

or $\qquad W^2 \quad + \quad 3W \quad - \quad 70 = 0 \qquad$ in standard quadratic form.

$$\downarrow \qquad\qquad \downarrow \qquad\qquad \downarrow$$

$$\boxed{a = 1} \qquad \boxed{b = 3} \qquad \boxed{c = -70}$$

To find the solution, substitute a, b, and c into the quadratic formula.

$$W = \frac{-(3) \pm \sqrt{(3)^2 - 4(1)(-70)}}{2(1)}$$

or $\qquad W = \dfrac{-3 \pm \sqrt{289}}{2}$

$$W = \frac{-3 \pm 17}{2}$$

The solution is

$$W = -10 \qquad \text{or} \qquad W = 7$$

Only the positive value makes sense. The answer is $W = 7$ in.

Substituting back into the equation $L = 3 + W$, we find that $L = 10$ in.

Check to see that the area is indeed 70 sq in.

EXAMPLE

In Chapter 9, the total surface area of a cylinder was found using the formula

$S = 2\pi r^2 + 2\pi rh$

Find the radius of a 15-in. high cylinder with a total surface area of 339 sq in. Use $\pi = 3.14$ and round to the nearest inch.

SOLUTION

First, substitute the given information into the formula:

$$S = 2\pi r^2 + 2\pi rh$$
$$339 = 2(3.14)r^2 + 2(3.14)r(15)$$
$$339 = 6.28r^2 + 94.2r$$

Then, subtract 339 from both sides and solve using the quadratic formula:

$$0 = 6.28r^2 + 94.2r - 339$$

Here $a = 6.28$, $b = 94.2$, and $c = -339$

$$r = \frac{-94.2 \pm \sqrt{(94.2)^2 - 4(6.28)(-339)}}{2(6.28)}$$

$r = 2.999\ldots \approx 3$ in.   or $r = -17.999\ldots \approx -18$ in.

A negative radius cannot exist, so the radius of the cylinder is approximately 3 in.

PROBLEMS

Solve the following problems. (Round each answer to two decimal places if necessary.)

(a) Find the side length of a square whose area is 200 sq m.

(b) **Sheet Metal Technology** The cross-sectional area of a rectangular duct must be 144 sq in. If one side must be twice as long as the other, find the length of each side.

(c) **Plumbing** Find the diameter of a circular pipe whose cross-sectional area is 3.00 sq in.

(d) **Auto Mechanics** The SAE rating of an engine is given by $R = \dfrac{D^2 N}{2.5}$, where D is the bore of the cylinder in inches, and N is the number of cylinders. What must the bore be for an eight-cylinder engine to have an SAE rating of 33.8?

(e) One side of a rectangular plate is 6 in. longer than the other. The total area is 216 sq in. How long is each side?

(f) **Metalworking** A man has a long strip of sheet steel 12 ft wide. He wishes to make an open-topped water channel with a rectangular cross section. If the cross-sectional area must be 16 ft^2, what should the dimensions of the channel be?
(*Hint:* $H + H + W = 12$ ft.)

(a) $A = s^2$

$200 = s^2$

$s = \pm\sqrt{200}$

$s \approx 14.14$ m, rounded   The negative solution is not possible.

(b) $L \cdot W = A$

and $L = 2W$

Therefore, $(2W)\,W = A$

or $2W^2 = A$

$2W^2 = 144$

$W^2 = 72$

$W = +\sqrt{72}$

$W \approx 8.485 \approx 8.49$ in.   rounded   for the positive root

Then $L = 2W$

or $L \approx 16.98$ in.   rounded

(c) $A = \dfrac{\pi d^2}{4}$

$3.00 = \dfrac{\pi d^2}{4}$

or $d^2 = \dfrac{4(3.00)}{\pi}$

$d^2 = 3.8197\ldots$

$d \approx +\sqrt{3.8197}$

$d \approx 1.95$ in.   rounded

  4 $\times$ **3** $\div$ π $=$ $\sqrt{}$ $\rightarrow$ `1.954410048`

(d) $R = \dfrac{D^2 N}{2.5}$

$33.8 = \dfrac{D^2 \cdot 8}{2.5}$

or $D^2 = \dfrac{(33.8)(2.5)}{8}$

$D^2 = 10.5625$

$D = +\sqrt{10.5625}$

or $D = 3.25$ in.

(e) "One side . . . must be 6 in. longer than the other."

$$L \quad = \quad 6 \quad + \quad W$$

For a rectangle, area $= LW$.

Then,

$216 = LW$

or $216 = (6 + W)W$

$216 = 6W + W^2$

$W^2 \; + \; 6W \; - \; 216 \; = 0$ in standard quadratic form

$\boxed{a = 1}$ $\boxed{b = 6}$ $\boxed{c = -216}$

When we substitute into the quadratic formula,

$$W = \frac{-(6) \pm \sqrt{(6)^2 - 4(1)(-216)}}{2(1)}$$

$$W = \frac{-6 \pm \sqrt{900}}{2}$$

$$W = \frac{-6 \pm 30}{2}$$

The solution is

$W = -18$ or $W = 12$

Only the positive value is a reasonable solution. The answer is $W = 12$ in.

Substituting into the first equation yields

$L = 6 + W$
$L = 6 + (12) = 18$ in.

(f)

$H + H + W = 12$

$2H + W = 12$

or $W = 12 - 2H$

Area $= HW$

or $16 = HW$ Substituting $W = \boxed{12 - 2H}$ into the area equation.

$16 = H \, \boxed{(12 - 2H)}$

$16 = 12H - 2H^2$ or $8 = 6H - H^2$ Divide each term by 2 to get a simpler equation.

$H^2 \; - \; 6H \; + \; 8 \; = 0$ in standard quadratic form

$\boxed{a = 1}$ $\boxed{b = -6}$ $\boxed{c = 8}$

Substituting a, b, and c into the quadratic formula yields

$$H = \frac{-(-6) \pm \sqrt{(-6)^2 - 4(1)(8)}}{2(1)}$$

$$H = \frac{6 \pm \sqrt{4}}{2}$$

$$H = \frac{6 \pm 2}{2}$$

The solution is $H = 4$ ft or $H = 2$ ft

For $H = 4$ ft, $W = 4$ ft, since $16 = HW$
For $H = 2$ ft, $W = 8$ ft

The channel can be either 4 ft by 4 ft or 2 ft by 8 ft. Both dimensions give a cross-sectional area of 16 sq ft.

Exercises 11-2   Quadratic Equations

A. Which of the following are quadratic equations?

1. $5x - 13 = 23$ 2. $2x + 5 = 3x^2$

3. $2x^3 - 6x^2 - 5x + 3 = 0$ 4. $x^2 = 0$

5. $8x^2 - 9x = 0$

Which of the following quadratic equations are in standard form? For those that are not, rearrange them into standard form.

6. $7x^2 - 5 + 3x = 0$ 7. $14 = 7x - 3x^2$

8. $13x^2 - 3x + 5 = 0$ 9. $23x - x^2 = 5x$

10. $4x^2 - 7x + 3 = 0$

B. Solve each of these quadratic equations. (Round to two decimal places if necessary.)

1. $x^2 = 25$ 2. $3x^2 - 27 = 0$

3. $5x^2 = 22x$ 4. $2x^2 - 7x + 3 = 0$

5. $4x^2 = 81$ 6. $6x^2 - 13x - 63 = 0$

7. $15x = 12 - x^2$ 8. $4x^2 - 39x = -27$

9. $0.4x^2 + 0.6x - 0.8 = 0$ 10. $0.001 = x^2 + 0.03x$

11. $x^2 - x - 13 = 0$ 12. $2x - x^2 + 11 = 0$

13. $7x^2 - 2x = 1$ 14. $3 + 4x - 5x^2 = 0$

15. $2x^2 + 6x + 3 = 0$ 16. $5x^2 - 7x + 1 = 0$

17. $3x^2 = 8x - 2$ 18. $x = 1 - 7x^2$

19. $0.2x^2 - 0.9x + 0.6 = 0$ 20. $1.2x^2 + 2.5x = 1.8$

C. Practical Problems (Round to the nearest hundredth.)

1. The area of a square is 625 mm^2. Find its side length.

2. **Manufacturing**   The capacity in gallons of a cylindrical tank can be found using the formula

$$C = \frac{\pi D^2 L}{924}$$

where C = capacity in gallons
    D = diameter of tank in inches
    L = length of tank in inches

   (a) Find the diameter of a 42-in.-long tank that has a capacity of 30 gallons.

   (b) Find the diameter of a 60-in.-long tank that has a capacity of 50 gallons.

3. **Plumbing**   The length of a rectangular pipe is three times longer than its width. Find the dimensions that give a cross-sectional area of 75 sq in.

4. **Sheet Metal Technology**   Find the radius of a circular vent that has a cross-sectional area of 250 cm^2.

5. **Sheet Metal Technology**   The length of a rectangular piece of sheet metal must be 5 in. longer than the width. Find the exact dimensions that will provide an area of 374 sq in.

6. **Construction**   The building code in a certain county requires that public buildings be able to withstand a wind pressure of 25 lb/ft^2. The pressure P is related to the wind speed v in miles per hour by the formula $P = v^2/390$. What wind speed will produce the maximum pressure allowed? (Round to two significant digits.)

7. **Plumbing**   Find the diameter of a circular pipe whose cross-sectional area is 40 sq in.

8. **Electrical Technology**   If $P = RI^2$ for a direct current circuit, find I (current in amperes) if

   (a)   The power (P) is 405 watts and the resistance (R) is 5 ohms.

   (b)   The power is 800 watts and the resistance is 15 ohms.

9. **Auto Mechanics**   Total piston displacement is given by the following formula:

   P.D. $= \frac{\pi}{4}D^2LN$

   where P.D. = piston displacement, in cubic inches
   D = diameter of bore of cylinder
   L = length of stroke, in inches
   N = number of cylinders

   Find the diameter if

   (a)   P.D. = 400 cu in., L = 4.5 in., N = 8

   (b)   P.D. = 392.7 cu in., L = 4 in., N = 6

10. **Sheet Metal Technology**   An open-topped channel must be made out of a 20-ft-wide piece of sheet steel. What dimensions will result in a cross-sectional area of 48 sq ft?

11. **Sheet Metal Technology**   A cylindrical water heater has a total surface area of 4170 in.^{2}. If the height of the water heater is 75 in., find the radius to the nearest inch.

Check your answers in the Appendix. Then turn to Problem Set 11 for a problem set covering both systems of equations and quadratic equations.

PROBLEM SET

11

Advanced Algebra

Answers are given in the Appendix.

A. Solve and check each of the following systems of equations. If the system is inconsistent or dependent, say so.

1. $x + 4y = 27$
  $x + 2y = 21$

2. $3x + 2y = 17$
  $x = 5 - 2y$

3. $5x + 2y = 20$
  $3x - 2y = 4$

4. $x = 10 - y$
  $2x + 3y = 23$

5. $3x + 4y = 45$
  $x - \frac{1}{3}y = 5$

6. $2x - 3y = 11$
  $4x - 6y = 22$

7. $2x + 3y = 5$
  $3x + 2y = 5$

8. $5x = 1 - 3y$
  $4x + 2y = -8$

9. $3x - 2y = 10$
  $4x + 5y = 12$

B. Solve and check each of the following quadratic equations.

1. $x^2 = 9$

2. $x^2 - 3x - 28 = 0$

3. $3x^2 + 5x + 1 = 0$

4. $4x^2 = 3x + 2$

5. $x^2 = 6x$

6. $2x^2 - 7x - 15 = 0$

7. $x^2 - 4x + 4 = 9$

8. $x^2 - x - 30 = 0$

9. $\frac{5x^2}{3} = 60$

10. $7x + 8 = 5x^2$

C. Practical Problems

For each of the following, set up either a system of equations in two variables or a quadratic equation and solve. (Round to two decimal places if necessary.)

1. The sum of two numbers is 38. Their difference is 14. Find them.

2. The area of a square is 196 sq in. Find its side length.

3. The difference of two numbers is 21. If twice the larger is subtracted from five times the smaller, the result is 33. Find the numbers.

4. Carpentry The perimeter of a rectangular door is 22 ft. Its length is 2 ft more than twice its width. Find the dimensions of the door.

5. Heating and Air Conditioning One side of a rectangular heating pipe is four times as long as the other. The cross-sectional area is 125 cm². Find the dimensions of the pipe.

Name _____

Date _____

Course/Section _____

6. **Carpentry** A mixture of 650 nails costs $43.50. If some of the nails cost 5 cents apiece, and the rest cost 7 cents apiece, how many of each are there?

7. **Electrical Technology** In the formula $P = RI^2$ find the current I in amperes if
   (a) The power P is 1352 watts and the resistance R is 8 ohms.
   (b) The power P is 1500 watts and the resistance is 10 ohms.

8. **Sheet Metal Technology** The length of a rectangular piece of sheet steel is 2 in. longer than the width. Find the exact dimensions if the area of the sheet is 168 sq in.

9. **Electrical Technology** A 42-in. piece of wire is to be cut into two parts. If one part is 2 in. less than three times the length of the other, find the length of each piece.

10. **Landscaping** The perimeter of a rectangular field is 750 ft. If the length is four times the width, find the dimensions of the field.

11. **Sheet Metal Technology** Find the diameter of a circular vent with a cross-sectional area of 200 in.^{2}.

12. **Painting and Decorating** A painter mixes paint worth $16 per gallon with paint worth $19 per gallon. He wishes to make 15 gallons of a mixture worth $18 per gallon. How many gallons of each kind of paint must be included in the mixture?

13. **Construction** According to Doyle's log rule, the volume V (in board feet) produced from a log of length L (in feet) and diameter D (in inches) can be estimated from the formula

    $$V = \frac{L(D^2 - 8D + 16)}{16}$$

    What diameter log of length 20 ft will produce 90 board feet of lumber?

14. **Plumbing** The length of a rectangular pipe is 3 in. less than twice the width. Find the dimensions if the cross-sectional area is 20 sq in.

15. **Landscaping** Fifty tons of a mixture of decorative rock cost $4200. If the mixture consists of rock costing $80 per ton and a more expensive rock costing $100 per ton, how many tons of each are used to make the mixture?

Answers to Previews

Answers to Preview 1

1. (a) 125   (b) 8607   (c) 37   (d) 2068   (e) 77

2. (a) 2368   (b) 74,115   (c) 640,140   (d) 334 remainder 2
   (e) 203

3. 445 lb

4. (a) 1, 2, 3, 4, 6, 12   (b) $2 \cdot 2 \cdot 3$

5. (a) 33   (b) 33   (c) 33   (d) 48

Answers to Preview 2

1. (a) $7\frac{3}{4}$ (b) $\frac{31}{8}$ (c) $\frac{20}{64}$ (d) $\frac{56}{32}$ (e) $\frac{5}{32}$ (f) $1\frac{7}{8}$

2. (a) $\frac{35}{256}$ (b) 3   (c) $\frac{9}{10}$ (d) $1\frac{1}{2}$ (e) $2\frac{3}{10}$ (f) 8

3. (a) $\frac{5}{8}$ (b) $1\frac{15}{16}$ (c) $\frac{11}{20}$ (d) $2\frac{11}{16}$

Answers to Preview 3

1. (a) 5.916   (b) 2.791   (c) 22.97   (d) 2.18225
   (e) 3256.25   (f) 225   (g) 3.045

2. 4.3

3. (a) 0.1875   (b) 106.027   (c) twenty-six and thirty-five thousandths
   (d) 3.452   (e) 9.225   (f) 0.305

4. (a) $474.60   (b) 0.34 lb

Answers to Preview 4

1. (a) 5:2 or 2:5   (b) $8\frac{2}{3}$ to 1

2. (a) $x = 10$ (b) $y = 14.3$

3. (a) 64 ounces   (b) 18 ft   (c) 12 machines

4. (a) 25%   (b) 46%   (c) 500%   (d) 7.5%

5. (a) 0.35   (b) 0.0025   (c) $1\frac{3}{25}$

6. (a) 225   (b) 54   (c) 6.25% or $6\frac{1}{4}$%   (d) 75
   (e) 0.90%   (f) 10.4 lb   (g) $94.09   (h) 80%
   (i) 11.4%

Answers to Preview 5

1. (a) nearest hundredth (b) nearest whole number (c) nearest ten
   (a) is most precise
   (d) 2 significant digits (e) 3 significant digits (f) 1 significant digit

2. (a) 44.2 sec (b) 5.94 lb

3. (a) 94 sq ft (b) 15 hr (c) 13,000 lb

4. (a) 1040 oz (b) 4.7 mi (c) 504 sq in.

5. (a) 100 lb (b) 85 qt

6. (a) 6.5 g (b) 450 m (c) 2850 sq m

7. (a) 12.7 gal (b) 72 km/hr (c) 108°F

Answers to Preview 6

1. (a) -14 (b)
   (c) -250

2. (a) -8 (b) $11\frac{3}{4}$ (c) -7.5 (d) -3

3. (a) -3 (b) -8 (c) 10.9 (d) $-1\frac{7}{8}$

4. (a) -72 (b) 6 (c) 31 (d) $-\frac{3}{20}$ (e) 1500

5. (a) 64 (b) 4.2025

6. (a) 11 (b) 500 (c) 22

7. (a) 13 (b) 3.81

Answers to Preview 7

1. (a) 7 (b) 5 (c) $A = 12$ (d) $T = 4$ (e) $P = 120$

2. (a) $6ax^2$ (b) $-3x - y$ (c) $4x - 2$ (d) $3x^2 + 7x + 12$

3. (a) $x = 5$ (b) $x = 9$ (c) $x = 12$ (d) $N = 2S - 52$
   (e) $A = \dfrac{8M + BL}{L}$

4. (a) $4A$ (b) i^2R (c) $E = \dfrac{100u}{I}$ (d) $R = \dfrac{12L}{D^2}$
   (e) $24.07 (f) 10.37 hr, or 11 hr, rounded

5. (a) $6y^2$ (b) $-12x^5y^4$ (c) $3xy - 6x^2$ (d) $-5x^5$ (e) $\dfrac{b^2}{3a^3}$
   (f) $4m^2 - 3m + 5$

6. (a) 1.84×10^{-4} (b) 2.13×10^5 (c) 1.44×10^{-3}
   (d) 6.5×10^{-8}

Answers to Preview 8

1. about $32°$

2. (a) $80°, 80°, 100°, 80°, 100°$ (b) $20°$

3. (a) Triangle   (b) Parallelogram   (c) Trapezoid
   (d) Square   (e) Hexagon

4. 1.5 in.

5. (a) 14 sq in.   (b) 24 cm^2 (c) 0.42 sq in., 2.4 in.
   (d) 22.9 sq in., 17.0 in.   (e) 554 sq in., 114 in.   (f) 19.3 sq in.

Answers to Preview 9

1. (a) Cube   (b) Frustum of a cone   (c) Rectangular prism

2. (a) 12.6 cu ft, 31.4 sq ft   (b) 38.8 cm^3, 55.4 cm^2 (c) 3.7 cu in.
   (d) 80.5 sq in., 75.4 cu in.   (e) 180 cm^2, 336 cm^3

3. (a) 40,400 gal   (b) 24.3 lb   (c) $4536

Answers to Preview 10

1. (a) Yes   (b) Yes   (c) Yes   (d) Yes   (e) No

2. (a) 0.438   (b) 0.105   (c) 0.943

3. (a) 14°0′   (b) 56°43′   (c) 75.3°   (d) 48.6°

4. (a) 0.80   (b) 0.15   (c) 1.40
   (d) 4.2 in., 12.6 sq in.   (e) 1.7 rad/sec

5. (a) $X \approx 10.5, Y \approx 13.4$ (b) $\angle m = 44°25′, X \approx 3.7$ in.
   (c) ≈ 39.8 in.   (d) 5°43′

6. (a) $A = 82°, B = 66°, C = 32°$ (b) $C = 33°, a = 7.6$ ft, $c = 4.4$ ft

Answers to Preview 11

1. (a) (4, 0)   (b) (−9, −3)   (c) Dependent   (d) Inconsistent, no solution

2. (a) 20 and 15   (b) 8 in. and 6 in.   (c) 300 of the 8¢ screws, and 500 of the 12¢ screws

3. (a) $x = 4$ or $x = -4$ (b) $x = 0$ or $x = 7$
   (c) $x = 7$ or $x = -2$ (d) $x = -2\frac{2}{3}$ or $x = 2$ (e) No solution

4. (a) 6 cm   (b) ≈ 8.4 in.   (c) 12.9 amp   (d) 6.5 in. by 9.5 in.

Answers

Chapter 1   *Practice, page 6*

A. | 10 | 11 | 11 | 12 | 16 | 7 | 9 | 15 | 12 | 13 |
|----|----|----|----|----|----|----|----|----|----|
| 10 | 13 | 15 | 8 | 14 | 9 | 13 | 18 | 9 | 11 |
| 16 | 8 | 10 | 17 | 9 | 14 | 12 | 7 | 10 | 11 |
| 13 | 13 | 12 | 16 | 9 | 11 | 10 | 10 | 14 | 17 |
| 11 | 14 | 11 | 10 | 11 | 14 | 13 | 12 | 13 | 15 |

B. | 11 | 12 | 14 | 17 | 18 | 21 | 11 | 20 | 18 | 15 |
|----|----|----|----|----|----|----|----|----|----|
| 15 | 14 | 15 | 18 | 22 | 17 | 14 | 14 | 18 | 16 |
| 12 | 19 | 8 | 10 | 22 | 12 | 19 | 15 | 14 | 16 |

Exercises 1-1, page 12

A.
1. 70   2. 65   3. 80   4. 103   5. 123
6. 124   7. 132   8. 136   9. 393   10. 1003
11. 1390   12. 831   13. 1009   14. 806   15. 861
16. 5525   17. 9461   18. 9302   19. 11,428   20. 15,715
21. 25,717   22. 47,111   23. 11,071   24. 14,711   25. 175,728

B.
1. 1042   2. 5211   3. 2442   4. 6441
5. 7083   6. 16,275   7. 6352   8. 7655
9. 6514   10. 9851   11. 64   12. 141
13. 55   14. 148   15. 357

C.
1. Three hundred fifty-seven
2. Two thousand three hundred four
3. Seventeen thousand ninety-two
4. Two hundred seven thousand six hundred thirty
5. Two million thirty-four
6. Ten thousand seven
7. Seven hundred forty thousand one hundred six
8. Five million fifty-five thousand five hundred fifty
9. One hundred eighteen million one hundred eighty thousand eighteen
10. Six thousand seven hundred nine

11. 3006   12. 17,024   13. 11,100
14. 3,002,017   15. 4,040,006   16. 720,000,010
17. 360   18. 4400   19. 4000
20. 5000   21. 230,000   22. 226,000

D.
1. 4861 ft   2. 11,365 fbm   3. 1636 screws
4. $19,488   5. 1129 minutes   6. 6670 shingles
7. (a) 3607 watts   (b) 1997 watts   (c) 850 watts   8. 111 qt
9. 3114   10. 97 in.   11. $1314
12. 580   13. 2929 ohms   14. 51 volts

E. 1. (a) *Daily Totals*   (b) *Machine Totals*   (c) yes

     1. 1919   A. 4074
     2. 2125   B. 4462
     3. 1958   C. 2185
     4. 1594   D. 5665
     5. 2192   E. 2883
     6. 1953
     7. 1974
     8. 1988
     9. 1667
    10. 1899

2. (a) $307,225   (b) $732,813   (c) $2,298,502   (d) $7156

3. (a) $8200   (b) $7735   (c) SELL

4. (a) #12 BHD: 11,453   (b) A3:3530
    #Tx: 258   A4:8412
    410 AAC: 12,715   B1:4294
    110 ACSR: 8792   B5:5482
    6B: 7425   B6:5073
                        C4:6073
                        C5:7779

Exercises 1-2, page 21

A. 1. 6   2. 7   3. 2   4. 8   5. 4   6. 9
   7. 3   8. 0   9. 3   10. 3   11. 8   12. 8
   13. 9   14. 9   15. 9   16. 9   17. 3   18. 6
   19. 8   20. 5   21. 7   22. 18   23. 7   24. 8
   25. 0   26. 5   27. 8   28. 9   29. 6   30. 8
   31. 6   32. 9   33. 5   34. 7   35. 4   36. 4

B. 1. 13   2. 29   3. 12   4. 19   5. 15
   6. 36   7. 38   8. 29   9. 46   10. 22
   11. 25   12. 38   13. 189   14. 85   15. 281
   16. 254   17. 408   18. 154   19. 273   20. 715
   21. 574   22. 29   23. 2809   24. 5698   25. 12,518
   26. 56,042   27. 4741   28. 9614   29. 47,593   30. 22,422

C. 1. $459   2. 2353 sq ft   3. 1758 ft   4. 29 cm
   5. $330,535   6. $3144   7. 3 drums, by 44 liters
   8. (a) 9 in.   (b) 4 in.   (c) 4 in.   (d) 5 in.
   9. 174 gal   10. 5840 lb   11. 7750   12. $A = 32$ in.; $B = 32$ in.
   13. $13,500 \, \Omega$

D. 1.

Truck No.	1	2	3	4	5	6	7	8	9	10
Mileage	1675	1167	1737	1316	1360	299	1099	135	1461	2081

Total mileage 12,330

2. First sum: 1,083,676,269
Second sum: 1,083,676,269
They are the same.
3. $24,431
4. $116,805

5. (a) $2065

(b)

Deposits	Withdrawals	Balance
		$6375
	$ 379	5996
$1683		7679
474		8153
487		8640
	2373	6267
	1990	4277
	308	3969
	1090	2879
	814	2065

6. 8467 cubic feet

Practice, page 26

A.
12	32	63	36	12	18	0	24	14	8
48	16	45	30	10	9	72	35	18	4
28	15	36	49	8	40	42	54	64	24
20	0	25	27	81	6	1	48	16	63

B.
16	30	9	35	18	20	28	48	12	63
32	0	18	24	9	25	24	45	10	72
15	49	40	54	36	8	42	64	0	4
25	27	7	56	36	12	81	0	2	56

Exercises 1-3, page 32

A.   1. 42   2. 56   3. 48   4. 72   5. 63
    6. 87   7. 54   8. 28   9. 45   10. 84
    11. 296   12. 168   13. 576   14. 423   15. 320
    16. 156   17. 290   18. 564   19. 416   20. 644
    21. 792   22. 1088   23. 1404   24. 153   25. 282
    26. 308   27. 720   28. 1728   29. 5040   30. 7138
    31. 1938   32. 1650   33. 4484   34. 928   35. 3822
    36. 8930

B.   1. 37,515   2. 375,750   3. 297,591   4. 38,023
    5. 378,012   6. 41,064   7. 30,780   8. 1,368,810
    9. 397,584   10. 60,241   11. 7281   12. 4263
    13. 25,000   14. 325,200   15. 3,532,536   16. 8954
    17. 10,112   18. 31,868   19. 89,577   20. 113,680

C.   1. $2600   2. 2100 ft   3. 1300 ft   4. $14,603
    5. 8000 envelopes   6. 225 in.   7. 2430   8. 132 in.
    9. 1200   10. 200,000 ohms   11. 7650   12. 44,280,000
    13. 23,040   14. 15,120 μA   15. 115 in. or   16. 234 V
                                            9 ft 7 in.

D.   1. $681,355 left   2. (a) Nets $36,500   (b) Nets $36,400
                          (c) Nets $31,200   (d) Nets $671,088.63
                          Therefore, (d) gives you the most money.
    3. (a) 111,111,111; 222,222,222; 333,333,333
       (b) 111,111; 222,222; 333,333

(c) 1; 121; 12,321; 1,234,321; 123,454,321

(d) 42; 4422; 444,222; 44,442,222; 4,444,422,222

4. 34,600 lb

5.

Alpha	Beta	Gamma	Delta	Tau
$28,080	$45,560	$37,120	$72,520	$56,160

Exercises 1-4, page 42

A.
1. 9		2. 11 rem. 4		3. Not defined
4. 7 rem. 2		5. 10 rem. 1		6. 1
7. 8		8. 4		9. 6
10. 35		11. 23 rem. 6		12. 57
13. 51 rem. 4		14. 1103 rem. 1		15. 21
16. 52		17. 37		18. 50 rem. 1
19. 23		20. 20 rem. 2		21. 39
22. 25		23. 9 rem. 1		24. 53
25. 22		26. 63 rem. 23		27. 8 rem. 35

B.
1. 120	2. 9 rem. 6	3. 56 rem. 8
4. 95 rem. 6	5. 96	6. 142 rem. 6
7. 222 rem. 2	8. 32	9. 305 rem. 5
10. 84 rem. 41	11. 119	12. 3001
13. 501	14. 8001 rem. 3	15. 604
16. 20,720	17. 200	18. 50 rem. 4
19. 108 rem. 4	20. 2009 rem. 2	21. 600
22. 61	23. 102 rem. 98	24. 81
25. 100 rem. 11	26. 19 rem. 66	27. 17 rem. 123

C. 1. (a) 1, 2, 3, 6   (b) 2×3 2. (a) 1, 2, 4, 8, 16   (b) $2 \times 2 \times 2 \times 2$
    3. (a) 1, 19   (b) 19   4. (a) 1, 3, 9, 27   (b) $3 \times 3 \times 3$
    5. (a) 1, 2, 4, 5, 8, 10, 20, 40   6. (a) 1, 2, 3, 4, 6, 8, 12, 16, 24, 48
        (b) $2 \times 2 \times 2 \times 5$ (b) $2 \times 2 \times 2 \times 2 \times 3$

D.
1. 27 in.	2. 6	3. 13 hr	4. $W = 12$ in., $H = 8$ in.
5. 28	6. 50	7. 7 in.	8. 373 HP
9. $68	10. 11 hr	11. 48	12. 13 cu ft
13. 27 reams	14. 36 days		

E. 1. (a) 21,021,731   (b) 449   (c) 93   (d) 27,270
    2. $4060
    3. 19.148255 or 20 rivets to be sure
    4. 123,091 lb   5. 51 hours 40 minutes, or 52 hr, rounded
    6. (a) 16,094   (b) 201,023   (c) 2283   (d) 357
    7. 42 hours

Exercises 1-5, page 46

A.
1. 50	2. 14	3. 36	4. 24	5. 57	6. 14
7. 4	8. 12	9. 42	10. 1	11. 6	12. 99
13. 61	14. 23	15. 112	16. 2	17. 31	18. 26
19. 17	20. 49	21. 84	22. 27	23. 4	24. 32
25. 6	26. 8	27. 36	28. 26	29. 2	30. 4
31. 8	32. 3	33. 13	34. 3	35. 7	36. 8
37. 6	38. 3	39. 5	40. 43		

B. 1. $3 \times \$18 + 5 \times \$20 = \$154$ 2. $\$468 + \$90 \times 8 = \$1188$
    3. $12 \times \$11 - \$9 \times 5 = \$87$ 4. $520 - 48 \times 5 + 300 = 580$

C.
1. 8347	2. 403	3. 7386	4. 1262
5. 5	6. 1,003,632	7. 1359	8. 1265
9. 1691	10. 12,998	11. 1458	12. 165
13. 13,920	14. 30	15. 63	16. 44

A.
1.	93	2.	83	3.	528	4.	860	
5.	934	6.	2980	7.	15	8.	26	
9.	649	10.	196	11.	195	12.	2615	
13.	1504	14.	3423	15.	1407	16.	3690	
17.	13,041	18.	290,764	19.	230,384	20.	1,575,056	
21.	37	22.	213	23.	57	24.	62	
25.	9	26.	43	27.	18	28.	69	
29.	6	30.	9	31.	115	32.	40	
33.	7	34.	627	35.	1245	36.	833	

B.
1. (a) 1, 2, 4, 8   (b) $2 \times 2 \times 2$ 2. (a) 1, 2, 4, 7, 14, 28   (b) $2 \times 2 \times 7$
3. (a) 1, 31   (b) 31   4. (a) 1, 5, 7, 35   (b) 5×7
5. (a) 1, 2, 3, 4, 6, 9, 12, 18, 36   (b) $2 \times 2 \times 3 \times 3$
6. (a) 1, 2, 3, 6, 7, 14, 21, 42   (b) $2 \times 3 \times 7$

C.
1.	43 ft	2.	64 rods	3.	1892 sq ft	4.	445 lb	
5.	6	6.	24 hr	7.	207 lb	8.	$754	
9.	$1742	10.	252 ft	11.	739,689 mi	12.	$165,359	
13.	650 gpm	14.	(a) 513		(b) 5068 ft			
15.	2839 lb	16.	263°	17.	24 hr			

18. $3 \times \$20 + 3 \times 6 \times \$9 + 3 \times 6 \times \$4$
or $3 \times \$20 + 3 \times 6 \times (\$9 + \$4) = \$294$

19.	87,780 cu in.	20.	646	21.	193 rpm	22.	101 lb
23.	6 ft	24.	1145 pF				

Chapter 2 *Exercises 2-1, page 69*

A.
1. $\frac{7}{3}$ 2. $\frac{15}{2}$ 3. $\frac{67}{8}$ 4. $\frac{17}{16}$ 5. $\frac{23}{8}$
6. $\frac{2}{1}$ 7. $\frac{8}{3}$ 8. $\frac{259}{64}$ 9. $\frac{29}{6}$ 10. $\frac{29}{16}$

B.
1. $8\frac{1}{2}$ 2. $1\frac{3}{5}$ 3. $1\frac{3}{8}$ 4. $2\frac{8}{16}$ or $2\frac{1}{2}$ 5. $1\frac{1}{2}$
6. $3\frac{2}{3}$ 7. $16\frac{4}{6}$ or $16\frac{2}{3}$ 8. $1\frac{1}{3}$ 9. $2\frac{16}{32}$ or $2\frac{1}{2}$ 10. $2\frac{1}{2}$

C.
1. $\frac{3}{4}$ 2. $\frac{2}{3}$ 3. $\frac{3}{8}$ 4. $\frac{9}{2}$ 5. $\frac{2}{5}$
6. $\frac{7}{6}$ 7. $\frac{4}{5}$ 8. $\frac{5}{2}$ 9. $4\frac{1}{4}$ 10. $\frac{17}{16}$
11. $\frac{21}{32}$ 12. $\frac{2}{7}$ 13. $\frac{5}{12}$ 14. $\frac{5}{2}$ 15. $\frac{19}{12}$

D.
1. 14   2. 12   3. 8   4. 24   5. 20   6. 92
7. 36   8. 34   9. 5   10. 24   11. 42   12. 34

E.
1. $\frac{3}{5}$ 2. $\frac{13}{8}$ 3. $1\frac{1}{2}$ 4. $\frac{13}{16}$ 5. $\frac{7}{8}$ 6. $2\frac{1}{2}$
7. $\frac{6}{4}$ 8. $\frac{25}{60}$ 9. $\frac{13}{5}$ 10. $2\frac{7}{4}$ 11. $\frac{5}{12}$ 12. $1\frac{1}{5}$

F.
1. $15\frac{3}{4}$ in.   2. $\frac{3}{4}$ 3. $\frac{19}{6}, \frac{25}{8}$ 4. $\frac{13}{64}$-in. fastener
5. No.   6. $2\frac{2}{3}$ in.   7. $\frac{3}{5}$ 8. $1\frac{1}{2}$ 9. $\frac{1}{5}$ 10. $\frac{3}{12}$

Exercises 2-2, page 73

A.
1. $\frac{1}{8}$ 2. $\frac{4}{15}$ 3. $\frac{2}{15}$ 4. 3   5. $2\frac{2}{3}$
6. $\frac{11}{45}$ 7. $1\frac{1}{9}$ 8. $\frac{13}{16}$ 9. $2\frac{1}{2}$ 10. 14
11. 3   12. 8   13. $3\frac{1}{4}$ 14. $1\frac{1}{21}$ 15. 69
16. $35\frac{3}{4}$ 17. 74   18. $9\frac{7}{8}$ 19. $10\frac{3}{8}$ 20. $21\frac{1}{3}$
21. $\frac{1}{8}$ 22. $\frac{3}{10}$ 23. $\frac{1}{15}$ 24. $1\frac{1}{3}$ 25. 2

B.   1. $\frac{1}{6}$ 2. $\frac{3}{32}$ 3. $\frac{1}{2}$ 4. $\frac{7}{16}$ 5. $\frac{3}{4}$

    6. $\frac{15}{16}$ 7. $1\frac{5}{16}$ 8. $2\frac{7}{9}$ 9. 1 10. $\frac{7}{10}$

    11. $1\frac{1}{20}$ 12. $2\frac{2}{5}$ 13. $2\frac{5}{8}$ 14. $\frac{5}{14}$ 15. 1 16. $1\frac{1}{2}$

C.   1. $137\frac{3}{4}$ in.   2. $99\frac{3}{4}$ in.   3. $11\frac{2}{3}$ ft

    4. $4\frac{5}{16}$ in.   5. 14 ft $1\frac{1}{2}$ in.   6. $110\frac{1}{4}$ in.

    7. $318\frac{1}{2}$ mi   8. $36\frac{3}{4}$ in.   9. $356\frac{1}{2}$ lb

    10. 118 in.   11. $\frac{9}{10}$ in.   12. 210 in. or 17 ft 6 in.

    13. $9\frac{3}{4}$ in.   14. $431\frac{41}{64}$ cu in.   15. $348\frac{3}{4}$ min

    16. 1001 cu in.   17. $10\frac{2}{3}$ hr   18. 126 in.

    19. $5\frac{2}{5}$ in.   20. 390 lb   21. $22\frac{1}{2}$ picas

    22. 24 MGD   23. $31\frac{1}{8}$ 24. $45°, 72°, 60°$

    25. $3\frac{1}{4}$ in.   26. $2\frac{1}{2}$ in.   27. $\frac{15}{32}$ in.

Exercises 2-3, page 80

A.   1. $1\frac{2}{3}$ 2. 9 3. $\frac{5}{16}$ 4. 32 5. $\frac{1}{2}$ 6. 1

    7. $\frac{1}{4}$ 8. $2\frac{2}{5}$ 9. 9 10. 4 11. $1\frac{1}{3}$ 12. $1\frac{3}{4}$

    13. $1\frac{1}{5}$ 14. $8\frac{1}{3}$ 15. 16 16. $\frac{1}{9}$ 17. 18 18. $\frac{6}{7}$

    19. $7\frac{1}{2}$ 20. $\frac{3}{5}$ 21. $\frac{5}{6}$ 22. $\frac{7}{8}$ 23. $\frac{1}{8}$ 24. $1\frac{1}{3}$

B.   1. 8 ft   2. 14 ft   3. 48   4. 12   5. 84

    6. $40\frac{1}{2}$ ft   7. 18   8. 8   9. 210   10. 29 ft by 34 ft

    11. 7 sheets   12. 108 rev   13. 45 threads   14. 14 ft by $18\frac{1}{2}$ ft

    15. 284 sq ft   16. $10\frac{7}{8}$ in.

Exercises 2-4, page 90

A.   1. $\frac{1}{4}$ 2. $1\frac{1}{3}$ 3. $\frac{3}{4}$ 4. $\frac{5}{6}$ 5. $\frac{1}{2}$ 6. $\frac{5}{8}$

    7. $\frac{2}{5}$ 8. $\frac{1}{4}$ 9. $\frac{15}{16}$ 10. $1\frac{3}{8}$ 11. $1\frac{1}{2}$ 12. $2\frac{1}{4}$

    13. $\frac{3}{4}$ 14. $\frac{13}{16}$ 15. $\frac{17}{24}$ 16. $\frac{29}{48}$ 17. $\frac{1}{8}$ 18. $\frac{7}{32}$

    19. $\frac{7}{16}$ 20. $\frac{13}{32}$ 21. $\frac{29}{40}$ 22. $1\frac{7}{15}$ 23. $\frac{19}{40}$ 24. $\frac{7}{36}$

    25. $\frac{5}{8}$ 26. $\frac{11}{48}$ 27. $1\frac{3}{4}$ 28. $3\frac{3}{16}$ 29. $4\frac{1}{8}$ 30. $3\frac{7}{20}$

    31. $3\frac{8}{15}$ 32. $2\frac{1}{8}$ 33. $2\frac{3}{8}$ 34. $3\frac{2}{3}$ 35. $1\frac{7}{60}$ 36. $2\frac{14}{15}$

B.   1. $5\frac{1}{8}$ 2. $1\frac{13}{16}$ 3. $2\frac{13}{16}$ 4. $1\frac{17}{60}$ 5. $\frac{7}{8}$ 6. $20\frac{3}{8}$

    7. $2\frac{1}{8}$ 8. $3\frac{3}{8}$ 9. $2\frac{11}{16}$ 10. $5\frac{5}{12}$ 11. $3\frac{9}{10}$ 12. $\frac{15}{56}$

C.   1. 9 in.   2. $\frac{13}{16}$ in.   3. $11\frac{5}{8}$ in.   4. $\frac{3}{32}$ in.

    5. $1\frac{9}{16}$ in.   6. $\frac{15}{16}$ in.   7. $25\frac{1}{4}$ c.i.   8. $2\frac{3}{8}$ ft

    9. $1\frac{1}{8}$ in.   10. $7\frac{31}{32}$ in.   11. $23\frac{5}{16}$ in.   12. $1\frac{5}{8}$ in.

    13. $2\frac{13}{16}$ in.   14. No   15. $\frac{3}{8}$ in.   16. $23\frac{3}{4}$ in.

    17. $7\frac{3}{4}$ in. by $6\frac{1}{2}$ in.   18. 5 in.   19. $\frac{9}{16}$ in.   20. $1\frac{1}{8}$ in.

    21. $4\frac{33}{64}$ in.   22. $9\frac{1}{4}$ in.

Problem Set 2, page 95

A. 1. $\frac{9}{8}$ 2. $\frac{21}{5}$ 3. $\frac{5}{3}$ 4. $\frac{35}{16}$ 5. $\frac{99}{32}$ 6. $\frac{33}{16}$

     7. $\frac{13}{8}$ 8. $\frac{55}{16}$ 9. $2\frac{1}{2}$ 10. $9\frac{1}{2}$ 11. $8\frac{1}{3}$ 12. $1\frac{1}{8}$

     13. $1\frac{9}{16}$ 14. $1\frac{5}{16}$ 15. $8\frac{3}{4}$ 16. $2\frac{1}{3}$ 17. $\frac{3}{16}$ 18. $\frac{1}{4}$

     19. $\frac{3}{8}$ 20. $\frac{3}{4}$ 21. $\frac{1}{6}$ 22. $1\frac{4}{7}$ 23. $1\frac{4}{5}$ 24. $3\frac{2}{5}$

     25. 9   26. 28   27. 44   28. 44   29. 68   30. 18

     31. 15   32. 26   33. $\frac{7}{16}$ 34. $\frac{2}{3}$ 35. $\frac{7}{8}$ 36. $1\frac{1}{4}$

     37. $\frac{3}{5}$ 38. $\frac{2}{10}$ 39. $\frac{7}{4}$ 40. $\frac{1}{9}$

B. 1. $\frac{3}{32}$ 2. $\frac{1}{2}$ 3. $\frac{7}{12}$ 4. $\frac{5}{256}$ 5. $1\frac{1}{4}$ 6. $\frac{49}{80}$

     7. $\frac{5}{64}$ 8. $5\frac{1}{4}$ 9. $7\frac{1}{2}$ 10. $\frac{2}{3}$ 11. 27   12. 34

     13. $11\frac{2}{3}$ 14. $7\frac{1}{2}$ 15. 2   16. $\frac{4}{5}$ 17. 32   18. $10\frac{2}{3}$

     19. $\frac{1}{6}$ 20. $\frac{3}{4}$ 21. $\frac{7}{10}$ 22. $\frac{5}{6}$ 23. $2\frac{4}{9}$ 24. $1\frac{13}{15}$

C. 1. $1\frac{1}{4}$ 2. $1\frac{1}{4}$ 3. $\frac{7}{32}$ 4. $1\frac{5}{8}$ 5. $1\frac{13}{30}$ 6. $\frac{29}{40}$

     7. $\frac{3}{8}$ 8. $\frac{3}{8}$ 9. $\frac{7}{16}$ 10. $\frac{19}{30}$ 11. $\frac{23}{40}$ 12. $1\frac{13}{32}$

     13. $3\frac{3}{8}$ 14. $2\frac{7}{16}$ 15. $4\frac{1}{2}$ 16. $1\frac{1}{8}$ 17. $1\frac{19}{24}$ 18. $1\frac{5}{12}$

     19. $1\frac{1}{30}$ 20. $3\frac{19}{20}$ 21. $1\frac{1}{6}$ 22. $1\frac{1}{3}$ 23. $\frac{2}{5}$ 24. $3\frac{1}{3}$

D. 1. $37\frac{1}{8}$ in.   2. 22   3. $22\frac{5}{16}$ in.; $22\frac{7}{16}$ in.   4. $1\frac{3}{8}$ in.

     5. $\frac{25}{32}$ in.   6. Yes   7. $4\frac{3}{5}$ cu ft   8. $15\frac{15}{16}$ in.

     9. $92\frac{13}{16}$ in.   10. 13 ft   11. $\frac{15}{16}$ in.

     12. A: $2\frac{11}{16}$ in.   B: $2\frac{5}{32}$ in.   C: $6\frac{9}{32}$ in.   D: $8\frac{7}{16}$ in.

     13. $4\frac{3}{32}$ in.   14. $21\frac{3}{8}$ in.   15. $244\frac{1}{2}$ in.   16. $859\frac{3}{8}$ in. or 71 ft $7\frac{3}{8}$ in.

     17. $1\frac{1}{16}$ in. and $\frac{11}{16}$ in.   18. $\frac{3}{5}$ in. and $\frac{9}{10}$ in.   19. $5\frac{3}{16}$ in.

     20. $24\frac{5}{8}$ in.; $24\frac{1}{8}$ in.   21. $6\frac{2}{3}$ min   22. $\frac{15}{32}$ in.

Chapter 3   *Exercises 3-1, page 109*

A. 1. Seventy-two hundredths
     2. Eight and seven tenths
     3. Twelve and thirty-six hundredths
     4. Five hundredths
     5. Three and seventy-two thousandths
     6. Fourteen and ninety-one thousandths
     7. Three and twenty-four ten-thousandths
     8. Six and eighty-three ten-thousandths
     9. 0.004   10. 3.4   11. 6.7
     12. 0.005   13. 12.8   14. 3.021
     15. 10.032   16. 40.7   17. 0.0116
     18. 0.0047   19. 2.0374   20. 10.0222

B. 1. 21.01   2. 78.17   3. $15.02   4. $151.11
     5. 1.617   6. 5.916   7. 828.6   8. 238.16
     9. 63.7305   10. 462.04   11. 6.97   12. 1.04
     13. $15.36   14. $6.52   15. 42.33   16. 36.18
     17. $22.02   18. $24.39   19. 113.96   20. 13.22
     21. 45.195   22. 245.11   23. $27.51   24. 151.402
     25. 95.888   26. 39.707   27. 15.16   28. 86.07
     29. 8.618   30. 18.6373   31. 31.23   32. 292.19

33. 17.608　　34. 29.395　　35. 0.0776　　36. 0.04262
37. 24.22　　38. 0.612　　39. 1.748　　40. 1.833

C. 1. 0.473 in.　　2. 11.85 lb
　　3. (a) 0.013 in.　　(b) smaller; 0.021 in.　　(c) #14
　　4. $2673.90　　5. A: 2.20 in.　　B: 0.45 in.　　C: 4.22 in.
　　6. 32.2 in.　　7. 2.267 in.　　8. $1667.05　　9. No
　　10. 68 ft　　11. 3.4 hr　　12. 0.013 cm　　13. 3.37 in.
　　14. 19.3 cm　　15. 0.843 in.　　16. 70.15 in.

D. 1. $308.24　　2. $24,869.19
　　3. (a) 0.7399　　(b) 4240.775　　(c) 510.436　　(d) 7.4262

Exercises 3-2, page 125

A. 1. 0.00001　　2. 21.5　　3. 4　　4. 0.09
　　5. 0.84　　6. 0.00006　　7. 0.00003　　8. 2.18225
　　9. 0.07　　10. 0.03　　11. 2.16　　12. 3.6225
　　13. 6.03　　14. 120　　15. 20　　16. 130
　　17. 0.045　　18. 126　　19. 60　　20. 10,000
　　21. 400　　22. 0.037　　23. 6.6　　24. 3256.25
　　25. 605　　26. 13.915　　27. 0.00378　　28. 0.29
　　29. 0.048　　30. 3200　　31. 45,000　　32. 0.000365
　　33. 0.00008　　34. 0.001128　　35. 0.000364　　36. 0.000048
　　37. 0.000096　　38. 55.53　　39. 1.705

B. 1. 42.88　　2. 0.6　　3. 6.5　　4. 117.625
　　5. 79.14　　6. 1463　　7. 3.6494　　8. 19.48
　　9. 0.216　　10. 56.83

C. 1. 3.33　　2. 0.83　　3. 10.53　　4. 37.04
　　5. 0.12　　6. 2.62　　7. 33.86　　8. 4.96
　　9. 33.3　　10. 0.2　　11. 0.3　　12. 11.1
　　13. 0.2　　14. 0.336　　15. 0.143　　16. 0.224
　　17. 65　　18. 13.268　　19. 2.999　　20. 1109.001

D. 1. $105　　2. $15.60　　3. 18.8 lb, or 19 lb rounded　　4. 7.8 hr
　　5.

	W	C
A	16.91 lb	$16.57
B	23.18	20.63
C	8.64	9.07
D	3.04	6.54

T = $52.81

　　6. 40.035 lb　　7. 48 lb　　8. 8.36 lb
　　9. 153.37 ft　　10. 6255 lb　　11. 0.075 volts
　　12. $1666.48　　13. 335　　14. 1600　　15. 27 gal
　　16. $67.99　　17. $576　　18. $869　　19. 2625 cycles

E. 1. 8.00000007
　　2. (a) 0.01234567　　(b) 0.00112233　　(c) 0.000111222
　　3. 4.2435 in.　　4. 0.0659 mm　　5. $580
　　6. (a) 0.031 lb　　(b) 12¢　　7. 26.2 psi　　8. $0.022 or 2.2¢
　　9. 23.343 mi　　10. 19.908 psi

Exercises 3-3, page 134

A. 1. 0.25　　2. 0.67　　3. 0.75　　4. 0.4　　5. 0.8
　　6. 0.83　　7. 0.29　　8. 0.57　　9. 0.86　　10. 0.38

11.	0.75	12.	0.1	13.	0.3	14.	0.17	15.	0.42
16.	0.19	17.	0.38	18.	0.56	19.	0.81	20.	0.15
21.	0.22	22.	0.65	23.	0.46	24.	0.61	25.	0.03
26.	0.213	27.	0.019	28.	0.34				

B.
1. 4.385   2. 1.77   3. 1.5   4. 0.7681
5. 7.88   6. 2.33   7. 1.43   8. 2.98
9. 3.64   10. 10.908   11. 7.65   12. 1.07

C.
1. 1.375 g; 5.2 tablets
2. (a) 420 sq ft   (b) 3712.5 sq ft of 4 in. and 4851 sq ft of 6 in.
   (c) $31,450.20
3. 2.3 squares   4. 2 in.
5. (a) $64.94   (b) $385.94   (c) $316.74
   (d) $295.83; Total: $1063.45
6. 523 full lots   7. 0.396 in.
8. 3.4775 in.   9. (a) 120.2 ft   (b) $12 \text{ ft } 6\frac{5}{8}$ in.
10. $19\frac{3}{8}$ in.

D.
1.

Gauge No.	Thickness (in.)	Gauge No.	Thickness (in.)
7–0	0.5	14	0.078
6–0	0.469	15	0.07
5–0	0.438	16	0.063
4–0	0.406	17	0.056
3–0	0.375	18	0.05
2–0	0.344	19	0.044
0	0.313	20	0.038
1	0.281	21	0.034
2	0.266	22	0.031
3	0.25	23	0.028
4	0.234	24	0.025
5	0.219	25	0.022
6	0.203	26	0.019
7	0.188	27	0.017
8	0.172	28	0.016
9	0.156	29	0.014
10	0.141	30	0.013
11	0.125	31	0.011
12	0.109	32	0.01
13	0.094		

2. $26,067.97; $20,540.38; $7999.92; $8246.40; Total: $62,854.67
3. $699.48

Problem Set 3, page 137

A.
1. Ninety-one hundredths
2. Eighty-four hundredths
3. Twenty-three and one hundred sixty-four thousandths
4. Sixty-three and two hundred nineteen thousandths
5. Nine and three tenths
6. Three and forty-five hundredths
7. Ten and six hundredths
8. Fifteen and thirty-seven thousandths
9. 0.07   10. 0.018   11. 200.8   12. 16.17
13. 63.063   14. 110.021   15. 5.0063   16. 11.0218

B.
1. 23.19   2. 174.96   3. $19.29   4. 26.06   5. 1.94
6. $3.95   7. 88.26   8. 12.55   9. 277.104   10. 5.311
11. 239.01   12. 253.01   13. 83.88   14. 3.985   15. 33.672
16. 4.71   17. 4.28   18. 0.34   19. 1.92   20. 6.77

C.   1. 0.00008   2. 0.003   3. 0.84   4. 37.68   5. 0.108

     6. 2.12   7. 19.866   8. 1.24   9. 61.7   10. 0.26

   11. 3.8556   12. 28.585   13. 0.006   14. 5.26   15. 18

   16. 0.2   17. 4.34   18. 4.36   19. 0.23   20. 1.50

   21. 2.78   22. 0.22   23. 0.04   24. 16.84   25. 0.526

   26. 0.503   27. 214.634 28. 0.028   29. 27.007   30. 1.182

   31. 1.5   32. 23.5   33. 4.6   34. 0.9   35. 47.2

   36. 123.8

D.   1. 0.0625   2. 0.875   3. 0.15625   4. 0.4375

    5. 1.375   6. 1.75   7. 2.55   8. $1.1\overline{6}$

    9. $2.\overline{6}$ 10. 1.03125   11. 2.3125

   12. 19.28   13. 30.67   14. 1.92   15. 78.57

   16. 0.96   17. 0.13

E.   1. 291   2. 2.75 in.   3. 9.772 in.   4. $588.80

    5. 0.0021 in.   6. 1.132 ohms   7. 489.4 lb   8. 0.02374 in.

    9. (a) 0.1875 in.   (b) 0.15625 in.   (c) 0.375 in.

        (d) 0.203125 in.

   10. (a) $\frac{3}{16}$ in.   (b) $\frac{2}{16}$ in.   (c) $\frac{5}{16}$ in.   (d) $\frac{13}{16}$ in.

   11. (a) 0.0089 in.   (b) 0.0027 in.   (c) 0.55215 in.   (d) 0.306175 in.

   12. $3808.74   13. 17.5 ft   14. 0.0423 in.

   15. Max: 2.525 in. Min: 2.475 in.   16. $823.13   17. $229.90

   18. $42.50   19. 56.7 cu ft   20. 27 in.   21. 18

   22. $27.97/hr   23. 11.4 hr   24. 1.4045 in.   25. 52 lines

   26. 359 gal   27. $22.12   28. $200.27

   29. Last year was greater by $0.83   30. 0.358 in.

Chapter 4   *Exercises 4-1, page 157*

A. (In each case only the missing answer is given.)

    1. (a) 7:1   (b) 12:7   (c) 6   (d) 6

      (e) 45   (f) 9   (g) 16   (h) 50

      (i) 3:2   (j) 2:5

    2. (a) 8:3   (b) 5:4   (c) 16 in.   (d) 6 cm

      (e) 40 cm   (f) $2\frac{1}{2}$:1 or 5:2   (g) 0.75 or 3:4   (h) 6.5 cm

      (i) 5.4 cm   (j) 17.8 cm

    3. (a) 16:12   (b) 8 ft   (c) 28 ft   (d) 6.88:12

      (e) 7.2:12   (f) 4 ft   (g) 20 ft   (h) 5 ft 1 in.

B.   1. $x = 12$ 2. $R = 86.4$ 3. $y = 100$ 4. $H = 60$

    5. $P = \frac{7}{6}$ 6. $x = 102$ 7. $A = 3.25$ 8. $M = 7.2$

    9. $T = 1.8$ 10. $N = 0.42$ 11. $x = 3$ 12. $A = 1.3$

   13. $x = 5.04$ 14. $R = 6\frac{1}{2}$ 15. $L = 5$ ft   16. $x = 58.9$ cm

   17. $x = 4$ cm   18. $W = \frac{1}{8}$ in. or 0.125 in.

C.   1. 5 cu in.   2. 15 gal   3. 21 pins

    4. 25.5 lb   5. 5.25 cu yd   6. 300 lb

    7. 288.75 gal   8. 224 ounces   9. 124.8 lb

   10. 12 ounces   11. $736   12. 72 parts

   13. 375 ft   14. $20\frac{5}{8}$ lb or 20.625 lb   15. $102

   16. $2\frac{3}{4}$ pints   17. 168   18. 20.4 in.

Exercises 4-2, page 167

A.   1. 32%   2. 100%   3. 50%   4. 210%

    5. 25%   6. 375%   7. 4000%   8. 67.5%

    9. 200%   10. 7.5%   11. 50%   12. $16\frac{2}{3}$%

   13. 33.5%   14. 0.1%   15. 0.5%   16. 30%

17.	150%	18.	7.5%	19.	330%	20.	20%
21.	40%	22.	10%	23.	95%	24.	3%
25.	30%	26.	1.5%	27.	60%	28.	775%
29.	120%	30.	400%	31.	604%	32.	900%
33.	25%	34.	20%	35.	35%	36.	37.5%
37.	$83\frac{1}{3}$%	38.	237.5%	39.	370%	40.	180%

B.
1.	0.06	2.	0.45	3.	0.01	4.	0.33
5.	0.71	6.	4.56	7.	0.005	8.	0.0005
9.	0.0625	10.	0.0875	11.	0.3	12.	0.021
13.	8	14.	0.08	15.	0.0025	16.	$0.16\overline{3}$
17.	0.07	18.	0.03	19.	0.56	20.	0.15
21.	0.01	22.	0.075	23.	0.90	24.	0.003
25.	1.5	26.	0.015	27.	0.0675	28.	0.005
29.	0.1225	30.	1.252	31.	0.012	32.	2.4

C.
1.	$\frac{1}{20}$	2.	$\frac{1}{5}$	3.	$2\frac{1}{2}$	4.	$\frac{4}{25}$	5.	$\frac{53}{100}$
6.	$\frac{1}{1000}$	7.	$\frac{23}{25}$	8.	$\frac{1}{50}$	9.	$\frac{9}{20}$	10.	$1\frac{3}{4}$
11.	$\frac{1}{12}$	12.	$\frac{3}{8}$	13.	$\frac{6}{25}$	14.	$1\frac{19}{100}$	15.	$\frac{1}{2000}$
16.	$\frac{3}{20}$	17.	$4\frac{4}{5}$	18.	$\frac{5}{6}$	19.	$\frac{5}{8}$	20.	$\frac{3}{250}$

Exercises 4-3, page 175

A.
1.	80%	2.	64%	3.	15	4.	100%
5.	54	6.	12.5%	7.	150	8.	80
9.	$66\frac{2}{3}$%	10.	500%	11.	100	12.	52%
13.	$21.25	14.	600%	15.	1.5	16.	$23\frac{1}{3}$
17.	43.75%	18.	$133\frac{1}{3}$%	19.	2.4	20.	375
21.	40	22.	3.25	23.	5000	24.	17.6

B.
1.	225	2.	25%	3.	20¢	4.	180%
5.	160	6.	$2.72	7.	150%	8.	2%
9.	17.5	10.	400%	11.	427	12.	22.1
13.	2%	14.	3.625	15.	460	16.	1400
17.	50	18.	2.475				

C.
1.	88%	2.	$48,200	3.	5.5%	4.	$96,900
5.	$165.60	6.	$130	7.	$2.14	8.	$141.14
9.	9.6 in. by 12 in.	10.	67%	11.	1414.5		
12.	4082 copies	13.	76%	14.	32 lb		
15.	38%	16.	8780 lineal ft.	17.	52,400 lineal ft		

Exercises 4-4, page 194

1. 64%   2. 3.78 kW   3. 63.6%
4. 5035 to 5565, 2695 to 2805, 6120 to 7480, 4536 to 6804
   tolerances: 265, 55, 680, 1134
5. 67.5%   6. 550 ohms   7. 5%
8. $504   9. 12.5%
10. $16\frac{2}{3}$% ($\frac{2}{3}$ lb) tin; 2% (0.08 lb) zinc; $81\frac{1}{3}$% (3.25 lb) copper
11. 0.9%   12. 1660 lb   13. 81.25 hp   14. Yes (1910)
15. 0.13%   16. $118.80
17. (a) 0.03%   (b) 0.44%   (c) ±0.007 in.
18. $16.41   19. 37.5%   20. $121.00   21. $16.69   22. $725
23. $554.25   24. (a) $10,625   (b) $460.42
25. $94.00   26. 21.1 mpg   27. B, by $1.74   28. $170.64
29. 25%   30. 8%   31. 15.6 MGD   32. 39%   33. 150 lb

34.

Measurement	Tolerance	Maximum	Minimum	Percent Tolerance
1.58 in.	±0.002 in.	1.582 in.	1.578 in.	0.13%
	±0.005 in.	1.585 in.	1.575 in.	0.32%
	±0.002 in.	1.582 in.	1.578 in.	0.15%
0.647 in.	±0.004 in.	0.651 in.	0.643 in.	0.62%
	±0.001 in.	0.648 in.	0.646 in.	0.15%
	±0.001 in.	0.648 in.	0.646 in.	0.20%
165.00 mm	±0.15 mm	165.15 mm	164.85 mm	0.09%
	±0.50 mm	165.50 mm	164.50 mm	0.30%
	±0.08 mm	165.08 mm	164.92 mm	0.05%
35.40 mm	±0.01 mm	35.41 mm	35.39 mm	0.03%
	±0.07 mm	35.47 mm	35.33 mm	0.20%
	±0.02 mm	35.42 mm	35.38 mm	0.05%
	±0.03 mm	35.43 mm	35.37 mm	0.08%
	±0.04 mm	35.44 mm	35.56 mm	0.10%

35. 28% 36. $1389.98 37. $105.22 38. $588,571.43
39. Approx. 2.8% 40. 20% 41. $68,090.88 42. $4367.30
43. (a) 8800 Btu/hr (b) 10,080 Btu/hr

Exercises 4-5, page 207

A. 1. (a) $\frac{5}{8}$ in. (b) 31.2 cm (c) 8 ft (d) 3 m
    2. (a) $2\frac{4}{5}$ in. (b) $12\frac{1}{2}$ ft (c) 1.4 m (d) 42 cm
    3. (a) 100 (b) 9 (c) 25 (d) 1500
    4. (a) 150 (b) 15 (c) 8 (d) 700

B. 1. 24 in. 2. (a) 43.5 ft (b) 48 to 1
    3. 139 HP 4. 216 ft 5. 109.1 lb
    6. 120 min 7. 980 rpm 8. 144 lb
    9. 23 lb 10. 2500 rpm 11. 500 rpm
    12. 21.6 watts 13. 6 hr 14. 33 turns
    15. 1470 ft 16. 3.8 hr, rounded 17. 0.96 sq in.
    18. 12.5 cu yd cement, 37.5 cu yd sand, 75 cu yd gravel
    19. 4 in. 20. $4\frac{11}{16}$ in. by $2\frac{3}{8}$ in. 21. 22.4 psi
    22. 4.766 ohms 23. $377.54
    24. $A = 5.95$ cm, $B = 5.10$ cm, $C = 4.25$ cm
        $D = 3.40$ cm, $E = 2.55$ cm, $F = 1.70$ cm
        $G = 0.85$ cm
    25. 14.4 hr, rounded 26. 86.1 psi 27. (a) 6 in. (b) 8 to 3
    28. (a) 5 in. (b) 4 29. 458 lb.

Problem Set 4, page 213

A. 1. (a) 30 in. (b) 5 to 2 (c) 25 in.
    2. (a) 1 to 3 (b) 65 (c) 42

B. 1. $x = 35$ 2. $x = 22.5$ 3. $x = 20$
    4. $x = 12$ 5. $x = 14.3$ 6. $x = 8.928$

C. 1. 72%   2. 6%   3. 60%   4. 35.8%   5. 130%
   6. 303%   7. 400%   8. 70%   9. $16\frac{2}{3}$%   10. 260%

D. 1. 0.04   2. 0.37   3. 0.11   4. 0.94   5. 0.0125
   6. 0.0009   7. 0.002   8. 0.017   9. 0.03875   10. 0.0802
   11. 1.15   12. 2.1

E. 1. $\frac{7}{25}$ 2. $3\frac{3}{4}$ 3. $\frac{81}{100}$ 4. $\frac{1}{25}$
   5. $\frac{1}{200}$ 6. $\frac{7}{10}$ 7. $\frac{7}{50}$ 8. $\frac{5}{12}$

F. 1. 60%   2. $6   3. 5.6   4. 200%   5. 42
   6. 120%   7. $28.97   8. 125   9. 8000   10. 0.099

G. 1. 29.92 in.   2. 0.9 lb of carbon in cast iron; 0.016 lb in wrought iron
   3. (a) $3827.25   (b) $399.50   (c) $1699   4. $107.87
   5. (a) $10.58   (b) $15.85   (c) 73¢   (d) $67.42   (e) $349.75
   6. $427.27   7. 1748 ft   8. 4%
   9. (a) 0.06%   (b) 2.82%   (c) ±0.009 in.   (d) 0.03%
      (e) ±0.62 in.
   10. 22%   11. $7216.25   12. 14.71 cm
   13. ±135 ohms; 4365 to 4635 ohms
   14. 90%   15. 0.9%   16. $35.06   17. 72%
   18. (a) $6412.50   (b) $1428.13   19. 9.12 HP
   20. 22.5%   21. 3.6%
   22. 1950 bricks   23. (a) 66 fps   (b) 15 mph
   24. (a) 0.07 in. per foot   (b) $1\frac{3}{4}$ in. per 25 feet
   25. $5\frac{1}{4}$ in.   26. 8.8 lb
   27. 150 seconds or $2\frac{1}{2}$ minutes
   28. $2170 per month   29. 450 parts
   30. (a) 12:5 or 2.4 to 1   (b) 3600 rpm
   31. $3\frac{3}{4}$ hr
   32. (a) 9 ft.   (b) 30 ft   (c) $2\frac{3}{4}$ ft   (d) $7\frac{1}{2}$ ft
   33. 6 sacks   34. 60 teeth   35. 960 lb
   36. 360 turns   37. 1.47 amperes   38. 12,632 copies
   39. $673.98   40. 546 rpm   41. 35 employees
   42. 1.984 in.   43. 83.96 psi   44. 74

Chapter 5   *Exercises 5-1, page 234*

A.

		Precision to nearest	Accuracy in significant digits
	1.	tenth	2
	2.	thousandth	1
	3.	whole number	3
	4.	hundred	2
	5.	thousandth	4
	6.	ten	2
	7.	hundred	3
	8.	hundredth	4
	9.	tenth	3
	10.	hundredth	2

B. 1. 21 in.   2. 1000 lb   3. 9.6 in.   4. 28.8 psi
   5. 1.91 gal   6. 51.8 mi   7. 1.19 oz   8. 10.216 sec
   9. 50 psi   10. 9.7 tons

C. 1. 40 sq ft   2. 5.3 sq in.   3. 7.2 sq ft   4. 260 miles
   5. 1.9 sq ft   6. 380 sq in.   7. 4.6 cu ft   8. 8.6 ft
   9. 63 mph   10. $0.43/lb   11. 1.6 hr   12. 1.9 ft

D. 1. (a) 1.88 in. (b) 4.05 in. (c) 3.13 sec (d) 0.06 in.
        (e) 0.09 in. (f) 1.19 lb (g) 2.27 in. (h) 0.59 in.
        (i) 0.38 lb

    2. (a) $\frac{15}{16}$ in. (b) $2\frac{9}{16}$ in. (c) $1\frac{13}{16}$ in. (d) $3\frac{11}{16}$ in.
        (e) $\frac{13}{16}$ in. (f) $\frac{5}{16}$ in. (g) $1\frac{15}{16}$ in. (h) $1\frac{9}{16}$ in.
        (i) $\frac{13}{16}$ in.

    3. (a) $1\frac{29}{32}$ in. (b) $\frac{27}{32}$ in. (c) $2\frac{11}{32}$ in. (d) $\frac{21}{32}$ in.
        (e) $2\frac{3}{32}$ in. (f) $\frac{9}{32}$ in. (g) $\frac{19}{32}$ in. (h) $\frac{22}{32}$ in.
        (i) $1\frac{17}{32}$ in.

    4. (a) $\frac{15}{64}$; error of 0.0006   (b) $\frac{33}{64}$; error of 0.0006
        (c) $1\frac{51}{64}$; error of 0.0031   (d) $2\frac{27}{64}$; error of 0.0019
        (e) $3\frac{11}{64}$; error of 0.0031   (f) $2\frac{55}{64}$; error of 0.0006
        (g) $1\frac{60}{64}$; error of 0.0025   (h) $\frac{41}{64}$; error of 0.0044
        (i) $\frac{31}{64}$; error of 0.0044

E. 1. Yes   2. $\frac{13}{32}$; error 0.0034   3. $\frac{30}{64}$ in.
    4. $\frac{24}{32}$ in.   5. 4.28 in.   6. 4.7 cu ft   7. $3\frac{5}{16}$ in.
    8. 17.5 ft   9. 354 mi   10. 23 ft   11. 0.013 sec

Box, page 245–246

1. (a) 2 bf   (b) 16 bf   (c) 8 bf   (d) $10\frac{2}{3}$ bf   (e) 7 bf   (f) 3 bf
2. 1280 bf   3. 5880 bf

Exercises 5-2, page 247

A. 1. 51 in.   2. 99 ft   3. 18,000 ft   4. 277
    5. 4 bbl   6. 15,000 yd   7. 3.1 atm   8. 120 oz
    9. 582.8 gal   10. $2.\overline{6}$ or 2.7 ft   11. 0.879   12. 3.5 rods
    13. 8.7 yd   14. 14.4 lb   15. 1180 cu in.   16. 26,300 cu in.
    17. 0.80 mi   18. 6.4 gal   19. 83 in.   20. 216 oz
    21. 20 lb 6 oz   22. 9 yd 2 ft

B. 1. 860   2. 109,000   3. 72.6   4. 42.8
    5. 0.563   6. 5.42   7. 0.03   8. 7.1
    9. $0.\overline{3}$ or $\frac{1}{3}$ 10. 500   11. 12,800   12. 95
    13. 2700   14. 380   15. 0.30   16. 220

C. 1. (a) 43,560   (b) 325,829   (c) 10,344   (d) 1613
    2. (a) $\frac{1}{12}$ (b) 144   3. 4
    4. (a) $12\frac{1}{2}$ gal   (b) $30.57   (c) 2000 sq ft   (d) 2.3 gal
        (e) $82.43
    5. 32.6 ± 0.6 in.
    6. (a) 5400 in. × 900 in. × 540 in. or 450 ft × 75 ft × 45 ft
        (b) 117 in. or 9 ft 9 in.
    7. 141 sq yd   8. 0.1 lb/cu in.   9. 10.2 rps   10. 28 pieces
    11. 1.93 cfs   12. 270 min   13. $466\frac{1}{2}$ in.   14. 18,000 oz
    15. 8.5 gal   16. (a) Carpetorium at $9.40 per sq ft   (b) $315.40
    17. 66'7"

Exercises 5-3, page 265

A. 1. (c)   2. (b)   3. (a)   4. (a)   5. (b)   6. (c)   7. (a)
    8. (c)   9. (a)   10. (b)   11. (c)   12. (c)   13. (a)   14. (c)
    15. (a)   16. (a)   17. (a)   18. (b)   19. (a)   20. (b)

B. 1. (b) 2. (c) 3. (a) 4. (a) 5. (b) 6. (c)
    7. (b) 8. (b) 9. (b) 10. (c) 11. (a) 12. (a)

C. 1. 56 mm 2. 80 mg 3. 2500 sq m 4. 64.6 liters
    5. 45 g 6. 347 cm/sec 7. 1250 cu cm 8. 720 kg
    9. 420 cm 10. 102.6 cm 11. 16 cm/sec 12. 5.98 m
    13. 9620 m 14. 0.785 m 15. 56.5 km 16. 7.8 ha
    17. 9.5 g 18. 52 km/h 19. 0.58 liters 20. 25 m/min
    (rounded) (rounded)

D. 1. 7.6 2. 130 3. 3.2 4. 6.25 5. 23
    6. 68 7. 68.9 8. 1.5 9. 19 10. 295
    11. 11.9 12. 2.6 13. 6.1 14. 98 15. 5.0
    16. 110 17. 50 18. 89 19. 19 20. 6.6
    21. 26.5 sq m 22. 8390 cu cm
    23. 3.59 cu yd 24. 7.16 sq in.

E. 1.

in.	mm
0.030	0.762
0.035	0.889
0.040	1.016
0.045	1.143

in.	mm
$\frac{1}{16}$	1.588
$\frac{5}{64}$	1.984
$\frac{3}{32}$	2.381
$\frac{1}{8}$	3.175

in.	mm
$\frac{5}{32}$	3.969
$\frac{3}{16}$	4.763
$\frac{3}{8}$	9.525
$\frac{11}{64}$	4.366

    2. (a) 2.2 lb (b) 28.3 kg (c) 1.0 kg
    3. 11.9 mph; 19.1 km/h
    4. (a) 103,000 (b) 689 (c) 2000 (d) 120
    5. (a) 0.6093 (b) 564 (c) 0.89 (d) 0.057 (e) 1.02
    6. 0.94 in. by 1.42 in.
    7. 12 km/liter 8. $4.16 9. 9.3 sq m
    10. (a) 2.4 m × 1.2 m × 1.2 m (b) 3.6 cu m
    11. (a) 1.6093 (b) 76.2 (c) 21.59 cm × 27.94 cm (d) 2.54
         (e) 804.65 (f) 0.0648 (g) 28.35; 453.6 (h) 0.0648
         (i) 7.57 (j) 37.85 (k) 7568
    12. 7 liters 13. (a) 18 kV (b) 0.435 mV 14. $2.51

Exercises 5-4, page 288

A. 1. (a) $\frac{5}{8}$ in. (b) $1\frac{7}{8}$ in. (c) $2\frac{1}{2}$ in. (d) 3 in.
        (e) $\frac{1}{4}$ in. (f) $1\frac{1}{16}$ in. (g) $2\frac{5}{8}$ in. (h) $3\frac{1}{2}$ in.

    2. (a) $\frac{5}{16}$ in. (b) $\frac{3}{4}$ in. (c) $1\frac{1}{32}$ in. (d) $1\frac{17}{32}$ in.
        (e) $\frac{1}{8}$ in. (f) $\frac{17}{32}$ in. (g) $\frac{53}{64}$ in. (h) $1\frac{7}{16}$ in.

    3. (a) $\frac{2}{10}$ in. = 0.2 in. (b) $\frac{5}{10}$ in. = 0.5 in. (c) $1\frac{3}{10}$ in. = 1.3 in.
        (d) $1\frac{6}{10}$ in. = 1.6 in. (e) $\frac{25}{100}$ in. = 0.25 in. (f) $\frac{72}{100}$ in. = 0.72 in.
        (g) $1\frac{49}{100}$ in. = 1.49 in. (h) $1\frac{75}{100}$ in. = 1.75 in.

B. 1. 0.650 in. 2. 0.705 in. 3. 0.287 in. 4. 0.061 in.
    5. 0.850 in. 6. 0.441 in. 7. 0.4068 in. 8. 0.5997 in.
    9. 0.2581 in. 10. 0.7796 in. 11. 0.0888 in. 12. 0.8451 in.
    13. 11.57 mm 14. 3.41 mm 15. 22.58 mm 16. 18.21 mm
    17. 18.94 mm 18. 13.27 mm

C. 1. 3.256 in. 2. 0.836 in. 3. 2.078 in. 4. 0.609 in. 5. 2.908 in.
    6. 1.682 in. 7. 2.040 in. 8. 0.984 in. 9. 0.826 in. 10. 2.656 in.

D. 1. 62°21′   2. 19°38′   3. 34°56′   4. 65°10′
    5. 35°34′   6. 63°48′   7. 20°26′   8. 44°15′

E. 1. 2.000 in. + 0.200 in. + 0.050 in. + 0.057 in. + 0.0503 in.
    2. 1.000 in. + 0.500 in. + 0.200 in. + 0.140 in. + 0.056 in. + 0.0507 in.
    3. 0.200 in. + 0.120 in. + 0.053 in. + 0.0502 in.
    4. 2.000 in. + 1.000 in. + 0.500 in. + 0.200 in. + 0.050 in. + 0.058 in. +
        0.0507 in.
    5. 1.000 in. + 0.500 in. + 0.070 in. + 0.058 in. + 0.0509 in.
    6. 2.000 in. + 0.500 in. + 0.100 in. + 0.050 in. + 0.058 in. + 0.0508 in.
    7. 0.050 in. + 0.053 in. + 0.0509 in.
    8. 2.000 in. + 0.500 in. + 0.300 in. + 0.120 in. + 0.057 in. + 0.0506 in.
    9. 2.000 in. + 0.140 in. + 0.0505 in.
    10. 1.000 in. + 0.400 in. + 0.056 in. + 0.0506 in.

Problem Set 5, page 293

A. | | **Precision to nearest** | **Accuracy in significant digits** |
    | --- | --- | --- |
    | 1. | Tenth | 2 |
    | 2. | Ten | 2 |
    | 3. | Hundredth | 3 |
    | 4. | Whole unit | 4 |
    | 5. | Hundredth | 1 |
    | 6. | Whole unit | 2 |
    | 7. | Thousand | 1 |
    | 8. | Thousandth | 4 |

B. 1. 3.5 sec   2. 6.8 sq ft   3. 0.75 in.   4. 40 sq cm
    5. $16\frac{7}{8}$ in.   6. 3 lb 14 oz   7. 21.1 mi/gal   8. 9.7 cm
    9. 17.65 psi   10. 3.2 ft   11. 3.87 in.   12. 404 lb
    13. 21 ft 3 in.   14. 140 cu in.

C. 1. 15 cm   2. 79 ft/sec   3. ≈72°F   4. 2.7 qt
    5. 820 g   6. 16.2 mm   7. 45,800 cu in. 8. 5°C
    9. 31.5 lb   10. 8.415 atm   11. 667 cm/sec   12. 12.7 ha
    13. 23.3 km   14. 5700 cu cm   15. 60.2 cu ft   16. 15 sq m
    17. 1.4 liters   18. 339 km/h   19. 11.48 bu   20. 1.8 gal
    21. 980 sq in.   22. 11.59 liters   23. 95 mg   24. 596 km
    25. 104 cm   26. 114 cm/sec   27. 1850 cu cm   28. 4000 g
    29. 53 mph   30. 13 kg

D. 1. (b)   2. (b)   3. (a)   4. (a)   5. (b)   6. (c)

E. 1. (a) $\frac{5}{10}$ in. = 0.5 in.   (b) $1\frac{3}{10}$ in. = 1.3 in.   (c) $\frac{17}{100}$ in. = 0.17 in.
        (d) $1\frac{35}{100}$ in. = 1.35 in.   (e) $\frac{3}{4}$ in.   (f) $3\frac{3}{4}$ in.   (g) $\frac{5}{16}$ in.
        (h) $1\frac{19}{32}$ in.   (i) $\frac{3}{8}$ in.   (j) $1\frac{27}{64}$ in.

    2. (a) 0.807   (b) 0.550   (c) 12.13 mm   (d) 19.07 mm
        (e) 0.5469 in.   (f) 0.8179 in.
    3. (a) 2.156 in.   (b) 3.030 in.   (c) 0.612 in.   (d) 1.925 in.
        (e) 1.706 in.   (f) 1.038 in.

F. 1. $22\frac{9}{16}$ in.   2. 53 mph   3. 6.1 mph
    4. 0.0505 in., 0.057 in., 0.130 in., 0.200 in., 2.000 in.
    5. $\frac{15}{32}$ in.   6. 500°F   7. 936 sq in.   8. 18.9 gallons
    9. 293°C   10. 164 cu in.   11. 7.623 in.   12. (a) 3.85 cu m/hr
        (b) 7.62 m   (c) 6.35 cm   13. 0.28 lb/cu in.
    14. 28.3 liter   15. 3.81 cm × 8.89 cm or 4 cm by 9 cm   16. 1.52 m
    17. 3.4 sq mi

Chapter 6   *Exercises 6-1, page 305*

A.   1. −4   2. −17   3. 18   4. −15
     5. −10   6. −31   7. −65   8. 23
     9. −47   10. −3   11. 0   12. −1.4
     13. −2.8   14. −3¼   15. −11.4   16. −0.1

B.   1.

     2.

     3.

     4.

C.   1. −6   2. −300   3. +12,000   4. +240
     5. −9   6. −7   7. −80   8. +15
     9. −5   10. −10   11. +6000   12. −30
     13. −$17.50   14. −26

D.   1. 5   2. −4   3. −15   4. −25
     5. 17   6. −5   7. 46   8. 31
     9. −16   10. −16   11. 9   12. −27
     13. 29   14. −65   15. −30   16. −80
     17. −0.9   18. −2.2   19. −28.09   20. −133.84
     21. −1¼   22. 2½   23. −14   24. −15⅓
     25. −¾   26. 3⅛   27. 1⁹⁄₁₆   28. −22¹⁄₁₀
     29. −10   30. 3   31. −16   32. 0
     33. 2   34. −3   35. 18   36. −15
     37. −230   38. 10   39. 7   40. 23.69
     41. 3599   42. −4.89 43. −26,620   44. −0.0655

E.   1. +22 (22 above quota)   2. 6 yd; $\frac{6}{7}$ yards per carry
     3. −9° (9 degrees below zero)   4. $884
     5. −$136,000 (a loss of $136,000)   6. −2$\frac{15}{16}$ in.

Exercises 6-2, page 311

A.   1. −2   2. −13   3. −16   4. −24
     5. 9   6. 21   7. −6   8. −12
     9. 8   10. 14   11. 11   12. 3

13.	-18	14.	-35	15.	-31	16.	-29
17.	8	18.	27	19.	26	20.	45
21.	9	22.	-11	23.	-19	24.	-26
25.	-17	26.	-55	27.	8	28.	14
29.	-87	30.	-112	31.	-67	32.	-27
33.	-10	34.	4	35.	-42	36.	25
37.	6	38.	-13	39.	-1	40.	-9
41.	$-\frac{5}{8}$	42.	$5\frac{3}{4}$	43.	$-6\frac{1}{2}$	44.	$-\frac{1}{6}$
45.	9.2	46.	-1.65	47.	-63.5	48.	-0.0725
49.	41.02	50.	-6.14	51.	-0.85	52.	-1.575

B. 1. 14,698 ft 2. 263° 3. 35°F
    4. 4423 ft 5. $1817 6. -9 (9 under par)
    7. lost $5211 8. -0.625 9. 420 V

Exercises 6-3, page 316

A.	1.	-63	2.	-40	3.	77	4.	36
	5.	-4	6.	-4	7.	12	8.	4
	9.	-4	10.	9	11.	$12\frac{3}{8}$	12.	$-\frac{1}{2}$
	13.	5.1	14.	$-\frac{1}{4}(-0.25)$	15.	$\frac{1}{5}(0.2)$	16.	52
	17.	-90	18.	-54	19.	84	20.	-20
	21.	-28	22.	20	23.	5	24.	-3.45
	25.	-8	26.	8	27.	$\frac{1}{6}$	28.	-15
	29.	$-2\frac{11}{20}$	30.	$1\frac{1}{9}$	31.	0.4	32.	$11\frac{3}{4}$
	33.	$-19\frac{1}{2}$	34.	0.077	35.	-4.48	36.	-0.06
	37.	-90	38.	224	39.	-20	40.	700
	41.	-0.4	42.	50				

B. 1. -19.3 2. 57 3. 0.0019
    4. 160,000 5. -3.16 6. 3.87
    7. 0.23 8. -0.030 9. 46
    10. 215 11. $-7\overline{0},000,000$ 12. $-8,500,000$

C. 1. $-8\frac{1}{3}$°C 2. $-$$23.6 million 3. (a) -120; (b) 40
    4. -3.2 mm 5. -2000 ft/min 6. 17.6°F

Exercises 6-4, page 326

A.	1.	16	2.	9	3.	64	4.	125
	5.	1000	6.	49	7.	256	8.	36
	9.	512	10.	81	11.	625	12.	100,000
	13.	-8	14.	-243	15.	729	16.	1
	17.	5	18.	1	19.	32	20.	64
	21.	196	22.	441	23.	3375	24.	256
	25.	108	26.	576	27.	1125	28.	2744
	29.	2700	30.	9216	31.	18	32.	6
	33.	216	34.	49	35.	19	36.	2
	37.	116	38.	22	39.	2	40.	111
	41.	71	42.	37				

B.	1.	9	2.	12	3.	6	4.	4
	5.	5	6.	3	7.	16	8.	20
	9.	15	10.	7	11.	18	12.	11
	13.	2.12	14.	22.36	15.	3.52	16.	26.46
	17.	14.49	18.	17.92	19.	28.46	20.	9.62
	21.	31.62	22.	44.72	23.	158.11	24.	50
	25.	12.25	26.	17.32	27.	54.77	28.	173.21

29. 1.12   30. 1.00   31. 13.42   32. 8.47
33. 51.26   34. 7.35   35. 4.80   36. 11.45

C.   1. 13.6 ft   2. 35.6 ft   3. 89,000 psi
    4. 19,000 ohms   5. 96 fps   6. 43 mph
    7. 26.6   8. 13.2 ft   9. 73 ft

Problem Set 6, page 329

A.   1. $-13, -8, -2, 4, 7$
    2. $-0.96, -0.6, -0.55, -0.138, 0, 0.4$
    3. $-180, -160, -150, -140, -120, 0$
    4. $-\frac{3}{4}, -\frac{5}{8}, -\frac{1}{2}, \frac{3}{8}, \frac{1}{2}, \frac{3}{4}$

B.   1. -10 2. -26 3. -6 4. 0
    5. -15 6. -50 7. 25   8. 8
    9. 2.3   10. $-2\frac{1}{2}$ 11. $-\$9.15$ 12. $-9\frac{7}{8}$
   13. -16 14. -27.35

C.   1. -60 2. 35   3. 12   4. -40
    5. $7\frac{1}{32}$ 6. $-\frac{3}{4}$ 7. -0.4 8. -36
    9. -120 10. 54   11. -16 12. 0.318

D.   1. -43.75 2. 15   3. -4 4. $-1\frac{1}{32}$
    5. 64   6. 243   7. 289   8. 12,167
    9. 0.25   10. 1.728   11. 0.0004   12. 0.000027
   13. 0.000000001   14. 4.0401   15. 16.1604
   16. -27 17. 16   18. 37   19. 28
   20. 2.1   21. 0.4   22. 2.3   23. 16.29
   24. 34.201   25. 121.68   26. 2.2   27. 432
   28. 22   29. 10   30. 20.52   31. 0.5
   32. 6.7   33. 8.94   34. 10.30   35. 17.61
   36. 1.34   37. 2.05   38. 1.74   39. 1.04
   40. 0.28   41. 2.76   42. 0.21   43. 0.78
   44. 7.71

E.   1. $+7$ amp   2. 15 ft   3. $-22°$F
    4. $-\$848$ 5. 163 A   6. 539 gal/min
    7. 700   8. $\$61,780$ 9. 3200 nanosec
   10. -300 volts   11. (a) $-20°$C   (b) $-4°$C
   12. (a) $-5.8°$F   (b) 10.4°F

Chapter 7   *Exercises 7-1, page 344*

A.   1. $A = 6$ 2. $D = 7$ 3. $T = 13$ 4. $H = 9$
    5. $K = 8$ 6. $Q = 38$ 7. $F = 19$ 8. $W = 6$
    9. $L = 1$ 10. $A = 8$ 11. $B = 13$ 12. $F = 9$

B.   1. $A = 11$ 2. $V = 250$ sq ft   3. $I = 15$ 4. $H = 10$ cm^2
    5. $T = 23$ m^2 6. $V = 221$ cu in.   7. $P = 40$ 8. $W = 40$ lb
    9. $V = 121$ m^3 10. $V = 14$ cu ft

C.   1. $P = 39$ in.   2. $i = 12$ amp   3. $P = 96$ watts
    4. $A = 1260$ cm^2 5. $F = 104°$ 6. $V = 140$ cu in.
    7. $A = \$12,500$ 8. $R = 120$ ohms   9. $T = 19$ cm^2
   10. (a) 0.3125 in.   (b) 0.1185 in.   (c) 0.2625 cm   (d) $\frac{5}{32}$ in.
   11. $L = 18\frac{1}{4}$ in.   12. $L = 16.5$ in.   13. $L = 5700$ in.
   14. $L = 120$ in.   15. $X = 1000$ ft   16. $C = 179$ ft/min
   17. $P = 2.64$ kW   18. $V = 600$ mph   19. $L = 920$ cm
   20. $F = 1250$ lb   21. 36 lines   22. 3 pipes

23. $V = 4.2850$ cu in.   24. $T = 39.3$ min
25. (a) 0.849 sq in.   (b) 0.601 sq in.
    (c) 13.4 sq in.   (d) 3.33 sq cm
26. 79.84 ft   27. 6.5 cu ft
28. (a) $-8.5°$C   (b) $-18.1°$C or $-0.6°$F
29. $3.1\ \mu$F

Exercises 7-2, page 354

A. 1. $9y$ 2. $9x^2y$ 3. $6E$
    4. $-4ax$ 5. $7B$ 6. 0
    7. $-2x^2$ 8. $10x + 16y$ 9. $2R^2 + 5R$
    10. $1.45A - 0.8A^2$ 11. $\frac{1}{8}x$ 12. $-2x$
    13. $1\frac{1}{2} - 3.1W$ 14. $q - 2\frac{1}{2}p$ 15. $3x + 6xy$
    16. $4ab$ 17. $5x^2 + x^2y + 3x$ 18. $4.6p + 0.3pq$

B. 1. $3x^2 + 2x - 5$ 2. $6 - 3a + 8b$
    3. $4m^2 + 10m$ 4. $11x - 5x^2$
    5. $2 - x - 5y$ 6. $7x - 4 - 2y$
    7. $3a - 8 + 6b$ 8. $5 - w + 6z$
    9. $-10x^2 - 3x$ 10. $8m - 6n$
    11. $23 - 3x$ 12. $-3 - 5y$
    13. $x + 8y$ 14. $8 - 9m$
    15. $-14 - 7w + 3z$ 16. $-14x - 4y + 8$
    17. $9x - 12y$ 18. $20a + 24b$
    19. $-56m - 48$ 20. $-12x + 6$
    21. $-9x - 15$ 22. $-2y - 16$
    23. $23m - 42$ 24. $-19w + 15$
    25. $3 - 8x - 12y$ 26. $6 - 6a + 10b$
    27. $28 - 6w$ 28. $-75 - 55a$
    29. $6x + 8y - 24x^2 + 20y^2$ 30. $-20a + 24b + 14ab - 28b^2$
    31. $4x + 10y$ 32. $18w - 31z$
    33. $3x + 9y$ 34. $33x^2 + 6x$
    35. $-2x^2 - 42x + 62$ 36. $-10a - 18b + 44ab$

Exercises 7-3, page 364

A. 1. $x = 9$ 2. $A = 17$ 3. $x = -24$ 4. $z = 47$
    5. $a = 8\frac{1}{2}$ 6. $x = -62$ 7. $y = 24.66$ 8. $R = 7.9$
    9. $x = -13$ 10. $y = -13$ 11. $a = 0.006$ 12. $x = 21$
    13. $z = 0.65$ 14. $N = \frac{3}{4}$ 15. $m = -12$ 16. $x = -27$
    17. $y = 6.5$ 18. $a = -36$ 19. $T = -18$ 20. $y = 36$
    21. $Q = 0.7$ 22. $Z = 10$ 23. $x = 25$ 24. $y = 24$
    25. $K = 4.24$ 26. $z = -0.3$

B. 1. 360 volts   2. 5.64 hr or   3. 28.8 parts per   4. 12.8 ft
                         5 hr 38 min   million
    5. 1.715 m   6. 18 psi   7. 21 windings   8. 2900 rpm

Exercises 7-4, page 372

A. 1. $x = 10$ 2. $x = -1$ 3. $x = 35$ 4. $y = -2.5$
    5. $m = 2.5$ 6. $x = \frac{1}{6}$ 7. $n = 10$ 8. $a = 5$
    9. $z = -6$ 10. $q = 10$ 11. $x = 48$ 12. $m = -5.5$
    13. $n = -9$ 14. $y = 25$ 15. $x = \frac{1}{4}$ 16. $x = 10$
    17. $a = 7$ 18. $x = 14$ 19. $z = -5\frac{1}{2}$ 20. $x = -3$
    21. $x = \frac{2}{5}$ 22. $x = -24$ 23. $P = 3$ 24. $x = 8$
    25. $x = 10$ 26. $x = 6$ 27. $x = 5\frac{1}{3}$ 28. $x = -7$

1. 5 hr   2. 22,000 ft   3. 32.5 years   4. 10%
       5. 0.125 in. 6. 0.25 in.   7. 28.75 hr   8. 18 in.

Exercises 7-5, page 385

A.   1. $x = 9$ 2. $y = -17$ 3. $n = 4\frac{5}{6}$ 4. $a = 1\frac{1}{6}$
     5. $x = -7$ 6. $m = 11$ 7. $y = 7\frac{1}{2}$ 8. $z = -2$
     9. $x = 1$ 10. $w = -\frac{1}{2}$ 11. $c = 5$ 12. $y = 2\frac{1}{3}$
    13. $x = -7$ 14. $a = -\frac{3}{4}$ 15. $t = \frac{1}{5}$ 16. $y = 1\frac{1}{2}$
    17. $x = -2$ 18. $n = -3$ 19. $y = 2$ 20. $p = 4$
    21. $x = 2$ 22. $y = 6$ 23. $t = -3\frac{3}{4}$ 24. $A = \frac{8}{9}$
    25. $x = 4\frac{2}{7}$ 26. $x = -1\frac{2}{3}$ 27. $x = 3$ 28. $x = 1$

B.   1. $L = \dfrac{S}{W}$ 2. $B = \dfrac{2A}{H}$ 3. $I = \dfrac{V}{R}$

     4. $D = 2H + R$

     5. $T = \dfrac{2S - WA}{W}$ 6. $H = \dfrac{V}{\pi R^2}$ 7. $B = \dfrac{P - 2A}{2}$

     8. $R = 2H - 0.1$ 9. $P = \dfrac{T(R + 2)}{R}$ 10. $V = IR - E$

C.   1. (a) $a = \dfrac{360L}{2\pi R}$ (b) $R = \dfrac{360L}{2\pi a}$ (c) $L = 5.23$ in.

     2. (a) $a = \dfrac{360A}{\pi R^2}$ (b) $A = 57$ sq in.

     3. (a) $R_1 = \dfrac{V - R_2 i}{i}$ (b) $R_2 = 50$ ohms

     4. (a) $N = \dfrac{3.78P}{t^2 d}$ (b) $P = 0.05$ HP

     5. (a) $V_1 = \dfrac{V_2 P_2}{P_1}$ (b) $V_2 = \dfrac{V_1 P_1}{P_2}$ (c) $P_1 = \dfrac{V_2 P_2}{V_1}$

         (d) $P_2 = \dfrac{V_1 P_1}{V_2}$ (e) $P_1 = 300$ psi

     6. (a) $L = \dfrac{6V}{\pi T^2}$ (b) $T^2 = \dfrac{6V}{\pi L}$

     7. (a) $D = \dfrac{CA + 12C}{A}$ (b) 0.14 gram

     8. $H = PF - T$

     9. (a) $i_L = \dfrac{i_s T_p}{T_s}$ (b) $i_s = \dfrac{i_L T_s}{T_p}$ (c) $i_L = 22.5$ amp

    10. (a) $P = IV$ (b) $I = \dfrac{P}{V}$ (c) $V = 375$ volts

    11. $L = 4138$ microhenrys
    12. (a) 1.11 in.   (b) 0.989 in.   (c) 0.741 in.
    13. 7732 parts   14. 3.125 years   15. 110.7°

Exercises 7-6, page 397

A.   1. $H = 1.4\,W$ 2. $W_1 + W_2 = 167$ 3. $V = \frac{1}{4}h\pi d^2$
     4. $K = 0.454\,P$ 5. $V = AL$ 6. $V = \frac{1}{3}\pi h r^2$

     7. $D = \dfrac{N}{P}$ 8. $T = \dfrac{L}{FR}$ 9. $W = 0.785\,hDd^2$

    10. $V = iR$

B. 1. $133\frac{1}{3}$ gal and $266\frac{2}{3}$ gal
   2. 36 lb and 48 lb
   3. $13\frac{5}{7}$ ft, $6\frac{6}{7}$ ft, $3\frac{3}{7}$ ft
   4. 10 V, 12 V, 16 V
   5. 22, 22, 22, 20, 18, 16, 14, 12, 10 blocks
   6. $6\frac{3}{4}$ cu ft of cement, $13\frac{1}{2}$ cu ft of water, $13\frac{1}{2}$ cu ft of aggregate, $20\frac{1}{4}$ cu ft of sand
   7. Jo: $48,800; Ellen: $36,600
   8. Pavement: $640,000; Base material: $320,000
      Sidewalk: $160,000
   9. 21 (with some left over)
   10. 64 ft   11. $6\frac{2}{3}$ hr   12. 36 avocados   13. 5000 parts
   14. $35.16   15. $50.02   16. 16,667 pieces

Exercises 7-7, page 403

A. 1. $20x$ 2. a^5 3. $6R^2$ 4. $9x^2$
   5. $-8x^3y^4$ 6. $8x^3$ 7. $0.6a^2$ 8. $2x^4A^3$
   9. $\frac{1}{16}Q^4$ 10. $1.2p^4q^4$ 11. $24M^6$ 12. $6x^2 - 2x$
   13. $-2 + 4y$ 14. $a^3b - ab^3$ 15. p^6 16. $6x^2$
   17. $5x^7$ 18. $8x^6$ 19. $10y^7$ 20. $4a^4b^3$
   21. x^4y^4 22. $x^2 + 2x$ 23. $3p^3$ 24. $x^2y + xy^2$
   25. 5^{10} 26. 3^9 27. 10^{-3} 28. 2^3
   29. $-6x^4 + 15x^5$ 30. $24x^7 + 18x^5$
   31. $6a^3b^2 - 10a^2b^3 + 14ab^4$ 32. $-20x^3y^3 + 30x^2y^4 - 40xy^5$

B. 1. 4^2 2. 10^6 3. x^3 4. y^4

   5. $\dfrac{1}{10^2}$ 6. $\dfrac{1}{m^5}$ 7. $2a^4$ 8. $-3m^6$

   9. $-\dfrac{2}{3y}$ 10. $\dfrac{4}{t^2}$ 11. $3ab^3$ 12. $-4x^2y^4$

   13. $-\dfrac{3}{m^4n}$ 14. $\dfrac{3c^4}{4d^2}$ 15. $-\dfrac{3}{5b^3}$ 16. $-\dfrac{10}{xy^3}$

   17. $3x^2 - 4$ 18. $3y^2 + 2y$ 19. $-2a^5 + a^3 - 3a$
   20. $-3m^4 + 2m + 1$

Exercises 7-8, page 410

A. 1. 5×10^3 2. 4.5×10^2 3. 9×10^1
   4. 4.07×10^4 5. 3×10^{-3} 6. 7.1×10^{-2}
   7. 4×10^{-4} 8. 5.9×10^{-3} 9. 6.77×10^6
   10. 3.82×10^4 11. 2.92×10^{-2} 12. 9.901×10^{-3}
   13. 1.001×10^3 14. 2×10^{-3} 15. 1.07×10^{-4}
   16. 8.1×10^5 17. 3.14×10^1 18. 6×10^{-1}
   19. 1.25×10^2 20. 7.4×10^{-1} 21. 2.9×10^7 lb/sq in.
   22. 2.4×10^{-4} cal/cm · sec   23. 9.55×10^7 watts
   24. 6.588×10^8 mph

B. 1. 200,000   2. 7,000,000   3. 0.00009   4. 0.0003
   5. 0.0017   6. 0.037   7. 51,000   8. 870
   9. 40,500   10. 0.00000701   11. 0.003205   12. 1007
   13. 2,450,000   14. 0.000319   15. 647,000   16. 0.000000826

C. 1. 8×10^7 2. 8.9×10^{-9} 3. 7.8×10^2
   4. 9.9×10^2 5. 2.8×10^{-9} 6. 2.5×10^{-8}
   7. 5×10^6 8. 1.5×10^6 9. 1.4×10^2
   10. 1.5×10^{-2} 11. 6.3×10^{-2} 12. 9.5×10^{-2}

13. 2.3×10^{10} 14. 6.5×10^{-1} 15. 8×10^{-11}
16. 5.9×10^{-3} 17. 2.8×10^{-2} 18. 1.2×10^{-6}
19. 5.9×10^{-12} 20. 2.0×10^{15} 21. 6.6×10^{13}
22. 6.5×10^{-10} 23. 4.0×10^{2} 24. 4.1×10^{-4}

D.   1.   13.3 cal/sec   2. 2×10^{-5} 3. $1.53 \times 10^{-4}\ \mu F$
     4. 3.98×10^{-19} Joules

Problem Set 7, page 413

A.   1. $10x$ 2. $12y$ 3. $12xy$ 4. $15xy^3$
     5. $1\frac{11}{24}x$ 6. 0 7. $0.4G$ 8. $16y - 8y^2$
     9. $21x^2$ 10. $8m^3$ 11. $-10x^2y^2z$ 12. $27x^3$
     13. $12x - 21$ 14. $6a^3b - 10ab^3$ 15. $9x + 11$ 16. $5y - 7$
     17. $9x - 3y$ 18. $-5x^2 + 9$ 19. 6^4 20. $\dfrac{1}{x^5}$
     21. $-3m^8$ 22. $-\dfrac{3}{5y^2}$ 23. $\dfrac{4a}{b}$ 24. $2x^2 - x + 3$

B.   1. $L = 13$ 2. $M = 8$ 3. $I = 144$ 4. $I = 183,333\frac{1}{3}$
     5. $V = 42\frac{21}{64}$ 6. $N = -95$ 7. $L = 115.2$ 8. $f = \frac{1}{48}$
     9. $t = 4$ 10. $V = 270.64$

C.   1. $x = 12$ 2. $m = 0.35$ 3. $e = 44$ 4. $a = 0.9$
     5. $x = -5$ 6. $x = 10\frac{1}{2}$ 7. $x = 4$ 8. $m = -6$
     9. $y = 32$ 10. $x = 3$ 11. $x = -6$ 12. $f = -1.15$
     13. $x = -4$ 14. $y = 2$ 15. $g = 10$ 16. $m = 1.5$
     17. $x = 16$ 18. $x = -4$ 19. $x = 3$ 20. $x = 6$
     21. $y = 3.75$ 22. $m = -1$ 23. $n = 9$ 24. $x = 3.5$
     25. $y = 2$ 26. $m = 10$ 27. $b = \dfrac{A}{H}$ 28. $P = R - S$
     29. $L = \dfrac{P - 2W}{2}$ 30. $F = \dfrac{w}{P}$
     31. $g = \dfrac{2S + 8}{t}$ 32. $A = \dfrac{\pi R^2 H - V}{B}$

D.   1. 7.5×10^3 2. 1.28×10^4 3. 4.1×10^{-2}
     4. 2.36×10^{-4} 5. 5.72×10^{-3} 6. 4.82×10^{-6}
     7. 4.47×10^5 8. 2.127×10^6 9. 8.02×10^7
     10. 4.671×10^4 11. 7.05×10^{-6} 12. 1.006×10^{-3}
     13. $930,000$ 14. 6020 15. 0.00029
     16. 0.00000305 17. 0.0000005146 18. 0.006203
     19. $90,710$ 20. $4,006,000$ 21. 5.7×10^{10}
     22. 7.5×10^{12} 23. 3.0×10^{-6} 24. 2.1×10^{-7}
     25. 4.2×10^{-1} 26. 5.6×10^{-1} 27. 1.2×10^2
     28. 8.1×10^{-3} 29. 2.8×10^{-2} 30. 2.2×10^3
     31. 2.4×10^{-8} 32. 1.6×10^8

E.   1.   (a) 31.25% (b) $L = \dfrac{100(H - X)}{CG}; L = 166.7$ in.

     (c) $H = \dfrac{L \cdot CG + 100x}{100}; H = 140$ in.

     (d) $x = \dfrac{100H - L \cdot CG}{100}; x = 156$ in.

   2.   (a) $L = \dfrac{FDU}{A}$ (b)   42 foot candles   (c) $A = \dfrac{FDU}{L}$

     (d)   66.7 sq ft
   3. $8W + 2W = 80; W = 8$ cm, $L = 32$ cm   4. $L = 225$ tons
   5. $2x + x = 0.048; x = 0.016$ in., $2x = 0.032$ in.

6. Sal: $8,763.64
   Gloria: $13,145.45
   Martin: $26,290.91
7. 9.7×10^6 m/sec   8. about 16 pipes
9. $51\frac{9}{16}$ in.   10. 999,600 BTU
11. 0.617   12. 0.1139 in. per in.
13. 3,770,000   14. (a) 1.08   (b) lower it 6 in.
15. (a) 70%   (b) 4.1 dS/m

Chapter 8   *Exercises 8-1, page 430*

A. 1. (a) $\angle B, \angle x, \angle ABC$ (b) $\angle O, \angle POQ$ (c) $\angle a$
    (d) $\angle T$ (e) $\angle 2$ (f) $\angle s, \angle RST$
   2. (a) $\angle AOB, \angle AOC$ (b) $\angle HOG, \angle HOF$ (c) $\angle AOF, \angle AOG$
    (d) $\angle AOE, \angle HOE$
    (e) $\angle AOB = 25°, \angle GOF = 40°, \angle HOC \approx 130°$
       $\angle FOH = 70°, \angle BOF \approx 87°, \angle COG \approx 100°$
   4. (a) $\angle 1 = 110°, \angle 2 = 139°, \angle 3 = 69°, \angle 4 = 150°$
    (b) $\angle 5 = 70°, \angle 6 = 109°$
    (c) $\angle 7 = 148°, \angle 8 = 89°, \angle 9 = 67°, \angle 10 = 150°$

B. 1. (a) $a = 55°, b = 125°, c = 55°$ (b) $d = 164°, e = 16°, f = 164°$
    (c) $p = 110°$ (d) $x = 75°, t = 105°, k = 115°$
    (e) $s = 60°, t = 120°$ (f) $p = 30°, q = 30°, w = 90°$
   2. (a) $A = 68°$ (b) $E = 137°$
    (c) $G = 42°$ (d) $133°$
    (e) $a = 70°$ (f) $53°$
    (g) $63°$ (h) $44°45'$ (i) $60°2'$
   3. (a) $w = 80°, x = 100°, y = 100°, z = 100°$
    (b) $a = 60°, b = 60°, c = 120°, d = 120°, e = 120°, f = 120°, g = 60°$
    (c) $n = 100°, q = 80°, m = 100°, p = 100°, h = 80°, k = 100°$
    (d) $z = 140°, s = 40°, y = 140°, t = 140°, u = 40°, w = 140°, x = 40°$

C. 1. $\angle ACB \approx 63°$ 2. $50°$ and $130°$ 3. $B \approx 122°, A \approx 28°, C \approx 30°$
   4. $a \approx 35°$ 5. $2''$ 6. $45°$ 7. $29°5'36''$
   8. $53°$ 9. $a = 112°, b = 68°$ 10. $22\frac{1}{2}°$
  11. $\angle 2 = 106°$ 12. $\angle y = 48°$

Exercises 8-2, page 446

A. 1. 20.0 m   2. 206 ft   3. 115 in.   4. 42 ft
   5. 48 in.   6. 104.8 m   7. 39 cm   8. 0.252 in.
   9. 22.81 in.   10. 23.8 ft   11. 64 ft   12. 26.4 in.

B.   *Perimeter*   *Area*   *Perimeter*   *Area*
   1. 20 in.   25 sq in.   2. 22.7 yd   29.6 sq yd
   3. 70.6 in.   285 sq in.   4. 98 ft   420 sq ft
   5. 56 m   160 m^2 6. 17.3 yd   18.8 sq yd
   7. 124 m   468 m^2 8. 42.5 ft   109.4 ft^2
   9. 86 ft   432 sq ft   10. 23 m   33.1 m^2
  11. 114.9 m   722.2 m^2 12. 847.5 ft   39496.7 sq ft

C. 1. 960   2. $35\frac{7}{8}$ sq ft   3. $25\frac{1}{3}$ sq yd, $1012.07
   4. 1840 sq   5. 11 bundles   6. 4385 sq in.
   7. 640 sq in.   8. 450 blocks
   9. 10 ft 8 in.   10. 56.25 lb
  11. $6\frac{2}{3}$ in.   12. 19.5 lb
  13. Cut the 6-in. card along the 38-in. side; 18 cards, 86 sq in. of waste
  14. 3500 BTU   15. $9729

Exercises 8-3, page 465

A.   1.   10.6 in.   2.   13.4 ft   3.   20 cm   4.   28.6 yd
     5.   46.8 mm   6.   39 in.   7.   27.7 in.   8.   4.5 ft
     9.   28 ft   10.   4.4 ft   11.   21.2 in.   12.   28 in.
     13.   6 in.   14.   8 cm   15.   17.4 ft   16.   5.4 in.
     17.   9.369 cm   18.   8.288 in.   19.   0.333 in.   20.   21.35 in.
     21.   118 ft   22.   43.91 m

B.   1. 173.2 mm^2 2. 439.1 cm^2 3. 336 ft^2 4. 53.2 ft^2
     5. 36 yd^2 6. 84.9 cm^2 7. 259.8 in.^2 8. 162 in.^2
     9. 520 ft^2 10. 77 in.^2 11. 5.1 cm^2 12. 161 m^2
     13. 433 ft^2 14. 234 in.^2 15. 354 ft^2 16. 2250 m^2
     17. 364.8 ft^2 18. 120 in.^2 19. 39 in.^2 20. 19.4 in.^2
     21. 0.586 in.^2 22. 0.039 cm^2 23. 7.94 m^2 24. 0.117 in.^2

C.   1.   7 gal   2.   268 sq in.   3.   5 squares   4.   74 ft
     5.   15.2 ft   6.   11.3 in.   7.   174.5 sq ft   8. 318 ft^2
     9. y = 36 ft 5 in., x = 64 ft 11 in.

Box, page 477

1.   4 in.   2.   8.5 in.   3.   5 in.   4.   10.4 in.   5.   5.1 in.

Exercises 8-4, page 477

A.   *Circumference*   *Area*
     1.   88.0 in.   616 sq in.
     2.   66.0 ft   346 sq ft
     3.   19.2 m 29.2 m^2
     4.   44.0 cm 154 cm^2
     5.   75.4 ft 452 ft^2
     6.   1.57 m 0.196 m^2
     7.   3.77 cm 1.13 cm^2
     8.   2.51 in. 0.503 in.^2

B.   1.   17.1 sq in.   2. 238.8 cm^2 3. 5.2 in.^2 4.   5.8 sq ft
     5. 15.1 in.^2 6. 15.9 m^2 7. 26.7 cm^2 8. 247.8 ft^2
     9.   0.7 sq in.   10. 135.3 cm^2 11. 246.1 in.^2 12.   8037.5 sq ft
     13.   25.1 in.   14.   55.1 in.   15.   52.3 cm

C.   1.   30.1 in.   2.   (a) 908 ft^2 (b)   226 sq ft   3.   0.6 in.
     4. 0.05 in.^2 5. 163 in.^2 6.   316 sq in.   7.   103.7 sq in.
     8. $1\frac{3}{4}$ in.   9.   314   10.   4.24 cm   11. $103.11
     12.   99.52 m   13.   174 in.   14.   10 ft 4.5 in.   15.   2.4 in.
     16.   67.0 sq yd   17.   2.6 cm   18.   200 mm
     19.   2.79 in.   20.   About $1\frac{1}{2}$ turns
     21.   2.1 fps   22.   393 lb
     23.   133 cm   24.   7

Problem Set 8, page 481

A.   1. $\angle 1$, acute   2. $\angle ABC$ or $\angle B$, obtuse   3. $\angle m$, acute
     4. $\angle E$, right   5. $\angle PQR$ or $\angle Q$, obtuse   6. $\angle G$, right
     7.   63° approx.   8.   90° approx.   9.   155° approx.
     10.   60° approx.   11.   70° approx.   12.   107° approx.
     14.   (a)   55°   (b) a = 95°, b = 85°, c = 95°
           (c) a = 130°, b = 50°, c = 130°, d = 50°, e = 130°, f = 50°, g = 130°
           (d)   131°   (e)   100°   (f)   All equal 60°   (g)   58°
           (h)   80°   (i)   35°

B.
	Perimeter	Area		Perimeter	Area
1.	12 in.	6 in.^{2}	2.	30 ft	48 ft^2
3.	31.4 cm	78.5 cm^2	4.	3 in.	$\frac{9}{16}$ in.^{2}
5.	12 ft	10.4 ft^2	6.	74.4 in.	289.8 in.^{2}
7.	17 yd	8 yd^2	8.	100 mm	561 mm^2
9.	36 in.	53.7 sq in.	10.	42 ft	102 sq ft
11.	12 in.	6.9 sq in.	12.	94.2 cm	706.9 cm^2
13.	17.2 ft	12.0 sq ft	14.	74 in.	145 sq in.
15.	51.7 ft	169.3 ft^2	16.	70 m	284 m^2
17.	28.3 in.	51.1 sq in.	18.	7.5 in.	2.3 sq in.
19.	5.77 mm		20.	15.59 in.	
21.	3.46 in.		22.	23.56 ft	
23.	$\frac{3}{4}$ in.		24.	11.70 in.	

C.
1.	$28,985	2.	7.07 in.	3.	45°	4.	640 sq ft
5.	250 ft	6.	105°	7.	$6981.45	8.	32.5 in.
9.	18 in.	10.	36 ft 5 in.	11.	36	12.	34.7 in.
13.	110°	14.	15 oz	15.	94 ft	16.	0.707 in.
17.	$100.62	18.	153 ft^2	19.	69.1 cm	20.	68.0 in.
21.	$753.60	22.	118°	23.	27.3 lb	24.	3.005 in.
25.	12.17 in.	26.	4 mi				

Chapter 9

Exercises 9-1, page 497

A.
1.	729 cu in.	2.	931.5 cu ft	3.	13,800 cm^3	4.	35 in.^{3}
5.	132.4 mm^3	6.	2.81 cu in.	7.	144 cm^3	8.	108 cm^3

B.
	Lateral Surface Area	Volume		Lateral Surface Area	Volume
1.	280 sq ft	480 cu ft	2.	576 cm^2	665.1 cm^3
3.	784 mm^2	2744 mm^3	4.	608 sq ft	1120 cu ft
5.	360 in.^{2}	480 in.^{3}	6.	28,800 yd^2	288,000 yd^3

C.
	Total Surface Area	Volume		Total Surface Area	Volume
1.	150 sq yd	125 cu yd	2.	311.3 m^2	357.9 m^3
3.	1457.8 ft^2	2816 ft^3	4.	11.8 in.^{2}	2.3 in.^{3}
5.	5290 cm^2	23,940 cm^3	6.	396 sq yd	288 yd^3

D.
1.	333.3 cu ft	2.	27.2 lb	3.	94,248 gal	4.	8 gal
5.	450 trips	6.	2012 lb	7.	71.3 cu yd	8.	249.1 lb
9.	16.9 lb	10.	63.9 lb	11.	3.8 ft	12.	5670 lb
13.	238 trips	14.	143,000 gal	15.	8.3 cu yd	16.	8100 lb
17.	3.5 yd^3	18.	2.4 gal	19.	7000 lb	20.	Yes. 5740 gal/24 hr
21.	2244 gal	22.	9.2 gal	23.	525 ft^3	24.	19.6 yd^3

Exercises 9-2, page 506

A.
	Lateral Surface Area	Volume		Lateral Surface Area	Volume
1.	384 sq in.	617 cu in.	2.	454 sq cm	520 cu cm
3.	127 sq in.	105 cu in.	4.	6000 sq cm	48,000 cu cm
5.	48.6 sq ft	46.7 cu ft	6.	72 sq ft	69.3 cu ft
7.	55.2 sq m	31.2 cu m	8.	397 sq in.	522 cu in.

B.
	Total Area	Volume		Total Area	Volume
1.	80.7 m^2	33.2 m^3	2.	1440 cm^2	3200 cm^3
3.	3200 ft^2	1020 ft^3	4.	59.2 cm^2	22.3 cm^3
5.	2500 in.^{2}	7340 in.^{3}	6.	3260 m^2	12,100 m^3
7.	173 ft^2	144 ft^3	8.	2240 cm^2	3460 cm^3

C. 1. $3825   2. 562 lb   3. 3 hr
   4. 322 bu   5. 384 sq in.   6. 66.9 cu yd

Exercises 9-3, page 513

A. | | Lateral Surface Area | Volume | | Lateral Surface Area | Volume |
|---|---|---|---|---|---|
| 1. | 24,800 cm^2 | 572,000 cm^3 | 2. | 1490 mm^2 | 6350 mm^3 |
| 3. | 245 sq in. | 367 cu in. | 4. | 301 sq yd | 603 cu yd |

B. | | Total Area | Volume | | Total Area | Volume |
|---|---|---|---|---|---|
| 1. | 7.07 in.^{2} | 1.77 in.^{3} | 2. | 4070 ft^2 | 24,400 ft^3 |
| 3. | 1280 ft^2 | 3530 ft^3 | 4. | 17.0 in.^{2} | 4.83 in.^{3} |
| 5. | 6520 m^2 | 39,900 m^3 | 6. | 3450 yd^2 | 13,200 yd^3 |

C. 1. 14 gal   2. 12.3 lb   3. 0.24 ft   4. 6283.2 liters
   5. 2.6 lb   6. 57.4 in.   7. 3.8 gal   8. 4071.5 sq in.
   9. 49.6 lb   10. 7.5 gal   11. 582 gal   12. 8.9 cu yd
   13. 0.43 lb   14. 11 gal   15. 28.1 in.   16. 53,300 gal
   17. 866 in.^{3} 18. 36.7 gal   19. 16 gal

Exercises 9-4, page 521

A. | | Lateral Surface Area | Volume | | Lateral Surface Area | Volume |
|---|---|---|---|---|---|
| 1. | 319 cm^2 | 581 cm^3 | 2. | 1580 ft^2 | 7070 ft^3 |
| 3. | 338 m^2 | 939 m^3 | 4. | 1460 in.^{2} | 6800 in.^{3} |

B. | | Total Area | Volume | | Total Area | Volume |
|---|---|---|---|---|---|
| 1. | 628 in.^{2} | 1010 in.^{3} | 2. | 961 ft^2 | 1880 ft^3 |
| 3. | 1110 in.^{2} | 2610 in.^{3} | 4. | 429 ft^2 | 462 ft^3 |

C. 1. 240 m^3 2. 1100 cu in.   3. 49 cm^3 4. 4.7 ft^3

D. 1. 104.7 m^3 2. 746.1 cm^2 3. 253.2 bu   4. $602
   5. 5.6 lb   6. 1073.6 cm^2 7. 4.2 cm   8. 6.6, or about 7
   9. 2.0 lb   truckloads

Problem Set 9, page 525

A. | | Lateral Surface Area | Volume | | Lateral Surface Area | Volume |
|---|---|---|---|---|---|
| 1. | 226.2 sq in. | 452.4 cu in. | 2. | 240 sq in. | 504 cu in. |
| 3. | 43.2 sq ft | 26.7 cu ft | 4. | 816.9 sq ft | 1638 cu ft |
| 5. | 259.2 mm^2 | 443.5 mm^3 | 6. | 630 sq in. | 1909.5 cu in. |
| 7. | 377.0 ft^2 | 980.2 ft^3 | 8. | 300 ft^2 | 912 ft^3 |
| 9. | 384 ft^2 | 512 ft^3 | 10. | 107.8 in.^{2} | 56.6 in.^{3} |
| 11. | 1767.2 cm^2 | 5301.5 cm^3 | 12. | 256.2 sq in. | 259.8 cu in. |
| 13. | 450 sq ft | 396 cu ft | 14. | 879.6 in.^{2} | 1570.8 in.^{3} |
| 15. | 1256.6 cm^2 | 4188.8 cm^3 | 16. | 648 in.^{2} | 1008 in.^{3} |
| 17. | 530.9 sq in. | 1150.3 cu in. | 18. | 1381.6 sq in. | 3479.1 cu in. |
| 19. | 1263.6 in.^{2} | 2520 in.^{3} | 20. | 35.3 in.^{2} | 14.7 in.^{3} |
| 21. | 329.2 sq in. | 312 cu in. | | | |

B. 1. 1125 boxes   2. 1204 yd^3 3. 2992 lb   4. 423.0 gal
   5. 40 qt   6. 84 trips   7. 1.5 ft^3 8. 8 ft
   9. 294 lb   10. 22,000 in.^{2} 11. 115 yd^3 12. 738 in.^{2}
   13. 2.8 gal   14. 8400 g   15. 6.3 ft

Chapter 10   *Exercises 10-1, page 544*

A. 1. Acute   2. Right   3. Obtuse   4. Right
   5. Straight   6. Obtuse   7. Acute   8. Acute
   9. Right   10. Acute

B. 1. 36°15′   2. 73°36′   3. 65°27′   4. 84°14′
    5. 17°30′   6. 47°23′   7. 16°7′   8. 165°22′
    9. 27.5°   10. 80.25°   11. 154.65°   12. 131.1°
   13. 57.05°   14. 44.33°   15. 16.88°   16. 0.27°
   17. 0.43   18. 3.24   19. 0.02   20. 0.26
   21. 1.18   22. 0.17   23. 1.41   24. 3.78
   25. 5.73°   26. 48.13°   27. 120.32°   28. 200.54°

C. 1. 26   2. 9.7   3. 24   4. 18.6
    5. 12 ft   6. 3.6 in.   7. 1.5   8. 3.3 cm
    9. 8.0 ft   10. 1.7 in.   11. (a) 45°   (b) 4.2 ft
   12. (a) 6.9 in.   (b) 8 in.   13. (a) 4   (b) 53.1°   (c) 36.9°
   14. (a) 3 cm   (b) 5.2 cm   (c) 60°   15. (a) 7   (b) 45°
   16. (a) 5 in.   (b) 53.1°   17. (a) 11.5 ft   (b) 5.8 ft   (c) 30°
   18. (a) 0.4 in.   (b) 0.3 in.   (c) 90°   19. (a) 10 in.   (b) 45°
   20. (a) 3.48 cm   (b) 60°   (c) 30°
   21. $S = 21$ cm, $A = 314$ cm^2 22. $S = 50$ ft, $A = 603$ sq ft
   23. $S = 88$ ft, $A = 4840$ sq ft   24. $S = 36$ m, $A = 324$ m^2
   25. $S = 6$ in., $A = 16$ sq in.   26. $S = 240$ cm, $A = 9600$ cm^2
   27. $S = 5$ m, $A = 25$ m^2 28. $S = 94$ in., $A = 1414$ in.^{2}

D. 1. (a) 325.2 in.^{2} (b) 36.1 in.   2. 1256.6 ft^2
    3. 21   4. $8670   5. 166 yd^3
    6. 58 mph   7. 829 in./sec   8. 1480 cm/sec
    9. 0.08 rad/sec   10. 157 rad/sec   11. 0.39 rad/sec
   12. 7.5 in., 32.0 in., 10.6 in.

Exercises 10-2, page 557

A. 1. 0.54   2. 0.74   3. 0.42   4. 0.53   5. 0.44   6. 0.69

B. 1. 0.454   2. 0.788   3. 0.213   4. 0.998
    5. 0.191   6. 0.105   7. 0.052   8. 0.994
    9. 1.206   10. 0.969   11. 0.752   12. 0.149
   13. 0.771   14. 3.522   15. 6.357   16. 0.218
   17. −0.139   18. 0.961   19. 0.585   20. −0.701

C. 1. 76°54′   2. 44°2′   3. 64°0′   4. 4°25′
    5. 74°49′   6. 22°0′   7. 9°40′   8. 69°45′
    9. 49.4°   10. 48.5°   11. 64.1°   12. 12.7°
   13. 41.1°   14. 28.2°   15. 37.2°   16. 81.1°

Exercises 10-3, page 563

A. 1. $X = 11.0$ in., $Y = 8.6$ in.   2. $X = 4.9$ ft, $Y = 7.7$ ft
    3. $X = 29.7$ cm, $Y = 32.0$ cm   4. $a = 32°28′, X = 26.1$ ft
    5. $a = 60°, X = 1.3$ in.   6. $X = 1.8$ mm, $Y = 3.3$ mm
    7. $X = 44.4$ in., $Y = 14.5$ in.   8. $a = 50°36′, X = 4.7$ yd
    9. $a = 36.0°, X = 1.54$ in.   10. $X = 33.50$ cm, $Y = 22.85$ cm
   11. $X = 1.16$ in., $Y = 1.47$ in.   12. $a = 61.3°, X = 1.47$ in.
   13. $m = 41.6°, X = 27.67$ cm   14. $X = 28.39$ in., $Y = 15.75$ in.
   15. $X = 1.27$ in., $Y = 1.42$ in.   16. $m = 16.5°, X = 17.97$ cm

B. 1. 31.1 ft   2. 43.8°   3. (a) 72 ft 9 in.   (b) No.   4. 4478 ft
    5. 51°41′   6. 4.3°   7. 115 in.   8. $A = 3.08$ in., $B = 4.66$ in.
    9. 3.74 in.   10. Yes   11. 1.1 in.   12. 68.75°
   13. 274 ft   14. 0.6308 in.   15. 33.4 cm   16. 6°37′
   17. 1.55 in.   18. 5 ft   19. 21 ft 1 in.   20. 14 ft 7 in.

Exercises 10-4, page 577

A. 1. $C = 75°, c = 9.2$ ft, $b = 8.4$ ft
    2. $c = 12.6$ in., $A = 65.0°, a = 16.5$ in.

3. $A = 39.0°, a = 202$ m, $c = 301$ m
4. $b = 1500$ yd, $a = 1100$ yd, $A = 25°$
5. $A = 55.9°, C = 94.1°, c = 116$ in. or $A = 124.1°, C = 25.9°, c = 50.6$ in.
6. $A = 108.6°, B = 44.4°, a = 359$ ft or $A = 17.4°, B = 135.6°, a = 113$ ft
7. $c = 5.6$ m, $A = 92°, B = 47°$ 8. $a = 24.3$ m, $B = 41.7°, C = 81.3°$
9. $b = 690$ ft, $A = 62.4°, C = 15.6°$
10. $a = 3890$ yd, $B = 14.0°, C = 41.0°$
11. $C = 94.2°, A = 37.5°, B = 48.3°$ 12. $A = 102°, B = 19°, C = 59°$
13. $b = 10.3$ m, $A = 45.4°, C = 34.9°$
14. $a = 0.865$ in., $B = 23.0°, C = 38.5°$
15. $B = 149.9°, A = 7.1°, C = 23.0°$
16. $A = 47.6°, B = 82.2°, C = 50.2°$

B. 1. 41 ft 2. 36.3 ft 3. 27.2°, 52.1°, 100.7°
    4. 182 ft 5. 19.1 ft 6. 26° and 130°
    7. 25.4 in.

Problem Set 10, page 579

A. 1. Yes 2. Yes 3. No 4. No 5. Yes
    6. Yes 7. No 8. No 9. Yes 10. No

B. 1. 0.391 2. 6.314 3. 0.070 4. 0.451
    5. 0.720 6. 0.488 7. 0.956 8. −0.545
    9. 14° 10. 57° 11. 67°58′ 12. 72°
    13. 10°59′ 14. 56°42′ 15. 39°27′ 16. 42°13′
    17. 83°13′ 18. 26.8° 19. 43.7° 20. 29.2°
    21. 8.9° 22. 74.2° 23. 63.9° 24. 18.7°
    25. 52.6° 26. 50.5° 27. 0.611 28. 0.374
    29. 1.300 30. 1.958 31. 25.8° 32. 97.4°
    33. 131.8° 34. 48.1°

C. 1. $X = 37.1, Y = 32.5$ 2. $X = 20.2, Y = 16.4$
    3. $a = 46°, X = 4.9$ 4. $X = 5.3, Y = 8.4$
    5. $a = 54.2°, X = 5.6$ 6. $X = 70.2$ ft, $Y = 77.2$ ft
    7. $t = 39.3°, X = 9.8$ in.   8. $X = 11.7$ ft, $Y = 10.9$ ft

D. 1. $C = 76.0°, c = 19.2$ in., $b = 12.4$ in.
    2. $A = 155.5°, B = 14.8°, C = 9.7°$
    3. $C = 52°, B = 91°, b = 430$ yd or $C = 128°, B = 15°, b = 110$ yd
    4. $a = 101$ in., $B = 49.7°, C = 11.8°$
    5. $B = 49.1°, C = 66.6°, c = 52.4$ m
    6. $b = 190$ ft, $A = 39.8°, C = 93.2°$
    7. $A = 24.9°, B = 38.9°, b = 1730$ m
    8. $A = 52°, C = 98°, c = 9.5$ in. or $A = 128°, C = 22°, c = 3.6$ in.

E. 1. 1.3 in. 2. $a = 5°13′, V = 219$ mph 3. 2 in.
    4. 6°2′ 5. (a) $X = 28.9$ ft (b) 57.7 ft
    6. (a) 80.9 ft (b) 8°32′ 7. (a) 1°8′ (b) 0.3 in.
    8. $x = 16.5$ ft, $y = 25.7$ ft 9. 0.394 in. 10. 90° 11. 17.2 cm
    12. 1.08 in.

Chapter 11 *Exercises 11-1, page 605*

A. 1. $x = 2, y = 8$ 2. $x = 4, y = 7$ 3. $x = 4, y = 5$
    4. Inconsistent   5. $x = 2, y = 4$ 6. $x = -\frac{1}{9}, y = -\frac{1}{9}$
    7. $x = -1, y = 2$ 8. $x = \frac{1}{2}, y = -3$

B. 1. $x = 9, y = -4$ 2. $x = 4, y = 1$ 3. Inconsistent
    4. Inconsistent   5. $x = -2, y = -3$ 6. $x = -3, y = 1$

7. $x = -1, y = -1\frac{1}{2}$ 8. $x = 4, y = 7$ 9. Dependent

10. $x = \dfrac{a + b}{2}, y = \dfrac{a - b}{2}$ 11. $x = 1, y = 5$ 12. $x = 3, y = \frac{1}{2}$

C. 1. $x + y = 39$ Solution: $x = 23$
$x - y = 7$ $y = 16$

2. $x + y = 14$ Solution: $x = 11$
$x = 2 + 3y$ $y = 3$

3. $x + y = 20$ Solution: $x = 12$
$2x = 3y$ $y = 8$

4. $\frac{1}{2}(x + y) = 25$ Solution: $x = 29$
$x - y = 8$ $y = 21$

5. $4b + 3f = 11$ Solution: $b = \$2$ A bleeb costs $2.00.
$3b + 4f = 10$ $f = \$1$

6. $2L + 2W = 14$ Solution: $L = 4$ ft
$L = 2W - 2$ $W = 3$ ft

7. $10d + 5n = 100$ Solution: $n = 2$
$d = n + 7$ $d = 9$

8. $4L + 3S = 26$ Solution: $L = 5$
$3L - 2S = 11$ $S = 2$

9. $5p + 4c = 841$ Solution: $c = 99¢$
$p = c - 10$ $p = 89¢$

10. $L = 2W$ Solution: $L = 40$ in.
$L - W = 20$ $W = 20$ in.

11. $x + y = 30$ Solution: $x = 24$ in.
$x = 4y$ $y = 6$ in.

12. $x + y = 12$ Solution: $x = 5$ gal (of $15 paint)
$15x + 18y = 201$ $y = 7$ gal (of $18 paint)

13. $x + y = 6$ Solution: $x = 1.5$ liter (10%)
$0.1x + 0.02y = 0.24$ $y = 4.5$ liter (2%)

14. $2L + 2W = 520$ Solution: $L = 200$ ft
$L = 20 + 3W$ $W = 60$ ft

15. $x + y = 24$ Solution: $x = 16$ tons ($30 rock)
$30x + 40y = 800$ $y = 8$ tons ($40 rock)

Exercises 11-2, page 619

A. 1. No   2. Yes   3. No
4. Yes   5. Yes   6. $7x^2 + 3x - 5 = 0$
7. $3x^2 - 7x + 14 = 0$ 8. OK   9. $x^2 - 18x = 0$
10. OK

B. 1. $x = 5$ or $x = -5$ 2. $x = 3$ or $x = -3$
3. $x = 0$ or $x = 4\frac{2}{5}$ 4. $x = \frac{1}{2}$ or $x = 3$
5. $x = 4\frac{1}{2}$ or $x = -4\frac{1}{2}$ 6. $x = 4\frac{1}{2}$ or $x = -2\frac{1}{3}$
7. $x \approx 0.76$ or $x \approx -15.76$ 8. $x = 9$ or $x = \frac{3}{4}$
9. $x \approx -2.35$ or $x \approx 0.85$ 10. $x = 0.02$ or $x = -0.05$
11. $x \approx 4.14$ or $x \approx -3.14$ 12. $x \approx 4.46$ or $x \approx -2.46$
13. $x \approx 0.55$ or $x \approx -0.26$ 14. $x \approx 1.27$ or $x \approx -0.47$
15. $x \approx -2.37$ or $x \approx -0.63$ 16. $x \approx 1.24$ or $x \approx 0.16$
17. $x \approx 2.39$ or $x \approx 0.28$ 18. $x \approx -0.46$ or $x \approx 0.31$
19. $x \approx 3.69$ or $x \approx 0.81$ 20. $x \approx 0.57$ or $x \approx -2.65$

C.　1.　25 mm　　2.　(a)　14.49 in.　　(b)　15.66 in.
　　3.　5 in. by 15 in.　　4.　8.92 cm　　5.　17 in. by 22 in.
　　6.　99 mph　　7.　7.14 in.　　8.　(a)　9 amp　　(b)　7.30 amp
　　9.　(a)　3.76 in.　　(b)　4.56 in.　　10.　6 ft by 8 ft or 4 ft by 12 ft
　　11.　8 in.

Problem Set 11, page 621

A.　1.　$x = 15, y = 3$　　2.　$x = 6, y = -\frac{1}{2}$　　3.　$x = 3, y = 2\frac{1}{2}$
　　4.　$x = 7, y = 3$　　5.　$x = 7, y = 6$　　6.　Dependent
　　7.　$x = 1, y = 1$　　8.　$x = -13, y = 22$　　9.　$x = \frac{74}{23}, y = -\frac{4}{23}$

B.　1.　$x = 3$　or　$x = -3$　　　　　2.　$x = 7$　or　$x = -4$
　　3.　$x \approx -1.43$　or　$x \approx -0.23$　　4.　$x \approx 1.18$　or　$x \approx -0.43$
　　5.　$x = 0$　or　$x = 6$　　　　　6.　$x = 5$　or　$x = -1\frac{1}{2}$
　　7.　$x = 5$　or　$x = -1$　　　　　8.　$x = 6$　or　$x = -5$
　　9.　$x = 6$　or　$x = -6$　　　　　10.　$x \approx 2.15$　or　$x \approx -0.75$

C.　1.　$x + y = 38$　　2.　$s^2 = 196$　　3.　$x - y = 21$
　　　　$x - y = 14$　　　　$s = 14$ in.　　　　$5y - 2x = 33$
　　　　$x = 26$　　　　　　　　　　　　$x = 46$
　　　　$y = 12$　　　　　　　　　　　　$y = 25$
　　4.　$2L + 2W = 22$　　5.　$L = 4W$　　　　6.　$x + y = 650$
　　　　$L = 2 + 2W$　　　　$4W^2 = 125$　　　　$5x + 7y = 4350$
　　　　$W = 3$ ft　　　　　$W \approx 5.59$ cm　　$x = 100$ (at 5¢ each)
　　　　$L = 8$ ft　　　　　$L \approx 22.36$ cm　　$y = 550$ (at 7¢ each)
　　7.　(a)　$I = 13$ amp　　(b)　$I = 12.25$ amp
　　8.　$L = 2 + W$　　　　9.　$x + y = 42$　　10.　$2L + 2W = 750$
　　　　$W(2 + W) = 168$　　　$x = 3y - 2$　　　　$L = 4W$
　　　　$W = 12$ in.　　　　　$y = 11$ in.　　　　$W = 75$ ft
　　　　$L = 14$ in.　　　　　$x = 31$ in.　　　　$L = 300$ ft
　　11.　$d = 15.96$ in.　　12.　$x + y = 15$　　　　　　13.　12.5 in.
　　　　　　　　　　　　　　$16x + 19y = 270$
　　　　　　　　　　　　　　$x = 5$ gal of \$16 paint
　　　　　　　　　　　　　　$y = 10$ gal of \$19 paint
　　14.　$L = 2W - 3$　　　15.　$x + y = 50$
　　　　$W(2W - 3) = 20$　　　$80x + 100y = 4200$
　　　　$W = 4$ in.　　　　　$x = 40$ tons (of \$80 rock)
　　　　$L = 5$ in.　　　　　$y = 10$ tons (of \$100 rock)

×	0	1	2	3	4	5	6	7	8	9	10
0	0	0	0	0	0	0	0	0	0	0	0
1	0	1	2	3	4	5	6	7	8	9	10
2	0	2	4	6	8	10	12	14	16	18	20
3	0	3	6	9	12	15	18	21	24	27	30
4	0	4	8	12	16	20	24	28	32	36	40
5	0	5	10	15	20	25	30	35	40	45	50
6	0	6	12	18	24	30	36	42	48	54	60
7	0	7	14	21	28	35	42	49	56	63	70
8	0	8	16	24	32	40	48	56	64	72	80
9	0	9	18	27	36	45	54	63	72	81	90
10	0	10	20	30	40	50	60	70	80	90	100

TABLE OF SQUARE ROOTS

Number	Square root	Number	Square root	Number	Square root	Number	Square root
1	1.0000	51	7.1414	101	10.0499	151	12.2882
2	1.4142	52	7.2111	102	10.0995	152	12.3288
3	1.7321	53	7.2801	103	10.1489	153	12.3693
4	2.0000	54	7.3485	104	10.1980	154	12.4097
5	2.2361	55	7.4162	105	10.2470	155	12.4499
6	2.4495	56	7.4833	106	10.2956	156	12.4900
7	2.6458	57	7.5498	107	10.3441	157	12.5300
8	2.8284	58	7.6158	108	10.3923	158	12.5698
9	3.0000	59	7.6811	109	10.4403	159	12.6095
10	3.1623	60	7.7460	110	10.4481	160	12.6491
11	3.3166	61	7.8102	111	10.5357	161	12.6886
12	3.4641	62	7.8740	112	10.5830	162	12.7279
13	3.6056	63	7.9373	113	10.6301	163	12.7671
14	3.7417	64	8.0000	114	10.6771	164	12.8062
15	3.8730	65	8.0623	115	10.7238	165	12.8452
16	4.0000	66	8.1240	116	10.7703	166	12.8841
17	4.1231	67	8.1854	117	10.8167	167	12.9228
18	4.2426	68	8.2462	118	10.8628	168	12.9615
19	4.3589	69	8.3066	119	10.9087	169	13.0000
20	4.4721	70	8.3666	120	10.9545	170	13.0384
21	4.5826	71	8.4261	121	11.0000	171	13.0767
22	4.6904	72	8.4853	122	11.0454	172	13.1149
23	4.7958	73	8.5440	123	11.0905	173	13.1529
24	4.8990	74	8.6023	124	11.1355	174	13.1909
25	5.0000	75	8.6603	125	11.1803	175	13.2288
26	5.0990	76	8.7178	126	11.2250	176	13.2665
27	5.1962	77	8.7750	127	11.2694	177	13.3041
28	5.2915	78	8.8318	128	11.3137	178	13.3417
29	5.3852	79	8.8882	129	11.3578	179	13.3791
30	5.4772	80	8.9443	130	11.4018	180	13.4164
31	5.5678	81	9.0000	131	11.4455	181	13.4536
32	5.6569	82	9.0554	132	11.4891	182	13.4907
33	5.7446	83	9.1104	133	11.5326	183	13.5277
34	5.8310	84	9.1652	134	11.5758	184	13.5647
35	5.9161	85	9.2195	135	11.6190	185	13.6015
36	6.0000	86	9.2736	136	11.6619	186	13.6382
37	6.0828	87	9.3274	137	11.7047	187	13.6748
38	6.1644	88	9.3808	138	11.7473	188	13.7113
39	6.2450	89	9.4340	139	11.7898	189	13.7477
40	6.3246	90	9.4868	140	11.8322	190	13.7840
41	6.4031	91	9.5394	141	11.8743	191	13.8203
42	6.4807	92	9.5917	142	11.9164	192	13.8564
43	6.5574	93	9.6437	143	11.9583	193	13.8924
44	6.6332	94	9.6954	144	12.0000	194	13.9284
45	6.7082	95	9.7468	145	12.0416	195	13.9642
46	6.7823	96	9.7980	146	12.0830	196	14.0000
47	6.8557	97	9.8489	147	12.1244	197	14.0357
48	6.9282	98	9.8995	148	12.1655	198	14.0712
49	7.0000	99	9.9499	149	12.2066	199	14.1067
50	7.0711	100	10.0000	150	12.2474	200	14.1421

TABLE OF TRIGONOMETRIC FUNCTIONS

Angle	Sine	Cosine	Tangent	Angle	Sine	Cosine	Tangent
0°	0.0000	1.0000	0.0000	45°	0.7071	0.7071	1.000
1	0.0175	0.9998	0.0175	46	0.7193	0.6947	1.036
2	0.0349	0.9994	0.0349	47	0.7314	0.6820	1.072
3	0.0523	0.9986	0.0524	48	0.7431	0.6691	1.111
4	0.0698	0.9976	0.0699	49	0.7547	0.6561	1.150
5	0.0872	0.9962	0.0875	50	0.7660	0.6428	1.192
6	0.1045	0.9945	0.1051	51	0.7771	0.6293	1.235
7	0.1219	0.9925	0.1228	52	0.7880	0.6157	1.280
8	0.1392	0.9903	0.1405	53	0.7986	0.6018	1.327
9	0.1564	0.9877	0.1584	54	0.8090	0.5878	1.376
10	0.1736	0.9848	0.1763	55	0.8192	0.5736	1.428
11	0.1908	0.9816	0.1944	56	0.8290	0.5592	1.483
12	0.2079	0.9781	0.2126	57	0.8387	0.5446	1.540
13	0.2250	0.9744	0.2309	58	0.8480	0.5299	1.600
14	0.2419	0.9703	0.2493	59	0.8572	0.5150	1.664
15	0.2588	0.9659	0.2679	60	0.8660	0.5000	1.732
16	0.2756	0.9613	0.2867	61	0.8746	0.4848	1.804
17	0.2924	0.9563	0.3057	62	0.8829	0.4695	1.881
18	0.3090	0.9511	0.3249	63	0.8910	0.4540	1.963
19	0.3256	0.9455	0.3443	64	0.8988	0.4384	2.050
20	0.3420	0.9397	0.3640	65	0.9063	0.4226	2.145
21	0.3584	0.9336	0.3839	66	0.9135	0.4067	2.246
22	0.3746	0.9272	0.4040	67	0.9205	0.3907	2.356
23	0.3907	0.9205	0.4245	68	0.9272	0.3746	2.475
24	0.4067	0.9135	0.4452	69	0.9336	0.3584	2.605
25	0.4226	0.9063	0.4663	70	0.9397	0.3420	2.747
26	0.4384	0.8988	0.4877	71	0.9455	0.3256	2.904
27	0.4540	0.8910	0.5095	72	0.9511	0.3090	3.078
28	0.4695	0.8829	0.5317	73	0.9563	0.2924	3.271
29	0.4848	0.8746	0.5543	74	0.9613	0.2756	3.487
30	0.5000	0.8660	0.5774	75	0.9659	0.2588	3.732
31	0.5150	0.8572	0.6009	76	0.9703	0.2419	4.011
32	0.5299	0.8480	0.6249	77	0.9744	0.2250	4.331
33	0.5446	0.8387	0.6494	78	0.9781	0.2079	4.705
34	0.5592	0.8290	0.6745	79	0.9816	0.1908	5.145
35	0.5736	0.8192	0.7002	80	0.9848	0.1736	5.671
36	0.5878	0.8090	0.7265	81	0.9877	0.1564	6.314
37	0.6018	0.7986	0.7536	82	0.9903	0.1392	7.115
38	0.6157	0.7880	0.7813	83	0.9925	0.1219	8.144
39	0.6293	0.7771	0.8098	84	0.9945	0.1045	9.514
40	0.6428	0.7660	0.8391	85	0.9962	0.0872	11.43
41	0.6561	0.7547	0.8693	86	0.9976	0.0698	14.30
42	0.6691	0.7431	0.9004	87	0.9986	0.0523	19.08
43	0.6820	0.7314	0.9325	88	0.9994	0.0349	28.64
44	0.6947	0.7193	0.9657	89	0.9998	0.0175	57.29
45	0.7071	0.7071	1.0000	90	1.0000	0.0000	∞

Index

Index of Applications